W0234493

LOW-DIMENSIONAL ELECTRONIC PROPERTIES OF MOLYBDENUM BRONZES AND OXIDES

Physics and Chemistry of Materials with Low-Dimensional Structures

Editor-in-Chief

F. LÉVY, *Institut de Physique Appliquée, EPFL,*
Département de Physique, PHB-Ecublens, CH-1015 Lausanne, Switzerland

Honorary Editor

E. MOOSER, *EPFL, Lausanne, Switzerland*

International Advisory Board

J. V. ACRIVOS, *San José State University, San José, CA, U.S.A.*

S. BARISIC, *University of Zagreb, Department of Physics, Zagreb, Yugoslavia*

J. G. BEDNORZ, *IBM Forschungslaboratorium, Rüschlikon, Switzerland*

C. F. van BRUGGEN, *University of Groningen, Groningen, The Netherlands*

R. GIRLANDA, *Università di Messina, Messina, Italy*

D. HAARER, *University of Bayreuth, F.R.G.*

A. J. HEEGER, *University of California, Santa Barbara, CA, U.S.A.*

H. KAMIMURA, *Dept. of Physics, University of Tokyo, Japan*

W. Y. LIANG, *Cavendish Laboratory, Cambridge, U.K.*

P. MONCEAU, *CNRS, Grenoble, France*

J. ROUXEL, *CNRS, Nantes, France*

M. SCHLÜTER, *AT & T, Murray Hill, N.J., U.S.A.*

I. ZSCHOKKE, *Universität Basel, Basel, Switzerland*

LOW-DIMENSIONAL ELECTRONIC PROPERTIES OF MOLYBDENUM BRONZES AND OXIDES

Edited by

CLAIRE SCHLENKER

Laboratoire d'Etudes des Propriétés Electroniques des Solides
C.N.R.S., Grenoble, France

KLUWER ACADEMIC PUBLISHERS
DORDRECHT / BOSTON / LONDON

Library of Congress Cataloging-in-Publication Data

ISBN-13: 978-94-010-6685-3   e-ISBN-13: 978-94-009-0447-7
DOI: 10.1007/ 978-94-009-0447-7

Published by Kluwer Academic Publishers,
P.O. Box 17, 3300 AA Dordrecht, The Netherlands.

Kluwer Academic Publishers incorporates the publishing programmes
of D. Reidel, Martinus Nijhoff, Dr W. Junk and MTP Press.

Sold and distributed in the U.S.A. and Canada
by Kluwer Academic Publishers,
101 Philip Drive, Norwell, MA 02061, U.S.A.

In all other countries, sold and distributed
by Kluwer Academic Publishers Group,
P.O. Box 322, 3300 AH Dordrecht, The Netherlands.

All Rights Reserved
© 1989 by Kluwer Academic Publishers
Softcover reprint of the hardcover 1st edition 1989
No part of the material protected by this copyright notice may be reproduced or
utilized in any form or by any means, electronic or mechanical,
including photocopying, recording or by any information storage and
retrieval system, without written permission from the copyright owner.

TABLE OF CONTENTS

Preface   xiii

MARTHA GREENBLATT / Transition Metal Oxide Bronzes with Quasi Low-Dimensional Properties   1
1. Introduction   1
2. Band Structure and Electronic Properties of Oxide Bronzes   2
3. Molybdenum Bronzes   3
   3.1. Introduction   3
   3.2. Preparation   4
   3.3. The Blue Bronzes   5
      3.3.1. Crystal Structures   5
      3.3.2. Physical Properties — The Peierls Transition   6
   3.4. The Purple Bronzes   12
      3.4.1. Crystal Structure   12
      3.4.2. Charge Density Wave Instabilities   16
   3.5. The Red Bronzes, $A_{0.33}MoO_3$ 26
   3.6. $La_2Mo_2O_7$ — A Rare-Earth Molybdenum Bronze-Like Phase   30
4. Vanadium Bronzes with Quasi Low-Dimensional Properties — β-$A_xV_2O_5$ 32
   4.1. Introduction   32
   4.2. Preparation   33
   4.3. Crystal Structure   33
   4.4. Is There CDW Instability or Bipolaron Ordering in β-Vanadium Bronzes?   33
5. Tungsten Bronzes with Quasi Low-Dimensional Properties   38
   5.1. Hexagonal Tungsten Bronzes (HTB)   38
   5.2. Phosphate Tungsten Bronzes   41
Acknowledgements   43
References   43

HENRI VINCENT AND MASSIMO MAREZIO / On Structural Aspects of Molybdenum Bronzes and Molybdenum Oxides in Relation to Their Low-Dimensional Transport Properties   49
1. Introduction   49
2. Molybdenum Bronzes   50
   2.1. Molybdenum Bronzes Containing Infinite Single ReO_3 Octahedral Chains   54
      2.1.1. The Red $Na_{0.9}MoO_3$ and $K_{0.9}MoO_3$ Bronzes   54
      2.1.2. The Blue Bronze $K_{0.5}MoO_3$ 54
      2.1.3. The Violet $Rb_{0.27}MoO_3$ and Blue-Black $Rb_{0.44}MoO_3$ Bronzes   55
   2.2. Molybdenum Bronzes Containing Infinite Two-Dimensional Slabs Comprised of Corner-Sharing Octahedra   55

2.2.1. The Violet Bronzes $A_{0.9}Mo_6O_{17}$ with $A = Li, Na, K, Tl$ 56
2.2.2. The Blue-Black $Cs_{0.25}Mo_{0.97}O_3$ Bronze   59
2.3. Molybdenum Bronzes Containing Infinite Double ReO_3 Chains   61
2.3.1. Hydrogen Molybdenum Bronzes, H_xMoO_3 62
2.3.2. Sodium Hydrated Molybdenum Bronzes   64
2.4. Molybdenum Bronzes Containing Infinite Single Octahedral
      Ribbons   65
2.4.1. The Triclinic Lithium Bronze, $Li_{0.33}MoO_3$ 66
2.5. Molybdenum Bronzes Containing Infinite Chains of Octahedral
      Clusters   67
2.5.1. The Red $A_{0.33}MoO_3$ Bronze with $A = K, Cs,$ and Tl 67
2.5.2. The Copper Colored $Cs_{0.3}MoO_3$ Bronze   68
2.5.3. The Blue Bronzes, $A_{0.3}MoO_3$ with $A = K, Rb,$ and Tl 71
3. Molybdenum Oxides   73
3.1. The Molybdenum Oxides Containing Infinite ReO_3 Slabs: The
      Crystal Structure of the Monoclinic η-Mo_4O_{11} and the Ortho-
      rhombic γ-Mo_4O_{11} 74
3.2. The Molybdenum Oxides Containing Infinite Chains of Clusters
      and of ReO_3-Type: $Mo_8O_{23}, Mo_9O_{26}, Mo_5O_{14},$ and $Mo_{17}O_{47}$ 76
3.2.1. The Crystal Structure of Mo_8O_{23} and Mo_9O_{26} 76
3.2.2. The Crystal Structure of Mo_5O_{14} and $Mo_{17}O_{47}$ 77
3.3. The Molybdenum Oxides having the Layer MoO_3-Type Structure:
      MoO_3 and $Mo_{18}O_{52}$ 80
3.3.1. The Crystal Structure of MoO_3 80
3.3.2. The Crystal Structure of $Mo_{18}O_{52}$ 80
3.4. The Dioxide MoO_2 81
Acknowledgements   83
References   84

JEAN-PAUL POUGET / Structural Instabilities in the Low Dimensional
Molybdenum Bronzes and Oxides   87
1. Introduction   87
2. The Charge Density Wave Instability   88
2.1. Basic Features   88
2.1.1. Response Functions   88
2.1.2. Spatial Correlations above T_c 92
2.1.3. Coupling between Charge Density Waves   94
2.1.4. The Periodic Lattice Distortion   94
2.1.5. Temporal Fluctuations   95
2.1.6. The Electron—Phonon Coupling   100
2.2. The One-Dimensional Case   101
2.2.1. Intrachain Thermal Fluctuations   101
2.2.2. Interchain Coupling   104
2.2.3. Amplitude and Phase Excitations   109
3. The Blue Bronzes $A_{0.3}MoO_3$ $(A = K, Rb, Tl)$ 112
3.1. Room Temperature Structure   112

3.2. Basic Aspects of the Peierls Instability   114
3.3. Critical Wave Vector   119
   3.3.1. The Chain Component   119
   3.3.2. The Transverse Components   124
3.4. Spatial Fluctuations ($T > T_c$)   125
3.5. The Modulated Structure ($T < T_c$)   126
   3.5.1. The CDW Order   126
   3.5.2. Atomic Displacements   127
3.6. CDW Disorder   128
   3.6.1. Substitutional Disorder   129
   3.6.2. Electric Field Induced Disorder   129
3.7. Dynamics   131
   3.7.1. Pretransitional Temporal Fluctuations ($T > T_c$)   131
   3.7.2. Phase and Amplitude Excitations of the Incommensurate
       Structure ($T < T_c$)   132
3.8. Microscopic Parameters   135
   3.8.1. In-Chain Parameters   135
   3.8.2. Transverse Coupling   138
4. Other Oxides   138
4.1. The Titanium Bronze $Na_{0.25}TiO_2$ 139
4.2. The Layer Type Mo Oxides and Bronzes   141
   4.2.1. Phase Diagrams   141
   4.2.2. Mo_4O_{11} 142
   4.2.3. Purple Bronzes and Magneli Phases   145
5. Conclusion   148
Appendix A   149
Appendix B   151
Notes   154
References   155

CLAIRE SCHLENKER, JEAN DUMAS, CLAUDE ESCRIBE-FILIPPINI
AND HERVÉ GUYOT / Charge Density Wave Instabilities and Transport
Properties of the Low Dimensional Molybdenum Bronzes and Oxides   159
1. Introduction   159
2. Theoretical Background   160
2.1. The Peierls Transition and the CDW State   160
2.2. Charge Density Wave Transport   165
   2.2.1. The Fröhlich Mechanism   165
   2.2.2. Rigid CDW: Phenomenological Description of the Motion   165
   2.2.3. Deformable CDW: Microscopic Models of Pinning by
       Impurities   166
   2.2.4. Model Based on the Coupling of CDW with Lattice
       Phonons   168
   2.2.5. Models Involving CDW Structural Defects: Discommen-
       surations and Phase Dislocations   168
   2.2.6. Quantum Models   170

3. Quasi One-Dimensional Compounds: The Blue Bronzes $A_{0.30}MoO_3$ 171
  3.1. Introduction   171
  3.2. The Peierls Transition   173
    3.2.1. Ohmic Transport   173
    3.2.2. Magnetic Susceptibility   176
    3.2.3. Thermal and Elastic Properties   177
    3.2.4. Effect of Impurities and Point Defects   178
  3.3. Band Structure: Experiment and Theory   179
  3.4. Nonlinear Transport   180
    3.4.1. Introduction   180
    3.4.2. Threshold Electric Field   182
    3.4.3. Broad Band Noise   191
    3.4.4. Periodic Voltage Oscillations   192
    3.4.5. Very Low Frequency Phenomena   194
    3.4.6. High Velocity Sliding of the CDW at Low Temperature   196
  3.5. Hysteresis and Metastability Phenomena   197
    3.5.1. Introduction   197
    3.5.2. Low Field Regime   197
    3.5.3. Nonlinear Regime   200
    3.5.4. Pulse Memory Effects   201
    3.5.5. Remanent CDW Polarization   202
    3.5.6. Field Induced Deformation of the CDW   203
  3.6. Local Properties   204
    3.6.1. Ion Channeling Technique   204
    3.6.2. Mössbauer Effect   204
    3.6.3. Electron Paramagnetic Resonance Studies   207
    3.6.4. Nuclear Magnetic Resonance Studies   209
  3.7. Summary   210
4. Quasi Two-Dimensional Compounds: The Molybdenum Purple Bronzes and the Molybdenum Oxides   212
  4.1. Introduction   212
  4.2. Charge Density Wave Instabilities and Transport Properties 212
    4.2.1. Electrical Resistivity   213
    4.2.2. Thermopower   221
    4.2.3. Galvanomagnetic Properties   224
  4.3. Magnetic Susceptibility   231
    4.3.1. Experimental Results   231
    4.3.2. Discussion   236
  4.4. Specific Heat   237
  4.5. Band Structure   241
  4.6. Quantum Transport   244
    4.6.1. Experimental Results   244
    4.6.2. Discussion   244
  4.7. Superconductivity in $Li_{0.9}Mo_6O_{17}$ 247
5. Conclusion   250

Acknowledgements   252
References   252

ROBERT M. FLEMING AND ROBERT J. CAVA / Frequency-Dependent
Conductivity in $K_{0.30}MoO_3$ 259
1. Introduction   259
   1.1. Conceptual Models   259
   1.2. Uniform Pinning Model   260
   1.3. Random Pinning Model   262
2. Dielectric Relaxation Regime   263
   2.1. Phenomenological Description   264
   2.2. Dielectric Relaxation: Zero dc Bias   267
   2.3. Chemical Doping: Zero Bias   273
   2.4. Dielectric Relaxation: Finite dc Bias   277
   2.5. Dielectric Relaxation Temperature Dependence: Finite dc Bias   279
3. Phase Mode Regime   285
4. Far Infrared Regime   289
5. Normal-Electron Screening   289
6. Summary   291
Acknowledgements   292
References   292

SERGE AUBRY AND PASCAL QUEMERAIS / Breaking of Analyticity
in Charge Density Wave Systems: Physical Interpretation and Consequences   295
1. Introduction: The Peierls Instability in 1D Conductors   295
2. The Transition by Breaking of Analyticity (TBA) in the Discrete
   Frenkel—Kontorova (FK) Model   301
   2.1. Commensurate Ground States   303
   2.2. Incommensurate Ground States   308
   2.3. Critical Behavior at the TBA   312
   2.4. Incommensurate Structure as an Array of Equidistant Discom-
      mensurations   321
   2.5. Ising Representation of a Nonanalytic Incommensurate Structure   323
   2.6. Extended FK Models and Thermal Fluctuations   326
3. Another Transition by Breaking of Analyticity: The Localization
   Transition of Electrons in an Incommensurate Potential   327
   3.1. Description of the Breaking of Analyticity of the Eigenwaves of a
      Quasi-Periodic Schroedinger Equation   328
   3.2. Exact Results for a Self-Dual Model   330
   3.3. Some Numerical Investigations of the Self-Dual Model and Other
      Non-Self-Dual Models in One Dimension   334
   3.4. Other Self-Dual Models in One and Several Dimensions   338
   3.5. Questions and Remarks Concerning Discontinuous Quasi-Periodic
      Potentials   339

3.6. Comparison Between Extended States in Quasi-Periodic and Random Potentials   340
   3.6.1. The Kubo—Greenwood Formula   341
   3.6.2. The Numerical Technique   342
   3.6.3. Results for a Quasi-Periodic Potential in One and Two Dimensions   344
   3.6.4. Results for a Random Potential in One and Three Dimensions   348
4. The Transition by Breaking of Analyticity in One-Dimensional Peierls Chains   352
   4.1. The Holstein Model   354
   4.2. The Fröhlich—SSH Model   356
   4.3. Numerical Observation of the Transition by Breaking of Analyticity in Peierls Chains   358
   4.4. Critical Behavior at the TBA of Peierls Chains   365
      4.4.1. Coherence Length   365
      4.4.2. Peierls—Nabarro Energy Barrier   367
      4.4.3. Phason Gap and Phonon Spectrum   369
   4.5. Electronic Behavior at the TBA   376
   4.6. A Classical Lattice Gas Model for the Holstein Model in the Large Electron—Phonon Coupling Limit   378
5. Future Prospects: Quantum Lattice Effects and Thermal Effects   384
   5.1. Quantum Lattice Effects   385
      5.1.1. A Quantum Lattice Gas Model for the Holstein Model in the Large Electron—Phonon Coupling Limit   386
      5.1.2. Effects of the Quantum Lattice Fluctuations on an Incommensurate CDW   388
   5.2. Thermal Effects on a Nonanalytic Incommensurate CDW Ground State   396
      5.2.1. Order—Disorder CDW   397
      5.2.2. Displacive CDW   398
      5.2.3. Ohmic Conductivity of a Nonanalytic CDW   399
      5.2.4. Nonlinear Electric Conductivity of a Nonanalytic CDW   400
Acknowledgements   403
References   403

DENIS FEINBERG AND JACQUES FRIEDEL / Imperfections of Charge Density Waves in Blue Bronzes   407
1. Introduction   407
2. Properties of Static Imperfections of CDWs   409
   2.1. Incommensurate CDW   409
      2.1.1. Long Wavelength Distortions; the 3D Elastic Limit   410
      2.1.2. Short Wavelength Distortions. 2D Ridges and Walls   413
      2.1.3. 1D Perfect Dislocations of the CDW   414
      2.1.4. Disclinations and Point Singularities of CDWs   418

2.2. Nearly Commensurate CDW 420
2.3. Commensurate CDW 421
3. Interaction of CDW with Lattice Defects 422
  3.1. Surfaces and Interfaces 423
    3.1.1. Brute Force Processes 423
    3.1.2. Dislocation Multiplication 424
  3.2. Lattice Dislocations 429
  3.3. Point Defects 430
    3.3.1. Interactions with CDW 430
    3.3.2. Interaction with Dislocations of the CDW 432
4. Elastic and Anelastic Responses of CDW 435
  4.1. Elastic Equilibrium of a CDW under an Electric Field 435
  4.2. Anelastic Response 438
5. Plastic Properties of CDW 439
  5.1. Amplitude-Dependent Internal Friction. Approach to Critical Current 440
  5.2. Critical Field for Fröhlich Current 442
  5.3. Remanent Polarisation 444
6. Conclusions 444
Acknowledgements 446
References 446

INDEX 449

PREFACE

The history of low dimensional conductors goes back to the prediction, more than forty years ago, by Peierls, of the instability of a one dimensional metallic chain, leading to what is known now as the charge density wave state. At the same time, Fröhlich suggested that an "ideal" conductivity could be associated to the sliding of this charge density wave. Since then, several classes of compounds, including layered transition metal dichalcogenides, quasi one-dimensional organic conductors and transition metal tri- and tretrachalcogenides have been extensively studied. The molybdenum bronzes or oxides have been discovered or rediscovered as low dimensional conductors in this last decade. A considerable amount of work has now been performed on this subject and it was time to collect some review papers in a single book.

Although this book is focused on the molybdenum bronzes and oxides, it has a far more general interest in the field of low dimensional conductors, since several of the molybdenum compounds provide, from our point of view, model systems. This is the case for the quasi one-dimensional blue bronze, especially due to the availability of good quality large single crystals.

This book is intended for scientists belonging to the fields of solid state physics and chemistry as well as materials science. It should especially be useful to many graduate students involved in low dimensional oxides. It has been written by recognized specialists of low dimensional systems. However, the non-specialized reader will find in all chapters useful introductions. The points of view of solid state chemistry and crystallography are given in the first chapters. The physical properties, including non-linear transport due to the depinning of the charge density wave and the related frequency dependent properties are then considered. Finally, theoretical aspects in relation with strong electron-phonon coupling on one side and with charge density wave structural defects such as phase dislocations on the other, are developed.

Acknowledgments are due to the authors for their contribution. I wish also to thank warmly F. Lévy, Managing Editor of the Series "Physics and Chemistry of Materials with Low Dimensional Structures" for his suggestions and his constant encouragements.

Grenoble, March 1989   CLAIRE SCHLENKER

TRANSITION METAL OXIDE BRONZES WITH QUASI LOW-DIMENSIONAL PROPERTIES

MARTHA GREENBLATT

Chemistry Department, Rutgers, The State University of New Jersey,
New Brunswick, NJ 08903, U.S.A.

1. Introduction

The term bronze, originally given to the Na_xWO_3 compounds by Wohler in 1825 [1], is now applied to a variety of crystalline phases of the transition metal oxides; usually ternary compounds of the type $A_xM_zO_y$, but binary oxides with intense color and metallic luster, metallic or semiconducting properties, and resistance to attack by non-oxidizing acids also qualify. Ternary bronzes have been prepared in which M is Ti, V, Mn, Nb, Ta, Mo, W, or Re, and A is H, NH_4^+, alkali, alkaline earth, rare-earth, Group IB, IIB, or other metal ion.

The alkali tungsten bronzes, A_xWO_3 have been studied most intensively and have been reviewed in the past [2—5]. Extensive reviews including all bronze phases studied have been presented by Banks and Wold [3] and Hagenmuller [5] nearly twenty years ago. A rather brief review of the molybdenum bronzes (results up to ~1980) was given more recently in a chapter on 'Lower Valence Molybdenum Oxides' [6].

Interest in all types of transition metal oxide bronzes continued in recent years. A computer search of 'tungsten bronzes' and 'vanadium bronzes' in *Chemical Abstracts* came up with 670 and 331 references, respectively, since 1967. However, in the last five years the greatest renewal of interest and an explosion of publications has been in the study of molybdenum bronzes and especially of the so-called 'blue bronzes' $A_{0.3}MoO_3$ (A = K, Rb, Tl). This interest is primarily concerned with the physical properties, unique to some of these bronzes which show highly anisotropic transport properties characteristic of quasi one- and two-dimensional (1D and 2D) metallic systems, metal-to-semiconductor transition driven by a charge-density wave (CDW) state, and superconductivity. Nonlinear behavior of the dc electric field dependence on the conductivity above a small threshold potential (E_T) and a collection of related phenomena previously associated with a moving or 'sliding' CDW in the chalcogenides $NbSe_3$ [7], TaS_3 [8], and $(TaSe_4)_2I$ [9] have also been observed in this structurally different class of materials and are intensely studied in several laboratories around the world [10]. These new findings on the nonlinear transport and related properties of $K_{0.3}MoO_3$ and $Rb_{0.3}MoO_3$ have recently been reviewed in an earlier work in this series by Schlenker and Dumas [11a] and elsewhere by Fleming [11b] and are further treated in other chapters in this volume.

This article will review primarily the molybdenum bronzes, although some new

1

C. Schlenker (Ed.), Low-Dimensional Electronic Properties of Molybdenum Bronzes and Oxides,
1—48.
© 1989 *Kluwer Academic Publishers.*

results on other transition metal bronzes, particularly some of the vanadium and tungsten phosphate bronzes with quasi-low dimensional behavior will also be discussed.

2. Band Structure and Electronic Properties of Oxide Bronzes

Figure 1 shows a schematic diagram of the band structure developed by Goodenough [12] for ReO_3 and first applied to the molybdenum bronzes by Dickens and Neild [13]. It is assumed that the local octahedral coordination of the metal atom by six oxygen atoms largely determines the common bonding pattern in the bronzes (A_xMO_3) and the parent oxide (e.g. MoO_3, WO_3, ReO_3). For a discrete MO_6 unit (in the $4d$ transition metal series) $5p$, $5s$, and $4d(e_g)$ orbitals overlap with six sp hybrid orbitals of the oxygen atom to give a set of six bonding σ and six antibonding σ^* molecular orbitals. In the extended lattice, the discrete energy levels arising from this structure unit will broaden into bands. The metal $4d(t_{2g})$ orbitals can overlap with three of the surrounding oxygen $p\pi$ orbitals per octahedron to form bonding π and antibonding π^* bands. The σ and π bands are filled and constitute the valence bands separated by a large energy gap (3—4 eV) from the π^* conduction band. This band is empty in WO_3 and MoO_3, so these materials are insulators, but is partially filled in ReO_3, which is a metal. In the bronzes, $A_xM_yO_z$, the A cations transfer their valence electrons to the usually empty π^*-levels with strong d-character. The extent of delocalization of the d-

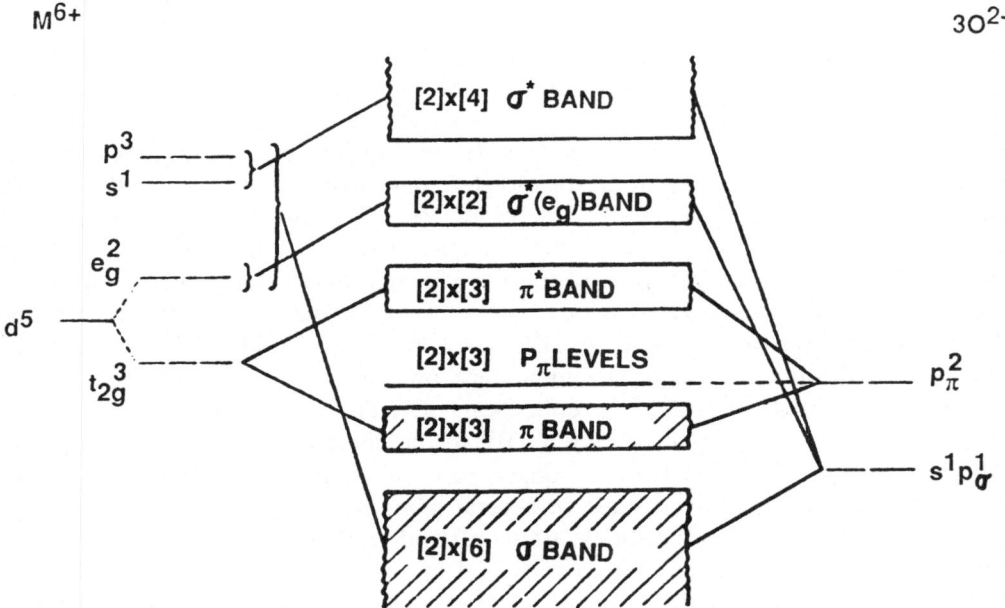

Fig. 1. Schematic Band Structure applicable to the bronzes. Above the Peierls Transition (after Ref. [13]).

electrons characterize the transport and optical properties of any specific bronze. The orbitals of the A cation do not contribute to the formation of these levels, as is confirmed by several experiments [14, 15].

In the vanadium bronzes, $A_xV_2O_5$ the $3d$ electrons are mostly localized, and these compounds are semiconductors although, some of the β-$A_xV_2O_5$ phases show quasi 1D metallic behavior, which will be discussed later. In contrast, the $5d$ electrons of the tungsten bronzes, A_xWO_3, are delocalized because of the greater radial extension of the $5d$ orbitals [16] and these compounds are metals at all temperatures. In the molybdenum bronzes the $4d$ orbitals have a radial extension intermediate between the $3d$ (localized) and $5d$ (delocalized) states, and these materials show complex electronic behavior; the blue, $A_{0.3}MoO_3$ and purple, $A_{0.9}Mo_6O_{17}$ bronzes are metallic at ambient temperature, but exhibit a metal-to-semiconductor and a metal-to-metal transition, respectively, while the red $A_{0.33}MoO_3$ bronzes are semiconducting. A recent band calculation using a tight-binding band scheme by Whangbo and Schneemeyer gives essentially the same band structure as shown in Figure 1 for $K_{0.3}MoO_3$ [17]; however, this calculation shows that there are three closely spaced $d\pi^*$ bands, two of which are partially filled and a third d band which is empty, but lies very close (~ 0.012 eV) above. The metal-to-semiconductor transition is predicted from the nesting properties of the Fermi surfaces of the two d bands.

3. Molybdenum Bronzes

3.1. INTRODUCTION

Molybdenum bronzes may be classified on the basis of similarity of color, stoichiometry, and structure. The more common alkali bronzes, which can be prepared by fused salt electrolysis at atmospheric pressure or by solid state reactions in vacuo are classified as: the blue bronzes, $A_{0.3}MoO_3$ with A = K, Rb and Tl; the red bronzes, $A_{0.33}MoO_3$ with A = Li, K, Rb, Cs, Tl, and the purple bronzes, $A_{0.9}Mo_6O_{17}$ with A = Li, Na, K, Tl. $Cs_{0.19}MoO_3$, a blue bronze, is unique both in stoichiometry and structure, but similar phases with other ternary cations may exist.

In contrast to the W bronzes which have simple three-dimensional (3D) structures of corner sharing WO_6 octahedra with the A cations located in cubooctahedral cavities, the Mo bronzes have complex, layered-type structures with edge and corner sharing MoO_6 octahedra forming infinite two-dimensional sheets which are held together by A cations (with the exception of $Li_{0.33}MoO_3$ and $Li_{0.9}Mo_6O_{17}$ as will be discussed later). While the W bronzes A_xWO_3 are nonstoichiometric ($0 < x \leqslant 1$), these Mo bronzes are stoichiometric or very nearly so.

The bronzes, $Na_{0.90-0.97}MoO_3$, $K_{0.89-0.93}MoO_3$, $K_{0.5}MoO_3$ and $Rb_{0.27}MoO_3$, which are prepared by high pressure synthesis and are isostructural with their W bronze analogs have not been studied extensively, but are known to be metallic [3, 5]. A relatively new class of compounds are the hydrogen molybdenum bronzes

H_xMoO_3 ($0 < x \leqslant 2.0$), in which the hydrogen is topotactically inserted into the MoO_3 matrix to yield four unique phases: $H_{0.23-0.40}MoO_3$, blue, orthorhombic; $H_{0.85-1.04}MoO_3$, blue, monoclinic; $H_{1.55-1.72}MoO_3$, red, monoclinic and H_2MoO_3, green, monoclinic; the structure of each is closely related to that of the structure of the MoO_3 host. The properties of these materials have been reviewed recently [6].

3.2. PREPARATION

The first molybdenum bronzes $K_{0.3}MoO_3$, $K_{0.33}MoO_3$ and $Na_{0.9}Mo_6O_{17}$ were prepared by Wold et al. [18] by electrolytic reduction of A_2MoO_4—MoO_3 melts [18] under carefully controlled conditions of temperature and composition. This technique has been used in slight variations to produce a variety of alkali metal molybdenum bronzes [19—31] as well as $Tl_{0.3}MoO_3$ [32] in single crystal form. Most of the crystals grown in this way are usually platelet-like, elongated along one of the crystal axis, typically $5 \times 2 \times 1$ mm^3 in size. The crystals can easily be cleaved parallel to the platelet plane.

Polycrystalline samples of a variety of alkali metal bronzes have been prepared from stoichiometric mixtures of A_2MoO_4, MoO_3 and MoO_2 in evacuated sealed gold tubes at selected temperatures [33—35]; all of the known Mo bronzes may be prepared similarly in polycrystalline form in evacuated quartz tubes at the appropriate temperature for each [36—42].

More recently a temperature gradient flux technique has been developed for the growth of good quality single crystal molybdenum bronzes from a stoichiometric mixture of reactants according to the equation: $n A_2MoO_4 + 2(1 - n)MoO_3 + n MoO_2 \rightarrow 2 A_nMoO_3$ [36—42]. In a typical experiment the reactant mixture is pelletized and sealed in an evacuated quartz tube which is placed in a two-zone furnace. The charge end of the ampule is placed in the hot zone and a temperature gradient exists throughout the length of the tube. For compositions $0.2 \leqslant n \leqslant 0.5$, melting of the charge throughout the tube is facilitated by appropriate adjustment of the temperature of the two zones. Single-crystal bronzes grow in the melt; different types of bronzes may grow at different regions of temperature in the same ampule. The type of bronze and the quality of the crystals in each case is highly dependent on the value of n and on the temperature gradient of the furnace. Optimal conditions of these parameters have been determined for the growth of large, good quality single crystals of most of the common ternary molybdenum bronzes [36—42]. Large single crystals of the binary phases of the Mo_4O_{11} and MoO_2 oxides can also be obtained by the temperature gradient flux technique.

A number of alkali metal molybdenum bronzes isostructural with known tungsten bronzes have been prepared by high pressure reaction between either alkali metal molybdate, MoO_3 and Mo metal powder [43] or alkali metal azide and MoO_3 [44] at 65 Kbar pressure and 770—1270 K. These phases have been reviewed in detail before [3, 5, 6].

Chemical or electrochemical reduction of MoO_3 in acidic media was shown to yield a series of solid phases H_xMoO_3 $0 < x \leqslant 2.0$ [45—52]. Deuterated molybdenum bronzes, D_xMoO_3 have also been prepared for various values of x [51, 52].

3.3. THE BLUE BRONZES

3.3.1. *Crystal Structures*

The structure of $K_{0.3}MoO_3$ was first solved by Graham and Wadsley [53]. More recent single crystal X-ray diffraction structural refinement of $K_{0.3}MoO_3$, $Rb_{0.3}MoO_3$ [54] and $Tl_{0.3}MoO_3$ [32] show that all of these blue bronzes are isostructural and form with monoclinic symmetry, in space group $C2/m$ with 20 molecules per unit cell, in an arrangement essentially as originally determined by Graham and Wadsley. Unit cell parameters of the known $A_{0.3}MoO_3$ phases are given in Table I. The structure is built of infinite sheets of distorted MoO_6 octahedra held together by the A cations as shown in Figure 2. The MoO_6 layers consist of clusters of 10 edge sharing octahedra linked by corners in the [010] and [102] directions. The structure can also be viewed as infinite chains of MoO_6 octahedra sharing corners along the monoclinic b direction. There are three crystallographically different Mo sites per cluster; only two of them, Mo(2) and Mo(3) [Refs. 32, 54] are involved in these infinite chains. Furthermore, it has been established by Zachariasen's method [55] that over 80% of the $4d$ electron density is found on the Mo(2) and Mo(3) sites [32, 54]. The data in Table I show significant variations in a, c and β with A ion, however, the unique monoclinic b axis remains practically unchanged with increasing effective size of the A cation. This suggests that the chain like coupling of MoO_6 octahedra which corner share along the b-axis is independent of the nature of the A ion in the bronze. If the stoichiometry corresponds exactly to the formula $A_{0.3}MoO_3(A_3Mo_{10}O_{30})$, all crystallographic sites are occupied and therefore no crystallographic disorder is expected. Next it will be shown how the low-dimensional electronic, magnetic, and optical properties are accounted for by the structure.

TABLE I
Unit cell parameters of $A_{0.3}MoO_3$ bronze phases.

Compound	$K_{0.30}MoO_3$ [a]	$Tl_{0.30}MoO_3$ [b]	$Rb_{0.30}MoO_3$ [a]
Unit Cell	Monoclinic	Monoclinic	Monoclinic
Space Group	$C2/m$	$C2/m$	$C2/m$
a (Å)	18.2587(7)	18.486(1)	18.6354(3)
		(18.543(3))	
b (Å)	7.5502(4)	7.5474(6)	7.555(1)
		(7.567(4))	
c (Å)	9.8614(4)	10.0347(7)	10.094(2)
		(10.067(2))	
β (°)	117.661(4)	118.377(6)	118.842(5)
		(118.39(2))	

[a] The values given in Ref. [58] have been transformed to a $C2/m$ unit cell.
[b] Ref. [38]; the values in parenthesis are given by Ref. [32]. The small differences between the two values may reflect small deviations in Tl stoichiometry or other concentrations of lattice or impurity defects.

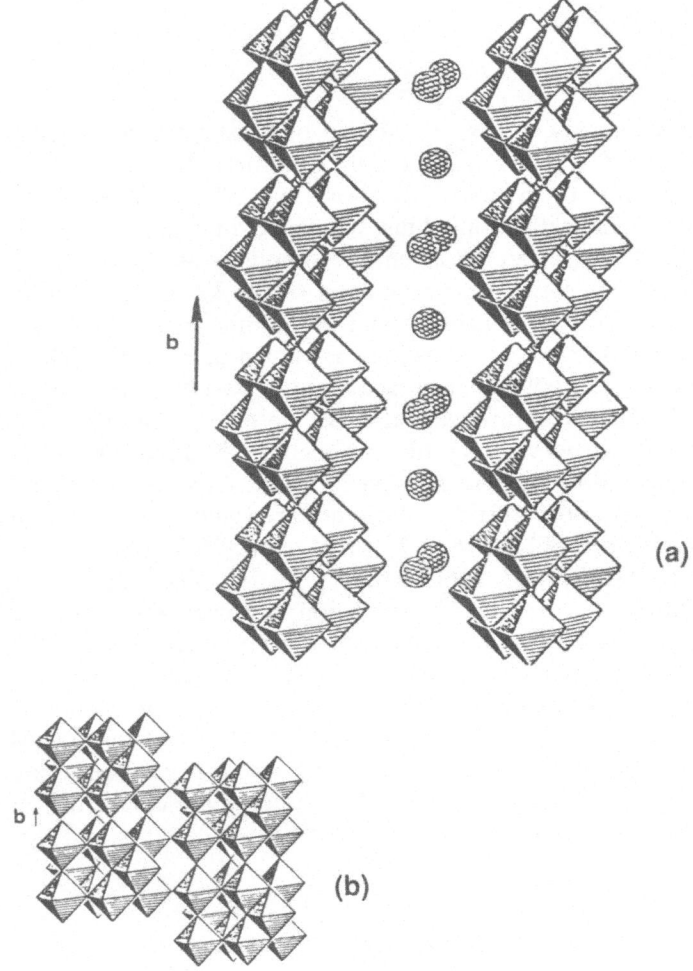

Fig. 2. (a) Crystal structure of the $A_{0.3}MoO_3$ blue bronzes showing infinite chains of Mo(2) and Mo(3) oxygen octahedra parallel to b. (b) Infinite layers of MoO_6 octahedra made up by corner sharing of the basic ten cluster MoO_6 edge/corner sharing octahedra along the b and [102] directions (after Ref. [11]).

3.3.2. *Physical Properties — The Peierls Transition*

Wold *et al.* [18] and Bouchard *et al.* [19] have studied the electrical properties of unoriented samples of $K_{0.3}MoO_3$ about twenty years ago and established the existence of a metal to semiconductor transition at ~ 180 K. Later, it was show by Perloff *et al.* [20] that the electrical conductivity is highly anisotropic. Subsequently Fogle and Perlstein [56] suggested several models to account for the metal-to-semiconductor transition including the Mott—Hubbard model [57] with a short-range electron—hole attraction and an excitonic-insulator model [58] for the semiconductor state. Furthermore, they noted that for $K_3Mo_{10}O_{30}$ with three electrons per lattice point, a new Brillouin zone could form at the Fermi surface (E_F) by a doubling of the periodicity due to a small distortion of the $Mo_{10}O_{30}$ clusters, which could also account for the metal-to-semiconductor transition. However, they failed to find a lattice distortion by X-ray diffraction [20, 59]. In addition they have observed nonohmic behavior above a critical voltage at low temperature, although the source of that behavior was then not recognized.

More recently Brusetti *et al.* [25] reexamined the conductivity, thermopower and magnetic properties of $K_{0.3}MoO_3$ and have shown that the resistivity of $K_{0.3}MoO_3$ is about an order of magnitude larger along the plane of the layers [102] than along b, and even higher along the [201] direction perpendicular to the layers of MoO_6 octahedra. Similar anisotropy of the temperature variation of ρ are found for $Rb_{0.3}MoO_3$ and $Tl_{0.3}MoO_3$ which is illustrated for the latter in Figure 3 [11, 29, 32, 38]. The optical reflectivity data obtained by Travaglini *et al.* [60] for $K_{0.3}MoO_3$ and Jandl *et al.* for $Rb_{0.3}MoO_3$ [61] established that above 180 K these bronzes are truly quasi one-dimensional (1D) metals. At 300 K, the blue bronzes show a metallic behavior with a plasma edge for light polarized along b, and semiconducting type behavior with an absorption edge of light polarized along [102]. The plasma frequency, $\omega_p = 2.7$ eV along b corresponds to a carrier concentration resulting from a complete charge transfer of the valence s electrons of the alkali metal into the π^* conduction band (Figure 1). $\omega_p \ll 0.03$ eV along [102] indicates that the carrier concentration in the semiconducting direction is 4—5 orders of magnitude less than in the metallic direction [60]. These results are in excellent agreement with the structural data discussed above which show that $>80\%$ of the $4d$ electrons are on Mo sites involved in infinite chains of corner sharing MoO_6 octahedra parallel to the b-axis of the monoclinic unit cell. In terms of the schematic band structure (Figure 1), the $4d$ electrons are delocalized in the π^* conduction band which form primarily by the overlap of Mo(2) and Mo(3) $4d(t_{2g})$ and oxygen $p(\pi)$ orbitals. Thus while the *bonding* is clearly 3D in the blue bronzes, the transport and optical properties are quasi one-dimensional, because the distribution of unpaired electrons is quasi one-dimensional. Once the one-dimensional metallic behavior of the $A_{0.3}MoO_3$ bronzes at $T > 180$ K was established, the nature of the metal-to-semiconductor transition observed at ~ 180 K could be clarified.

The idea that the electronic energy of a 1D metal could be lowered by a charge-density wave (CDW) was first predicted by Peierls [62] and Frohlich [63]. In one-dimensional conductors a coupled instability at $2k_f$ of the conduction electrons and phonons at low temperature leads to a modulation of the charge density, coupled to a lattice distortion or a CDW state which is accompanied by an opening of a gap at the Fermi surface as shown in Figure 4. The states near the gap below E_F, which are pushed down in energy, are occupied while those that are raised are empty; the total electronic energy is therefore lowered. If the decrease in electronic energy is greater than the potential energy cost of the lattice distortion, a charge-density wave will spontaneously appear. Subsequent work has shown that

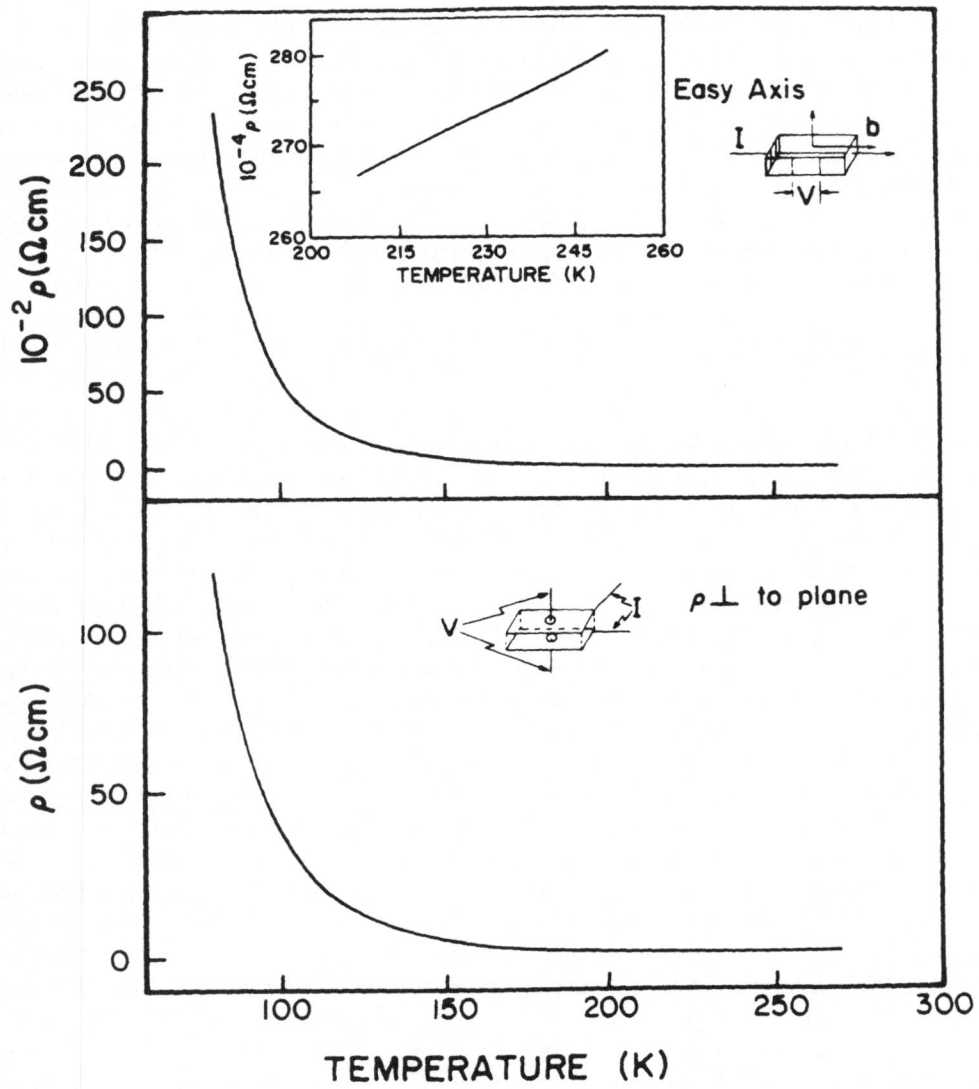

Fig. 3. Temperature variation of the resistivity of $Tl_{0.3}MoO_3$. (a) Current and voltage in the plane of the platelet crystals parallel to the b-axis. (b) Current and voltage perpendicular to b and the (201) cleavage plane. Inset shows detail of decreasing resistivity with decreasing temperature in the range ~ 250–200 K.

this instability is not restricted to one-dimensional systems. In the last 15 years CDW has been experimentally verified in a number of quasi low-dimensional organic and inorganic compounds including $NbSe_3$, TaS_2, $TaSe_2$ and $(TaSe_4)_2I$ and are fairly well understood in the transition metal di- and trichalcogenides [65, 66].

The blue bronzes are new examples of chemically and structurally different class of compounds to exhibit a CDW-driven metal-to-semiconductor transition.

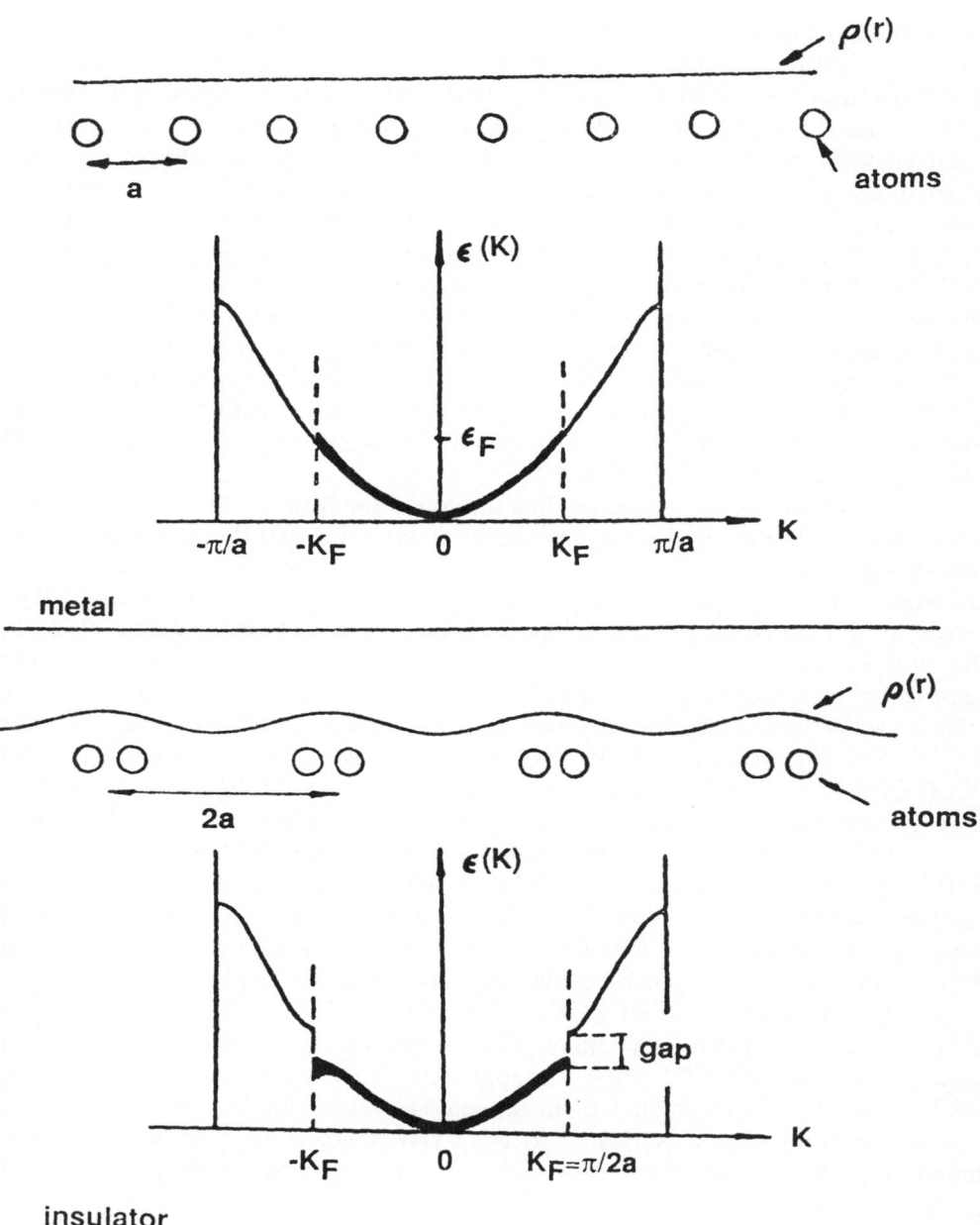

Fig. 4. Peierls distortion in a one-dimensional metal with a half-filled band (after Ref. [64]). (a) Undistorted metal. (b) Peierls insulator with period doubling.

Although such a transition generally can be detected as an anomaly in the temperature variation of the resistivity, magnetic susceptibility and specific heat, unambiguous evidence of the CDW state is the observation of superlattice lines by electron, X-ray or neutron diffraction which confirm the modulation of the lattice sites. X-ray diffuse scattering studies by Pouget et al. [29] have shown conclusively that the transition at 180 K is a Peierls transition leading to an incommensurate semiconducting CDW state. When the wave vector Q of the distortion is commensurate with the lattice, the distortion of the charge and atomic displacement simply gives the crystal a larger unit cell. Specifically if $Q = a^*/3$ then the unit cell dimension a is three times the old unit cell dimension and the crystal is periodic. When Q is incommensurate, no unit cell can contain the exact period of both the wave and the underlying crystal lattice. The crystal is no longer periodic; the whole sample is the unit cell. In $K_{0.3}MoO_3$ and $Rb_{0.3}MoO_3$ ($Tl_{0.3}MoO_3$ has not been studied by diffraction as yet below the transition) even at room temperature diffuse streaks normal to the b^* direction are observed which condense into well-defined superlattice reflections below 180 K [29]. The satellite reflections are located at $\pm(1 - q_b)b^*$ from the main reflections, where the component of Q along b, $q_b = 0.26 \pm 0.01$ at 110 K [29].

Analysis of the X-ray data shows that motion of the $Mo(2)$—O_6 and $Mo(3)$—O_6 octahedra are mainly involved in the structural distortion and that the alkali metals are not significantly affected. Similar X-ray diffraction studies by Sato et al. [67] of the superlattice reflections of the blue bronzes confirmed these findings. The temperature dependence of the Q vector studied by various investigators [68, 69, 70] indicate that $1 - q_b$ is incommensurate with decreasing temperature and approaches a commensurate value of 0.75 near 100 K. According to Fleming et al. [69] a true lock-in incommensurate-to-commensurate transition occurs near 100 K. This is consistent with the data of Pouget et al. [70], and with recent [87]Rb NMR data of $Rb_{0.3}MoO_3$ which shows a narrowing of the NMR line around 90 K corresponding to a finite number of sites with substantial hysteresis [71]. However, the existence and the nature of the incommensurate-commensurate transition near 100 K remains controversial; recent NMR results seem to indicate that the system remains incommensurate down to the lowest temperatures [72, 73]. Careful heat capacity measurements of $Rb_{0.3}MoO_3$ between 50 and 150 K show a kink at 100 K, but there is no conclusive evidence of a phase transition [74].

Iron-57 Mössbauer studies of Fe substituted for Mo in $K_{0.3}MoO_3$ also give evidence of local distortions taking place at the Peierls transition [74]. Raman scattering studies [75—77] show the appearance of a line at ~ 50 cm^{-1} in the semiconducting phase which has an anomalous temperature dependence below 180 K suggesting that it is related to the CDW state. A specific heat anomaly, found in $K_{0.3}MoO_3$ and $Rb_{0.3}MoO_3$ between about 150 K and 190 K is additional evidence of the CDW-driven phase transition [72].

The electronic structure of the blue bronzes has been deduced partly from optical reflectivity [60, 61] and X-ray diffuse scattering [29, 78]. More recently an X-ray photoemission spectroscopy study (XPS) of $K_{0.3}MoO_3$ shows a conduction band extending 2 eV below the Fermi level with large Mo $4d$ character [79]. The band structure is essentially in agreement with the model of Travaglini et al. [60]. Angular resolved UV photoemission studies measured at 300 K on $K_{0.3}MoO_3$ crystals establish that 0.27 Å$^{-1} < k_f < 0.36$ Å^{-1} [80]. This is in excellent agreement with the value obtained from the position of the X-ray satellite reflections in the low temperature CDW phase ($2k_f = 0.75b^*$, $k_f = 0.312$ Å^{-1}). This direct measurement of k_f corroborates the model of $2k_f$ instability responsible for the phase transition at 180 K and this may be the first example where this is so clearly established.

The magnetic susceptibility, χ of $K_{0.3}MoO_3$ has been measured by several investigators [19, 25, 28, 30, 81, 82] and is similar to the temperature variation of the susceptibility of $Tl_{0.3}MoO_3$ [38] shown in Figure 5. The very weak temperature dependence of χ between ~240 K and room temperature is consistent with metallic behavior. The observed decrease in the magnetic susceptibility from ~180 K is due to the opening of a gap at the Fermi surface associated with the CDW transition. A small anomaly of unknown origin is observed near 50 K which was also seen in the susceptibility of $K_{0.3}MoO_3$ [28].

The similarities of the structural and transport properties of $K_{0.3}MoO_3$, $Rb_{0.3}MoO_3$ and $Tl_{0.3}MoO_3$ suggest that a similar mechanism of transport is operative in all three systems. The nearly identical T_c values close to 180 K in each indicate that the electronic properties are determined primarily by the MoO_6 octahedral network which is nearly the same in each bronze as evidenced by the almost identical values of the crystallographic b dimension (Table I).

Additional aspects of the specific heat, optical magnetic, and lattice dynamic properties as well as some substitutional studies of the blue bronzes have been reviewed in greater detail recently in a previous volume [11] and are treated further in this volume.

The effect of impurities on the electrical properties of the blue bronzes has been

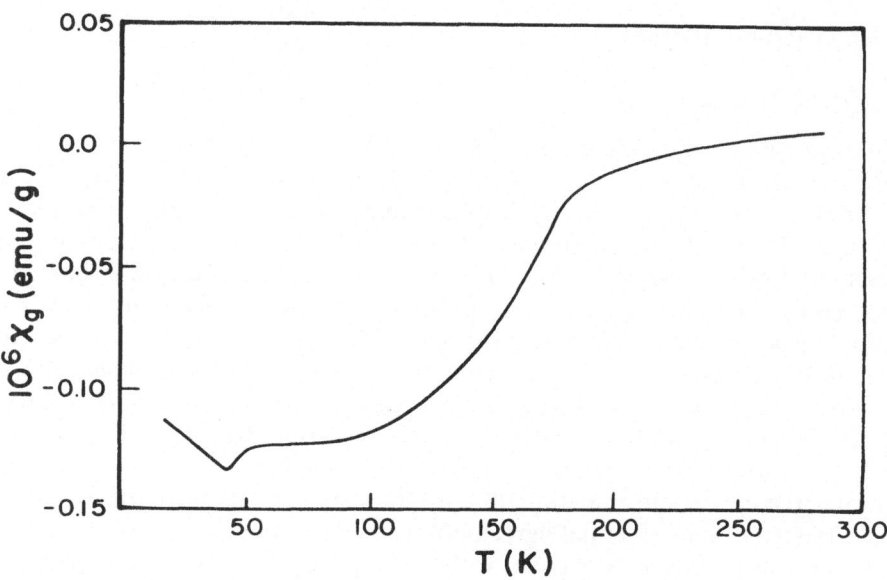

Fig. 5. Temperature dependence of the magnetic susceptibility of $Tl_{0.3}MoO_3$.

investigated. Substitution of nonisoelectronic vanadium [25] in $K_{0.3}Mo_{1-x}V_xO_3$ with $x = 0.04$ stabilizes the semiconducting state and no transition is found in the resistivity. It is possible that the smaller V^{5+} (0.46 Å) ions substituting for Mo^{6+} (0.59 Å) produce some structural distortion which stabilizes the semiconducting phase. Alternatively, the V ions may interfere with the overlap of Mo t_{2g} and O $p\pi$ orbitals and induce localization. The substitution of Rb for K even in large concentrations (50 at. %) does not modify the physical properties strongly as expected, since the properties of the K and Rb blue bronzes are very similar and the alkali metal does not participate in the band structure [30]. In contrast, substitution of isoelectronic W for Mo has a dramatic effect and decreases the transition temperature sharply even at doping levels of W ($K_{0.3}Mo_{1-x}W_xO_3$) with $x = 0.004$); the metallic-type behavior vanishes at $x = 0.03$ W content at room temperature and below [30]. Iron ions appear to substitute for the alkali metal ions and act as impurity traps pinning the CDW [80]. A recent study of cesium substitution indicates that large concentration of Cs (~ 50 at. %) in $(Rb_{1-x}Cs_x)_{0.3}MoO_3$ yields the blue bronze with little change in the electronic properties except for a shift in the metal to semiconductor transition to lower temperature; this is somewhat surprising, since the Cs analog of $A_{0.3}MoO_3$ does not appear to form [83].

A variety of new phenomenon associated with electrical conduction by a moving or 'sliding' CDW in the blue bronzes including nonlinear conductivity, high frequency voltage oscillations, frequency dependent conductivity and very low frequency phenomena have been reviewed by Schlenker et al. [11, 31], and will be treated in another chapter of this book.

3.4. THE PURPLE BRONZES

3.4.1. Crystal Structure

The first purple molybdenum bronze, $Na_{0.9}Mo_6O_{17}$ was prepared more than twenty years ago by Wold et al. by fused-salt electrolysis [18]. The crystal structure of this bronze was partly solved by Stephenson on a twinned crystal who reported an 'average' distorted perovskite-type structure [84]. Reau et al. have prepared polycrystalline specimens of $A_{0.9}Mo_6O_{17}$ with A = Li, Na and K and indicated that all three compounds are isostructural, pseudo-hexagonal, monoclinic phases [34]. Gatehouse et al. prepared untwinned single crystals of the Li, Na and K purple bronzes and attempted to solve the structure of each by X-ray analysis [85, 86]. They solved the structure of the Na and K phases based on centered monoclinic symmetry, however, they were unable to solve the structure of the Li compound, which they reported to have primitive monoclinic symmetry.

A more recent redetermination [87] of the crystal structure of $K_{0.9}Mo_6O_{17}$ shown in Figure 6 is in essential agreement with that reported by Gatehouse et al. [86]. The crystal structure is trigonal with space group $P\overline{3}$. The ideal structure of hexagonal $K_{0.9}Mo_6O_{17}$ can be described in terms of slabs of Mo—O corner-sharing polyhedra. Each slab of Mo—O polyhedra consists of four layers of ReO_3-

like MoO_6 octahedra sharing corners which are terminated on either side by a layer of MoO_4 tetrahedra which share corners with adjacent MoO_6 octahedra. These slabs are perpendicular to the c axis and are separated from each other by a layer of potassium ions in a KO_{12} icosahedral environment of oxygens. The MoO_4 tetrahedra in adjacent layers do not share corners, so that the Mo—O—Mo bonding, infinite in the a and b directions, is disrupted in the c direction. The effective Mo valences are $+6$ on the Mo in tetrahedral sites and $+5.1$ and $+5.8$ on the two crystallographically nonequivalent Mo in octahedral sites. Thus the $4d$ electrons of molybdenum atoms are located in the two-dimensional slabs of octahedra and the structural properties should lead to a very anisotropic Fermi surface which is consistent with the observed quasi two-dimensional conductivity [26]. The structure is very similar to the monoclinic form of Mo_4O_{11} [88].

Very recently the structure of $Li_{0.9}Mo_6O_{17}$ has been solved by Onoda et al. [89]. In Figure 7 the crystal structure projected in the ac plane shows significant differences from that of the $K_{0.9}Mo_6O_{17}$ structure (Figure 6). There are six crystallographically unique molybdenum ions in the unit cell; Mo(3) and Mo(6) are tetrahedrally coordinated while the other molybdenums occupy octahedral sites. Chains of four MoO_6 octahedra corner share along the [102] direction and ReO_3-like slabs are formed by three corner-sharing MoO_6 octahedra. The slabs are interconnected by corner-sharing octahedra and tetrahedra (Figure 7) along the

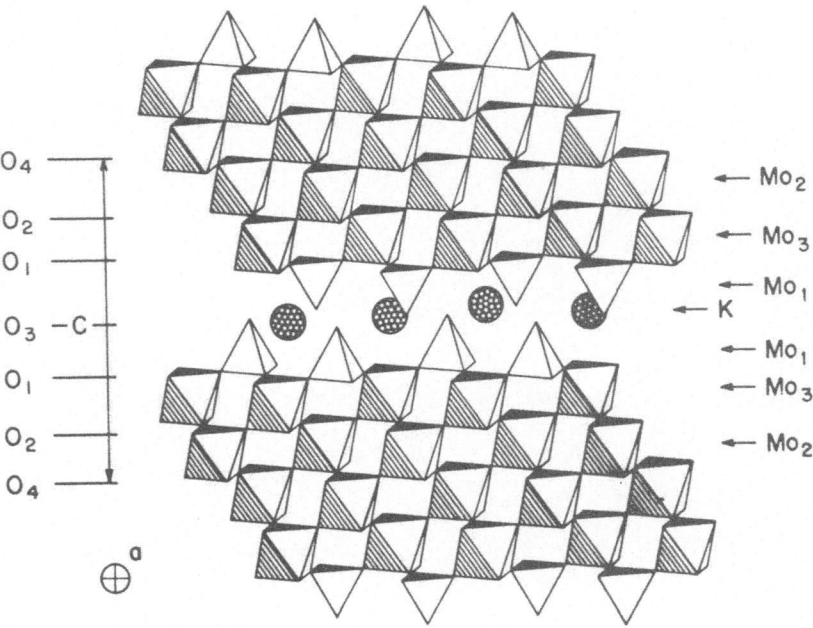

Fig. 6. Idealized structure of $K_{0.9}Mo_6O_{17}$ viewed along the a-axis showing the infinite layers of corner-sharing molybdenum-oxygen polyhedra stacked along c and held together by K ions located in icosahedral sites (after Ref. [31]).

c direction. Thus, in contrast to the $K_{0.9}Mo_6O_{17}$ structure, the Mo—O—Mo interactions along the *c* direction are uninterrupted. However, the Mo—O—Mo network has a much higher degree of crosslinking in the *ab* plane than parallel to *c*, so a quasi low-dimensional character is maintained. The Li^+ ions occupy large nine oxygen coordinated cavities created by the 3D molybdenum polyhedral network structure. The effective Mo valences are 5.05, 5.72, 5.76, 5.01, 5.76, and 5.76 for Mo(1), Mo(2), Mo(3), Mo(4), Mo(5), and Mo(6), respectively. Thus, in contrast to $K_{0.9}Mo_6O_{17}$, all the molybdenums have appreciable conduction electrons. Nevertheless, most of the electrons are located on Mo(1) and Mo(4) octahedral sites, which are associated in pairs to form —Mo(1)—O(11)—Mo(4)—O(11)— double zig-zag chains, extending along the *b*-axis. The Mo ions involved in MoO_4 and MoO_6 polyhedra interconnecting the slabs each have a valence of 5.76 and fewer unpaired electrons. Therefore the structural properties should lead to highly anisotropic electronic transport.

The crystal structure of $Na_{0.9}Mo_6O_{17}$ has been worked out in detail very

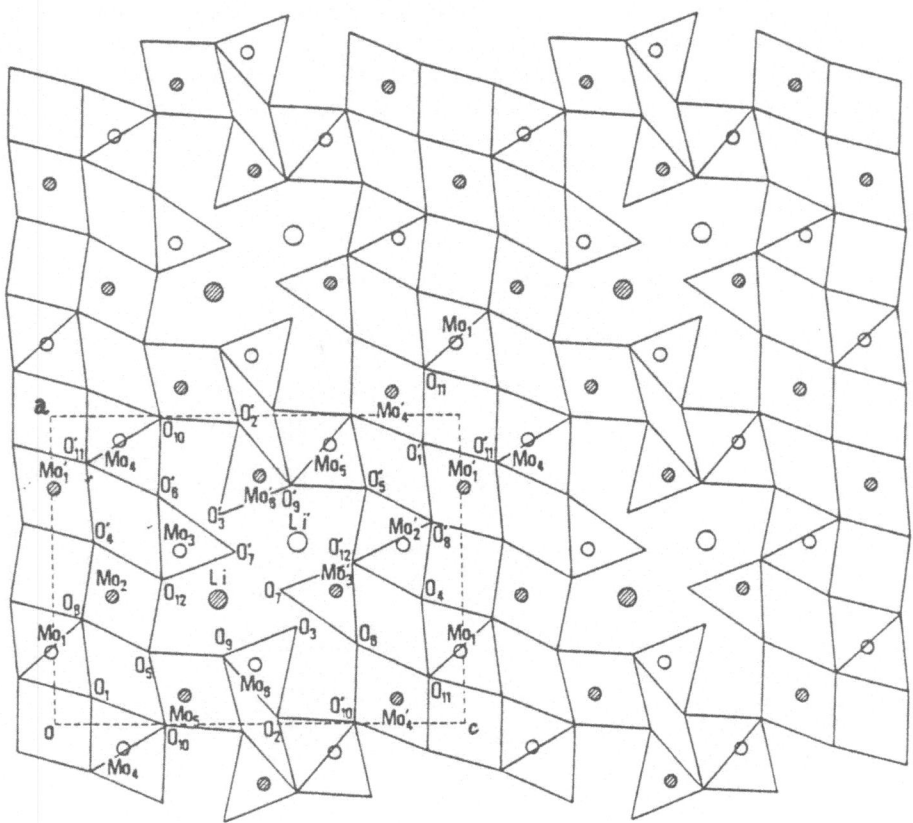

Fig. 7. Crystal structure of $Li_{0.9}Mo_6O_{17}$ projected in the *ac* plane. Open and hatched circles indicate atomic positions at $y = 1/4$ and $3/4$, respectively (after Ref. [89]).

recently [90]; the monoclinic space group of $C2$ symmetry as previously suggested by Gatehouse *et al.* [85, 86] has been confirmed. The structure of the Na compound is essentially identical with that of the $K_{0.9}Mo_6O_{17}$ phase [90].

Very recently single crystals of the thallium analog of the purple bronzes, $TlMo_6O_{17}$, have been prepared by the gradient flux technique [39]. Based on single crystal X-ray determination of the unit cell parameters, this phase appeared to be isostructural with the K compound [39]. This has been confirmed by a very recent structural investigation [91], which shows that although the space group of $TlMo_6O_{17}$ is $P\bar{3}m1$, different from that of $K_{0.9}Mo_6O_{17}$, which is $P\bar{3}$, the structural differences are very small and affect only two of the oxygen positions. Furthermore, the monovalent site is fully occupied only in the thallium bronze [91].

In Table II crystallographic parameters of the $A_{0.9}Mo_6O_{17}$ phases are summarized. The compounds in Table II are listed in order of increasing effective ionic radii for A^+ in a 12-coordinated oxygen polyhedron [92]. When the monoclinic cell parameters of the Li and Na phases are approximated to a pseudo hexagonal cell with $a_{hex} \simeq b$ and $c_{hex} \simeq c$ for easier comparison with $K_{0.9}Mo_6O_{17}$ and $TlMo_6O_{17}$, it is seen that a_{hex} is practically unchanged with increasing effective size of the A cations. This indicates that the 2D network structure of the Mo_6O_{17} infinite slabs is independent of the nature of the A cation (except, of course, in the case of the Li bronze, where the polyhedral Mo network is significantly different. However, even in the Li compound the Mo—O—Mo interactions in the slabs along b are very similar to that in the other purple bronzes). In contrast, c_{hex} increases dramatically with increasing ionic size of A^+ in the bronze. Despite the similarity of the effective ionic radii of Tl^+ (1.70 Å) and Rb^+ (1.72 Å), efforts in our group to synthesize the Rb analog have not been successful so far.

TABLE II
Unit cell parameters of $A_{0.9}Mo_6O_{17}$ phases.

Compound	$Li_{0.9}Mo_6O_{17}$[a]	$Na_{0.9}Mo_6O_{17}$[b]	$K_{0.9}Mo_6O_{17}$[c]	$TlMo_6O_{17}$[d]
Unit Cell	Monoclinic	Monoclinic	Trigonal	Trigonal
Space Group	$P2_1/m$	$C2$	$P\bar{3}$	$P\bar{3}m1$
a (Å)	9.499(1)	9.591(2)	5.538(1)	5.543(2)
	(9.487)	(9.565)		(5.557(3))
b (Å)	5.523(1)	5.518(1)	5.538(1)	5.543(2)
	(5.519)	(5.525)		(5.557(3))
c (Å)	12.762(2)	12.983(2)	13.656(2)	14.000(2)
	(12.748)	(12.978)		(14.030(4))
β (°)	90.61(1)	89.94(1)	—	—
	(90.59)	(90.09)		

[a] Ref. [89]; in parenthesis from NDPPA data Ref. [93].
[b] Ref. [90]; in parenthesis are values for a $Na_{0.88}Mo_6O_{17}$ composition obtained by least-squares refinement of powder data (Ref. [37]).
[c] Ref. [87].
[d] Ref. [39]; values given in parenthesis are from Ref. [91].

It appears that $TlMo_6O_{17}$, like $K_{0.9}Mo_6O_{17}$ has a fixed composition, unlike $Na_{0.9}Mo_6O_{17}$ which has a small range of nonstoichiometry in Na ($0.84 \leqslant$ Na $\leqslant$ 0.96) [34, 37, 39]. Thus, while in $TlMo_6O_{17}$ all the alkali sites are fully occupied, in the Na and K phases these sites are about 0.1 mol^{-1} vacant.

3.4.2. *Charge Density Wave Instabilities*

The results of four-probe resistivity measurements for $Li_{0.9}Mo_6O_{17}$, $Na_{0.9}Mo_6O_{17}$, $K_{0.9}Mo_6O_{17}$, and $TlMo_6O_{17}$ single crystals are shown in Figures 8, 9, 10 and 11

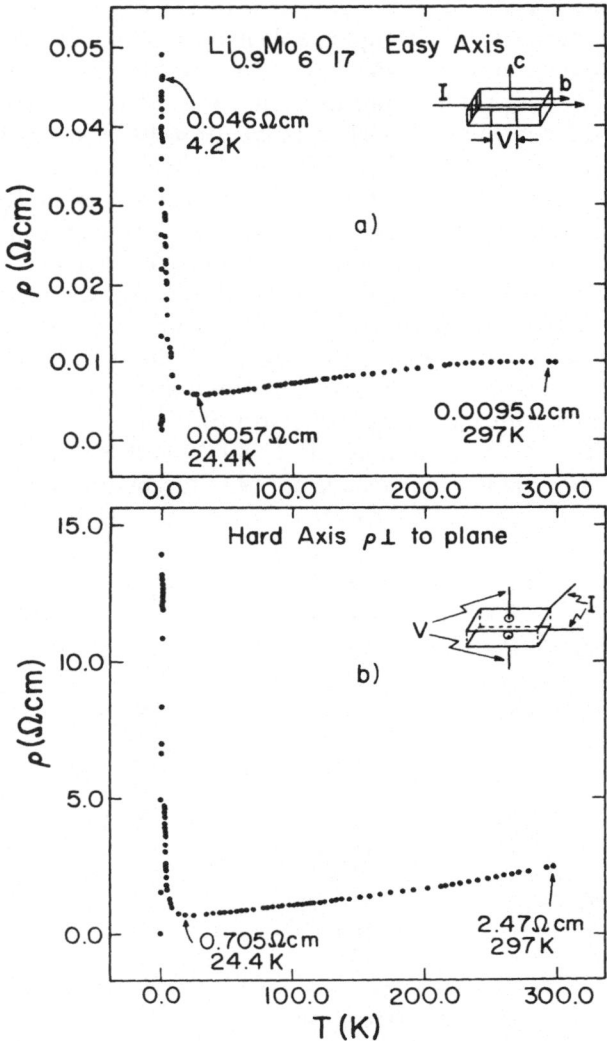

Fig. 8. Temperature variation of the resistivity of $Li_{0.9}Mo_6O_{17}$ from 1.5 K to 300 K. (a) Current in the plane of the platelet crystal. (b) Current perpendicular to plane of the platelet parallel to *c*.

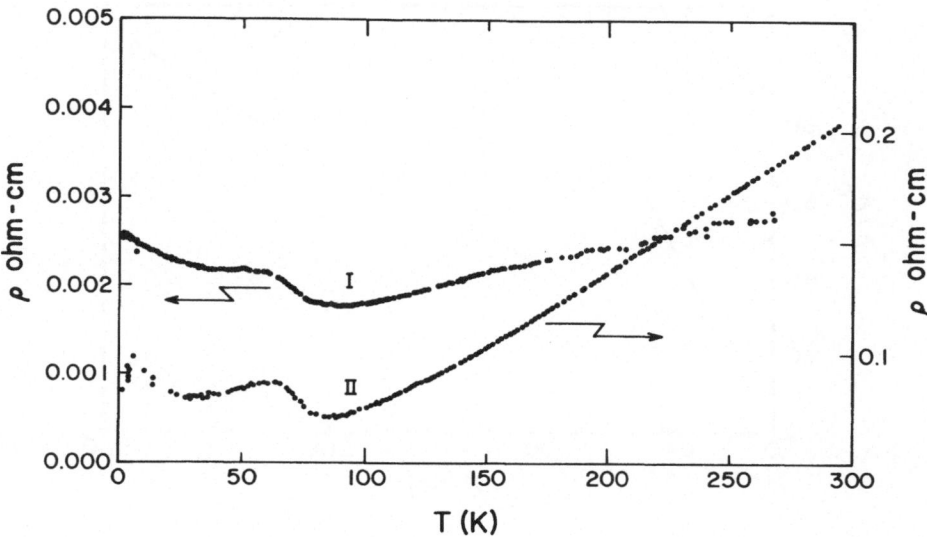

Fig. 9. Temperature variation of the resistivity of $Na_{0.9}Mo_6O_{17}$ from 1.5 to 300 K. (I) Current in the plane of the platelet crystal. (II) Current perpendicular to plane of the platelet parallel to c.

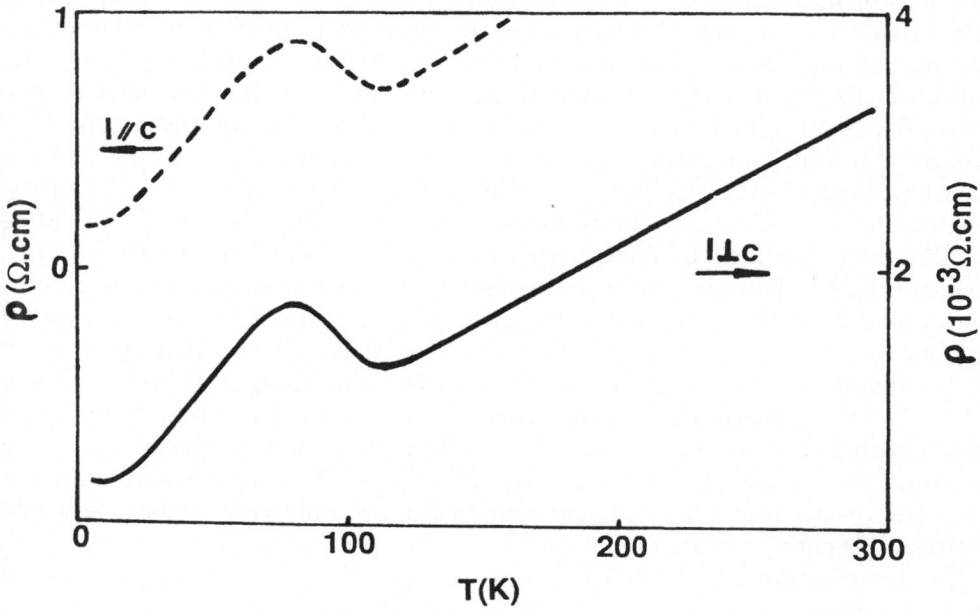

Fig. 10. $K_{0.9}Mo_6O_{17}$: Resistivity vs temperature measured with the current along or perpendicular to the c-axis (after Ref. [26]).

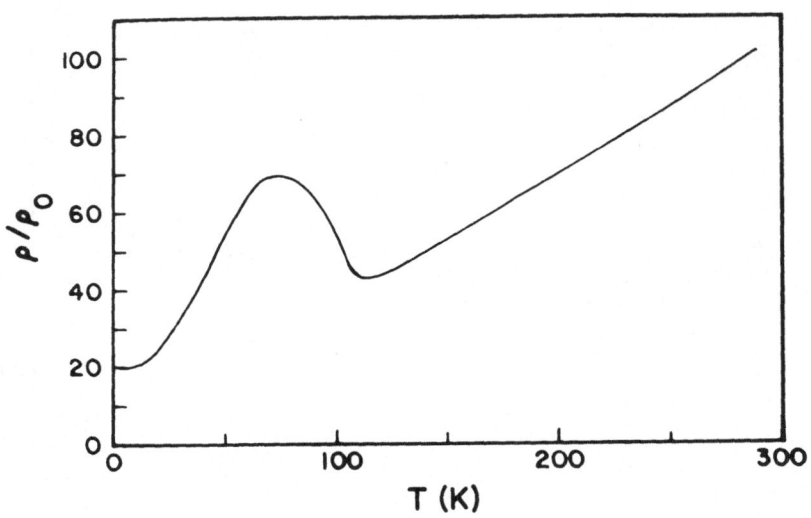

Fig. 11. Temperature variation of the resistivity of $TlMo_6O_{17}$.

respectively [94, 95, 26, 39]. The resistivity is strongly anisotropic and orders of magnitude higher along the c axis than in the plane $\perp$ c, where it is nearly isotropic in each case, except for $Li_{0.9}Mo_6O_{17}$ for which a large anisotropy is present in the ac plane. The purple molybdenum oxide bronzes are quasi two-dimensional metals and exhibit a metal-to-metal transition (T_c) at low temperature. However, the transition temperature at the transport anomaly varies over a wide range, depending on the nature of the alkali metal ion, in contrast to the blue bronzes, for which the metal-to-semiconductor transition, T_{ms} ~ 180 K for the reported three phases ($A_{0.3}MoO_3$, A = K, Rb, Tl). A change in the sign of the temperature dependence of the resistivity is found at approximately 24 K, 80 K, 120 K, and 113 K for the lithium, sodium, potassium and thallium purple bronzes, respectively. Furthermore, while the resistivity of the Na, K and Tl compounds show a maximum at the transition temperature, similar to that found for the quasi one-dimensional compound $NbSe_3$ [7], the lithium phase shows a sharp increase in the resistivity at ~ 24 K and a transition to a superconducting state at T_c ≃ 1.9 K [95—97]. This suggests that the electronic properties of the purple bronzes are somehow affected by the nature of the alkali metal ion, probably by subtle structural variations of the respective compounds. The significantly different structural properties of the Li compound, which was only very recently reported [89] bear this out.

The temperature variation of magnetic susceptibility, χ of $Li_{0.9}Mo_6O_{17}$ shows Pauli paramagnetic behavior and no anisotropy down to 4 K [95, 97]. In $K_{0.9}Mo_6O_{17}$ the susceptibility exhibits large anisotropy similar to the behavior seen in the resistivity and a sharp decrease in the susceptibility at ~ 120 K corresponding to the onset of the anomaly in the resistivity (Figure 12). At T < 70 K

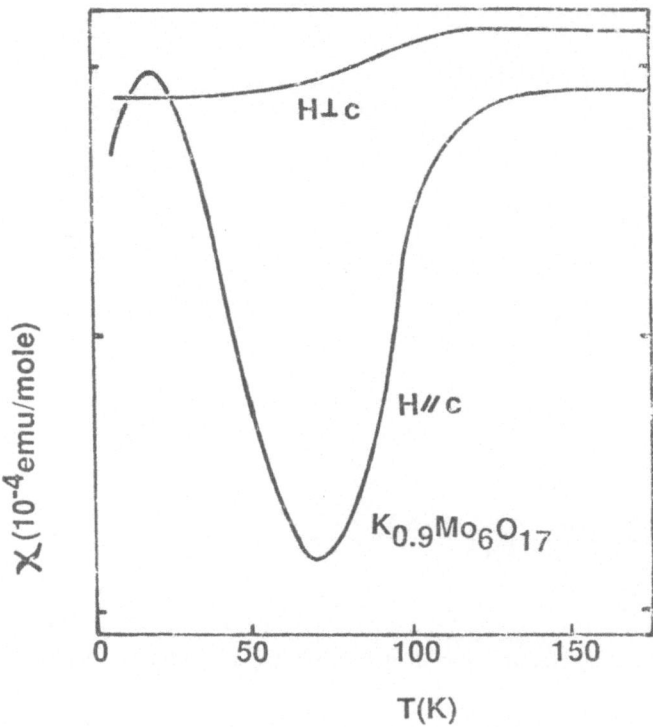

Fig. 12. Temperature variation of the susceptibility of $K_{0.9}Mo_6O_{17}$ measured with the magnetic field parallel or perpendicular to the c axis (after Ref. [98]).

the susceptibility increases and another transition is evident in the vicinity of 30 K [98]. This last transition may be related to the onset of a spin density-wave (SDW) [99]. The sodium purple bronze shows similar complex behavior (Figure 13): there appears to be a small drop in the susceptibility near ~ 80 K, where ρ begins to increase, a definite transition at $T < 100$ K and another transition at $T \sim 40$ K similar to the 30 K transition observed in the K compound. The temperature dependence of the susceptibility of the thallium purple bronze (Figure 14) shows a transition at 110 K which is also seen in the resistivity; the transition near ~ 30 K is not observed in the resistivity (Figure 11) [39, 91].

As was discussed above, the structure of $A_{0.9}Mo_6O_{17}$ bronzes with A = Na, K and Tl consist of layers of corner-sharing MoO_6 octahedra and MoO_4 tetrahedra that make up the Mo_6O_{17} slabs, which are connected by the alkali metal ions. The $4d$ electrons are delocalized primarily on the octahedral molybdenums in the two-dimensional ab planes, which are well separated along the c-direction. This gives rise to the observed quasi two-dimensional metallic behavior. Although the structure of $Li_{0.9}Mo_6O_{17}$ is considerably different from the Na, K and Tl analogs as reflected in its markedly different electronic transport properties, nevertheless, the quasi 2D (or perhaps quasi 1D) character is maintained.

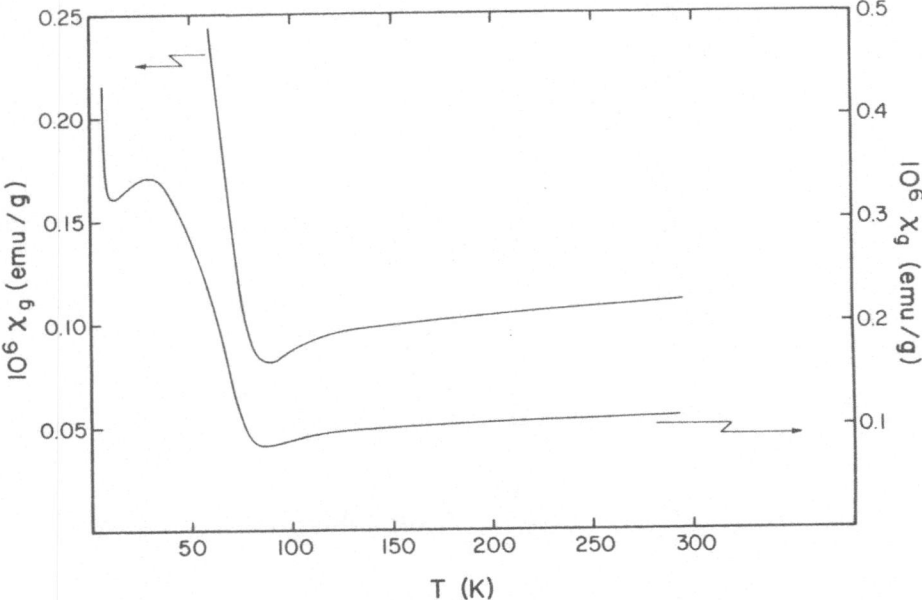

Fig. 13. Temperature variation of the magnetic susceptibility of $Na_{0.9}Mo_6O_{17}$.

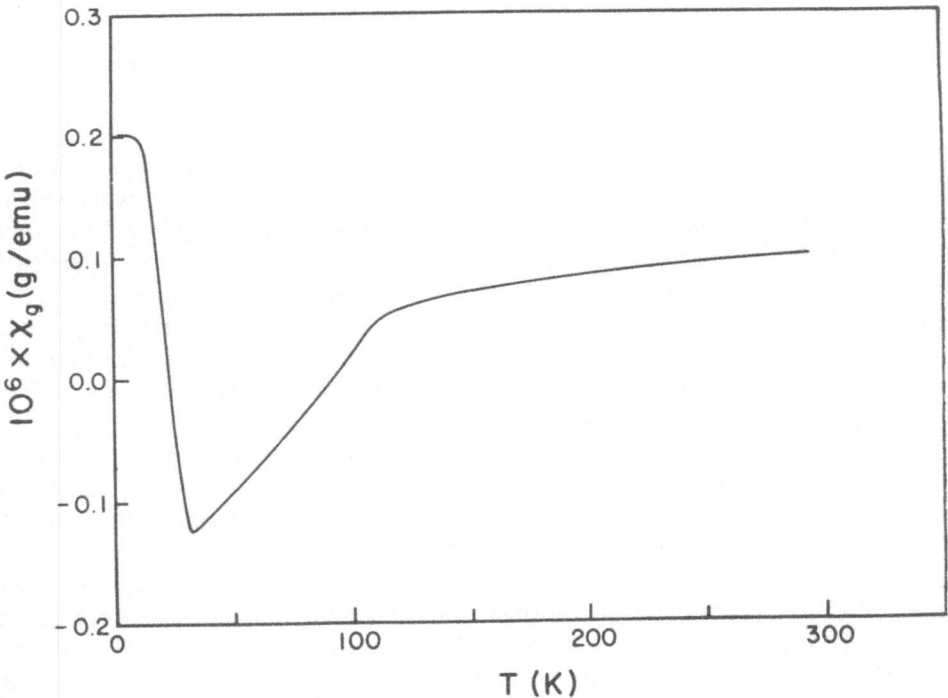

Fig. 14. Temperature dependence of the magnetic susceptibility of $TlMo_6O_{17}$.

In quasi two-dimensional metals the Fermi surface may be quasi-cylindrical with respect to an axis perpendicular to the two-dimensional plane, and will often show nesting with a wave vector Q parallel to this plane. Such materials are unstable toward lattice distortions of the wave vector Q, which can lead to an opening of a gap at the Fermi surface. This decreases the electronic energy of the system.

In quasi one-dimensional metals the gap opening at the Fermi surface is normally complete, and the Peierls transition is a metal-non-metal transition. In a quasi 2D metal the Peierls-like transition is associated with partial gap openings only, and therefore a metal-metal transition. In the $A_{0.9}Mo_6O_{17}$ phases the increase in the resistivity and decrease in the magnetic susceptibility at the metal-metal transition results from the loss of Fermi surface due to formation of gaps associated with the CDW state. Both X-ray diffuse scattering and electron diffraction studies have established that the transition seen in $K_{0.9}Mo_6O_{17}$ and $Na_{0.9}Mo_6O_{17}$ at ~ 120 K and ~ 80 K, respectively is due to a CDW instability which leads to a structural transition of a commensurate $2a \times 2a \times c$ state [100, 101]. Diffuse scattering seen at room temperature in the single crystal Weissenberg X-ray photographs of $TlMo_6O_{17}$ may also be an indication of lattice instability [39]. A broad endothermic peak ~ 120 K in the DSC is further evidence of a structural distortion in this compound [39].

Assuming that the basic band structure is primarily determined by the Mo—O polyhedra and in view of the similarity of the Mo_6O_{17} network structure of the Na, K and Tl phases, similar electronic behavior might be expected. We have already noted that the onset of the CDW instability varies from, 80 K for Na, 113 K for Tl to 120 K for K. Although the qualitative shape of the temperature dependence of resistivity at the CDW transition is 'sigmoidal' in these three purple bronzes, the temperature variation of the transport properties show significant differences in detail (Figures 8—11). For example, $K_{0.9}Mo_6O_{17}$ is metallic between ~ 80—2 K, but shows an anomly in χ below ~ 30 K; in $Na_{0.9}Mo_6O_{17}$ the resistivity is metallic in the range ~ 65—30 K beyond the Peierls-like transition, but ρ increases again below ~ 30 K to 1.5 K. A plot of χ vs T indicates even more complex behavior of $Na_{0.9}Mo_6O_{17}$ as discussed above (Figure 13). In $TlMo_6O_{17}$ the resistivity is temperature independent below ~ 5 K down to ~ 1.6 K, the limit of temperature measurement. An unambiguous transition to a superconducting state has been observed only in $Li_{0.9}Mo_6O_{17}$ thus far and nonlinear transport properties attributed to the sliding of the CDW, although sought for, have not been detected in these compounds. The significant differences seen in the transport properties of the $A_{0.9}Mo_6O_{17}$, A = Na, K, Tl purple bronzes as a function of the A cation is in contrast to the $A_{0.3}MoO_3$ blue bronzes (A = K, Rb and Tl) where the transition temperature of the CDW state and general behavior of the transport properties are nearly identical. There is structural evidence that the quasi 2D nature of the $K_{0.9}Mo_6O_{17}$ structure is maintained in all of these analogs. However, the nature of the A metal ion subtly effects the molybdenum environment and valency in the slabs. This appears to have a considerable effect on the band structure.

The anomalous behavior of $Li_{0.9}Mo_6O_{17}$ compared to the Na, K and Tl analogs has been clarified to some extent by the structural determination. Based on the distribution of electrons, which are located primarily on the double chains of

Mo(1) and Mo(4) along the b axis, the large anisotropy observed in the conductivity is accounted for ($\sigma_b : \sigma_a : \sigma_c \simeq 250 : 10 : 1$). The question whether the sharp upturn in the resistivity at 24 K is due to a CDW driven transition, or to a localization effect, has not been answered unambiguously. Sato and co-workers cite the absence of a decrease in the magnetic susceptibility in the vicinity of 24 K as argument against the formation of a CDW state [97]. However, similar effects have been observed in the susceptibility of $NbSe_3$ where the CDW onset at 32 K affects only the resistivity [102]. So whether CDW or localization is responsible for the anomaly in the resistivity at 24 K can only be resolved by diffraction experiments below T_c. Matsuda *et al.* attempted to find evidence for a competition of CDW and superconductivity in mixed $(Li_{1-x}K_x)_{0.9}Mo_6O_{17}$ and $(Li_{1-x}Na_x)_{0.9}Mo_6O_{17}$ phases, but the transition from superconductivity to CDW behavior is abrupt at $x = 0.48$ and $x = 0.4$, respectively, due to the structural phase transitions [97].

The temperature variation of the thermoelectric power (TEP) (Figure 15) shows that S is negative in the whole range of measured temperature. In the high temperature region TEP is temperature independent; there is a downturn at ~ 60 K and a minimum near the temperature where the resistivity begins to increase [104]. Interpretation of this data is not yet available.

The specific heat anomaly found for $Li_{0.9}Mo_6O_{17}$ between 22 K and 34 K [31] is in agreement with the transition temperature seen in the resistivity (Figure 8) at ~ 24 K and shows that a considerable amount of short range order persists above T_c. Similar specific heat data of $Li_{0.9}Mo_6O_{17}$ down to 1.6 K shows a deviation from linearity below $T^2 \sim 2.5$ for $Li_{0.9}Mo_6O_{17}$ which most likely corresponds to the

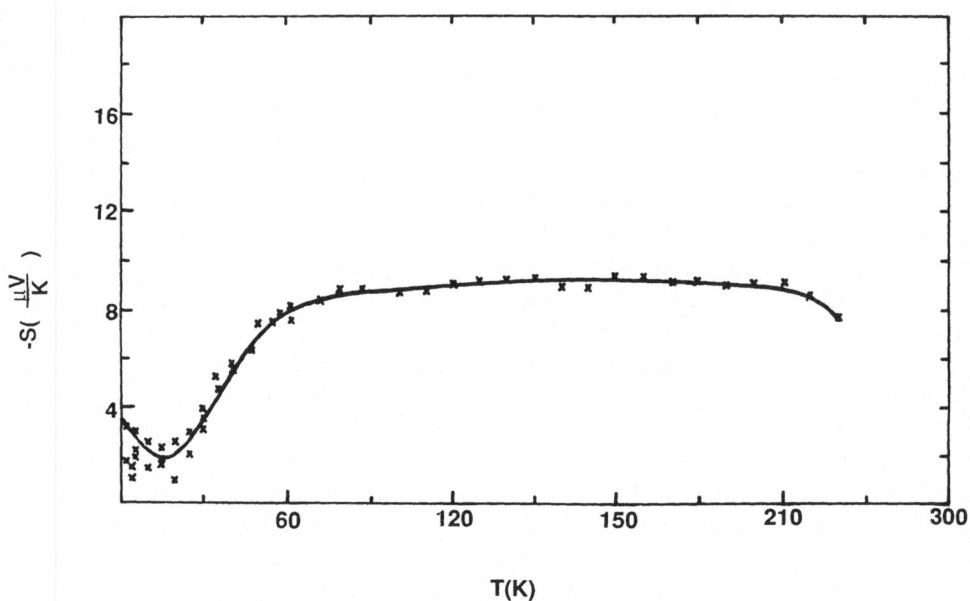

Fig. 15. Thermopower vs temperature measured in the bc plane of $Li_{0.9}Mo_6O_{17}$.

superconducting transition [31, 97], which is seen in the resistivity (Figure 8) and which was also confirmed by a.c. susceptibility measurements at ~ 2 K [95, 96].

The thermopower measured for $K_{0.9}Mo_6O_{17}$ is small and negative above 120 K, becomes positive at 120 K and shows a maximum at 70 K [26]. This indicates that at $T > 120$ K the major carriers are electrons while at $T < 120$ K the dominant carriers are holes. This behavior is consistent with a structural transition occurring below 120 K due to the partial opening of a gap at the Fermi surface, which leads to a change in the concentration of both types of charge carriers.

In $K_{0.9}Mo_6O_{17}$ the electrical resistivity (Figure 10), the specific heat [100] and the intensity of X-ray diffraction superlattice spots [100] as a function of temperature strongly suggest that the transition is due to the opening of gaps at the Fermi surface, which start at ~ 120 K and are fully opened at ~ 80 K. The satellite intensity is temperature independent at $T < 80$ K. The resistivity decreases as the temperature is further lowered $T < 80$ K, presumably because the gaps begin to close. The electronic energy gained by the distortion of the Fermi surface can no longer compensate the energy needed to distort the lattice as the temperature is lowered.

The specific heat of $Na_{0.9}Mo_6O_{17}$ in the temperature range 1.5—40 K (Figure 16) shows an anomaly at ~ 30 K, also seen in the resistivity and susceptibility,

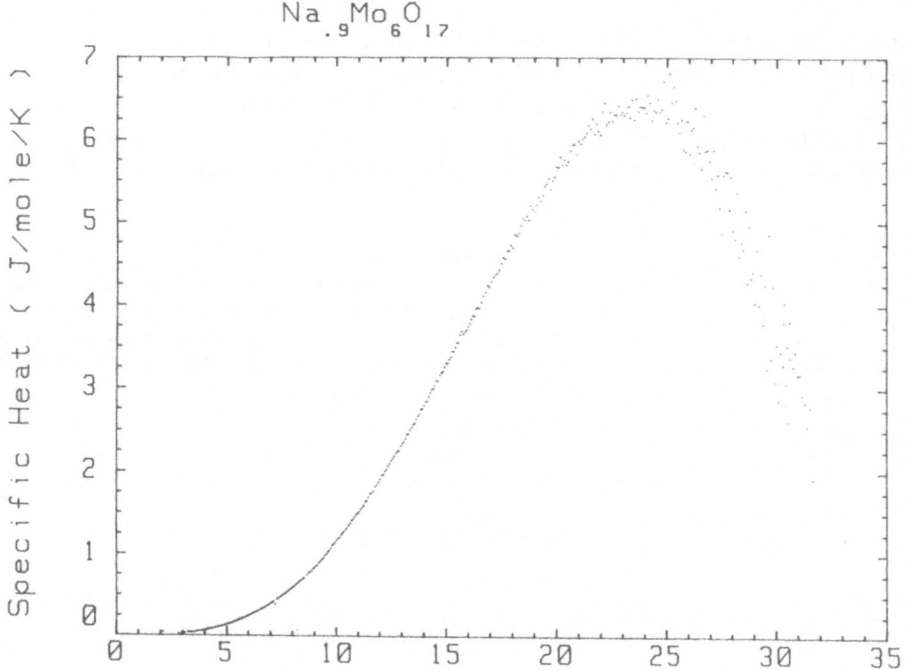

Fig. 16. Specific heat, C_p of $Na_{0.9}Mo_6O_{17}$ as a function of temperature (polycrystalline samples).

which suggests a second transition [104]. The TEP of $TlMo_6O_{17}$, above ~ 120 K, is negative and increases linearly with decreasing temperature as expected for metallic behavior. An anomaly is observed at ~ 120 K and S increases sharply and becomes positive indicating that at low temperatures the carriers are mainly holes [91]. The specific heat measured from 300 to 90 K shows a soft maximum at ~ 117 K confirming a phase transition near this temperature [91], which leads to the partial opening of gaps at the Fermi surface suggested by the transport and magnetic properties (Figures 11, 14).

The thermal data of the purple bronzes are summarized in Table III. γ, the coefficient of the linear term in the low-temperature heat capacity is significantly larger for the Li and Na compounds. This may be related to electron correlations and in $Li_{0.9}Mo_6O_{17}$ is consistent with the larger resistivity below T_c. The largest Debye temperature, θ_D of $Li_{0.9}Mo_6O_{17}$ reflects the increased three-dimensionality as evidenced by the 3D network structure, and is related to the smaller size of the Li^+ ion and the tighter coordination of Li^+ vs Na^+, K^+ and Tl^+, as indicated by the structural data. The onset of the CDW instability, T_c increases from Li to K with increasing effective ionic radius of the alkali metal (in 12-coordination); T_c of Tl is anomalously low in terms of the larger effective ionic radius of Tl compared to K. However in the Tl phase, the alkali metal sites are fully occupied, which may give it the additional stability against the Peierls-like distortion.

The magnetoresistance studies of $K_{0.9}Mo_6O_{17}$ are consistent with a quasi-cylindrical Fermi surface and with open orbits along the c axis [105].

Dramatic effects are observed in the electronic properties of tungsten doped $Li_{0.9}Mo_6O_{17}$ [40, 90, 97]. Even at the lowest W-doping levels (~ 0.18 at. wt %) the superconducting state near 2 K disappears. A rounding and broadening of the metal—metal transition temperature is seen with increasing tungsten content which

TABLE III

Thermal data for the purple bronzes and molybdenum oxides Mo_4O_{11}: T_c is the onset temperature of the charge density-wave transition. γ is the coefficient of the linear term in the low-temperature heat capacity. θ_D is the Debye temperature, ΔH is the enthalpy variation associated with the transition and ΔS is the entropy of the transition obtained as $\Delta H/T_c$. The accuracy of ΔH and ΔS is limited by the difficulty in estimating the background heat capacity.

	T_c (K)	ΔS (mJ mol^{-1} K^{-1})	ΔH (J mol^{-1})	γ (mJ mol^{-1} K^{-2})	θ_D (K)	Ref.
$Li_{0.9}Mo_6O_{17}$	24	240	2.6	5	365	[31]
				6	410	[97]
$Na_{0.9}Mo_6O_{17}$	80	—	—	10	290	[31]
$K_{0.9}Mo_6O_{17}$	120	530	64	1.6	320	[31]
				3.9		[97]
$TlMo_6O_{17}$	120	314 ± 17	36 ± 2	—	—	[91]
	117					[39]
η-Mo_4O_{11}	109	560	70	3	330	[31]
	25	37	0.9	—	—	
γ-Mo_4O_{11}	100	—	—	2.6	410	[31]

may be attributed to local distortions around the substitutional impurity. In contrast to $K_{0.3}Mo_{1-x}W_xO_3$ in which T_{ms} decreases with increasing W content, in $Li_{0.9}Mo_{6-x}W_xO_{17}$ the transition temperature increases with increasing tungsten content. In addition to the metal—metal transition at ~ 24 K probably associated with a CDW state, another activated conduction state is observed at lower temperature as shown in the log ρ vs $1/T$ plots in Figure 17. This low temperature anomaly is attributed to impurity defect levels in the gap region. Phases containing more than 6% tungsten are semiconducting over the measurement range of 298—1.5 K. These strong effects are similar to what was observed in $K_{0.3}Mo_{1-x}W_xO_3$ [30] and indicates that W substitution in the Mo polyhedra interferes with overlap of Mo $t_{2g}4d$ and oxygen $\pi 2p$ orbitals which determine the band structure.

The effect of alkali metal ion substitution on the electronic properties of $Li_{0.9}Mo_6O_{17}$ was studied in $(Li_{1-x}K_x)_{0.9}Mo_6O_{17}$ and $(Li_{1-x}Na_x)_{0.9}Mo_6O_{17}$ phases [40, 90, 97]. Substantial Na and K content merely broadens the metal-to-metal transition at 24 K and the superconducting state is still observed at ~ 2 K for $x \leqslant$ 0.4 and $x \leqslant$ 0.48, respectively (Figure 18). This is taken as strong evidence that the transition at 24 K is electron-phonon mediated and driven by a CDW instability primarily due to the anisotropic Fermi surface of the quasi two-dimensional Mo_6O_{17} lattice, rather than a localization due to disorder (Anderson type).

The only study of pressure effects on the electronic properties of the purple bronzes was carried out on $K_{0.9}Mo_6O_{17}$ [106]. This data indicated that the room temperature resistivity decreased under pressure, which may be consistent with a semimetallic behavior; the pressure increases the overlap of the highest occupied bands and therefore the carrier concentration. The temperature of the onset of the

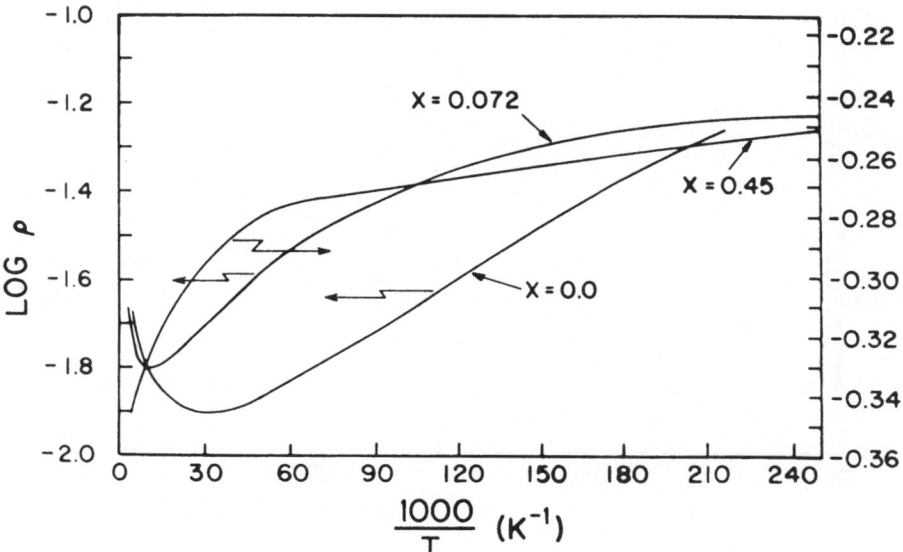

Fig. 17. Log(resistivity) vs inverse temperature for $Li_{0.9}Mo_{6-x}W_xO_{17}$.

CDW appears to be decreased by low applied pressure ($P \sim 0.50$ GPa) similar to what has been found in layered dichalcogenides [107] and where it has been attributed to the distortion of the Fermi surface under pressure. The nesting of the Fermi surface, responsible for the CDW would be reduced by the pressure and therefore the onset temperature decreased.

At low temperature, the increase of the resistivity under pressure (Figure 19) suggests a semiconducting type behavior which may indicate a weak localization of carriers due to pressure induced defects. This is similar to the low temperature properties of the Fe-doped $K_{0.9}Mo_6O_{17}$ under zero pressure [106] and also to that of $Na_{0.9}Mo_6O_{17}$ where impurities and defects may be responsible for the increase in ρ and χ seen at low temperature, due to weak Anderson localization.

3.5. THE RED BRONZES, $A_{0.33}MoO_3$

Wold *et al.* reported on the preparation of the first red molybdenum bronze $K_{0.33}MoO_3$ by fused salt electrolysis [18]. They have found $K_{0.33}MoO_3$ semiconducting with a positive temperature dependence and a room temperature resistivity of $\sim 10^4$ Ω cm. Later Bouchard *et al.* confirmed these results and showed that the susceptibility was nearly temperature independent and positive [19]. They attributed this behavior to spin-pairing of the d electrons or the formation of Mo^{4+} ions (d^2) with low spin configuration ($S = 0$). Stephenson and Wadsley determined the crystal structure and showed that the monoclinic structure (space group $C2/m$) is built of infinite sheets of corner/edge-sharing distorted MoO_6 octahedra which are joined together by the K^+ ions which are in

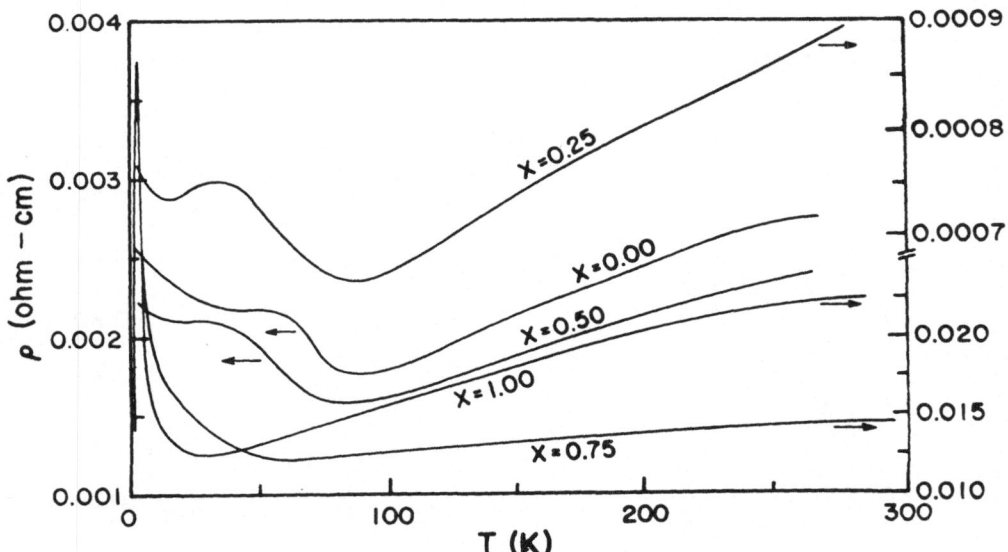

Fig. 18. Temperature variation of the resistivity of $(Na_{1-x}Li_x)_{0.9}Mo_6O_{17}$.

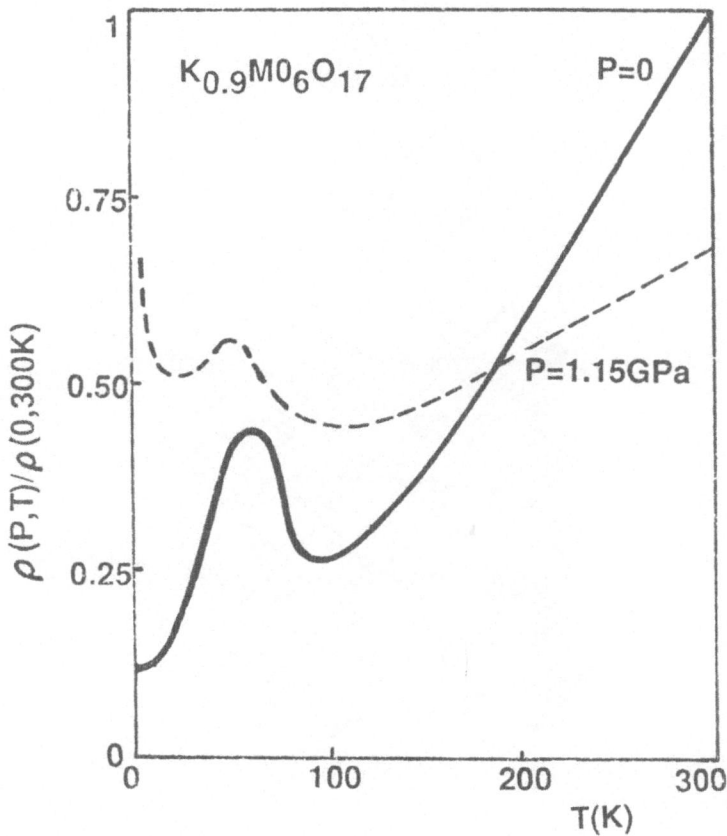

Fig. 19. Resistivity vs temperature at zero pressure and at 1.15 GPa for $K_{0.9}Mo_6O_{17}$. The resistivity is measured in the plane perpendicular to the c-axis (after Ref. [106]).

an irregular eightfold coordination [108]. The structure is very similar to that of the blue bronze except that in the red bronze the unit of structure in six edge-shared octahedra which then corner share along b and the [102] direction to form the infinite layers (Figure 20). If all the potassium sites are occupied, $K_{0.33}MoO_3 = K_2Mo_6O_{18}$, hence two electrons per Mo_6O_{18} cluster. Dickens and Neild [13] measured the electronic spectra and showed the donor levels to be at least 0.75 eV below the bottom of the conduction band (Figure 1). They also measured the ESR spectra and found a well-resolved signal at $g \sim 1.96$ which they attributed to Mo^{5+} centers. ESR studies by Bang and Sperlich [109] indicated two types of resonances; one of which was assigned to unpaired d electrons delocalized over six MoO_6 octahedra (i.e. in one cluster) while the other to pairs of exchanged coupled d-electrons [109].

More recently Travaglini et $al.$ [110] measured the optical reflectivity of $K_{0.33}MoO_3$ single crystals in the infrared and visible range in the temperature

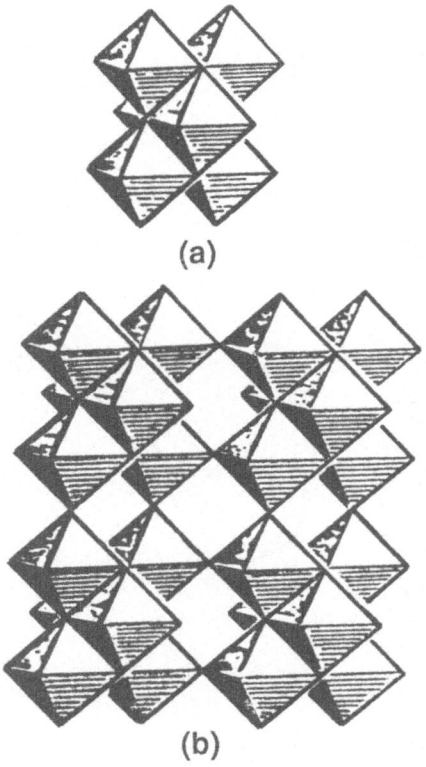

(a)

(b)

Fig. 20.   Idealized structure of $K_{0.33}MoO_3$ (after Ref. [108]).

range 4—300 K using polarized light. These data indicated that this compound is a
0.5 eV energy gap semiconductor with very strong anisotropy in the infrared and
visible range. These investigators derived a microscopic model to explain the
metal—semiconductor 'transition' between the blue bronze and the red bronze in
terms of a formal Mott transition [111]. This is attributed to the substantial
elongation of Mo—O bond distances along the b-axis in $K_{0.33}MoO_3$ compared to
those in $K_{0.3}MoO_3$ such that the dt_{2g}—$p\pi$, Mo—O—Mo overlap is less than
necessary for delocalization. It would be interesting to see the effect of pressure
along the b-axis on the electrical properties.

Reau et $al.$ reported on the red bronze $Rb_{0.33}MoO_3$ which, based on its unit cell
parameters, appears to be isostructural with the K analog [35]. In another
communication these authors reported on a blue-violet monoclinic phase,
Li_xMoO_3 with $0.31 \leqslant x \leqslant 0.39$ which they prepared by solid state reaction. Later
single crystals of a blue-violet bronze, $Li_{0.33}MoO_3$ prepared by electrolysis was
shown to be the same phase as the one reported by Reau et $al.$ and shown to be a
semiconductor with room temperature resistivity of 10^5 Ω cm [24]. More recently,

large single crystals of $Li_{0.33}MoO_3$ have been prepared by the gradient flux technique [36]. An X-ray crystal structure determination by Tsai *et al.* [112] showed that this bronze is triclinic, not monoclinic as previously reported. The structure is completely different from that of the K and Rb red bronzes, forming a 3D network of corner and edge sharing distorted MoO_6 and LiO_6 octahedra. The location of lithium atoms at completely occupied octahedral sites establishes the stoichiometric composition $Li_{0.33}MoO_3 = LiMo_3O_9$. Along the *c* direction MoO_6 octahedra are connected by corner sharing to form infinite chains and the Mo—O—Mo bond distances are regular within the critical distance [112] for good dt_{2g}—$p\pi$ overlap. Furthermore, bond length, bond strength type of calculations [113] show the unpaired $4d$ electrons on molybdenums in this chain. Although the electrical conductivity is highest along this direction, the behavior is semiconducting [41].

The structure of the thallium analog of $A_{0.33}MoO_3$ red bronzes has recently been determined [114]. $Tl_{0.33}MoO_3$ is isostructural with the K and Rb phases; it is a *p*-type semiconductor and ρ at room temperature is $\sim 10^5$ Ω cm as in the other red bronzes [38, 114]. Unit cell parameters of the known $A_{0.33}MoO_3$ phases are tabulated in Table IV. As in the blue bronzes, *a*, *c*, and β increase fairly regularly with increasing size of the A^+ cation in the isostructural monoclinic phases (K, Tl, Rb, Cs) while *b* remains remarkably constant but significantly larger than *b* in $K_{0.3}MoO_3$ (Table I). The *c* dimension in $Li_{0.33}MoO_3$, corresponding to the direction of chains of corner sharing MoO_6 octahedra in this bronze is surprisingly

TABLE IV
Unit cell parameters of $A_{0.33}MoO_3$ phases.

Compound	$Li_{0.33}MoO_3$[a]	$K_{0.33}MoO_3$[b]	$Tl_{0.33}MoO_3$[c]	$Rb_{0.33}MoO_3$[d]	$Cs_{0.33}MoO_3$[e]
Color	Blue	Red	Red	Red	Red
Unit Cell	Triclinic	Monoclinic	Monoclinic	Monoclinic	Monoclinic
Space Group	$P1$	$C2/m$	$C2/m$	$C2/m$	$C2/m$
a (Å)	13.079(2)	14.278(8)	14.537(1) (14.541(3))	14.809(8)	15.862(2)
b (Å)	15.453(2)	7.723(5)	7.7230(5) (7.780(2))	7.726(5)	7.728(2)
c (Å)	7.476(1)	6.387(4)	6.4096(4) (6.414(2))	6.410(4)	6.4080(7)
α (°)	96.97(2)	—	—	—	—
β (°)	106.56(2)	92.34	93.096(5) (93.09(2))	93.8(5)	94.37(1)
α (°)	103.368(9)	—	—	—	—

[a] Ref. [112].
[b] Ref. [57].
[c] Ref. [38]; values in parenthesis reported by Ref. [114].
[d] Ref. [35].
[e] Ref. [115].

short, in fact shorter than b in the blue bronzes. However, metallic behavior is not observed along this direction probably because of band filling.

Reid and Watts prepared monoclinic, copper colored Cs_xMoO_3 bronzes by fused salt electrolysis [23]. The X-ray crystal structure of this phase corresponding to a $Cs_{0.25}MoO_3$ composition was shown to be similar but not identical to that of $K_{0.33}MoO_3$ by Mumme and Watts [22]. Strobel and Greenblatt prepared a red bronze, $Cs_{0.33}MoO_3$ also by fused salt electrolysis and showed it to be semi-conducting with $\rho \sim 10^5 \; \Omega$ cm at room temperature [24]; upon grinding this bronze into a powder, it turned blue and the powder pattern could not be indexed with the cell parameters reported for a red Cs bronze by Mumme and Watts [22]. More recently $Cs_{0.33}MoO_3$ red bronze crystals were grown by the gradient flux technique by Ramanujachary et al. [37]. An X-ray structure analysis showed that this phase is isostructural with $K_{0.33}MoO_3$ [115]; moreover the X-ray powder pattern of the polycrystalline blue phase which results upon grinding of the red bronze crystals could be indexed with the lattice parameters determined by this structural investigation.

Differential scanning calorimeter (DSC) measurements at high pressure indicate a phase transition at 553 °C and 100 psi pressure; T_c decreases with increasing pressure [116]. The pressure variation of resistivity at room temperature, with the pressure applied perpendicular to the platelet direction of a single crystal of $Cs_{0.33}MoO_3$ indicates a phase transition at about the pressure expected from the DSC results. The resistivity of $Cs_{0.33}MoO_3$ decreases by several orders of magnitude with increasing pressure [116]. Measurements of ρ vs P at various temperatures are in progress [116].

A blue colored $Cs_{0.19}MoO_3$ semiconducting phase appears to be yet another structurally different Cs—Mo—O phase [27]. A redetermination of the composition by elemental analysis as well as a structural determination by X-ray diffraction indicates that the actual stoichiometry is $Cs_{0.25}MoO_3$ [117].

This phase is monoclinic with a unit cell that bears no obvious relationship to the unit cells of other alkali molybdenum bronzes (Table IV). The structural determination confirmed that $Cs_{0.25}MoO_3$ has a completely different structure than the $A_{0.3}MoO_3$ blue bronzes [117]. The anomalies in the susceptibility and the resistivity vs temperature suggest that this material undergoes a phase transition near 200 K. The susceptibility anomaly is similar in shape and even the temperature at which it occurs to the anomaly observed in the $A_{0.3}MoO_3$ blue bronzes [27]. However, the resistivity of $Cs_{0.25}MoO_3$ is semiconducting above as well as below the anomaly. Consequently, if the sample is single phase, this transition cannot be due to a CDW. Unit cell parameters of the different semiconducting Cs—Mo—O phases are summarized in Table V. The question of whether or not there are two structurally different red Cs—Mo—O bronze phases remains unresolved.

3.6. $La_2Mo_2O_7$ — A RARE-EARTH MOLYBDENUM BRONZE-LIKE PHASE

Recently single crystals of the purple, bronze-like $La_2Mo_2O_7$ were grown by fused

TABLE V
Crystallographic data for Cs_xMoO_3 bronzes.

	Color	a (Å)	b (Å)	c (Å)	β (°)	Space Group	Ref.
$Cs_{0.33}MoO_3$	Red	15.862(2)	7.728(2)	6.4080(7)	94.37(1)	C2/m	[115]
$Cs_{0.25}MoO_3$	Copper	6.425(5)	7.543(5)	8.169(5)	96.50(5)	$P2_1$ or $P2_1/m$	[22]
$Cs_{0.25}MoO_3$	Blue	19.198(4)	5.519(2)	12.213(2)	119.44(2)		[27]

salt electrolysis [118]. A single crystal structure determination showed that the basic building blocks of the structure are two MoO_6 octahedra which share a common edge to form a fused Mo_2O_{10} unit. These units then share corners to form Mo—O layers as shown in Figure 21. The layers are held together by the lanthanum ions. Although the structure is 3D overall, it can be considered as 2D with respect to the Mo—O network. The Mo—Mo distance in the Mo_2O_{10} groups is 2.478 Å, the shortest Mo—Mo distance thus far reported in molybdenum oxide systems.

$La_2Mo_2O_7$ has a room temperature resistivity of 3×10^{-4} Ω cm [119]. The positive coefficient of the temperature dependence of the resistivity above 200 K indicates metallic behavior. A large increase in the resistivity (Figure 22) at 125 K is probably associated with a phase transition. This anomaly shows up in the magnetic susceptibility, which indicates weakly temperature dependent Pauli paramagnetic behavior above the transition temperature (Figure 22) [119]. The metallic conductivity of $La_2Mo_2O_7$ probably arises from electron delocalization via Mo—O π bond formation as in the bronzes. The quasi two-dimensional nature of electron density leads to a quasi cylindrical Fermi surface and the opening of gaps at the low temperature transition. Further studies are needed to confirm the anisotropy of the transport properties and the CDW instability by X-ray diffraction.

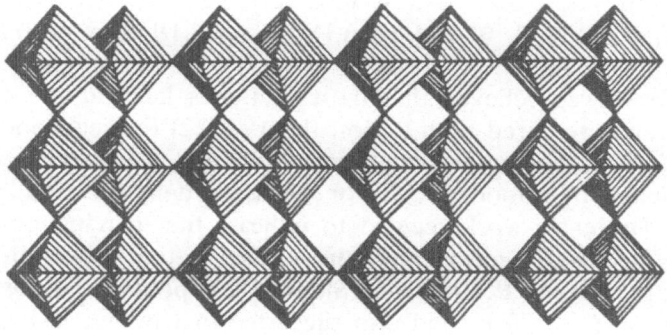

Fig. 21. A section of the $La_2Mo_2O_7$ structure showing corner sharing of Mo_2O_{10} in the crystallographic ac plane (after Ref. [119]).

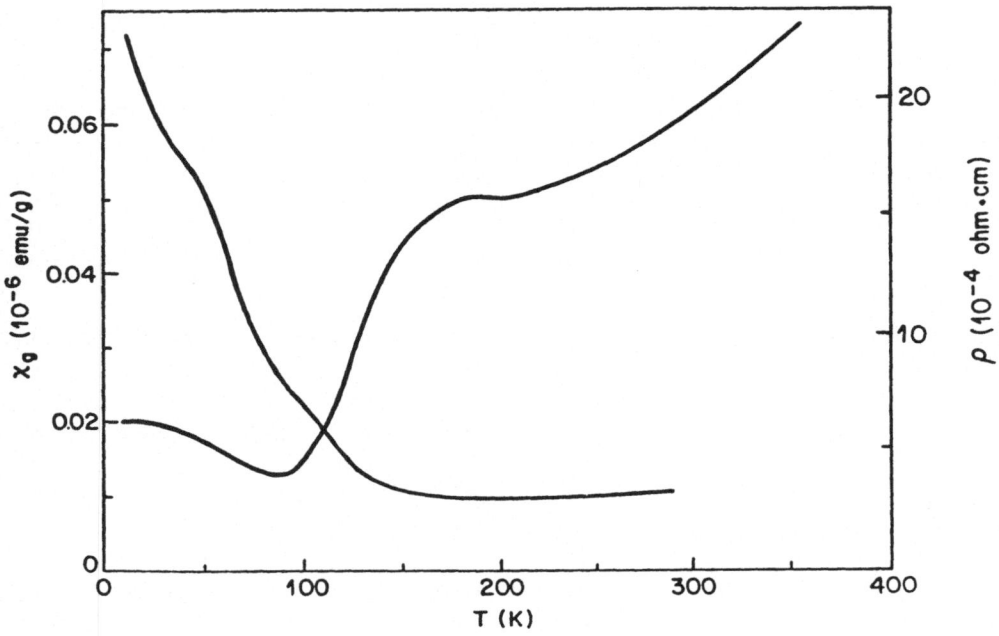

Fig. 22. Temperature variation of the resistivity and magnetic susceptibility of $La_2Mo_2O_7$ (corrected for paramagnetic impurities) (after Ref. [119]).

4. Vanadium Bronzes with Quasi Low-Dimensional Properties — β-$A_xV_2O_5$

4.1. INTRODUCTION

The β-phase vanadium bronzes, β-$A_xV_2O_5$ with A = Li, Na, Cu, Ag, K form in the range of composition: $0.22 \leqslant x \leqslant 0.62$ (Li), $0.22 \leqslant x \leqslant 0.40$ (Na), $0.26 \leqslant x \leqslant 0.64$ (Cu), $0.29 \leqslant x \leqslant 0.41$ (Ag) [120] and $0.19 \leqslant x \leqslant 0.27$ (K) [121], respectively.

In the 1960s it was believed that these materials lie near the critical donor concentration range required by the Mott theory [122] for semiconductor—metal transitions and that the V—V metal—metal spacing was close to the critical distance predicted by Goodenough for electron delocalization via d-orbital overlap [123]. The early work seemed to indicate that conduction occurred by small-polaron (localized electron and surrounding lattice distortion) hopping in these materials [123, 124]. Perlstein and Sienko [124] predicted a possible polaron band formation at $T < 77$ K, and also suggested that by increasing the electron density (e.g. adding more Na to $Na_{0.33}V_2O_5$) in the β-phase there should be a transition to a metallic state according to the theory of Mott [122]. Much of the early work up to ~ 1970 was reviewed by Hagenmuller [5] and by Wold and Banks [3] and up to ~ 1977 by Fotiev [121].

4.2. PREPARATION

Sienko and Sohn [125] prepared single crystals of $Na_{0.33}V_2O_5$ by heating a $6:1$ mixture of V_2O_5 and Na_2CO_3 for several days to $800\,°C$ in a platinum boat in a stream of purified Ar gas. The melt was then cooled at a rate of about $1\,°C/min$ yielding a mass of black needle-like crystals. Modifications of this method have been used for the preparation of $A_xV_2O_5$ single crystal specimens from a V_2O_5 melt and A_2CO_3 in most of the physical measurements published to date. Lack of control of the stoichiometry leads to poor reproducibility of the samples, therefore a technique for crystal growth of these phases with better control over A ion composition and oxygen loss would be important.

4.3. CRYSTAL STRUCTURE

The crystal structure of β-$Na_xV_2O_5$ (NaVB) with $x = 0.287$ was determined by Wadsley [126] and Ozerov et al. [127] in the 1950s and refined by Kobayashi [128] more recently. The structure shown in Figure 23a is made up of chains of vanadium polyhedra with Na^+ ions occupying tunnel sites. There are three crystallographically independent vanadium sites. V(1) and V(2) are each linked to six oxygens in a strongly distorted octahedral configuration. The octahedra are associated in pairs, with a shared edge and form a zig-zag chain along the b axis (Figure 23b). On the other hand, V(3) is five coordinated by oxygen to form trigonal bipyramids, which are also associated in pairs to form a double chain of edge-sharing polyhedra (Figure 23c).

The x electrons transferred by the A cations to the V—O network are located primarily on the V(1) site [123, 129]. The large anisotropy in the electrical conductivity, which is two orders of magnitude larger along the b axis than perpendicular to this direction is attributed to the delocalization of electrons either along the vanadium chains via V—V interactions or via V—O—V interactions (Figure 23d).

The sodium ions occupy two sites in the tunnels in an irregular seven coordination of oxygens. The distance between two sodium sites is only 1.95 Å; the occupancy probability is 50% at $x = 1/3$. Because of Coulombic interactions, sodium ions will tend to be arranged regularly.

The β-$Na_{0.33}V_2O_5$ structure originally determined by Wadsley forms in other $A_xV_2O_5$ systems: with larger A cations such as K, or with bivalent cations (Ca, Ba, Sr, Cd) this structure is stable for $x < 0.33$ [121]; systems which smaller cations $A^+ = Li^+, Na^+, Cu^+, Ag^+$ have single phase regions up to $x = 0.62, 0.40, 0.64$ and 0.41 respectively. Galy et al. [130] have shown that for the small cations alternative tetrahedral or octahedral or eight-coordinated sites in the tunnels are occupied at higher values of x without major changes in the basic β-phase V—O network.

4.4. IS THERE CDW INSTABILITY OR BIPOLARON ORDERING IN β-VANADIUM BRONZES?

In order to explain a linear γT term observed in the low-temperature specific-heat

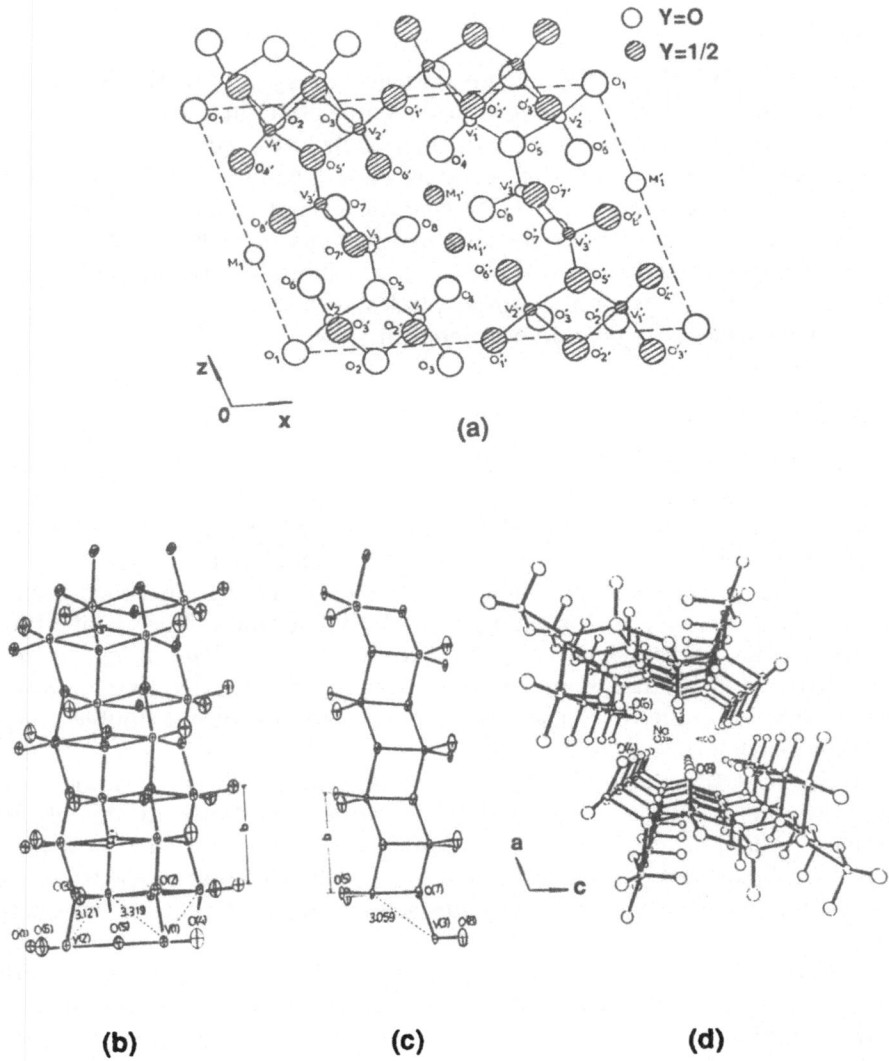

Fig. 23. (a) Projection of the β-Na$_{0.33}$V$_2$O$_5$ structure along [010] (after Ref. [126]). (b) Complex chain constructed of 6-coordinated V(1) and V(2) atoms along the b direction in β-A$_x$V$_2$O$_5$ (after Ref. [128]). (c) A double chain of the 5-coordinated vanadium atoms (after Ref. [128]). (d) Perspective view of the tunnel structure (after Ref. [128]).

behavior of nonmetallic β-A$_x$V$_2$O$_5$ phases, Chakraverty and Sienko [131] postulated that V^{4+} centers form near-neighbor pairs of bipolarons (phonon-induced electron pairs) and attributed the low temperature conducting mechanism to the tunneling motion of these bipolarons along the vanadium chains. The progressive breaking up with temperature of the singlet bipolarons was used to explain the magnetic susceptibility as well as the electrical conductivity.

Kapustkin *et al.* [132] in single crystal resistivity measurements of β-A$_x$V$_2$O$_5$ with M = Li, Na, K in the region 300—900 K, found a reversible semiconductor—metal transition along [010] in the vicinity of 340 K. This was attributed to a second-order phase transition which shifts the vanadium atoms along the *b*-axis such that the activation energy of conduction vanishes.

Further work by others suggested that the β-phase vanadium bronzes should be considered quasi one-dimensional metallic conductors. This view was supported by: (1) the anisotropy of the conductivity, $\sigma_\parallel/\sigma_\perp \sim 10^2$, where $\sigma_\parallel$ and $\sigma_\perp$ are the conductivities parallel and perpendicular to the *b* axis, respectively [133]; (2) optical reflectivity measurements which showed a metallic plasma edge with light polarized along the *b* axis, but a featureless behavior for perpendicular polarization [134]; (3) ESR measurements gave an asymmetric line shape when the electric-field vector (**E**) was parallel to the *b* axis but Lorentzian for $\mathbf{E} \perp b$ [135]; (4) the dielectric constant was highly anisotropic and very large parallel to the *b* axis [136]. These new findings appeared to conflict with the localized electron conduction mechanism.

More recently, Kobayashi showed that the room temperature resistivity of a

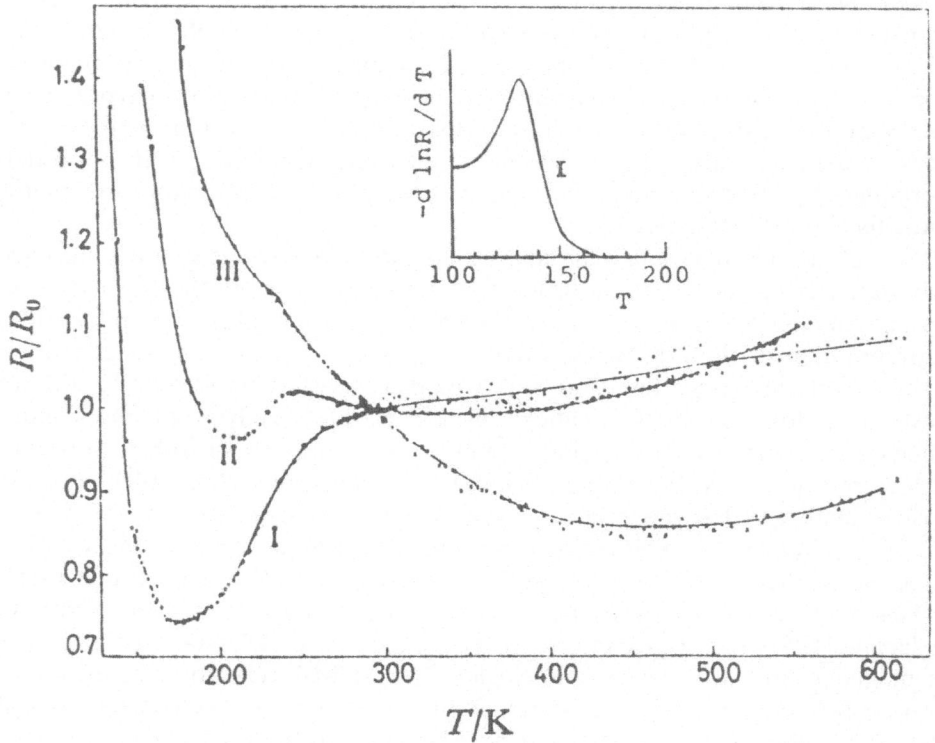

Fig. 24. Temperature dependence of the normalized resistivity of β-Na$_{0.33}$V$_2$O$_5$. Curves I—III indicate the resistivities of three different samples. Insert shows $-\mathrm{d}\ln R/\mathrm{d}T$ versus T (after Ref. [128]).

single crystal of β-NaVB is 6×10^{-3} Ω cm along b and that a metal—insulator (M—I) transition occurs at ~ 130 K (Figure 24) [128]. An anomaly at the same T in the DTA measurement confirmed the M—I transition. Kobayashi sought evidence of X-ray diffuse scatterings associated with one-dimensional charge-density wave instability and found broad diffuse scattering which suggested a doubling of the b axis. This, however, he attributed to the ordering of staggered strings of Na$^+$ ions which alternately occupy one of two possible sites in the tunnel [128].

The Seebeck coefficient of β-NaVB is almost constant at $T > 150$, and appears to be too high for a metallic state, but decreases abruptly in the vicinity of 130 K [124, 137]. Chock and Gruner [137] interpret this thermoelectric power behavior as being due to on-site correlation effects, which are important at all temperatures. Nearest neighbor interactions lead to a gradual ordering of charges and to a semiconducting gap only below ~ 130 K [137]. The magnetic susceptibility of β-NaVB follows a Curie—Weiss law from 150 to 450 K, but its temperature dependence becomes more complicated below T_c [138]. Nevertheless, a lowering of the susceptibility is indicated below the transition temperature [125].

Thus β-NaVB can be regarded as neither a simple metal nor a simple semiconductor, but a combination of both. One-dimensional metallic character originates from the delocalization of states along the vanadium chains parallel to b and a semiconductor behavior due to localization of state in the ac plane. The sample dependence of the resistivity seen in Figure 24 may be interpreted as a superposition of these two contributions. Below the M—I transition temperature R increases exponentially with decreasing T, thus the magnitude of R is primarily determined by a variable-range hopping process; at high temperature the 'metallic conductivity' predominates.

It is well known that small amounts of disorder and defects in a crystal have a large influence on the electrical properties and may smear out the M—I transition [139]. Small changes in the A ion stoichiometry and/or possible oxygen loss during the preparation of these bronzes is partly responsible for the sample dependence of the electrical properties and the varied properties reported by different investigators for seemingly identical phases. Careful analysis of the chemical composition is essential for meaningful interpretation and comparison of the data.

More recent ^{51}V NMR results of β-NaVB by Nagasawa et al. [140] confirmed previous evidence that the x electrons of β-A$_x$V$_2$O$_5$ are located on the Wadsley V(1) and V(3) sites and that a first-order phase transition occurs in the vicinity of 130 K. Subsequent ESR measurements of β-NaVB at 77 K indicated that most of the electrons are located on site V(1) and are coupled along the chain into bipolarons [129], which in contrast to the 'bipolaron' of Chakraverty et al. [131] was magnetic and not always spin-singlet. Their NMR relaxation measurements provided further microscopic evidence for the existence of electron paired states which move along the V(1) zig-zag chains parallel with the b axis giving rise to the one-dimensional conductivity [141b, c].

Nagasawa et al. [140] also found evidence of diffuse X-ray scattering in β-NaVB with wave vector (0, 0.5, 0) which grows below 200 K in agreement with

Kobayashi [128]. In β-Li$_{0.36}$V$_2$O$_5$ they observed two phase transitions by X-ray scattering, one at ~ 220 K with $q = 0.5$ and another below 170 K with $q = 0.53$ and 0.57. Their interpretation of these data appears to be conflicting. In one and the same paper these phase transitions are attributed to commensurate and incommensurate CDW instabilities, respectively [140], on the one hand, and to bipolaron-type electron hopping motion along b as the origin of quasi 1D conductivity, on the other hand. However, in three of their later papers the possibility that the observed phase transitions are due to the Peierls mechanism are ruled out because their magnetic resonance data support a bipolaron motion and not a one-dimensional band model as the mechanism of conductivity in β-A$_x$V$_2$O$_5$ phases [142]. The periodic lattice distortions observed in β-phases with A = Na, Li, Ag, Cu and Pb are explained in terms of a bipolaron ordering model (Figure 25) rather than by the Peierls transition. The fact that the satellite scattering intensity of β-NaVB is smaller than that of Li$_x$V$_2$O$_5$ is given as strong evidence that changes in the V$_2$O$_5$ network due to strong electron—phonon interactions and not A ion ordering are responsible for the superlattice lines.

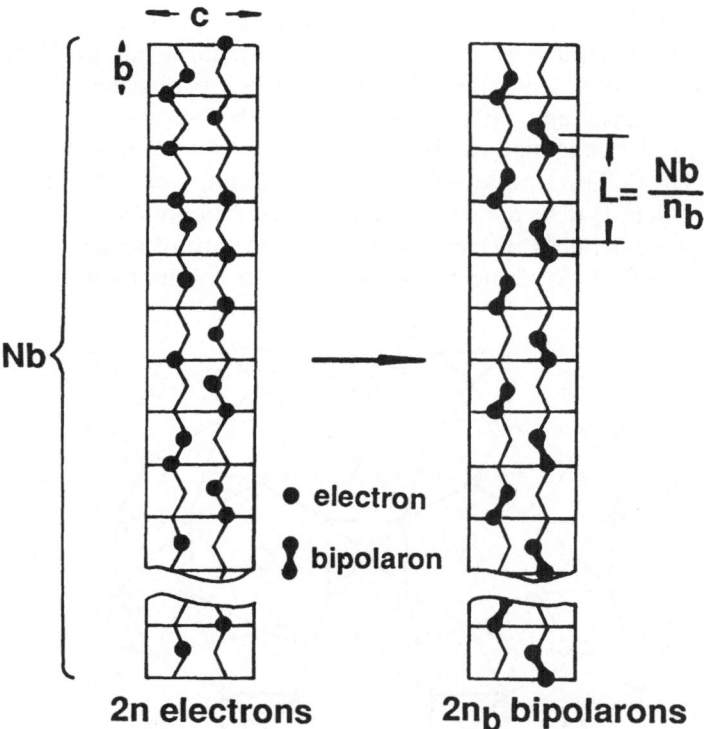

Fig. 25. Schematic sketch of bipolaron ordering on V(1) zig-zag chains. Left, $2n$ electrons distributed on two V(1) chains over N unit cells. Right, $2n_b$ bipolarons ordered, showing how this ordering must be accompanied by a period lattice distortion having a wave vector $\mathbf{Q}_b = (n_b/N)b^*$ where b^* is the reciprocal lattice vector (after Ref. [142b]).

It seems that much further work is needed in this field to sort out unambiguously the mechanism of conductivity in the β-phase vanadium bronzes.

5. Tungsten Bronzes with Quasi Low-Dimensional Properties

5.1. HEXAGONAL TUNGSTEN BRONZES (HTB)

The tungsten bronzes, compounds of composition A_xWO_3, where A is any of about 40 metallic elements and $0 < x < 1$, form a class of nonstoichiometric materials which crystallize in a number of phases, and many have been found to have interesting properties [3, 5, 143]. These compounds are metallic beyond a critical concentration of x as each A atom donates an electron to the WO_3 band. The structure of the compounds depends both on the value of x and on the size of the A cation. For example, Na_xWO_3 with $0.4 \leqslant x \leqslant 1$ has the cubic perovskite structure, and two tetragonal structures for $0.2 \leqslant x \leqslant 0.4$ (TI) and $x < 0.2$ (TII) [3, 5, 143].

A_xWO_3 with A = K, Rb, Cs, NH_4 have a hexagonal structure with large open channels along the c direction as shown in Figure 26. The A cations are located in the channels. Sweedler et al. [145] demonstrated that the hexagonal tungsten bronzes are superconducting. Since then much attention has been paid to these systems which revealed anomalies in the transport properties as a function of temperature. Wanlass and Sienko found that the superconducting transition temperature T_c of hexagonal Rb_xWO_3 $(0.16 < x < 0.33)$ is not a monotonic function of x [146]. The simple BSC theory, assuming constant electron—phonon interaction, predicts T_c to be a monotonic increasing function of x.

Recent measurements of the resistivities as a function of temperature of K_xWO_3 [147] and Rb_xWO_3 [148] along the c axis showed a minimum at a temperature

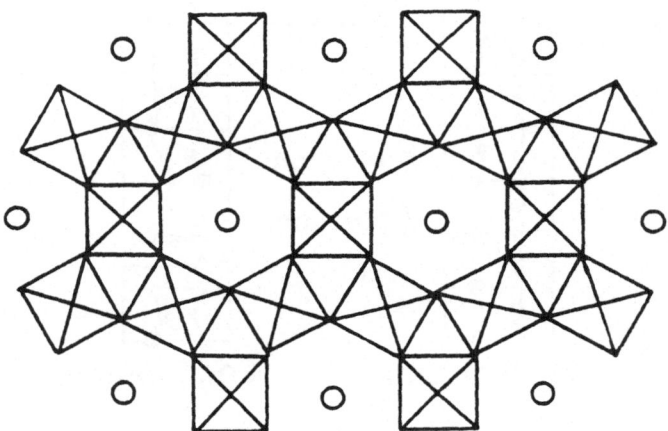

Fig. 26. Schematic figure of the high temperature phase of hexagonal A_xWO_3 projected along [001] (after Ref. [144]).

referred to as T_B, while none was found perpendicular to the c axis. A similar resistivity anomaly had been observed previously in In_xWO_3 [149]. The Hall coefficient is constant above T_B and increases linearly with temperature below T_B; the Seebeck coefficient shows a change in slope near T_B (Figure 27a). The

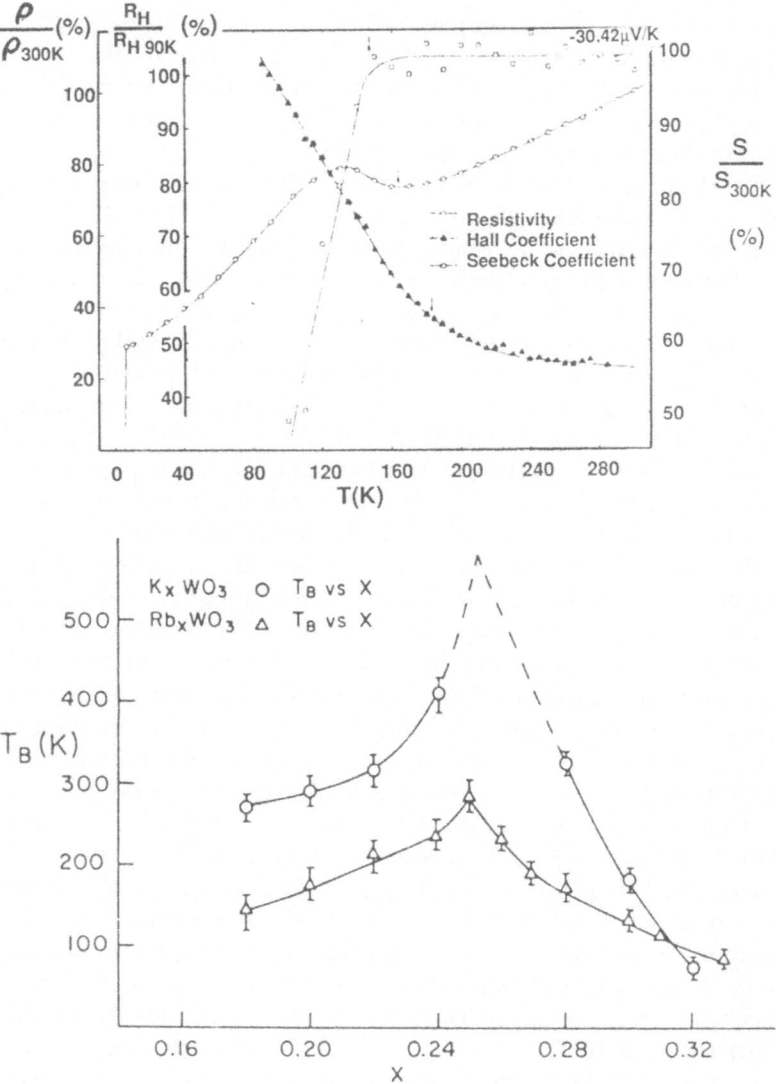

Fig. 27. (a) Temperature dependence of resistivity, Hall coefficient, and Seebeck coefficient for $Rb_{0.18}WO_3$. This figure shows that the high-temperature anomaly is reflected in all three of these parameters. Behavior shown is typical for all Rb concentrations < 0.33 (after Ref. [147b]). (b) Dependence of the onset temperature of the resistive anomaly, T_B, on composition in K_xWO_3 and Rb_xWO_3 (after Ref. [154]).

dependence of T_B on x shows a maximum at $x = 0.25$ for both K_xWO_3 and Rb_xWO_3 (Figure 27b) [147, 148]. This maximum in T_B coupled with an anomalous drop in T_c near $x = 0.25$ suggested a phase transition near $x = 0.25$. However a structural transformation could not be found by X-ray diffraction. Cs_xWO_3 shows no anomalies in the transport properties and the x dependence of T_c is a monotonic decreasing function of x [150].

The lattice dynamics of hexagonal A_xWO_3 have been investigated by specific heat [151] and neutron scattering experiments [152]. These studies established the existence of a low-lying phonon (LLP) with nearly dispersionless Einstein mode behavior in $Tl_{0.33}WO_3$ [151a] and in $Rb_{0.33}WO_3$ [152b] which appeared to be associated with the A ions.

Stanley *et al.* suggested the formation of CDW state as a possible explanation for the observed anomalous transport properties of HTB [147b]. However, if vacancies in the A lattice act as scattering centers, as the data seemed to indicate, T_B should decrease with decreasing concentration of A, as impurities are in general destructive to CDW. As Figure 27b shows, this is not the case. They speculated that a form of ordering of the A cations near $x = 0.25$ commensurate with a possible CDW might assist the formation of the CDW [147].

Sato *et al.*, by neutron diffraction and inelastic scattering experiments, observed two types of phase transition in K_xWO_3 and Rb_xWO_3 HTB [152, 153]. One is associated with a distortion of the WO_3 network at T_{c1}, accompanied by an optical phonon softening at the Γ point. This transition takes place above room temperature and appears to be second order [152]. The other transition at $T_{c2} \sim 250$ K is an order—disorder transition of the A atoms where the ordering scheme depends on x in a systematic way. In $Rb_{0.27}WO_3$ for example, there is a random distribution of the A cation vacancies in the planes; in $Rb_{0.24}WO_3$ on the other hand the vacancies order in two neighboring (001) planes while in the next two (001) planes the A cation sites are completely filled. As the number of vacancies increase (i.e. $Rb_{0.20}WO_3$) there is ordering of vacancies in every plane in a regular way as the temperature is lowered [153]. This structural transition temperature (T_{c2}) agrees reasonably well with T_B. There is some evidence of the coupling of the A atom motion with the host cage (i.e. distortion of the WO_3 cage due to A ion ordering [153]). Sato *et al.* attribute the resistivity anomaly at T_B as being due to the possible formation of a gap at the Fermi surface which results from the new periodicity along the [001] direction at T_{c2}. A local structural excitation model [153] is adopted to explain the x dependencies of the superconducting temperature (T_c) in A_xWO_3 in relation to the ordering schemes of the A ion [153a].

Very recently Krause *et al.*, by single crystal X-ray diffraction, found commensurate superlattices in the a^*b^* planes of K_xWO_3 corresponding to doubling or quadrupling of the volume of the original HTB unit cell [154]. They also observed incommensurate superlattice rows along c which indicated a long range periodicity of 50 to 250 Å depending on x and temperature. A clear relationship between layer periodicity and potassium concentration of the sample could not be established although trends were observed. The layer periodicity was highest at $x = 0.25$ and decreased both towards higher and lower concentrations. While the incommensurate reflections were most pronounced for compositions that do not allow a simple ordering of potassium atoms, the commensurate reflections were particularly strong near compositions where such ordering can take place: $x = 0.25$, $x = 0.29$ or $x = 0.20$. The temperature dependence of the periodicity for any given x is too scattered in most cases to be interpreted reliably. However, a very strong change with temperature was observed for x values much smaller and much larger than $x = 0.25$. The superlattice formation is attributed to the ordering of K ions and the data is consistent with that found by Sato et al. [152].

The exact nature of the structural instabilities in HTB and possible competition between CDW and superconductivity in these phases need further clarification.

5.2. PHOSPHATE TUNGSTEN BRONZES

Raveau and coworkers have recently reported on a series of new phosphate tungsten bronzes (PTB) which promise to have interesting anisotropic properties [155—166]. There are three major structural types: the monophosphate tungsten bronzes (MPTB), $(PO_2)_4(WO_3)_{2m}$ form one of the series whose framework is built up of ReO_3-type slabs of WO_6 octahedra interconnected by PO_4 tetrahedra. Empty pentagonal tunnels form at the junction of the ReO_3-type slabs and the slices of tetrahedra. For the $m = 6$ member the structure is very similar to that of γ-Mo_4O_{11} (Figure 28a) [159]. Closely related is the system of $A_x(PO_2)_4(WO_3)_{2m}$

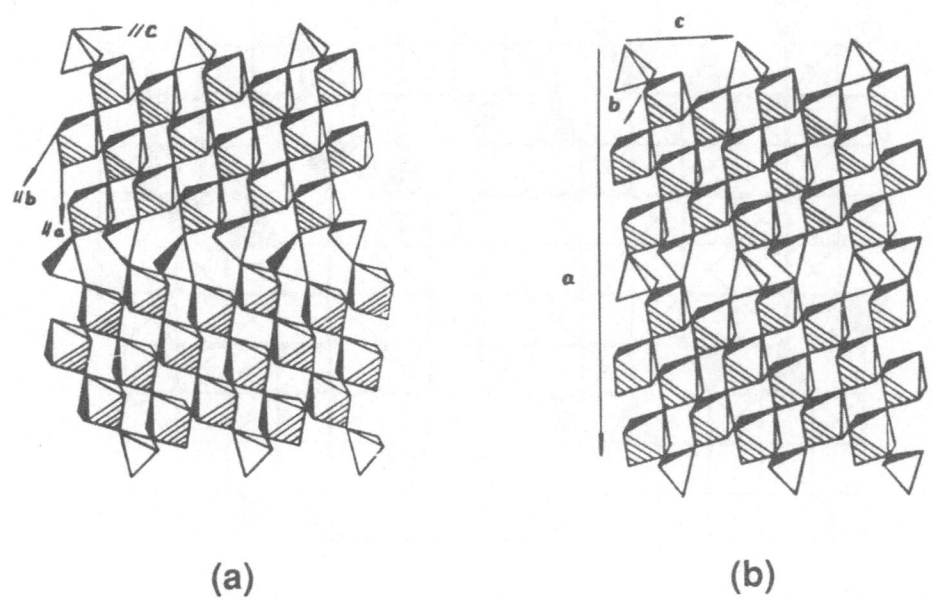

(a)   (b)

Fig. 28. Idealized structures of (a) $(PO_2)_4(WO_3)_{2m}$ and (b) $A_x(PO_2)_4(WO_3)_{2m}$ showing WO_6 octahedra slabs and PO_4 tetrahedra planes.

with A = K, Na, of MBTBs with hexagonal tunnels, which have the η-Mo_4O_{11} structure (Figure 28b); the A cations partially occupy the hexagonal cavities.

The diphosphate tungsten bronzes (DPTB), $A_x(P_2O_4)_2(WO_3)_{2m}$ A = Na, K, Rb, Tl form a closely related series in which the framework is built up from blocks of corner sharing WO_6 octahedra joined by P_2O_7 diphosphate tetrahedra creating distorted hexagonal tunnels in which the A cations are located (Figure 29) [155, 156, 165]. Different members of the series differ by the length of the ReO_3-like chains (indicated by the value of m) in all of these phases.

In general the phosphate tungsten bronzes are characterized by a three-dimensional network structure with interconnected large tunnels. In addition to the large pentagonal or hexagonal tunnels there are perovskite and perovskite-like cubooctahedral cavities which should allow diffusion of cations in these compounds. A recent investigation of the ion exchange properties of $K(P_2O_4)_2(WO_3)_{2m}$ in aqueous solutions confirm fast ionic motion [167]. Furthermore, topotactic reactions with n-butyllithium and sodium naphthalide showed that about one Li or Na/W, respectively, may be reversibly inserted into both the MPTB or DPTB [167], suggesting possible application of these materials as cathodes in secondary batteries.

Preliminary investigations of the electronic transport and magnetic properties

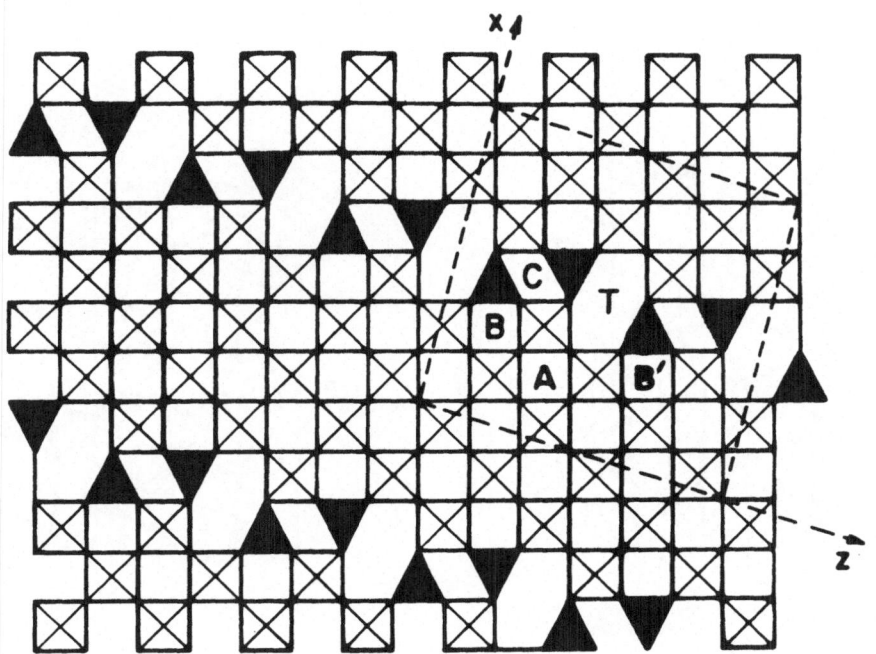

Fig. 29. Projection of the idealized structure of $AP_4W_{16}O_{56}$ ($A(PO_2)_4(WO_3)_{2m}$, $m = 8$) along the [010] direction. The different types of cavities in the structure are: tunnels (T) with distorted hexagonal geometry; cubooctahedral cages (A, B) related to perovskite-like cavities; cages (B′, C) created by 11 oxygen atoms (after Ref. [166]).

indicate highly anisotropic metallic behavior [162, 168]. Anomalies in the resistivity and susceptibility suggest CDW-like phase transitions [168]. The electronic properties of the PTB are expected to be similar to the $A_x WO_3$ phases. Because of the similarity of their structure to the $Mo_4 O_{11}$ phases quasi 2D behavior and a highly anisotropic Fermi surface is expected. Thus the anomalies in the transport properties observed at low temperature are likely to be due to structural instabilities which lead to partial openings of gaps at the Fermi surface.

Acknowledgements

The studies performed at Rutgers on the molybdenum bronzes would not have been possible without the many contributions of collaborators.

W. H. McCarroll, while here on a sabbatical from Rider College, N. J. worked out the gradient flux technique for the crystal growth of the lithium bronzes and continued to guide and contribute to all aspects of the work here. Mark Croft and Rick Neifeld of the physics department helped us start the low temperature resistivity measurements and were helpful in explaining the solid state physics aspect of the problem. K. V. Ramanujachary, a postdoctoral fellow here for three years, grew some fine bronze crystals, built a computer-controlled state-of-the-art device for the measurement of the temperature variation of the resistivity and specific heat, and made important contributions in all areas of research including the substitutional bronze studies. B. Collins, a Ph.D. candidate, worked out most aspects of Tl—Mo—O bronze results and contributed significantly to the substitutional studies. Many thanks to J. A. Potenza for his help in the single crystal X-ray studies. F. J. DiSalvo was a constant source of help in interpreting the magnetic and electronic properties of the materials studied and a willing listener to test ideas on. We are grateful to J. V. Waszczak for measuring all the reported magnetic susceptibility data.

Finally, the support of the National Science Foundation Solid State Chemistry Grants DMR—81—15977, DMR—84—04003, and DMR—84—08266 are gratefully acknowledged.

References

1. (a) F. Wohler, Ann. Chim. Phys. 29, 43 (1825). (b) F. Wohler, Phil. Mag. 66, 263 (1825) [This is the English translation of (a)].
2. M. J. Sienko, in Non-stoichiometric Compounds, Advances in Chemistry Series (ed. R. F. Gould), American Chem. Soc. Washington D.C., p. 224 (1963).
3. E. Banks and A. Wold, Preparative Inorganic Reactions, Vol. 4 (ed. William L. Jolly), Interscience (1968) p. 237.
4. P. G. Dickens and M. S. Whittingham, Quarterly Rev. 22, 30 (1968).
5. P. Hagenmuller, in Progress in Solid State Chemistry (ed. H. Reiss), Pergamon, Vol. 5, p. 71 (1971).
6. A. Manthiram and J. Gopalakrishnan, Rev. Inorg. Chem., Vol. 6 (ed. H. D. B. Jenkins), (Freund Publishing, 1984) p. 1.
7. (a) P. Monceau, N. P. Ong, A. M. Portis, A. Meerschaut, and J. Rouxel, Phys. Rev. Lett. 37, 602 (1976). (b) P. Monceau, J. Richard, and M. Renard, Phys. Rev. B25, 931 (1982). (c) ibid. B25, 948 (1982).

8. For reviews, see: (a) R. M. Fleming in *Physics in One Dimension*, Vol. 23 of Springer Series in Solid State Sciences (eds. J. Bernasconi and T. Schneider) Springer, Berlin (1981). (b) N. P. Ong, *Can. J. Phys.* **60**, 757 (1982). (c) P. Monceau, J. Richard, and M. Renard, *Phys. Rev.* **B25**, 1931 (1982). (d) J. Richard, P. Monceau, and M. Renard, *Phys. Rev.* **B25**, 948 (1982). (e) G. Gruner, *Comments in Solid State Phys.* **10** (1983). (f) G. Gruner, *Physica* **8D**, 1 (1983). (g) *Electronic Properties of Inorganic Quasi-One-Dimensional Compounds* (ed. P. Monceau) in *Physics and Chemistry of Materials with Low-Dimensional Structures, Series B: Quasi-One-Dimensional Structures*, Reidel, Dordrecht (1985). (h) G. Gruner and A. Zettl, *Phys. Rep.* **119**, 117 (1985).

9. (a) Z. Z. Wang, M. C. Saint-Lager, P. Monceau, M. Renard, P. Gressier, A. Meerschaut, and J. Rouxel, *Solid State Commun.* **46**, 325 (1983). (b) A. Meerschaut and J. Rouxel in *Crystal Chemistry and Properties of Materials with Quasi-One-Dimensional Structures* (ed. J. Rouxel) Reidel, Dordrecht (1986) p. 205. (c) M. Maki, M. Kaiser, A. Zettl, and G. Gruner, *Solid State Commun.* **46**, 497 (1983).

10. (a) Proc. Int. Symp. on Non-Linear Transport and Related Phenomena in Inorganic Quasi-One-Dimensional Conductors, Sapporo, Japan (Oct. 1983), Hokkaido University. (b) Proc. Int. Conf. on the Physics and Chemistry of Low Dimensional Synthetic Metals (ICSM 84), Abano Terme (Italy), June 1984, *Molecular Crystals and Liquid Crystals* (Vol. 121, 1985, Gordon and Breach). (c) Proc. Int. Conf. on Charge Density Waves in Solids, Budapest, Hungary, Sept. 1984, *Lecture Notes in Physics* **217** (ed. Gy Hutiray and J. Solyom) Springer-Verlag (1985). (d) Proc. of the Yamada Conf. XV on Physics and Chemistry of Quasi-One-Dimensional Conductors (ed. S. Tanaka) Lake Kawaguchi, Japan, May 1986.

11. (a) C. Schlenker and J. Dumas in *Crystal Chemistry and Properties of Materials with Quasi-One-Dimensional Structures* (ed. J. Rouxel) Reidel, Dordrecht (1986) p. 135. (b) R. M. Fleming, *Synthetic Metals* **13**, 241 (1986).

12. (a) J. B. Goodenough, *Bull. Soc. Chim.* 1200 (1965). (b) J. B. Goodenough, *Czech. J. Phys.* **B17**, 304 (1967).

13. P. G. Dickens and D. J. Neild, *Trans. Faraday Soc.* **64**, 13 (1968).

14. G. Sperlich, *Z. Physik* **250**, 335 (1972).

15. V. A. Joffe and I. V. Patrina, *Sov. Phys. Sol. State* **10**, 639 (1968).

16. J. B. Goodenough, *Progr. Solid State Chem.* **5**, 145 (1971).

17. M.-H. Whangbo and L. F. Schneemeyer, *Inorg. Chem.* **25**, 2424 (1986).

18. A. Wold, W. Kunnmann, R. J. Arnott, and A. Ferretti, *Inorg. Chem.* **3**, 545 (1964).

19. G. H. Bouchard, J. Perlstein, and M. J. Sienko, *Inorg. Chem.* **6**, 1682 (1967).

20. D. S. Perloff, M. Vlasse, and A. Wold, *J. Phys. Chem. Solids*, Suppl. No. 1, 361 (1967).

21. P. Hagenmuller, *Preparative Methods in Solid State Chemistry* (ed. P. Hagenmuller), Academic Press, 1972, p. 279.

22. W. A. Mumme and J. A. Watts, *J. Solid State Chem.* **2**, 16 (1970).

23. A. F. Reid and J. A. Watts, *J. Solid State Chem.* **1**, 310 (1970).

24. P. Strobel and M. Greenblatt, *J. Solid State Chem.* **36**, 331 (1981).

25. R. Brusetti, B. K. Chakraverty, J. Devenyi, J. Dumas, J. Marcus, and C. Schlenker, in *Recent Developments in Condensed Matter Physics*, Vol. 2 (eds. J. T. Deeveese, L. F. Lemmens, V. E. Van Doren, and J. Van Royen), Plenum, New York, Vol. 2, p. 181 (1981).

26. R. Buder, J. Devenyi, J. Dumas, J. Marcus, J. Mercier, C. Schlenker, and H. Vincent, *J. Physique Lett.* **43**, L59 (1982).

27. L. F. Schneemeyer, S. E. Spengler, F. J. DiSalvo, J. V. Waszczak, and C. E. Rice, *J. Solid State Chem.* **55**, 158 (1984).

28. L. F. Schneemeyer, F. J. DiSalvo, R. M. Fleming, and J. V. Waszczak, *J. Solid State Chem.* **54**, 358 (1984).

29. J. P. Pouget, S. Kagoshima, C. Schlenker, and J. Marcus, *J. Physique Lett.* **44**, L113 (1983).

30. L. F. Schneemeyer, F. J. DiSalvo, S. E. Spengler, and J. V. Waszczak, *Phys. Rev.* **30B**, 4297 (1984).

31. C. Schlenker, J. Dumas, C. Escribe-Filippini, H. Guyot, J. Marcus, and J. Fourcaudot, *Phil. Mag.* **521B**, 643 (1985).

32. M. Ganne, A. Boumaza, M. Dion, and J. Dumas, *Mat. Res. Bull.* **20**, 1297 (1985).

33. J.-M. Reau, C. Fouassier, C. Gleitzer, and M. Parmentier, *Bull. Soc. Chim. Fr.* 479 (1970).
34. J.-M. Reau, C. Fouassier, and P. Hagenmuller, *J. Solid State Chem.* 1, 326 (1970).
35. J.-M. Reau, C. Fouassier, and P. Hagenmuller, *Bull. Soc. Chim. Fr.* 2883 (1971).
36. W. H. McCarroll and M. Greenblatt, *J. Solid State Chem.* 54, 282 (1984).
37. K. V. Ramanujachary, M. Greenblatt, and W. H. McCarroll, *J. Cryst. Growth* 70, 476 (1984).
38. B. T. Collins, K. V. Ramanujachary, M. Greenblatt, and J. V. Waszczak, *Solid State Commun.* 56, 1023 (1985).
39. K. V. Ramanujachary, B. T. Collins, M. Greenblatt, and J. V. Waszczak, *Solid State Commun.* 59, 647 (1986).
40. K. V. Ramanujachary, B. T. Collins, M. Greenblatt, P. McNally, and W. H. McCarroll, *Solid State Ionics* 22, 105 (1986).
41. B. T. Collins, K. V. Ramanujachary, M. Greenblatt, W. H. McCarroll, P. McNally, and J. V. Waszczak, *J. Solid State Chem.* 76, 319 (1988).
42. B. T. Collins, K. V. Ramanujachary, and M. Greenblatt, to be published.
43. T. A. Bither, J. L. Gillson, and H. S. Young, *Inorg. Chem.* 5, 1559 (1966).
44. B. L. Chamberland, *Inorg. Chem.* 8, 1183 (1969).
45. O. Glemser and G. Lutz, *Z. Anorg. Allg. Chem.* 264, 17 (1951).
46. K.-A. Wilhelmi, *Acta Chem. Scand.* 23, 419 (1969).
47. O. Glemser, U. Hanschild, and G. Lutz, *Z. Anorg. Allg. Chem.* 269, 93 (1952).
48. O. Glemser, G. Lutz, and G. Meyer, *Z. Anorg. Allg. Chem.* 285, 173 (1956).
49. R. Schollhorn, R. Kuhlman, and J. O. Basenhard, *Mat. Res. Bull.* 11, 83 (1976).
50. J. J. Birtill and P. G. Dickens, *Mat. Res. Bull.* 13, 311 (1978).
51. P. G. Dickens and J. J. Birtill, *J. Electron Mater.* 7, 679 (1978).
52. P. G. Dickens, J. J. Birtill, and C. J. Wright, *J. Solid State Chem.* 28, 185 (1979).
53. J. Graham and A. D. Wadsley, *Acta Crystallogr.* 20, 93 (1966).
54. M. Ghedira, J. Chenavas, M. Marezio, and J. Marcus, *J. Solid State Chem.* 57, 300 (1985).
55. W. H. Zachariasen, *J. Less Comm. Met.* 62, 1 (1978).
56. W. Fogle and J. Perlstein, *Phys. Rev.* 6B, 1402 (1972).
57. D. Adler, *Solid State Phys.* 21, 1 (1968).
58. (a) W. Kohn, *Phys. Rev. Letters* 19, 439 (1967). (b) B. I. Halperin and T. M. Rice, *Rev. Mod. Phys.* 40, 755 (1968). (c) *Solid State Phys.* 21, 116 (1968).
59. D. S. Perloff, Ph.D. Thesis (Brown University, 1969), p. 52.
60. G. Travaglini, P. Wachter, J. Marcus, and C. Schlenker, *Solid State Commun.* 37, 599 (1981).
61. S. Jandl, M. Bonville, C. Pépin, J. Marcus and C. Schlenker, *Phys. Rev. B* (submitted).
62. R. E. Peierl, *Quantum Theory of Solids* (Clarendon, Oxford, 1955) p. 108.
63. H. Frohlich, *Proc. Roy. Soc.* A223, 296 (1954).
64. G. Gruner, *Perspectives in Solid State Physics*, to be published.
65. See, for example, the series *Physics and Chemistry of Materials with Layered Structures* (ed. F. Levy), Vol. 2 (1976); (eds. T. J. Wieting and M. Schluter), Vol. 3 (1979); (ed. P. A. Lee), Vol. 4 (1976) Reidel, Dordrecht.
66. G. A. Toombs, *Phys. Rep.* C40, 181 (1978).
67. M. Sato, *Lecture Notes in Physics*, Vol. 217 (ed. Gy. Hutiray and J. Solyom) Springer-Verlag (1985) p. 7.
68. M. Sato, H. Fujishita, and S. Hoshino, *J. Phys.* C16, L877 (1983).
69. R. M. Fleming, L. F. Schneemeyer, and D. E. Moncton, *Phys. Rev.* 31B, 899 (1985).
70. J. P. Pouget, A. H. Moudden, R. Moret, C. Escribe-Filippini, B. Hennion, J. Marcus, and C. Schlenker, *Mol. Crystals Liq. Crystals* 121, 111 (1985).
71. P. Butaud, P. Segransan, C. Berthier, J. Dumas, and C. Schlenker, *Phys. Rev. Lett.* 55, 253 (1985).
72. P. Segransan, A. Janossy, C. Berthier, J. Marcus, and P. Butaud, *Phys. Rev. Lett.* 56, 1854 (1986).
73. K. Nomura, K. Kume, and M. Sato, *Solid State Commun.* 57, 611 (1986).
74. J. Y. Veuillen, R. Chevalier, D. Salomon, J. Dumas, J. Marcus, and C. Schlenker, *Charge Density Waves in Solids*, Lecture Notes in Physics Vol. 217 (Berlin: Springer-Verlag), p. 129.
75. G. Travaglini, I. Morke, and P. Wachter, *Solid State Commun.* 45, 289 (1983).

76. G. Travaglini and P. Wachter, in Proc. 15th Conf. Physics of Semiconductors, Kyoto 1980, *J. Phys. Soc. Japan* **49**, 1980, Suppl. A, p. 869.

77. S. B. Dierker, K. B. Lyons, and L. F. Schneemeyer, *Bull. Phys. Soc.* **29**, 469 (1984).

78. J. P. Pouget, C. Nguera, A. H. Moudden, and T. Moret, *J. Physique* **46**, 1731 (1985).

79. G. K. Wertheim, L. F. Schneemeyer, and D. N. E. Buchanan, *Phys. Rev.* **32B**, 3568 (1985).

80. J. Dumas, C. Schlenker, J. Y. Veuillen, R. Chevalier, J. Marcus, R. Cinti, and E. Al Khoury Nemeh, Proceedings International Conference on Science and Technology of Synthetic Metals (ICMS, 86), *Synthetic Metals* **19**, 937 (1987).

81. G. Bang and G. Sperlich, *Z. Phys.* **B22**, 1 (1975).

82. D. C. Johnston, *Phys. Rev. Lett.* **52**, 2049 (1984).

83. B. T. Collins, K. Ramanujachary, M. Greenblatt, and J. V. Waszczak, *J. Solid State Chem.* **77**, 348 (1988).

84. N. C. Stephenson, *Acta Cryst.* **20**, 59 (1966).

85. B. M. Gatehouse and D. J. Loyd, *Chem. Commun.* 13 (1971).

86. B. M. Gatehouse, D. J. Loyd, and B. K. Miskin, NBS Special Publication 364 (1972) p. 15.

87. H. Vincent, M. Ghedira, J. Marcus, J. Mercier, and C. Schlenker, *J. Solid State Chem.* **47**, 113 (1983).

88. L. Kihlborg, *Arkiv Kemi* **21**, 471 (1963).

89. M. Onoda, K. Toriumi, Y. Matsuda, and M. Sato, *J. Solid State Chem.* **66**, 163 (1987).

90. M. Onoda, Y. Matsuda, and M. Sato, *J. Solid State Chem.* **69**, 67 (1987).

91. M. Ganne, M. Dion, A. Boumaza, and M. Tournoux, *Solid State Commun.* **59**, 137 (1986).

92. R. D. Shannon, *Acta Crystallogr.* **A32**, 751 (1976).

93. A. Santoro, M. Greenblatt, and W. H. McCarroll, unpublished data.

94. M. Greenblatt, W. H. McCarroll, R. Niefeld, M. Croft, and J. V. Waszczak, *Solid State Commun.* **51**, 671 (1984).

95. M. Greenblatt, K. V. Ramanujachary, W. H. McCarroll, R. Niefeld, and J. V. Waszczak, *J. Solid State Chem.* **59**, 149 (1985).

96. $T_c \simeq 1.15$K for an electrolytically grown crystal, Ref. |31|.

97. Y. Matsuda, M. Sato, M. Onoda, and K. Nakao, *J. Phys. C: Solid State Phys.* **19**, 6039 (1986).

98. J. Dumas, C. Escribe-Filippini, J. Marcus, and C. Schlenker, Proceedings of the NATO DAVY Advanced Study Institute, Cambridge U.K. (1983), *Physics and Chemistry of Electrons and Ions in Condensed Matter.*

99. J. Dumas, E. Bervas, J. Marcus, D. Salomon, C. Schlenker, and G. Fillion, *J. Magn. Materials* **31—44**, 535 (1983).

100. C. Escribe-Filippini, K. Konate, J. Marcus, C. Schlenker, R. Almairac, R. Ayroles, and C. Roucau, *Phil. Mag.* **B50**, 321 (1984).

101. C. Escribe-Filippini, J. Marcus, C. Schlenker, R. Ayroles, C. Roucau, S. Kagoshima, and J. P. Pouget (1985), unpublished data.

102. F. J. DiSalvo and J. V. Waszczak, *J. Phys. Chem. Solids* **41**, 1311 (1980).

103. L. Mihaly, G. Gruner, and M. Greenblatt, unpublished data.

104. K. V. Ramanujachary and M. Greenblatt, to be published.

105. E. Bervas, R. W. Cochrane, J. Dumas, C. Escribe-Filippini, J. Marcus, and C. Schlenker, *Charge Density Waves in Solids*, Lecture Notes in Physics Vol. 217 (Berlin: Springer-Verlag, 1985), p. 144.

106. J. Dumas, C. Escribe-Filippini, J. Marcus, J. Mercier, D. Salomon, C. Schlenker, and F. Razavi, *Physica* **117B** and **118B**, 602 (1983).

107. R. Delaplace, P. Molinie, and D. Jerome, *J. Phys.-Lettres* **37**, L13 (1976).

108. N. C. Stephenson and A. D. Wadsley, *Acta Crystallogr.* **19**, 241 (1965).

109. (a) G. Bang and G. Sperlich, *Phys. Lett.* **49A**, 21 (1974). (b) G. Sperlich, *J. Solid State Chem.* **12**, 360 (1975).

110. G. Travaglini, P. Wachter, J. Marcus, and C. Schlenker, *Solid State Commun.* **42**, 407 (1982).

111. G. Travaglini and P. Wachter, *Solid State Commun.* **47**, 217 (1983).

112. P. P. Tsai, J. A. Potenza, M. Greenblatt, and H. J. Schugar, *J. Solid State Chem.* **64**, 47 (1986).

113. I. D. Brown, *Structure and Bonding in Crystals* (Eds. M. O'Keefe and A. Navrotsky), Vol. II, p. 1 (Academic Press, New York, 1981).

114. M. Ganne, M. Dion, and A. Boumaza, *C.R. Acad. Sc. Paris.* **302**, 635 (1986).
115. P. P. Tsai, J. A. Potenza, and M. Greenblatt, *J. Solid State Chem.* **69**, 329 (1987).
116. K. V. Ramanujachary and M. Greenblatt, to be published.
117. S. C. Abrahams, P. Marsh, L. F. Schneemeyer, C. E. Rice, and S. E. Spengler, *J. Mater. Res.* **2**, 82 (1987).
118. W. H. McCarroll, C. Darling, and G. Jakubicki, *J. Solid State Chem.* **48**, 187 (1983).
119. A. Moini, M. Subramanian, A. Clearfield, F. J. DiSalvo, and W. H. McCarroll, *J. Solid State Chem.* **66**, 136 (1987).
120. P. Hagenmuller, J. Galy, M. Pouchard, and A. Casalot, *Mat. Res. Bull.* **1**, 45 (1966).
121. A. A. Fotiev, B. L. Volkov, and B. K. Kapustkin, *Oxide Vanadium Bronzes*, Academy Nauk, Moscow, 1978.
122. N. F. Mott, *Metal-Insulator Transitions*, Taylor and Francis, London (1974).
123. J. B. Goodenough, *J. Solid State Chem.* **1**, 349 (1970).
124. J. H. Perlstein and M. J. Sienko, *J. Chem. Phys.* **48**, 174 (1968).
125. M. J. Sienko and J. Sohn, *J. Chem. Phys.* **44**, 1369 (1966).
126. A. D. Wadsley, *Acta Crystallogr.* **8**, 695 (1955).
127. R. P. Ozerov, G. A. Gol'der, and G. S. Zhdanov, *Soviet Phys. Cryst.* **2**, 211 (1957).
128. H. Kobayashi, *Bull. Chem. Soc. Japan* **52**, 1315 (1979).
129. T. Takahashi and H. Nagasawa, *Solid State Commun.* **39**, 1125 (1981).
130. J. Galy, J. Darriet, A. Casalot, and J. B. Goodenough, *J. Solid State Chem.* **1**, 339 (1970).
131. B. K. Chakraverty, M. J. Sienko, and J. Bonnerot, *Phys. Rev.* **17B**, 3781 (1978).
132. V. K. Kapustkin, V. L. Volkov, and A. A. Fotiev, *J. Solid State Chem.* **19**, 359 (1976).
133. R. H. Wallis, N. Sol, and Zylbersztejn, *Solid State Commun.* **23**, 539 (1977).
134. D. Kaplan and A. Zylbersztejn, *J. de Phys.* **37**, L123 (1976).
135. G. Sperlich, W. D. Laze, and G. Bang, *Solid State Commun.* **16**, 489 (1975).
136. W. J. Gunning, A. J. Heeger, R. H. Wallis, N. Sol, and A. Zylbersztejn, *Solid State Commun.* **26**, 155 (1978).
137. E. P. Chock and G. Gruner, *Solid State Commun.* **36**, 1017 (1980).
138. A. Friedrich, D. Kaplan, and N. Sol, *Solid State Commun.* **25**, 633 (1978).
139. (a) R. A. Craven, Y. Tomkiewicz, E. M. Englar, and A. R. Taranko, *Solid State Commun.* **23**, 429 (1977). (b) S. Etemad, E. M. Englar, T. D. Schultz, T. Penny, and B. A. Scout, *Phys. Rev.* **17B**, 513 (1978).
140. H. Nagasawa, T. Takahashi, T. Erata, M. Onoda, Y. Kanai, and S. Kagoshima, *Mol. Cryst. Liq. Cryst.* **86**, 1935 (1982).
141. (a) K. Maruyama and H. Nagasawa, *J. Phys. Soc. Japan Lett.* **48**, 2159 (1980). (b) T. Erata, T. Takahashi, and H. Nagasawa, *Solid State Commun.* **39**, 321 (1981). (c) M. Onoda, T. Takahashi, and H. Nagasawa, *J. Phys. Soc. Japan* **51**, 3868 (1982).
142. (a) Y. Kanai, S. Kagoshima, and H. Nagasawa, *J. Phys. Soc. Japan* **51**, 697 (1982). (b) Y. Kanai, S. Kagoshima, and H. Nagasawa, *Synthetic Metals* **9**, 369 (1984). (c) H. Nagasawa, T. Erata, M. Onoda, H. Suzuki, S. Uji, Y. Kanai, and S. Kagoshima, *Mol. Cryst. Liq. Cryst.* **121**, 121 (1985).
143. P. G. Dickens and M. S. Whittingham, *Quart. Rev. Chem. Soc.* **22**, 30 (1968).
144. A. Magneli, *Acta Chem. Scand.* **7**, 315 (1953).
145. A. R. Sweedler, Ch. J. Raub, and B. T. Matthias, *Phys. Lett.* **15**, 108 (1965).
146. D. R. Wanlass and M. J. Sienko, *J. Solid State Chem.* **12**, 362 (1975).
147. (a) R. K. Stanley, R. C. Morris, and W. G. Moulton, *Solid State Commun.* **27**, 1277 (1978). (b) R. K. Stanley, R. C. Morris, and W. G. Moulton, *Phys. Rev.* **B20**, 1903 (1979).
148. L. H. Cadwell, R. C. Morris, and W. G. Moulton, *Phys. Rev.* **B23**, 2219 (1981).
149. J. Bouchard and J. L. Gillson, *Inorg. Chem.* **7**, 969 (1968).
150. M. R. Skokan, W. G. Moulton, and R. C. Morris, *Phys. Rev.* **B20**, 3670 (1979).
151. (a) A. J. Bevolo, H. R. Shanks, P. H. Sidles, and G. C. Danielson, *Phys. Rev.* **B9**, 3220 (1974). (b) C. N. King, T. A. Benda, R. L. Greene, and T. H. Gabelle, Proc. 13th Int. Conf. Low Temperature Physics (Boulder, Colorado) 1972, eds. W. J. O'Sullivan, K. D. Timmerhaus, and E. F. Hammel (New York, Plenum) (1974).
152. (a) W. A. Kamitakahara, K. Scharnberg, and H. R. Shanks, *Phys. Rev. Lett.* **43**, 1607 (1979).

(b) M. Sato, *J. Phys. C: Solid State Phys.* **13**, L481 (1980). (c) M. Sato, H. Fujishita, A. R. Moodenbaugh, S. Hoshino, and B. H. Grier, Proc. 6th Yamada Conf. Condensed Matter (Hakone) 1982. (d) M. Sato, B. H. Grier, and H. Fujishita, *Phys. Rev.* **B25**, 501 (1982).

153.   (a) M. Sato, B. H. Grier, H. Fujishita, S. Hoshino, and A. R. Moodenbaugh, *J. Phys. C: Solid State Phys.* **16**, 5217 (1983). (b) M. Sato, H. Fujishita, A. R. Moodenbaugh, S. Hoshino, and B. H. Grier, *Physica* **120B**, 275 (1983).

154.   H. B. Krause, W. G. Moulton, and R. C. Morris, *Acta Crystallogr.* **B41**, 11 (1985).

155.   J. P. Giroult, M. Goreaud, Ph. Labbe, and B. Raveau, *Acta Crystollgr.* **B37**, 2139 (1981).

156.   J. P. Giroult, M. Goreaud, Ph. Labbe, and B. Raveau, *Acta Crystollgr.* **B37**, 1163 (1981).

157.   J. P. Giroult, M. Goreaud, Ph. Labbe, and B. Raveau, *Acta Crystollgr.* **B36**, 2570 (1980).

158.   J. P. Giroult, M. Goreaud, Ph. Labbe, and B. Raveau, *J. Solid State Chem.* **44**, 407 (1984).

159.   B. Domenges, F. Studer, and B. Raveau, *Mat. Res. Bull.* **18**, 669 (1983).

160.   A. Benmoussa, D. Giroult, Ph. Labbe, and B. Raveau, *Acta Crystallogr.* **40**, 573 (1984).

161.   M. Hervieu and B. Raveau, *J. Solid State Chem.* **43**, 299 (1982).

162.   M. Bermoussa, D. Giroult, and B. Raveau, *Rev. Chim. Min.* **21**, 710 (1984).

163.   Ph. Labbe, D. Ouachee, M. Goreaud, and B. Raveau, *J. Solid State Chem.* **50**, 163 (1983).

164.   J. P. Giroult, M. Goreaud, Ph. Labbe, and B. Raveau, *Rev. Chim. Min.* **20**, 829 (1983).

165.   B. Domenges, M. Goreaud, Ph. Labbe, and B. Raveau, *J. Solid State Chem.* **50**, 173 (1983).

166.   J. P. Giroult, M. Goreaud, J. Provost, Ph. Labbe, and B. Raveau, *Mat. Res. Bull.* **16**, 811 (1981).

167.   E. Wang and M. Greenblatt, *J. Solid State Chem.* **68**, 38 (1987).

168.   Since this review has been submitted, the following papers have been submitted or are in press:

E. Wang, M. Greenblatt, I. E.-I. Rachidi, E. Canadell, and M.-H. Whangbo, *Inorg. Chem.*, in press.

E. Wang, M. Greenblatt, I. E.-I. Rachidi, E. Canadell, and M.-H. Whangbo, *J. Solid State Chem.*, in press.

E. Wang, M. Greenblatt, E. Canadell, I. E.-I. Rachidi, M.-H. Whangbo, and S. Vadlmannati, *Phys. Rev. B*, in press.

E. Canadell, I. E.-I. Rachidi, E. Wang, M. Greenblatt, and M.-H. Whangbo, *Inorg. Chem.*, in press.

E. Wang, M. Greenblatt, E. Canadell, I. E.-I. Rachidi, and M.-H. Whangbo, *J. Solid State Chem.*, in press.

ON STRUCTURAL ASPECTS OF MOLYBDENUM BRONZES AND MOLYBDENUM OXIDES IN RELATION TO THEIR LOW-DIMENSIONAL TRANSPORT PROPERTIES

HENRI VINCENT* and MASSIMO MAREZIO

Laboratoire de Cristallographie, C.N.R.S., B.P. 166X, 38042 Grenoble Cedex, France

1. Introduction

In 1823 Wöhler [1] observed that the reduction of Na_2WO_4 by hydrogen yielded compounds which were chemically inert, strongly colored and exhibited a metallic aspect. Because of these last two features they were called tungsten bronzes. Subsequently, a number of other bronzes, containing transition elements such as Ti, Mn, Nb, and Mo, were synthesized. Review articles on bronzes have been published by Hagenmuller [2], and by Dickens and Wiseman [3].

The oxide bronzes have the general formula A_xMO_n: where A is an electropositive, readily ionizable element, such as an alkali metal, and MO_n the transition metal oxide corresponding to the highest oxidation state of the M cation. The insertion of an A^+ cation in the MO_n oxide causes a decrease of the M cation oxidation state according to the equation:

$$M^{+2n}O_n + x\,A^+ = A_x^+M^{+(2n-x)}O_n.$$

This decrease corresponds to an increase of the number of d electrons which can be either localized around the M cations or delocalized in the MO_n lattice. In the former case the compounds are semiconductors (as $\beta - Na_xV_2O_5$) while in the latter they exhibit metallic conductivity (as Na_xWO_3, $x > 0.3$).

The number of A cations per unit cell which can be inserted in the oxide matrix can vary over a large range, and the bronze stoichiometry changes accordingly. For example, there exists only one phase for sodium tungsten bronze, Na_xWO_3 with $0.37 \leqslant x \leqslant 0.95$. The structure can be considered as a solid solution of Na cations into the WO_3 matrix.

In addition to their excellent chemical stability, their inertness to the non-oxidant acids and their conductivity properties, the oxide bronzes show the composition suppleness of liquid solutions.

It is understandable, therefore, that such compounds are potentially of high technological interest. In particular, they could be very useful in the field of electrochemical energy storage as they are excellent candidates as electrode materials [4].

Because of their variety and their relative complexity the oxide bronzes are also very interesting from the structural point of view. The M cations which have a high electric charge and a small ionic radius establish strong anisotropic interactions with the surrounding first nearest oxygen neighbors. Although the insertion of

C. Schlenker (Ed.), Low-Dimensional Electronic Properties of Molybdenum Bronzes and Oxides,
49—85.
© 1989 *Kluwer Academic Publishers.*

large A^+ induces the breaking of the host MO_n structure, the MO_p polyhedron array is maintained along certain directions and the A cations are accommodated to form either layers or chains. Such structural arrangements allow one to vary the amount of inserted cations over a large range without any structural change. The sublattices forming the structure may have a dimensionality lower than 3, and consequently the chemical and physical properties are also direction dependent. The oxide bronzes constitute, in fact, a very large class of compounds exhibiting low-dimensionality properties. For all these reasons they have been intensively investigated during the last ten years. We shall limit our discussion to the oxide molybdenum bronzes.

The molybdenum suboxides, $Mo_m O_n (9/26 \leqslant m/n \leqslant 4/11)$ can be obtained by reacting the appropriate mixture of MoO_3 and Mo metal in a vacuum-sealed tube. One can regard these suboxides as Mo-inserted MoO_3 compounds. Since the structural similarities between the Magneli phases and the oxide bronzes can be extended to the chemical and physical properties, the suboxides will be included herein, whereas the molybdenum mixed oxides other than the oxide bronzes will not be discussed. A comprehensive description of all these compounds can be found in the review article published by Gleitzer [7].

2. Molybdenum Bronzes

The molybdenum bronzes are less stable than their tungsten counterparts, consequently their preparation is more difficult. It was only in 1963 that Wold et al. [8] were able to synthesize the first oxide molybdenum bronzes: $Na_{0.9}Mo_6O_{17}$, $K_{0.33}MoO_3$, and K_xMoO_3 with $0.28 \leqslant x \leqslant 0.30$. They were called, according to their color, violet sodium bronze, red and blue potassium bronzes, respectively. They were prepared by electrolytic reduction of a mixture of A_2MoO_4 (A = Na, K) and MoO_3, held at about 560 °C.

In 1966 Bither et al. [9] obtained Na_xMoO_3 ($0.90 \leqslant x \leqslant 0.93$), $K_{0.5}MoO_3$, $Rb_{0.27}MoO_3$, and $Rb_{0.44}MoO_3$. The syntheses were carried out under pressure (65 kbar) and at high temperature (1200 °C). These compounds are isostructural with their tungsten counterparts.

A few years later Réau and Hagenmuller [10] carried out a systematic study of the alkali-metal molybdenum bronzes. Mixtures of A_2MoO_4, MoO_2 and MoO_3 were heat-treated at relatively low temperatures (550–590 °C) in gold-sealed tubes. Besides the bronzes prepared by Wold et al. new compounds were obtained by this procedure: Li_xMoO_3 ($0.31 \leqslant x \leqslant 0.39$), $Li_{0.9}Mo_6O_{17}$, and $K_{0.9}Mo_6O_{17}$. Almost at the same time Reid and Watts [11] prepared the reddish $Cs_{0.3}MoO_3$ bronze by electrolitic reduction of a Cs_2MoO_4 and MoO_3 mixture at 530 °C. By using the same method Strobel and Greenblatt [12] obtained the blue lithium bronze, $Li_{0.3}MoO_3$, a red cesium bronze, $Cs_{0.33}MoO_3$, and the rubidium bronze, $Rb_{0.3}MoO_3$. Recently Schneemeyer et al. [13] synthesized single crystals of a blue-black Mo-deficient cesium bronze by following a similar procedure.

A different preparation method for obtaining molybdenum bronzes has been developed by McCarroll and Greenblatt [14]. Single crystals of compounds such as $Li_{0.9}Mo_6O_{17}$, $Li_{0.33}MoO_3$, and $Li_{0.04}MoO_3$ were obtained. The starting products were sealed in a silica tube under vacuum and placed in a furnace with two temperature zones (640 °C and 590 °C). Of the three compounds only the last one is new. However $Li_{0.33}MoO_3$ reported by Réau and Hagenmuller seems to be different from that prepared by McCarroll and Greenblatt.

A large number of tungsten bronzes with A other than an alkali cation is known. On the other hand, the number of the corresponding molybdenum bronzes is somewhat smaller. In 1951 Glemser et al. [15] reported the synthesis of hydrated molybdenum oxides which were prepared by dehydrogenation under vacuum of H_2MoO_3 obtained by reacting MoO_3 with nascent hydrogen in acqueous solution. Subsequently, a compound whose formula was thought to be $H_{0.5}MoO_3$, but was in fact $H_{0.3}MoO_3$, was characterized by Kihlborg et al. [68]. In 1978, Birtill and Dickens [16] characterized the green H_2MoO_3, the red $H_{1.7}MoO_3$, and $H_{0.9}MoO_3$ and $H_{0.3}MoO_3$ both of blue color. These authors recognized that there compounds are bronze-like. The thallium bronzes, the red $Tl_{0.33}MoO_3$, the blue $Tl_{0.3}MoO_3$ and the violet $TlMoO_3$ were prepared by Ganne et al. [64, 17, 65] following the method developed by Réau and Hagenmuller [10]. Recently Marcus et al. [18] reported the synthesis of rare earth molybdenum bronzes, $RE_{0.1}MoO_3$.

Molybdenum bronzes containing both hydrogen and an alkali cation, or more precisely hydrated alkali molybdenum bronzes, were first synthesized by Schöllhorn et al. [19] and subsequently by Iwamoto et al. [20]. These compounds, which are obtained by reducing MoO_3 with $Na_2S_2O_4$ in aqueous solution, were characterized by Thomas and McCarren [21]. Their general formula is: $[A(H_2O)_n]MoO_3$ with A = Li, Na, K, Rb, and Cs. Crystallographic studies have been carried out in detail for two of such compounds, namely $[Na(H_2O)_2]_{0.25}MoO_3$ and $[Na(H_2O)_5]_{0.25}MoO_3$.

Table I lists the various molybdenum bronzes reported so far, together with the preparation method and the crystallographic data.

TABLE I
Crystal data of molybdenum bronzes.

Typical Formula	Preparation mode	Cell		Space group or Symmetry
H_2MoO_3	$M_0O_3 + H_2$ (nascent)	$a = 3.897$	$b = 13.554$	
		$c = 4.053$	$\beta = 94.6°$	Monoclinic
$H_{1.7}MoO_3$	$H_2M_0O_3$ dehydrogenation	$a = 3.770$	$b = 13.967$	
		$c = 4.055$	$\beta = 93.9°$	Monoclinic
$H_{0.9}MoO_3$		$a = 3.796$	$b = 14.531$	
		$c = 3.863$	$\beta = 93.7°$	Monoclinic
$H_{0.3}MoO_3$		$a = 3.896$	$b = 14.066$	$Cmcm$
		$c = 3.736$		
$Li_{0.35}MoO_3$	Silica tube (460 °C)	$a = 24.54$	$b = 7.450$	
		$c = 15.11$	$\beta = 106.5°$	Monoclinic

Table I (Continued)

Typical Formula	Preparation mode	Cell	Space group or Symmetry
$Li_{0.9}Mo_6O_{17}$	(560 °C)	$a = 12.762$ $b = 5.523$ $c = 9.499$ $\beta = 90.6°$	$P2_1/m$
$Li_{0.33}MoO_3$	Flux + temperature gradient	$a = 13.079$ $b = 15.453$ $c = 7.476$ $\alpha = 97.0$ $\beta = 106.6$ $\gamma = 103.4$	$P\bar{1}$
$Li_{0.04}MoO_3$		unknown	unknown
$Na_{0.9}Mo_6O_{17}$	Electrolytic réduction	$a = 12.983$ $b = 5.518$ $c = 9.591$ $\beta = 89.9°$	$A2$
$Na_{0.95}MoO_3$	High pressure (65 kB)	$a = 3.851$	$Pm3m$
$K_{0.33}MoO_3$	Electrolytic reduction	$a = 14.278$ $b = 7.723$ $c = 6.387$ $\beta = 92.3°$	$C2/m$
$K_{0.3}MoO_3$		$a = 16.231$ $b = 7.550$ $c = 9.861$ $\beta = 94.9°$	$I2/m$
$K_{0.9}MoO_3$	High pressure (65 kB)	$a = 3.918$	$Pm3m$
$K_{0.5}Mo_6O_{17}$		$a = 12.32$ $c = 3.859$	$P4/mbm$
$K_{0.9}Mo_6O_{17}$	Electrolytic reduction	$a = 5.538$ $c = 13.656$	$P\bar{3}$
$Rb_{0.3}MoO_3$	Electrolytic reduction	$a = 16.361$ $b = 7.555$ $c = 10.094$ $\beta = 93.9°$	$I2/m$
$Rb_{0.27}MoO_3$	High pressure (65 kB)	$a = 7.321$ $c = 7.683$	$C6/m\,cm$
$Rb_{0.44}MoO_3$		$a = 7.724$ $b = 37.624$ $c = 7.385$	$Pmmm$
$Cs_{0.33}MoO_3$	Electrolytic reduction	$a = 15.862$ $b = 7.728$ $c = 6.408$ $\beta = 94.4°$	$C2/m$
$Cs_{0.3}MoO_3$		$a = 16.425$ $b = 7.543$ $c = 8.169$ $\beta = 96.5°$	$P2_1/m$
$Cs_{0.25}Mo_{0.97}O_3$		$a = 19.063$ $b = 5.583$ $c = 12.115$ $\beta = 118.9°$	$C2/m$
$Tl_{0.33}MoO_3$	Silica tube (560 °C)	$a = 14.541$ $b = 7.780$ $c = 6.414$ $\beta = 93.1°$	$C2/m$
$Tl_{0.3}MoO_3$		$a = 18.543$ $b = 7.567$ $c = 10.067$ $\beta = 118.4°$	$C2/m$
$TlMo_6O_{17}$		$a = 5.557$ $b = 14.030$	$P\bar{3}m1$
$[Na(H_2O)_2]_{0.25}MoO_3$	MoO_3 reduction by $Na_2S_2O_4$	$a = 3.876$ $b = 19.093$ $c = 3.733$	$Ammm$
$[Na(H_2O)_5]_{0.25}MoO_3$		$a = 3.884$ $b = 22.618$ $c = 3.751$	$Cmcm$

Because of their interesting chemical and physical properties the molybdenum bronzes have been systematically investigated from the crystallographic point of view. In most cases the precise determination of the structure has been carried out along with that of the $4d$ electron distribution. For the latter determination the empirical formula of Brown and Shannon [22] or that of Zachariasen [23] has been used.

These formulae, which are generalizations of the second Pauling rule [24], allow one to determine the cation (anion) valence state by summing the bond strengths existing between a given cation (anion) and the surrounding anions (cations). The bond strengths are calculated directly from the bond lengths. For example, for an oxide Zachariasen proposed the following formula:

$$s(i) = \exp[(1/A)(1 - D_i/D_1)]$$

where D_i is the observed interatomic distance and A and D_1 are some constants characteristic of the metallic element.

As pointed out by Dickens and Wiseman [3] the structures of all molybdenum bronzes have a common basic unit, namely a corner-sharing single octahedral chain running along the 4-fold axis of the octahedra (see Figure 1). Following Wells' nomenclature [25] we shall refer to such a chain as being of ReO_3-type. By combining these units in different ways, all the structural arrangements existing for the molybdenum bronzes can be obtained. In the ReO_3 structure as well as in the ABO_3 perovskite compounds the basic units are linked to the adjacent ones via corner-sharing and form identical square channels. In the former structure these channels are empty while in the latter they accommodate the large A cations (see Figure 2). We adopt here Dickens and Wiseman's structure description mode.

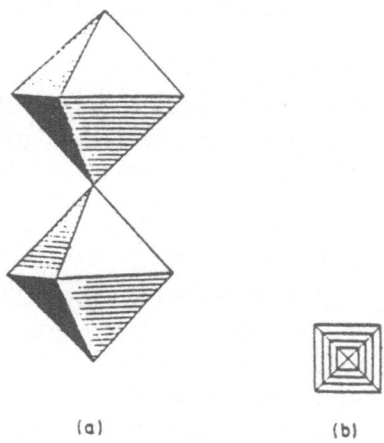

(a)   (b)

Fig. 1 (From Ref. [3]). (a) Single ReO_3 chain. (b) Projection along the chain axis.

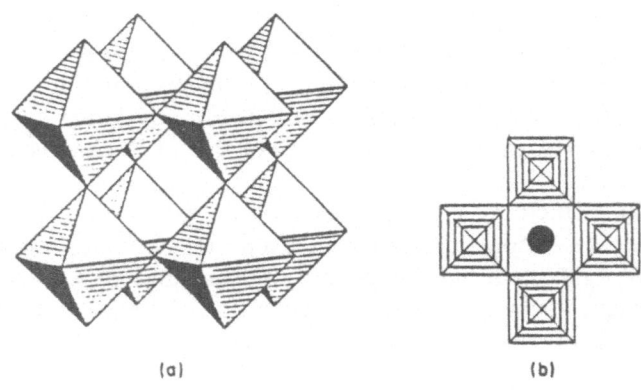

(a)   (b)

Fig. 2 (From Ref. [3]). Perovskite structure.

2.1. MOLYBDENUM BRONZES CONTAINING INFINITE SINGLE ReO_3 OCTAHEDRAL CHAINS

All these bronzes were obtained under pressure by Bither *et al.* [9]. Their structure has never been determined in detail; however, according to the powder photographs and the lattice parameters, they are isostructural with the corresponding tungsten bronzes whose structure was determined by Magneli [26—29]. In the structure of sodium tungsten bronze, Na_xWO_3 $0 \leqslant x \leqslant 1$, the infinite ReO_3 chains run along the octahedron 4-fold axis and are linked to each other by sharing all the other four corners with as many adjacent chains. The way these chains are linked together and the crystal symmetry depend upon the value of x. The latter increases with increasing x. Projected along the chain direction the structural arrangement can be considered as a polygonal network, where each point of connection is bounded by two chains and two channels.

2.1.1. *The Red* $Na_{0.9}MoO_3$ *and* $K_{0.9}MoO_3$ *Bronzes*

These two compounds are isostructural with Na_xWO_3 with $0.3 \leqslant x \leqslant 0.9$. Their symmetry is cubic (space group $Pm3m$) and the lattice parameter for both compounds is $a \sim 3.9$ Å. The structural arrangement is that of the perovskite. The Na or the K cations occupy partially the cuboctahedral sites (see Figure 2).

2.1.2. *The Blue Bronze* $K_{0.5}MoO_3$

This compound is isostructural with Na_xWO_3 with $0.28 \leqslant x \leqslant 0.38$ or with K_xWO_3 with $0.48 \leqslant x \leqslant 0.57$. Its symmetry is tetragonal (space group $P4/mbm$) with lattice parameters $a = 12.32$ Å and $c = 3.859$ Å. In this compound the infinite ReO_3 chains are linked together as shown in Figure 3. The linkage between chains is still via corner-sharing, But the resulting empty channels are of three types: square, triangular, and pentagonal. The K cations are equally distributed among the square and the pentagonal channels and have 12 and 15 first-nearest oxygen neighbors, respectively.

Fig. 3 (From Ref. [3]). $K_{0.5}MoO_3$ tetragonal structure.

2.1.3. *The Violet* $Rb_{0.27}MoO_3$ *and Blue-Black* $Rb_{0.44}MoO_3$ *Bronzes*

The violet bronze, $Rb_{0.27}MoO_3$, is isostructural with the corresponding tungsten bronze, $Rb_{0.29}WO_3$. Its symmetry is hexagonal (space group $C6/mcm$) with lattice parameters $a = 7.321$ Å and $c = 7.683$ Å. The infinite single ReO_3 chains are linked together to form hexagonal and triangular channels (see Figure 4). In the plane perpendicular to the 6-fold axis each octahedron is surrounded by 2 triangular and 2 hexagonal channels. The Rb cations which occupy partially the hexagonal channels, and surrounded by 18 oxygen atoms.

The blue-black bronze, $Rb_{0.44}MoO_3$ has a similar structural arrangement; however the larger amount of Rb cations distorts the structure and the resulting symmetry is orthorhombic (space group Pmnm). Its unit cell is 3 times the orthohexagonal cell of the violet bronze with $a = c_H = 7.724$ Å, $b = 3\sqrt{3}a_H = 37.624$ Å and $c = a_H = 7.385$ Å.

2.2. MOLYBDENUM BRONZES CONTAINING INFINITE TWO-DIMENSIONAL SLABS COMPRISED OF CORNER-SHARING OCTAHEDRA

The molybdenum bronzes having this type of structural arrangement can be separated into two categories: the violet bronzes $A_{0.9}Mo_6O_{17}$ with A = Li, Na, K, Tl and the black-blue cesium bronze $CsMo_{3.88}O_{12}$ recently prepared by L. F. Schneemeyer [13]. In the first type the slabs are ReO_3-like, namely the octahedra

Fig. 4 (From Ref. [3]). $Rb_{0.27}MoO_3$ hexagonal structure.

form a corner-sharing three-dimensional array. The slabs are infinite in two directions which are parallel to the (111) plane of the idealized cubic ReO_3 structure, while they are 4-octahedra in width. In $CsMo_{3.88}O_{12}$ the structural arrangement is more complex as certain MoO_6 octahedra inside the slabs are replaced by MoO_4 tetrahedra.

In this type of bronzes, whose formula can be written as $A_{0.16}Mo_{1.06}O_3$ and $Cs_{0.25}Mo_{0.97}O_3$, the alkali cation sublattice is not three-dimensional. The A cations are inserted between the ReO_3 slabs. Furthermore, the bronzes described in section 2.1. contain a three-dimensional network of corner-sharing octahedra, while the $A_{0.16}Mo_{1.06}O_3$ and $Cs_{0.25}Mo_{0.97}O_3$ bronzes are comprised of discrete slabs of corner-sharing octahedra for the former and of octahedra and tetrahedra for the latter. This structural difference is likely due to the smaller amount of alkali cations inserted in the MoO_3 structure.

2.2.1. The Violet Bronzes $A_{0.9}Mo_6O_{17}$ with A = Li, Na, K, Tl

The lithium bronze was found to be monoclinic (space group $P2_1/m$) with $a = 12.762$ Å, $b = 5.523$ Å, $c = 9.499$ Å and $\beta = 90.6°$. The structure has recently been determined by Onoda et al. [30]. The sodium bronze is also monoclinic (space group $A2$) with $a = 12.983$ Å, $b = 5.551$ Å, $c = 9.591$ Å, and $\beta = 89.9°$. Its structure was first determined by Stephenson [31] in 1966. Unfortunately, all his crystals were twinned and the structural determination was not entirely correct. Recently Onoda et al. [32] were able to obtain untwinned crystals of $Na_{0.9}Mo_6O_{17}$ and carried out a new structural determination. The potassium bronze is trigonal (space group $P\bar{3}$) with $a = 5.538$ Å and $c = 13.656$ Å. The structure has been determined by Vincent et al. [33]. The thallium bronze $TlMo_6O_{17}$ is also trigonal (space group $P\bar{3}m1$) with $a = 5.557$ Å and $c = 14.030$ Å. The structure has been determined by Ganne et al. [65]. Although the symmetry of these bronzes is different, their structural arrangement is quite similar, especially for the sodium, potassium and thallium bronzes. Figure 5 represents the idealized structure of

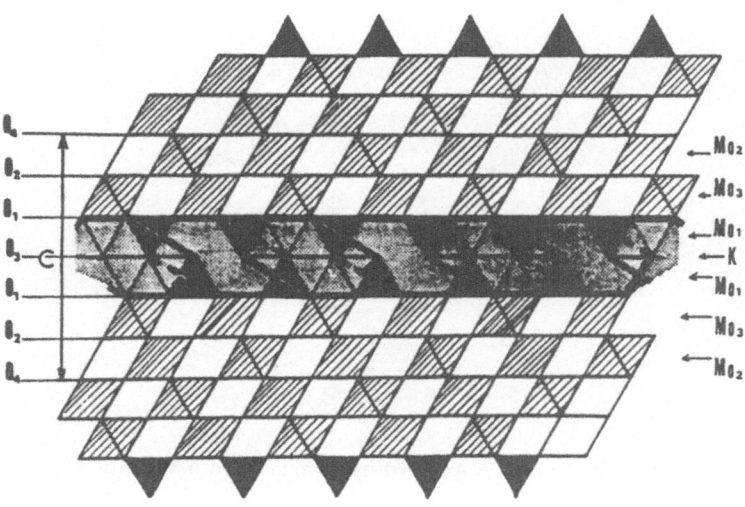

Fig. 5 (From Ref. [33]). View along the a axis of the $K_{0.9}Mo_6O_{17}$ structure.

$K_{0.9}Mo_6O_{17}$ along the a axis. It comprises corner-sharing MoO_6-octahedron slabs parallel to the (001) plane. These slabs, 4 octahedra (2 $Mo(2)O_6$ and 2 $Mo(3)O_6$) in width, are intercalated by one KO_{12}-icosahedron layer. $Mo(1)O_4$-tetrahedra are inserted between the slabs and the KO_{12}-icosahedron layer. Each icosahedron is linked via face-sharing to 2 $Mo(3)O_6$ octahedra belonging to the slabs above and below, respectively. Furthermore, it is bonded to 6 $Mo(1)O_4$ tetrahedra and 6 KO_{12} icosahedra as shown in Figures 6 and 7. Three of the 6 $Mo(1)O_4$ tetrahedra are bonded via corner sharing to 6 $Mo(3)O_6$ octahedra of the slab above while the other 3 are bonded in the same way to 6 $Mo(3)O_6$ octahedra of the slab below (see Figure 7).

The effective charge calculated by the Zachariasen formula, is +6 for the Mo(1) cations in the tetrahedral sites, while the values of +5.1 and +5.8 were obtained

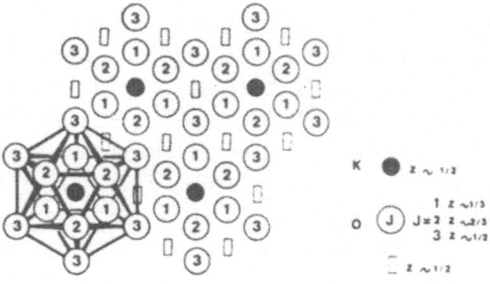

Fig. 6 (From Ref. [33]). Projection along the c axis of a K icosahedra layer; small rectangles represent oxygen positions usually occupied in h.c. stacking.

Fig. 7 (From Ref. [33]). K icosahedron surrounded by 6 Mo tetrahedra.

for the octahedrally-coordinated Mo(2) and Mo(3) cations, respectively. These valence states indicate that the $4d$ electrons of the Mo cations are located essentially towards the slab central plane. These electrons hop from one Mo site (Mo(2) and partially Mo(3)) to another via the Mo—O—Mo bonds. There exists, thus, a hybridized conduction band comprised of the Mo $4d$ states and the oxygen $2p$ states. Electron transfers along the direction perpendicular to the slabs do not take place because they are blocked at the Mo(1) sites which have a +6 valence state. Along this direction the electron transfers would take splace along the bonds Mo(3)—O—Mo(1)—O—O—Mo(1)—O—Mo(3) whose exchange integrals should be very small. The conduction band and the Fermi surface must be very anisotropic and the electrical conductivity is two-dimensional.

The sodium and thallium bronzes have an identical structural arrangement; The $Na_{0.9}Mo_6O_{17}$ symmetry is monoclinic which allows a greater distortion of the various coordination polyhedra (see Figure 8). The $4d$ electron distribution among the Mo cation sites is identical to that found in the K bronze. Thus, the electrical conductivity of the Na and Tl bronzes is also two-dimensional.

The lithium bronze has a somewhat similar arrangement. However, in this case, the difference is not only due to an additional distortion of the coordination polyhedra, but is the result of important structural changes. The MoO_6 octahedra forming the ReO_3-like slabs are much more distorted than those of the Na and K counterparts. The slabs can be considered as being 3-octahedra in width, in which every sequence of three octahedra is prolonged, along the sequence axis, by a forth octahedron alternately on each side of the slabs. This type of arrangement gives a spiked aspect to the slabs (see Figure 9).

The 4-coordinated Mo cations occupy two crystallographically independent sites. The two tetrahedra surrounding one of the two sites are linked via corner-sharing to two spike octahedra belonging to two adjacent slabs and facing each other. Therefore, contrary to the two-dimensional arrangement of the Mo—O—Mo—O bonds found in the violet Na and K bronzes, in the Li coun-

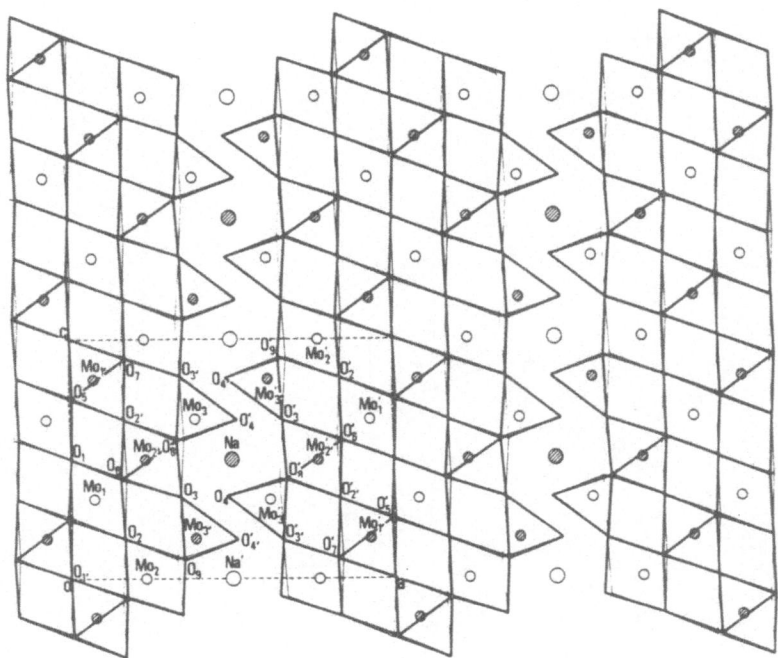

Fig. 8 (From Ref. [32]). $Na_{0.9}Mo_6O_{17}$ structure projected on the ac-plane; open and hatched circles indicate atoms at $y = 0$ and $y = 1/2$ respectively.

terpart at certain locations there exist Mo—O—Mo—O bonds which link together two adjacent slabs. Thus, the Mo—O—Mo—O bonding scheme is not entirely two-dimensional. The lithium cations are surrounded by 9 oxygen nearest neighbors arranged as a tetradecahedron (see Figure 10). These polyhedra are more distorted than those surrounding the Na and K cations in $Na_{0.9}Mo_6O_{17}$ and $K_{0.9}Mo_6O_{17}$, respectively.

The average effective charge, calculated by the Zachariasen formula, for the Mo(2), Mo(3), Mo(5), and Mo(6) cations, namely those located near the slab exterior and between slabs, is +5.7; while it is +5.03 for the Mo(1) and Mo(4) cations, that is for those located near the slab central plane. Although the probability that the $4d$ electrons are located near the slab central plane is much higher, the $4d$ electrons are located around all Mo cation sites. Therefore, the electrical conductivity is two-dimensional, but its anisotropy is less than that of the violet Na and K bronzes.

2.2.2. *The Blue-Black $Cs_{0.25}Mo_{0.97}O_3$ Bronze*

This bronze is monoclinic (space group $C2/m$) with lattice parameters $a = 19.063$ Å, $b = 5.583$ Å, $c = 12.115$ Å, and $\beta = 118.9°$. The crystal structure recently determined by S. C. Abrahams *et al.* [34] is somewhat related to that of the violet bronzes. On the other hand it is very different from the structure of the

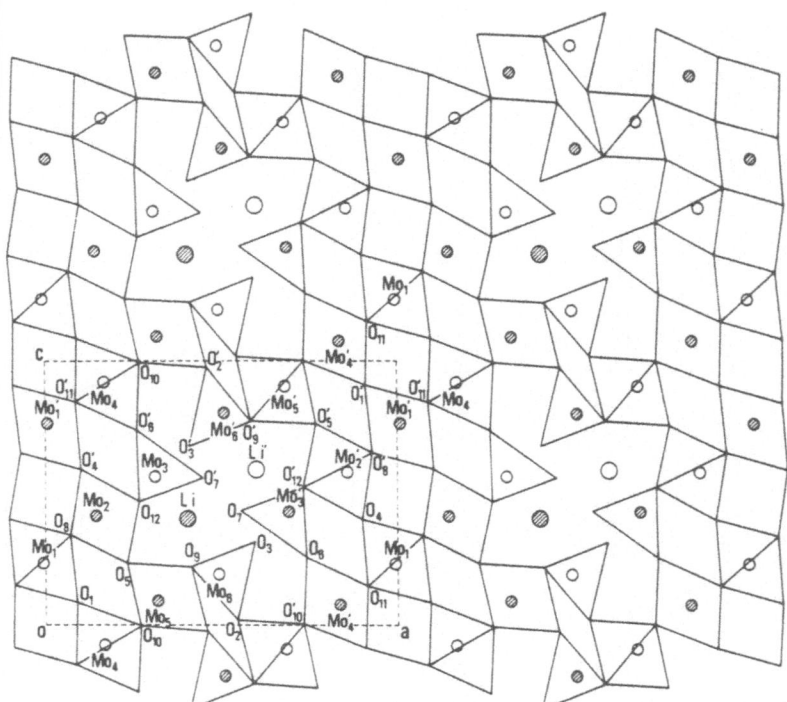

Fig. 9 (From Ref. [30]). $Li_{0.9}Mo_6O_{17}$ stucture projected on the ac-plane.

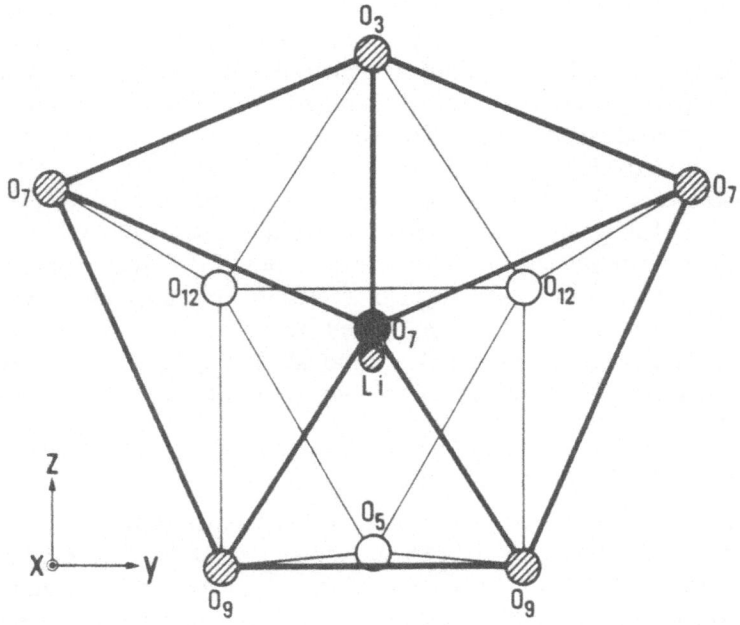

Fig. 10 (From Ref. [30]). Li cation tetradecahedron in $Li_{0.9}Mo_6O_{17}$.

reddish $Cs_{0.3}MoO_3$ reported by W. G. Mumme and J. A. Watts [35], which will be described in a subsequent section.

The structure of $Cs_{0.25}Mo_{0.97}O_3$ comprises slabs of MoO_4 tetrahedra and MoO_6 octahedra, parallel to the (100) plane and separated by chains of oxygen polyhedra centered around the cesium cations. These chains run parallel to the 2-fold axis (see Figure 11). A slab is built up of 3 layers of octahedra and tetrahedra. In contrast to the violet bronzes, where the MoO_4 tetrahedra are located at the slab edge, in the $Cs_{0.25}Mo_{0.97}O_3$ bronze the tetrahedra are located at the center of the slab. The polyhedral arrangement of a given slab is of ReO_3-type in which half the octahedra have been replaced by a tetrahedra. The zigzag chains running along the 2-fold axis are comprised only of corner-sharing octahedra. Two parallel chains are linked together by isolated tetrahedra whose unshared corners are pointing away from the chains (see Figure 12). The infinite Cs-polyhedron chains, which also run parallel to the 2-fold axis, are of two types. The first comprises edge-sharing CsO_{12} icosahedra, while the second comprises edge-sharing two-face-capped CsO_{10} tetragonal prisms (see Figure 13). The two types of chains are separated from each other by empty channels.

The Zachariasen formula, applied to this structure, gives effective charges of $+6$ and $+5.5$ for the tetrahedral and octahedral Mo cations, respectively. The $4d$ electrons are on the octahedral Mo cations, therefore they are located along the zigzag chains running along the 2-fold axis. The conduction mechanism described above for the violet bronzes also applies for $Cs_{0.25}Mo_{0.97}O_3$. The $Mo^{5+\varepsilon}$—O—$Mo^{5+\varepsilon}$ bonds are in this case monodimensional and consequently the metallic conductivity is also monodimensional.

2.3. MOLYBDENUM BRONZES CONTAINING INFINITE DOUBLE ReO_3 CHAINS

An infinite double ReO_3-chain is built up of two simple corner-sharing octahedral chains linked together by equatorial edge-sharing (see Figure 14). Double chains bonded in a zigzag fashion by equatorial edge-sharing form the slabs of the MoO_3 layer structure (see Figure 15).

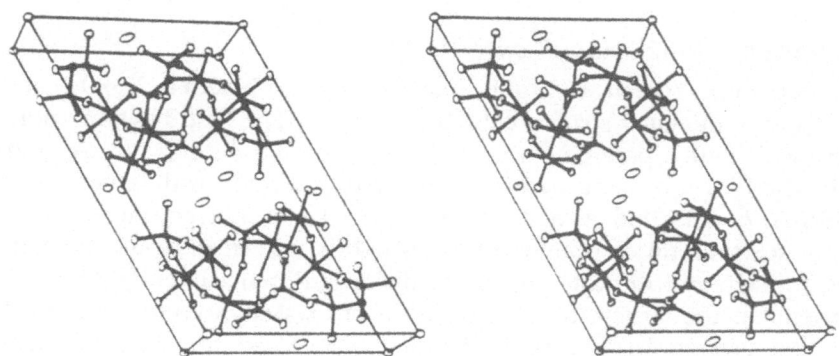

Fig. 11 (From Ref. [34]). Stereoview of one unit cell of $Cs_{0.25}Mo_{0.97}O_3$; a and c axis are vertical and horizontal respectively.

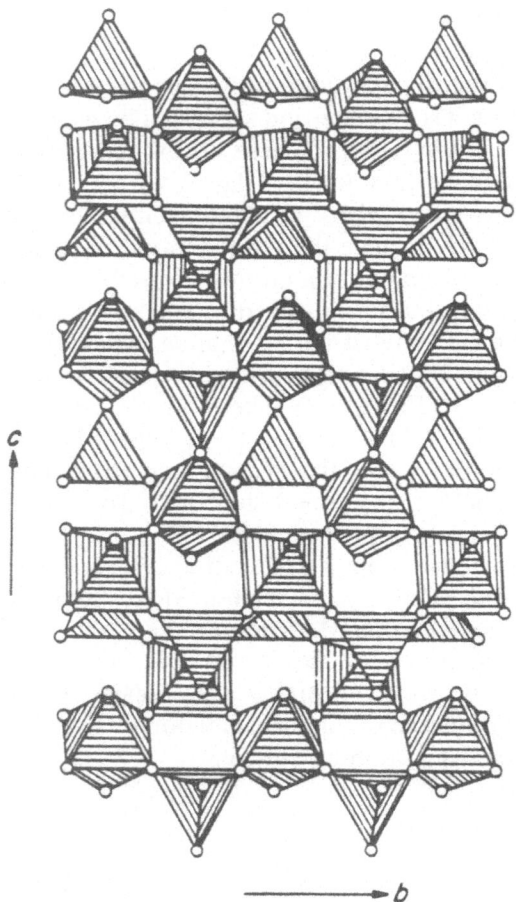

Fig. 12 (From Ref. [34]). Single polyhedral slab of $Cs_{0.25}Mo_{0.97}O_3$.

2.3.1. *Hydrogen Molybdenum Bronzes,* H_xMoO_3

These compounds (green H_2MoO_3, red $H_{1.7}MoO_3$, blue $H_{0.9}MoO_3$, and blue $H_{0.3}MoO_3$) have the layer structure described above. The first 3 exhibit monoclinic symmetry with lattice parameters $a \sim 14$ Å, $b \sim 3.8$ Å, $c \sim 4.0$ Å, and $\beta \sim 94°$, while the fourth is orthorhombic (space group *Cmcm*) with lattice parameters $a = 3.896$ Å, $b = 14.066$ Å, and $c = 3.736$ Å. Of the four crystal structures only that of an isostructural compound of $H_{0.3}MoO_3$ is known in detail. Wilhelmi [69] localized the Mo and O atoms by X-ray diffraction. Subsequently, Dickens *et al.* [36] carried out the structural determination of $D_{0.3}MoO_3$ using neutron diffraction. The deuteration was necessary in order to decrease the incoherent scattering due to hydrogen. Contrary to what one would expect, the deuterium atoms are not located between the slabs, but are bonded to two oxygen atoms belonging to the

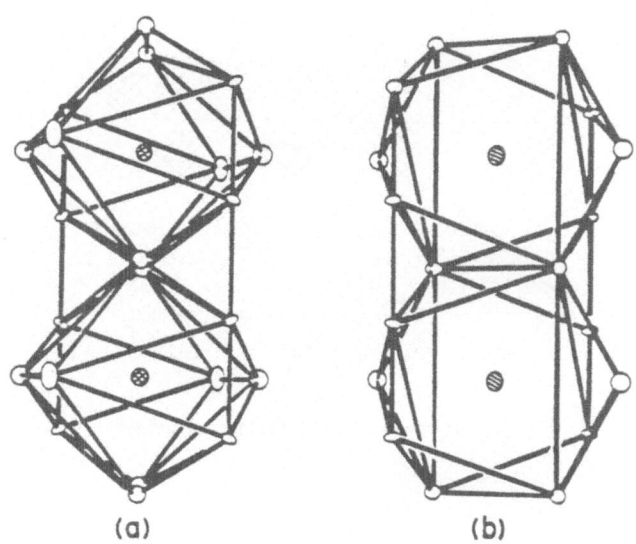

Fig. 13 (From Ref. [34]). (a) Cs(1) irregular icosahedra linked together in the *b* axis direction. (b) Cs(2) capped tetragonal prisms linked together in the *b* axis direction.

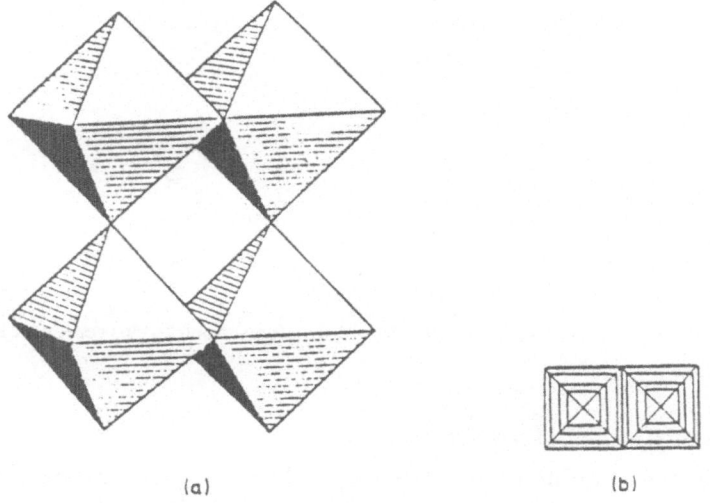

Fig. 14 (From Ref. [3]). Double ReO_3 chain.

same slab (see Figure 16). The asymmetric bonds O—D·····O are not ordered in the sense that the deuterium atoms form the short bond (1.09 Å) randomly with one of the two oxygen atoms. In $D_{0.3}MoO_3$ the MoO_6 octahedra are less distorted than those of the MoO_3 layered oxide.

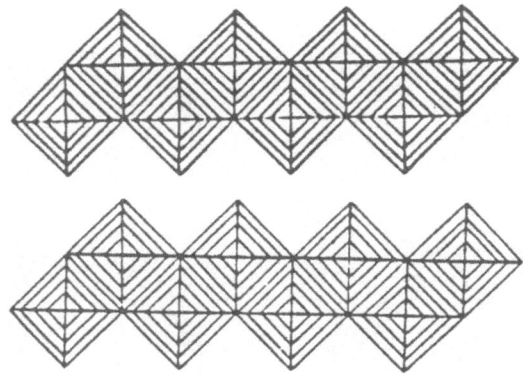

Fig. 15 (From Ref. [3]). MoO_3 layer structure.

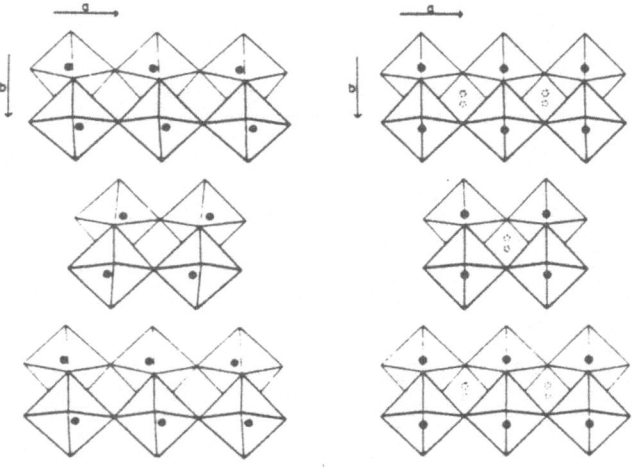

Fig. 16 (From Ref. [36]). MoO_3 and $D_{0.36}MoO_3$ structures; open circles represent D atoms.

2.3.2. Sodium Hydrated Molybdenum Bronzes

These bronzes also exhibit a layer structure of MoO_3-type. They are orthorhombic (space group *Ammm*) with lattice parameters $a = 3.876$ Å, $b = 19.093$ Å, $c = 3.733$ Å and $a = 3.884$ Å, $b = 22.618$ Å, $c = 3.751$ Å, for $[Na(H_2O)_2]_{0.25}MoO_3$ and $[Na(H_2O)_5]_{0.25}MoO_3$, respectively.

The sodium cations and the water molecules are inserted between MoO_3-type layers, whose separation increases with increasing amount of inserted cations. By analysing X-ray diffraction powder diagrams, Thomas *et al.* [21] proposed the arrangement shown in Figure 17. This arrangement should be confirmed by a neutron diffraction powder study.

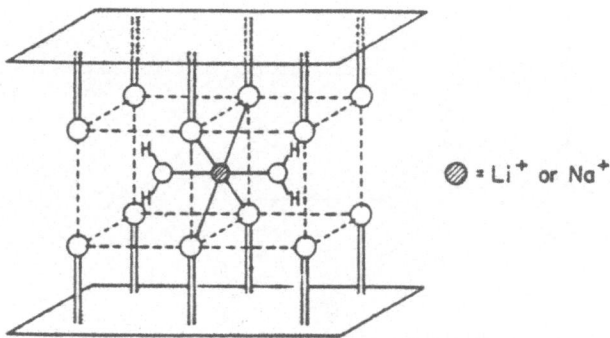

$\oslash = Li^+$ or Na^+

Fig. 17 (From Ref. [21]). Proposed model for the sodium hydrated molybdenum layer bronze.

2.4. MOLYBDENUM BRONZES CONTAINING INFINITE SINGLE OCTAHEDRAL RIBBONS

An infinite single octahedral ribbon is comprised of two ReO_3-chains which are interlocked in such a way that each octahedron shares two of its edges with two octahedra of the other chain (see Figure 18). Viewed from a perpendicular direction such a ribbon resembles a section of a MoO_3-type layer built by a zigzag juxtaposition of infinite double ReO_3-chains. The idealized three-dimensional V_2O_5 structure is generated by linking the ribbons via corner-sharing to obtain layers and by stacking parallel layers along the third direction via corner-sharing as well (see Figure 19).

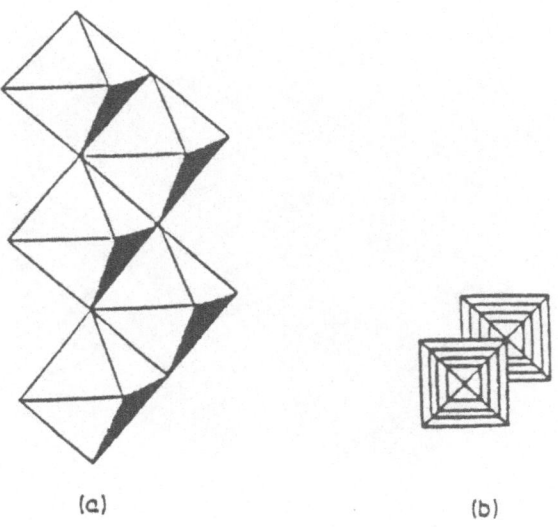

(a)   (b)

Fig. 18 (From Ref. [3]). Single octahedral ribbon.

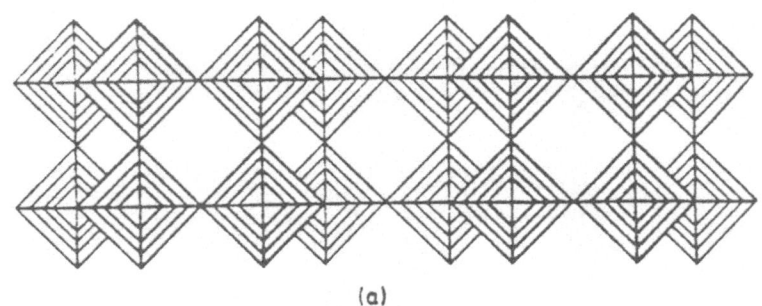

(a)

Fig. 19 (From Ref. [3]). V_2O_5 layer structure.

2.4.1. *The Triclinic Lithium Bronze,* $Li_{0.33}MoO_3$

The blue-violet stoichiometric lithium bronze, which has a triclinic symmetry (space group $P\bar{1}$), is different from the blue bronze reported by Réau and Hagenmuller [10], Li_xMoO_3 with $0.31 \leqslant x \leqslant 0.39$. The lattice parameters of $Li_{0.33}MoO_3$ were found to be $a = 13.079$ Å, $b = 15.453$ Å, $c = 7.476$ Å, $\alpha = 97.0°$, $\beta = 106.6°$, and $\gamma = 103.4°$. Its structure, which has recently been determined by Tsai *et al.* [37], is built up of V_2O_5-like layers stacked along the b^* axis. Every third octahedron throughout a layer is occupied by a lithium cation in an ordered fashion (see Figure 20).

The unit cell contains four layers. The first three (I, II, and III) are linked together by corner sharing, while layers III and IV are connected by edge-sharing (see Figure 21). The Mo cations are so displaced from the octahedron center that in some cases the coordination number should be taken as 4 + 2 or in others as

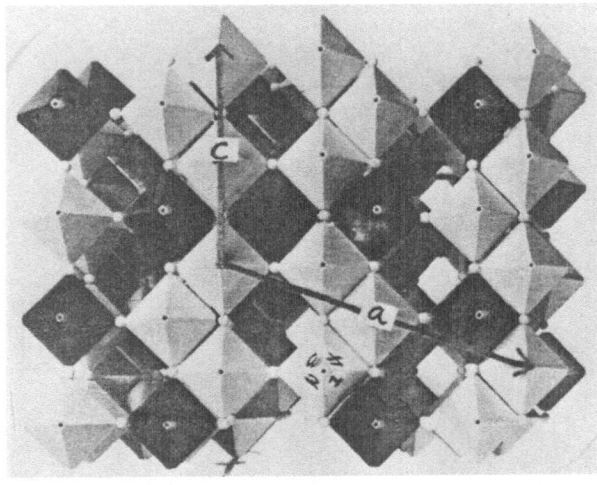

Fig. 20 (From Ref. [37]). V_2O_5-like plane of the $Li_{0.33}MoO_3$ structure; black octahedra represent LiO_6 octahedra.

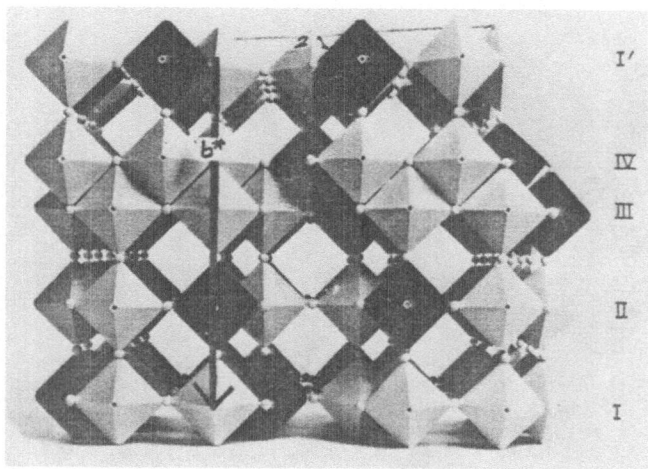

Fig. 21 (From Ref. [37]). Layer stacking along b^* of $Li_{0.33}MoO_3$.

5 + 1. Regular infinite ReO_3 chains can be outlined along the c axis (see Figure 20).

The calculated effective charge of the Mo cations is about +5.6 for the sites belonging to the chains running along the c axis and between +5.8 and +6 for the other sites. These values show that the $4d$ electrons are located on the average on all Mo sites, however they are preferentially along the chains. Therefore, the triclinic lithium bronze, $Li_{0.33}MoO_3$, should exhibit monodimensional electrical conductivity along the c axis.

2.5. MOLYBDENUM BRONZES CONTAINING INFINITE CHAINS OF OCTAHEDRAL CLUSTERS

The infinite chains of octahedral clusters are basically comprised of infinite double ReO_3 chains. Four contiguous octahedra of one double chain are linked to one or two additional octahedra by edge-sharing. Each of the four octahedra share one edge with each linking octahedron. This octahedron is either always on one side of the chains, or alternately on one side at a given level and on the other at the next level, or on both sides at the same level. The five or six octahedra constitute a cluster (see Figure 22). In some cases, as that of the blue bronzes, the infinite cluster chains are comprised of two double ReO_3 chains. These clusters are then made up of more octahedra.

2.5.1. The Red $A_{0.33}MoO_3$ Bronze with A = K, Cs and Tl

The red potassium bronze is monoclinic (space group $C2/m$) with lattice parameters $a = 14.278$ Å, $b = 7.723$ Å, $c = 6.387$ Å, and $\beta = 92.3°$. The structure, which was determined by Stevenson and Wadsley in 1965 [38], contains clusters of six octahedra. The two Mo(1) octahedra which are encased in the double ReO_3-

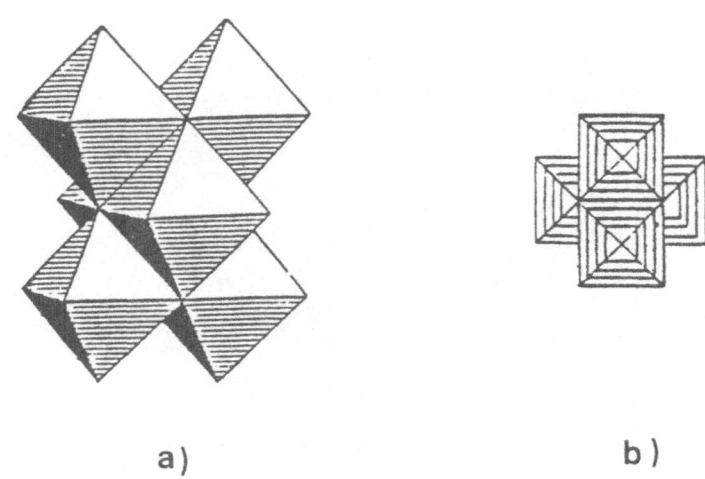

a)   **b)**

Fig. 22 (From Ref. [3]). The six octahedra cluster of the red $K_{0.33}MoO_3$ bronze.

chains, are at the same level on each side of the chain (see Figure 22). The cluster chains which run along the b axis, are linked together by corner-sharing to form parallel one cluster wide slabs. There are two such slabs per unit cell (see Figure 23). The slabs are bonded to each other by the K cations. Four oxygen atoms of one slab and four of the adjacent one form a two-face capped triangular prism whose center is occupied by the K cations. These polyhedra form infinite edge-sharing chains parallel to the b axis (see Figure 24). The red thallium and cesium bronzes are isomorphous of the former. The space group is the same one and the lattice parameters are: for $Tl_{0.33}MoO_3$, $a = 14.541$ Å, $b = 7.780$ Å, $c = 6.414$ Å, $\beta = 93.1°$ and for $Cs_{0.33}MoO_3$, $a = 15.862$ Å, $b = 7.728$ Å, $c = 6.408$ Å and $\beta = 94.4$ Å. The crystal structures have been determined by Ganne *et al.* [64] and by Tsai *et al.* [66] respectively.

The red bronzes are stoichiometric compounds whose formula is $AMo_3O_9(1/2$ $A_2O + 2\ MoO_3 + 1/2\ Mo_2O_5)$. They do not exhibit any metallic conductivity, mono- or two-dimensional, as one would predict from their structural arrangement, but are semiconductors at all temperatures. The $4d$ electrons are likely to be localized in the three compounds on only one Mo cation site. The determinations of the $4d$ electron distribution in $Tl_{0.33}MoO_3$ and $Cs_{0.33}MoO_3$ are consistent with this hypothesis. Approximately 90% of the $4d$ electrons are located on the Mo(2) cations of the ReO_3-chains [64, 66]. According to EPR measurements the $4d$ electrons are paired and lie in a cluster molecular orbital [64, 66].

2.5.2. *The Copper Colored $Cs_{0.3}MoO_3$ Bronze*

This bronze is also monoclinic, although, its space group is $P2_1/m$. The lattice parameters are $a = 6.425$ Å, $b = 7.453$ Å, $c = 8.169$ Å, and $\beta = 96.5°$. Its structure, which was determined by Mumme and Watts [35], is quite similar to that

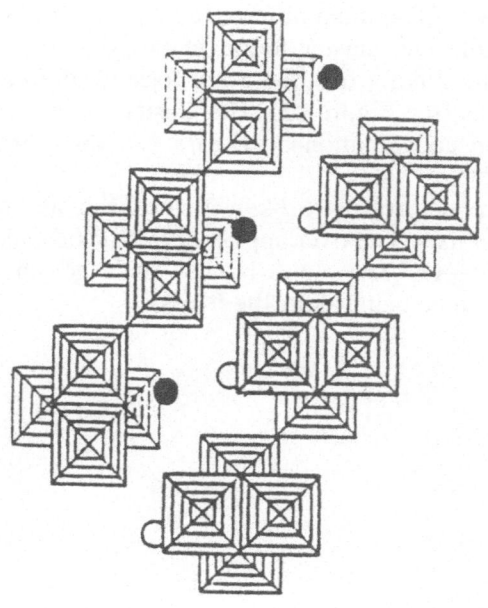

Fig. 23 (From Ref. [3]). Octahedral slabs linked by inter-layer K ions in $K_{0.33}MoO_3$; open and black circles represent K atoms at $y = 0$ and $y = 1/2$ respectively.

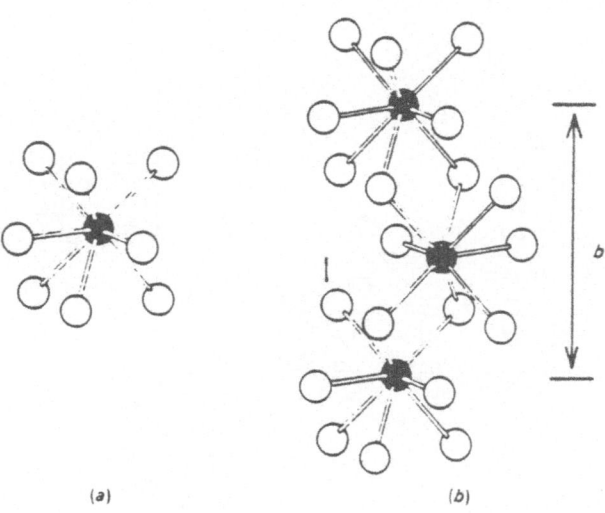

(a) (b)

Fig. 24 (From Ref. [38]). Zigzag chains of K—O polyhedra along the b axis.

of red $K_{0.33}MoO_3$ bronze. The main difference between the two structures is in the octahedral cluster itself. The cluster found in this Cs bronze still comprises six octahedra; however the linking octahedra are located alternately on each side of the double ReO_3 chains (see Figure 25). The cluster chain arrangement in forming slabs, the alkali cation coordination polyhedra and their array, are the same as those found in $K_{0.33}MoO_3$.

The Mo cation $4d$ electrons are likely delocalized along the infinite cluster chains parallel to the b axis. The overlapping of the $4d$ orbitals with the oxygen $2p$ ones along the —Mo—O—Mo—O— bonds, leads to an anisotropic metallic conductivity, the large value being along the b axis.

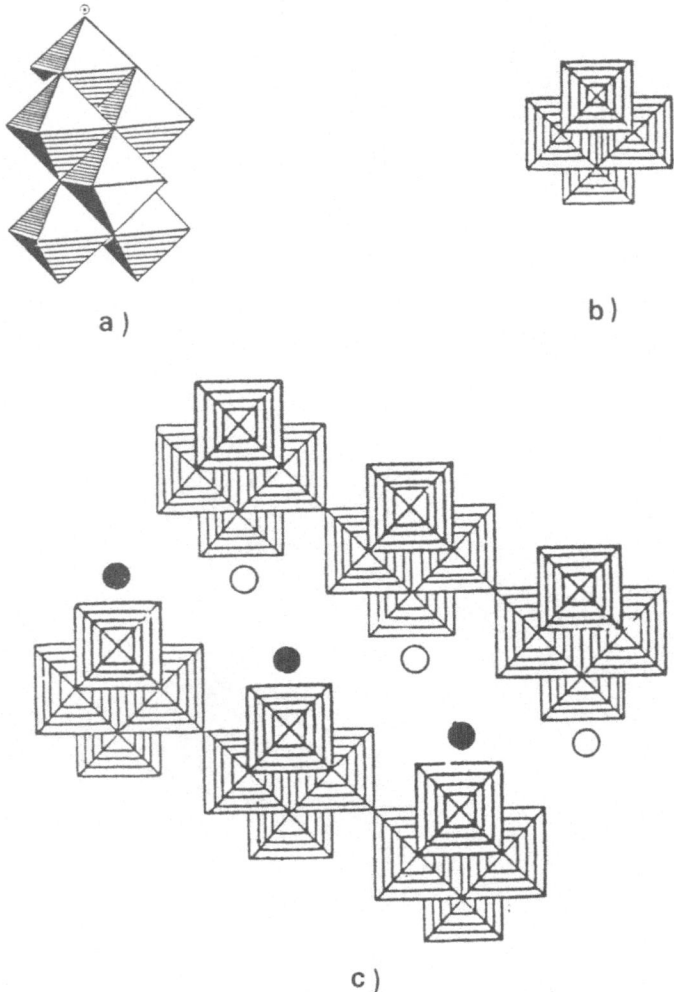

a)

b)

c)

Fig. 25 (From Ref. [3]). Structure of $Cs_{0.3}MoO_3$. (a, b) the six octahedra cluster; (c) the octahedral slabs linked by inter-layer Cs ions.

2.5.3. The Blue Bronzes, $A_{0.3}MoO_3$ with $A = K$, Rb, and Tl

The blue bronzes, $K_{0.3}MoO_3$, $Rb_{0.3}MoO_3$, and $Tl_{0.3}MoO_3$, have the same structure. Their symmetry is monoclinic (space group $C2/m$) with lattice parameters: for $K_{0.3}MoO_3$, $a = 18.249$ Å, $b = 7.560$ Å, $c = 9.855$ Å, $\beta = 117.3°$; for $Rb_{0.3}MoO_3$, $a = 18.635$ Å, $b = 7.555$ Å, $c = 10.094$ Å, $\beta = 118.8°$; for $Tl_{0.3}MoO_3$, $a = 18.543$ Å, $b = 7.567$ Å, $c = 10.087$ Å, $\beta = 118.4°$.

The crystal structures of $K_{0.3}MoO_3$ and $Tl_{0.3}MoO_3$ were solved by Graham and Wadsley in 1966 [39] and by Ganne et al. in 1985 [17], respectively. In order to determine the $4d$ electron distribution and relate it to the physical properties, Ghedira et al. [40] carried out a precise refinement of the $K_{0.3}MoO_3$ structure. Since the isomorphism of the corresponding Rb bronze had never been established, they refined the structure of $Rb_{0.3}MoO_3$ as well.

The description of the blue bronze structure is facilitated by using a monoclinic I-unit cell [40], related by the matrix: $(a\ b\ c)_I = (101, 0\bar{1}0, 00\bar{1})\ (a\ b\ c)_C$ to the C-centered one proposed by Graham and Wadsley. The new space group is $I2/m$. The two cells have the same volume, but β_I is closer to $90°$ than the β_C. Note, for example, that the I-cell contains only one slab running parallel to the diagonal (101) plane. The lattice parameters of the I-cells are: for $K_{0.3}MoO_3$ $a = 16.231$ Å, $b = 7.550$ Å, $c = 9.861$ Å, and $\beta = 94.9°$; for $Rb_{0.3}MoO_3$ $a = 16.361$ Å, $b = 7.555$ Å, $c = 10.094$ Å, $\beta = 93.9°$.

In these blue bronzes the infinite octahedral-cluster chains parallel to the b axis are more complex than those described above. They comprise two double ReO_3 chains bonded together by edge-sharing. The linking octahedra are encased at the same level to each double ReO_3 chain. The cluster is comprised of ten octahedra as shown in Figure 26.

As in the bronzes described above, the octahedral-cluster chains linked together by corner-sharing form one-cluster wide slabs between which the large cations (K or Rb or Tl) are inserted. These cations occupy two crystallographically independent sites. The first site is surrounded by ten oxygen atoms, eight of which are arranged as a distorted cube, while the other two cap two opposite faces. The second site is surrounded by seven oxygen atoms, six of which are arranged as a trigonal prism with the seventh capping one of the rectangular faces. Each type of polyhedra forms edge-sharing chains parallel to the b axis (see Figure 27), which link the slabs via corner-sharing.

The valence state calculations of the three crystallographically-independent Mo cation sites, as carried out by the Zachariasen formula, show that the $4d$ electrons are distributed over all Mo sites. However, the probability that they are located around the Mo(2) and Mo(3) site, where the average valence state is $+5.6$, is higher than the probability that they are located around the Mo(1) site, whose valence state is $+5.8$. The Mo(2) and Mo(3) sites form the infinite double ReO_3 chains, while the Mo(1) site corresponds to the linking octahedron.

The structure of the blue bronzes built up of octahedral slabs is definitely a two-dimensional structure with respect to the molybdenum and oxygen sublattices. The electrical conductivity should be two-dimensional as well if its mechanism is based on the overlapping of the $4d$—$2p$ orbitals of the Mo—O—Mo—O bonds. The

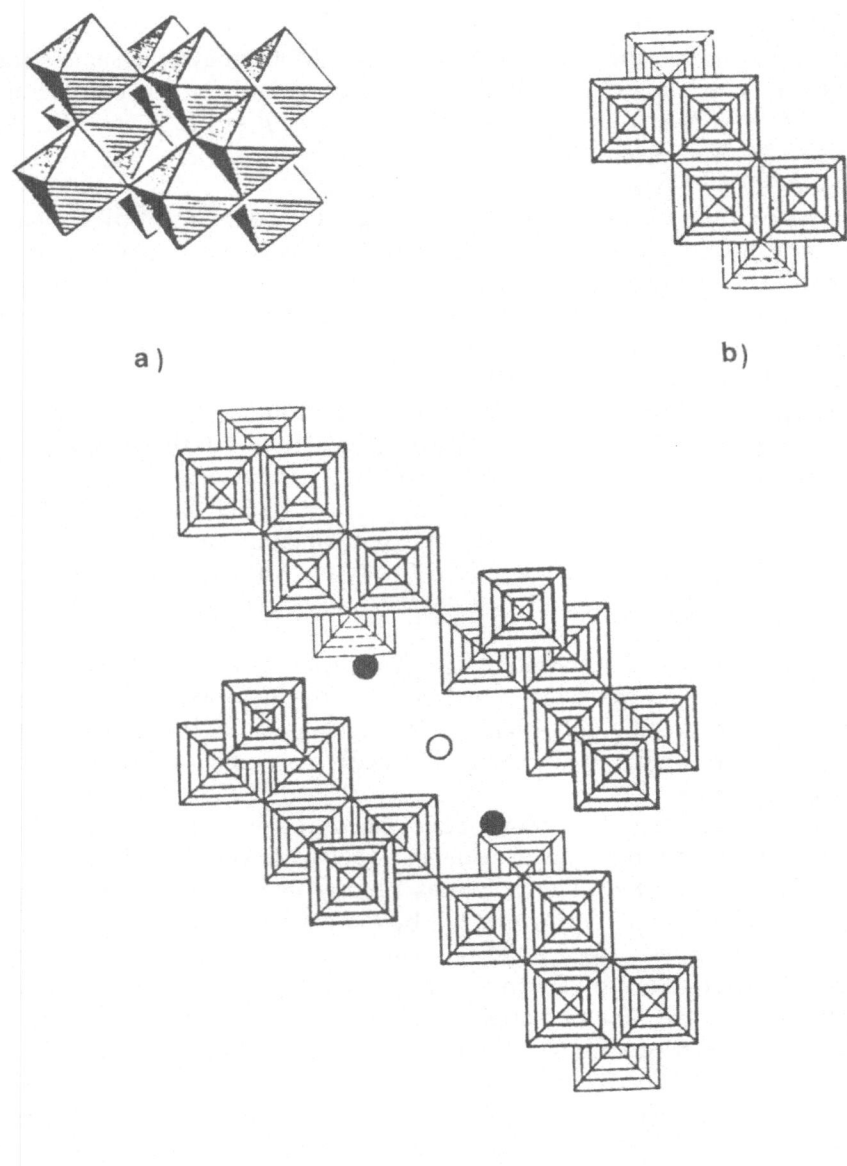

a)　　　　　　　　　　　　　　　　　　　b)

c)

Fig. 26 (From Ref. [3]). Structure of the blue $A_{0.3}MoO_3$ bronzes; (a, b) the ten octahedra cluster; (c) the octahedral slabs linked by inter-layer A ions.

experimental monodimensionality of the electrical conductivity can be explained if one takes into account the high density of the Mo—O—Mo—O bonds along the double ReO_3 chains with respect to that of the bonds between chains.

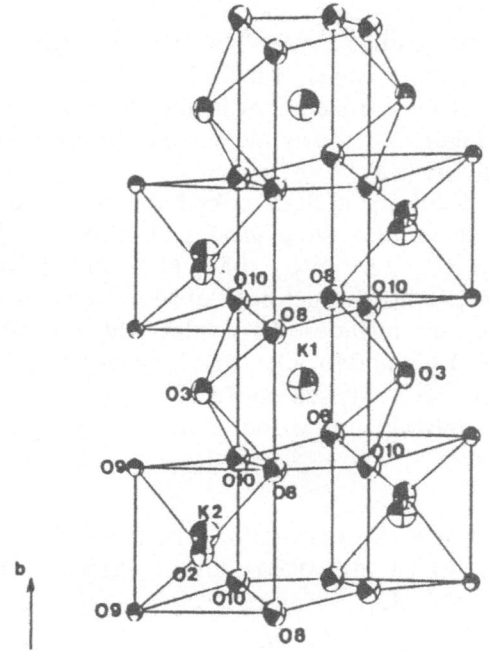

Fig. 27 (From Ref. [40]). Infinite chains of A cation polyhedra showing the two kinds of site.

3. Molybdenum Oxides

There are two molybdenum oxides, stable well beyond 1000 °C, with well defined valence states for the Mo cations, namely $Mo^{+4}O_2$ and $Mo^{+6}O_3$. Between these two end members there exists a series of intermediate oxides, which have been synthesized and characterized from the crystallochemical point of view by Hägg, Magnéli and his collaborators as early as 1944 [41—45]. A few years later Glemser and Lutz resumed these studies and virtually completed the pioneering work of the Swedish school [46]. More recently Kihlborg carried out a systematic structural study of the intermediate oxide series which includes seven members: $Mo_{18}O_{52}$, $Mo_{17}O_{47}$, Mo_9O_{26}, Mo_8O_{23}, Mo_5O_{14}, and two Mo_4O_{11}, the γ and η phases [47—54]. Except for the first two, the others belong to the series with the general formula Mo_nO_{3n-1}. As are the molybdenum bronzes, some of these oxides are metallic conductors of low dimensionality. Very recently their structures have been refined by different authors in order to localize the $4d$ electrons.

All intermediate oxides and MoO_2 can be obtained by H_2 reduction of MoO_3 at 500°. They can also be prepared by a solid state reaction of MoO_3 powder with either Mo metal or MoO_2 powder, in the proper stoichiometric ratio, in a vacuum-sealed tube at temperatures between 500—800 °C. Single crystals of MoO_2 and of some of the intermediate oxides have been prepared either by the electrolysis of molten mixtures of alkali molybdates and MoO_3 at about 600 °C, or by the vapor transport method at 800 °C using I_2 or $TeCl_4$ as transport agent [53], [55].

The list of molybdenum oxides, which have been characterized from the crystallographic point of view, together with their crystallographic constants, are reported in Table II. Figure 28 shows the stability domain for each intermediate oxide according to Kihlborg [52].

In a review article Kihlborg [52] described the structure of all molybdenum oxides and classified them into three groups. The first includes the oxides of ReO_3-type, namely Mo_8O_{23}, Mo_9O_{26}, γ-Mo_4O_{11}, and η-Mo_4O_{11}. The second group includes the oxides whose structure is of MoO_3 type, that is MoO_3 itself and $Mo_{18}O_{52}$. The third group includes the oxides which exhibit mixed polygonal networks, namely Mo_5O_{14} and $Mo_{17}O_{47}$. We have adopted for the molybdenum bronzes a classification which is based on ReO_3-type chains. Since these chains are also present in the molybdenum oxide structures, we present hereafter this alternative classification.

3.1. THE MOLYBDENUM OXIDES CONTAINING INFINITE ReO_3 SLABS: THE CRYSTAL STRUCTURE OF THE MONOCLINIC η-Mo_4O_{11} AND THE ORTHORHOMBIC γ-Mo_4O_{11}

The η phase of Mo_4O_{11} has a monoclinic symmetry (space group $P2_1/a$), with lattice parameters: $a = 25.54$ Å, $b = 5.439$ Å, $c = 6.701$ Å, and $\beta = 94.3°$. The crystal structure was determined by Kihlborg [53] and recently refined by Ghedira *et al.* [67].

The structural arrangement is similar to that of the violet molybdenum bronzes $A_{0.9}Mo_6O_{17}$ (A = Na or K) which has been described in a previous chapter. It comprises infinite ReO_3-type corner-sharing Mo-octahedron slabs. The slabs are three-octahedra thick along the direction of one of the three four-fold axes of the ReO_3 structure. The A icosahedra have disappeared and the slabs are linked

TABLE II
Crystal data of molybdenum oxides.

	a	b	c	β	Space group
$Mo_{18}O_{52}$	8.145	11.89	21.23	67°8	$P\bar{1}$
	$\alpha = 102°67$		$\gamma = 109°97$		
Mo_9O_{26}	16.75	4.030	14.45	96°0	$P2/a$
Mo_8O_{23}	16.88	4.052	13.39	106°1	$P2/a$
Mo_5O_{14}	46.00	23.00	3.937		$Pb2_1a$
$Mo_{17}O_{47}$	21.615	19.632	3.951		$Pba2$
$Mo_4O_{11}\,\eta$	24.54	5.439	6.701	94°3	$P2_1/a$
$Mo_4O_{11}\,\gamma$	24.49	5.457	6.752		$Pn2_1a$

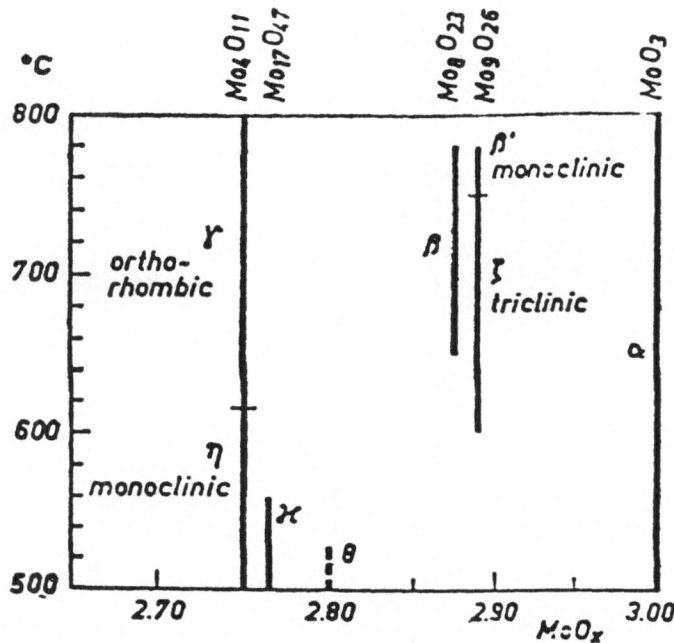

Fig. 28 (From Ref. [52]). Molybdenum oxides and their formation temperatures.

together by the MoO_4 tetrahedra, of which every other one is pointing either up or down (see Figure 29).

The comparison cannot be carried any further because the two structural types have their own symmetry elements and consequently their specific distortion.

The γ phase of Mo_4O_{11} is orthorhombic (space group $Pn2_1a$) with lattice parameters: $a = 24.49$ Å, $b = 5.457$ Å, and $c = 6.752$ Å. Its structure has been resolved by Magnéli [42], later refined by Kihlborg [53, 54] and recently by Ghedira *et al.* [55]. The structural arrangement is very similar to that of the monoclinic η structure. In the γ phase, as well as in the η, the slabs are three-octahedra thick along one of the three pseudo four-fold axes of the ReO_3 structure. The main difference between the two structures lies in the ReO_3-chain orientation in adjacent slabs. In the η phase the chains have the same orientation in all slabs, whereas in the γ phase the ReO_3-chains have the same orientation only in every other slab. This is due to existence of a diagonal glide mirror plane perpendicular to the a axis. In the γ phase the ReO_3-chains zigzag on going from one slab to the adjacent one (see Figure 29).

The Mo valence calculations carried out by using the Zachariasen formula for Mo_4O_{11}(O-rh) and Mo_4O_{11}(Mono.), led to a charge distribution which is quite similar to that found in the violet bronzes. The valence of the Mo(1) cations occupying the tetrahedral sites was found to be very close to +6, whereas that of the Mo cations occupying the octahedral sites was found to decrease on going

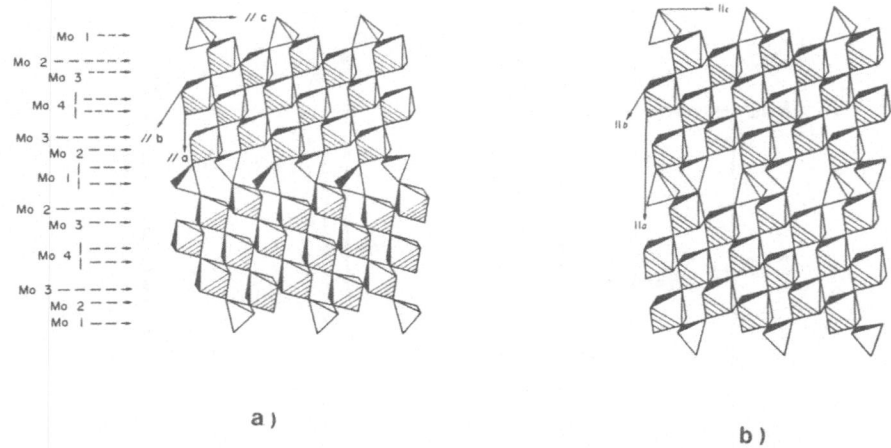

Fig. 29 (From Ref. [51]). Crystal structure of Mo_4O_{11} oxides; (a) orthorhombic form; (b) monoclinic form. MoO_6 octahedra slab are connected by MoO_4 tetrahedra.

from the edge to the center of the slabs. For example, for Mo_4O_{11}(O-rh) the following was found: $Mo(2) \equiv +5.8$, $Mo(3) \equiv +5.4$, $Mo(4) \equiv +5.0$ v.u. [55]. Kihlborg carried out the same type of calculations by using the Pauling's relation; the values obtained from his second refinement are in good agreement with the above values. The $4d$ electrons are located around the median plane of the slabs. As for the molybdenum bronzes, this $4d$ electron distribution brings about anisotropic conduction bands and anisotropic Fermi surfaces. Consequently these oxides exhibit a two-dimensional metallic conduction.

3.2. THE MOLYBDENUM OXIDES CONTAINING INFINITE CHAINS OF CLUSTERS AND OF ReO_3-TYPE: Mo_8O_{23}, Mo_9O_{26}, Mo_5O_{14}, AND $Mo_{17}O_{47}$

3.2.1. *The Crystal Structure of* Mo_8O_{23} *and* Mo_9O_{26}

Mo_8O_{23} and Mo_9O_{26} are monoclinic (space group $P2/a$) with lattice parameters: $a = 16.88$ Å, $b = 4.052$ Å, $c = 13.39$ Å and $\beta = 106.1°$ for the former, and $a = 16.75$ Å, $b = 4.030$ Å, $c = 14.45$ Å, and $\beta = 96.0°$ for the latter.

The crystal structures were determined by Magnéli *et al.* [41] and later Kihlborg refined the Mo_8O_{23} structure [51]. Recently, Fujishita *et al.* [56] have redetermined the structure of Mo_8O_{23} at 370 K and 100 K. These authors found that at room temperature the structure is incommensurate with a propagation vector $\mathbf{q} = [0.195, 0.5, 0.120]$. Therefore, the structural determination carried out by the previous authors at room temperature, is an average structure. However, it is stable above 360 °C.

Magnéli and Kihlborg described the structure of Mo_8O_{23} and Mo_9O_{26} as containing ReO_3-type slabs which are interlocked after adjacent slabs have undergone a shearing movement with respect to each other. Wadsley introduced the term 'crystallographic shear' (CS) for this structural phenomenon. In Mo_8O_{23} the slabs are respectively four and eight octahedra-thick along two of the three four-fold axes of the ReO_3 arrangement. In Mo_9O_{26} the slabs have a thickness of four or five octahedra along one of the four-fold axis and nine along the other (see Figure 30). Contrary to what has been observed in the structures of Mo_4O_{11}, the slabs of Mo_8O_{23} and Mo_9O_{26} are infinite along the third four-fold axis. The lattice parameter along this direction correspond to the octahedron height.

According to our description these oxides contain infinite cluster chains linked together via corner-sharing by two infinite simple ReO_3-chains. The cluster chains are comprised of two double ReO_3-chains linked together by equatorial edge-sharing (see Figure 31). As can be seen, these arrangements are closer to that of the blue bronzes than to that of the violet ones. This might explain why their electrical conductivity has a monodimensional character rather than a two-dimensional one.

3.2.2. The Crystal Structure of Mo_5O_{14} and $Mo_{17}O_{47}$

Mo_5O_{14} is, as first approximation, tetragonal (space group $P4/mbm$) with lattice parameters $a = 23.00$ Å, $c = 3.946$ Å, while $Mo_{17}O_{47}$ is orthorhombic (space group $Pba2$) with $a = 21.615$ Å, $b = 19.632$ Å, $c = 3.951$ Å. Their structures were determined by Kihlborg [49] and [47, 48]. A more recent study by Yamazoe and Kihlborg [71] showed that the tetragonal unit cell of Mo_5O_{14} is only an average symmetry due to twinning. The true symmetry is orthorhombic with lattice parameters $a' = 2a$, $b' = a$ and $c' = c$ (may be $2c$); cation displacements out of

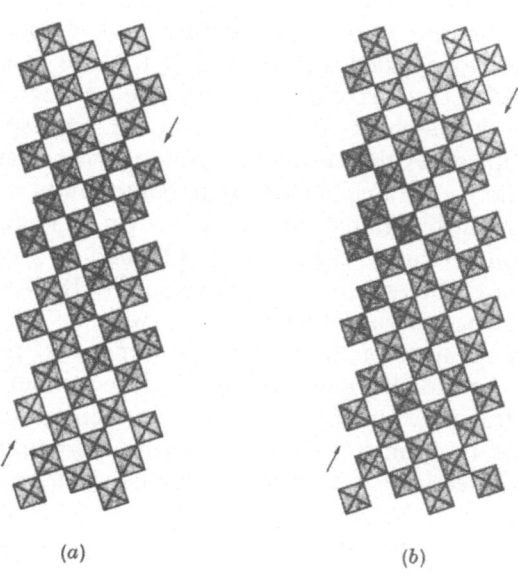

(a)   (b)

Fig. 30 (From Ref. [52]). ReO_3 slabs; (a) in Mo_8O_{23}; (b) in Mo_9O_{26}.

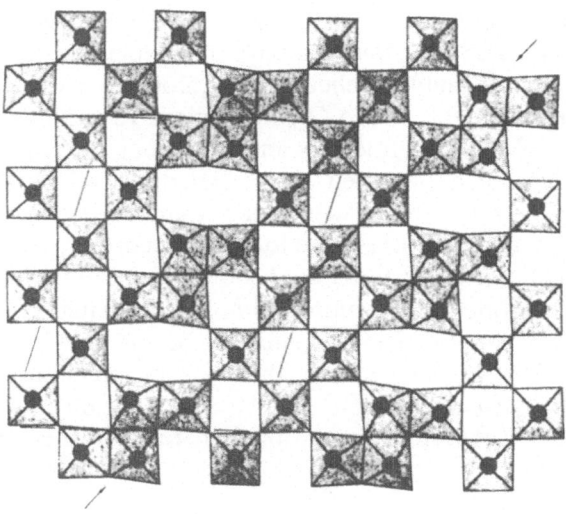

Fig. 31 (From Ref. [52]).   Crystal structure of Mo_8O_{23} viewed along the b axis.

the mirror plane perpendicular to the c axis are the causes of the lowering of symmetry.

In both arrangements, the Mo cations occupy, besides the usual MoO_6 octahedral sites, MoO_7 pentagonal bipyramidal sites as well. Each pentagonal bipyramid shares its five equatorial edges with five octahedra. This well known six-polyhedron cluster, sometimes called a 'pentagonal column' [70], is the basic unit of the Mo_5O_{14} structure. Two of such clusters linked together via equatorial edge-sharing between two octahedra (Oct_1 and Oct_2), one from each cluster, are the basic unit of the $Mo_{17}O_{47}$ structure. The distance between the two Mo atoms of Oct_1 and Oct_2 is quite short (2.64 Å); it could correspond to a metallic Mo—Mo bond. In both cases the clusters form via corner-sharing infinite chains along the c axis. The c parameter for both compounds corresponds to the octahedron or pentagonal bipyramid height. As shown in Figures 32 and 33, the infinite cluster and double-cluster chains in Mo_5O_{14} and in $Mo_{17}O_{47}$, respectively, are linked together via corner-sharing by infinite simple ReO_3-chains. The linking among the ReO_3-chains occurs via corner-sharing as well.

The structural arrangement of these oxides is similar to that of the tungsten bronzes or to that of the molybdenum bronzes obtained under high pressure. For example, the blue tetragonal bronze, $K_{0.5}MoO_3$, contains parallel infinite ReO_3-chains, arranged so as to form a polygonal network perpendicular to the chain axis. In this network there exist pentagonal channels where the K cations are accomodated (see Figure 3). In the oxide structures the channels contain alternatively oxygen and molybdenum atoms. It must be pointed out that the polygonal network of the oxides is different from that of the bronzes.

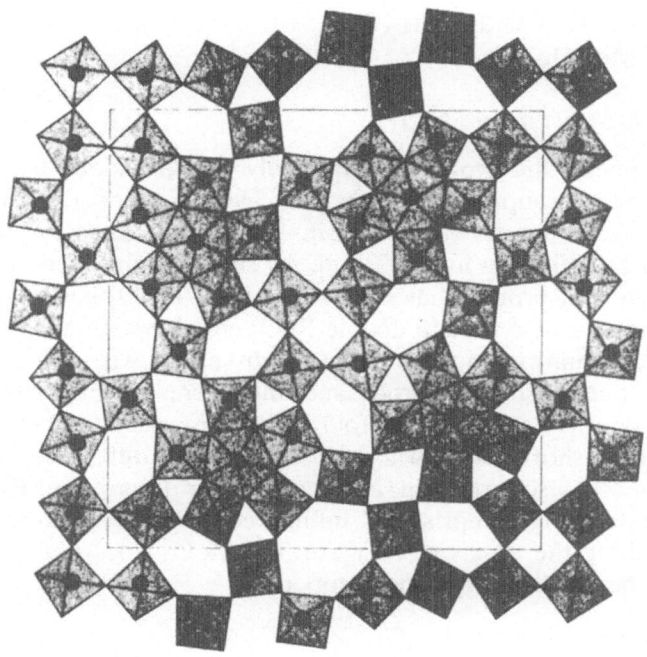

Fig. 32 (From Ref. [52]). Crystal structure of Mo_5O_{14} viewed along the c axis.

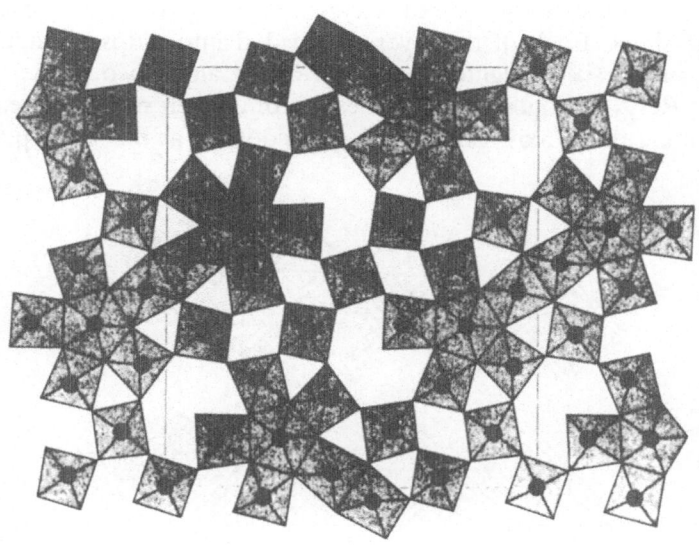

Fig. 33 (From Ref. [52]). Crystal structure of $Mo_{17}O_{47}$ viewed along the c axis.

3.3. THE MOLYBDENUM OXIDES HAVING THE LAYER MoO_3-TYPE STRUCTURE: MoO_3 AND $Mo_{18}O_{52}$

3.3.1. *The Crystal Structure of* MoO_3

MoO_3, which is one of the two most stable molybdenum oxides, is of orthorhombic symmetry (space group *Pbnm*) with $a = 3.963$ Å, $b = 13.855$ Å, and $c = 3.696$ Å. Its crystal structure was solved in 1931 by Bräkken [57] and Wooster [58] and refined by Kihlborg in 1963 [59]. As stated in the section in which the hydrogen molybdenum bronzes have been described, MoO_3 has a layer structure. Each layer is comprised of infinite double ReO_3-chains which are linked together by equatorial edge-sharing to form zigzag slabs as shown in Figure 34. Each octahedron has one unshared corner and the corresponding oxygen atom is bonded to only one Mo cation. The MoO_6 octahedra exhibit a Jahn—Teller type distortion with four short and two long distances. The difference between these two sets of distances is so large that, in an alternative description, Kihlborg refers to this structure as being comprised of infinite corner-sharing MoO_4-tetrahedron chains running along the c axis (see Figure 35). MoO_3 does not contain any $4d$ electrons, and it therefore has insulating properties.

3.3.2. *The Crystal Structure of* $Mo_{18}O_{52}$

$Mo_{18}O_{52}$ is triclinic (space group $P\bar{1}$) with lattice parameters $a = 8.145$ Å, $b = 11.89$ Å, $c = 21.23$ Å, $\alpha = 102.7°$, $\beta = 67.8°$, and $\gamma = 110.0°$. Its crystal structure, which was determined by Kihlborg [50], contains two-octahedron thick layers, geometrically identical to those found in MoO_3. However, inside the layers of $Mo_{18}O_{52}$ there exist well-ordered oxygen stacking defects caused by the crystallographic shear. Each of the layers is divided into strips, parallel to the b axis, which are either six or eighteen octahedra wide along two of the three four-fold axes of the ReO_3 arrangement. The shearing of a given strip with respect to its adjacent one, along the b axis, generates oxygen defects at the shear plane. These

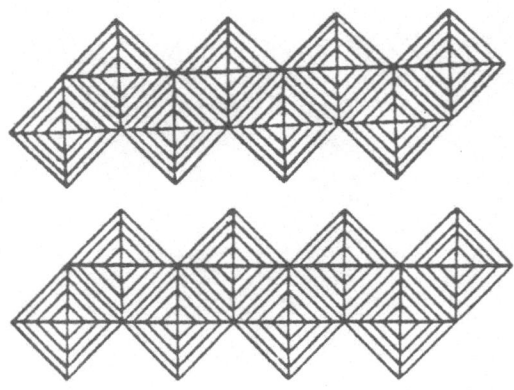

Fig. 34 (From Ref. [3]). MoO_3 layer structure viewed along the c axis.

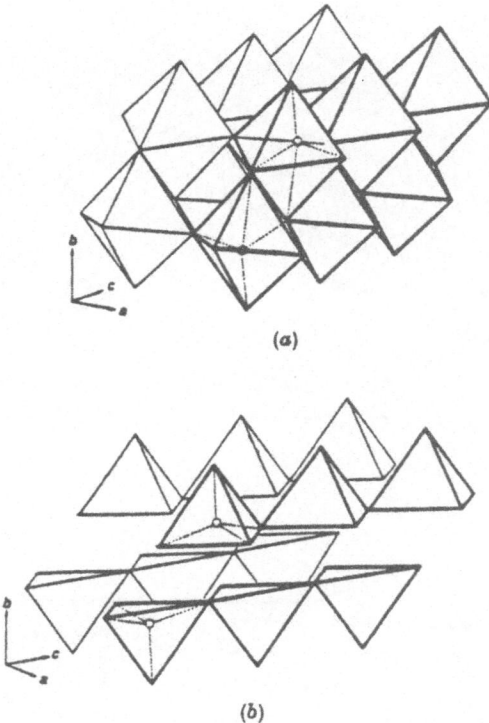

(a)

(b)

Fig.35 (From Ref. [52]). MoO_3 layer; (a) visualized as comprising regular octahedra; (b) visualized as built up of tetrahedra.

defects cause a decrease in coordination from octahedral to tetrahedral for some of the molybdenum cations (see Figure 36). It must be pointed out that in the structure of $Mo_{18}O_{52}$ the strips have a finite thickness (two octahedra) and not an infinite one as in the case, for example, of Mo_4O_{11}(Mono. and O-rh).

By taking into account the structural arrangement, $Mo_{18}O_{52}$ oxide could be a monodimensional (along the b axis) metallic conductor. By studying a powder sample Kihlborg showed that it exhibits resistive semiconducting properties [72]. This seems to indicate that the $4d$ electrons are localized only on some of the different Mo cation sites.

The existence of other molybdenum oxides, such as $Mo_{13}O_{38}$ and $Mo_{26}O_{75}$, has been evidenced by Kihlborg. The structural arrangements of these oxides are very similar to that of $Mo_{18}O_{52}$. The difference lies in the strip width; however their crystal structures have not been determined in detail.

3.4. THE DIOXIDE MoO_2

This is the only molybdenum oxide structure which cannot be related to the

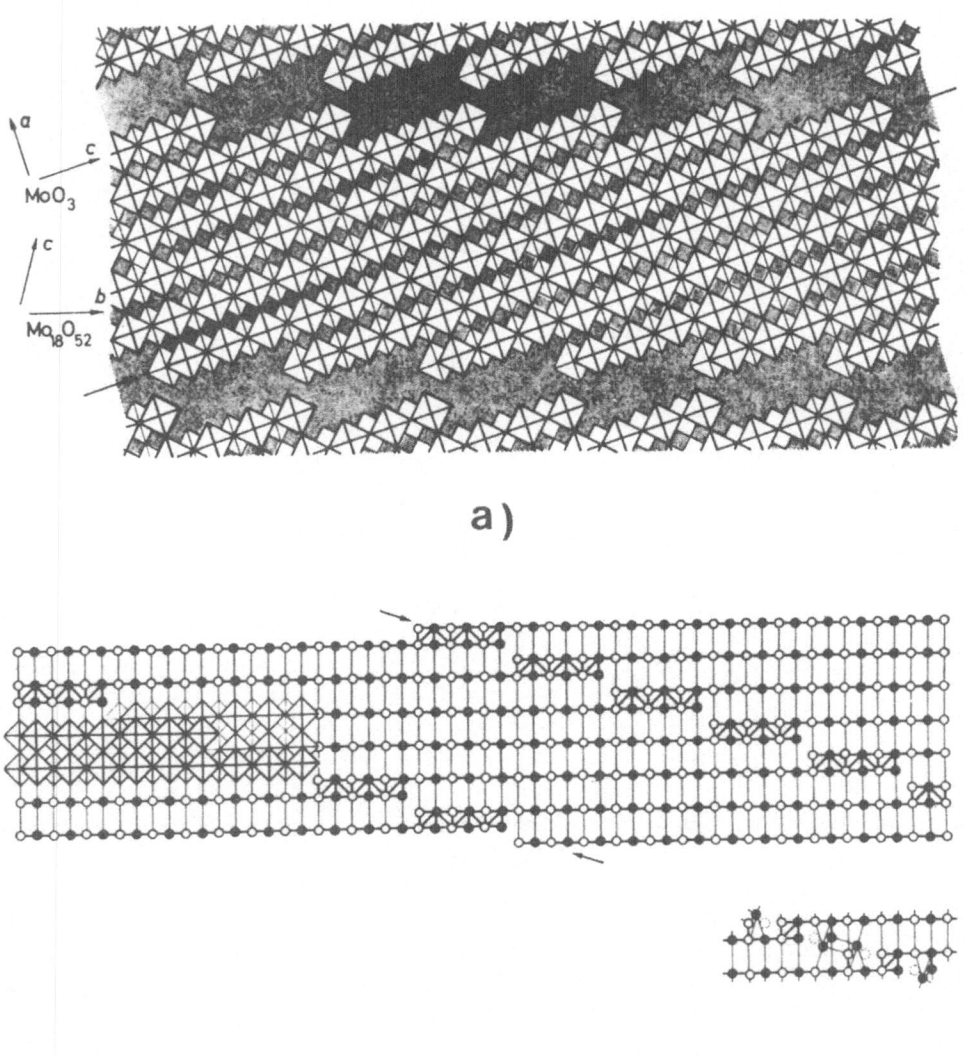

Fig. 36 (From Ref. [50]). $Mo_{18}O_{52}$ crystal structure; (a) MoO_3-type layers are divided into strips parallel to the b axis; (b) after shearing two adjacent strips are connect by MoO_4 tetrahedra; filled and open circles indicate molybdenum atoms.

ReO_3-chains. In order to describe the MoO_2 structure a different type of chains must be introduced, namely the rutile chains. These are infinite linear chains of edge-sharing octahedra, shown in Figure 37. The tetragonal structure of TiO_2 is comprised of such chains, which run along the four-fold axis and are linked together via corner-sharing along the other two directions (see Figure 38). At room temperature MoO_2 has a monoclinic-distorted rutile structure (space group $P2_1/c$) with $a = 5.611$ Å, $b = 4.856$ Å, $c = 5.628$ Å, and $\beta = 120.9°$. It was

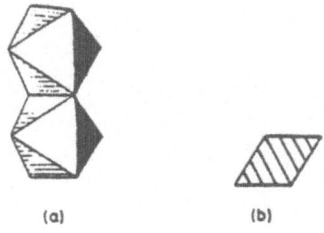

(a)   (b)

Fig. 37 (From Ref. [3]). The single rutile chain.

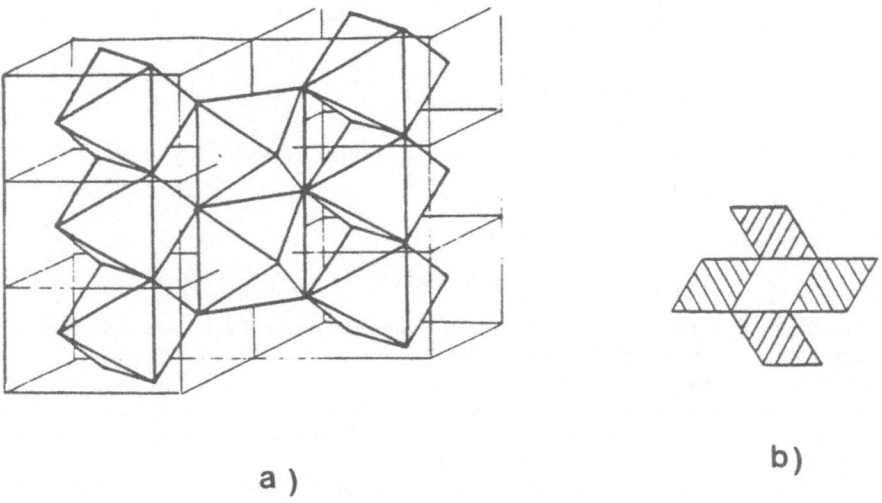

a)

b)

Fig. 38 (From Ref. [62]). The rutile structure of TiO_2.

determined by Magnéli [60] and refined by Brandt *et al.* [61]. Quite recently, in order to determine the Mo—Mo bond character, Ghedira *et al.* [62] have studied the structure of Ti-doped MoO_2 as a function of temperature from 24 to 900 °C.

The Mo—Mo distance across the shared edge is very short, which indicates that the Mo cations are paired to form covalent bonds. This pairing causes a symmetry lowering from tetragonal to monoclinic, which allows the octahedra to have a sizeable distortion (see Figure 39). The fact that MoO_2 has a metallic conductivity at all temperatures, indicates that only one $4d$ electron of Mo^{4+} participates in the Mo—Mo covalent bonds (σ and π bonds). The other must participate in a Mo—O π conduction band responsible for the metallic conductivity [62, 63].

Acknowledgements

The authors would like to thank A. S. Cooper (A. T. & T.), C. Gleitzer (University of Nancy) and L. Kihlborg (University of Stockholm) for their critical reading of the manuscript.

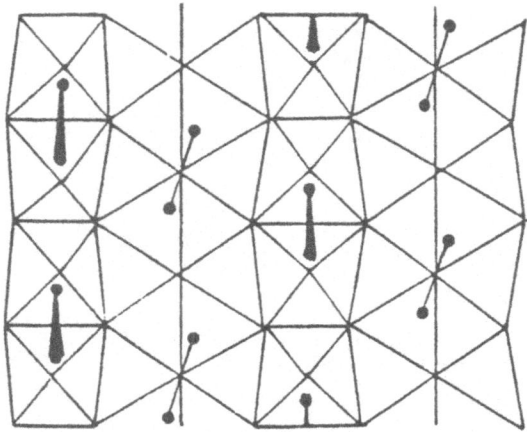

Fig. 39 (From Ref. [62]). The monoclinic-distorted rutile structure of MoO_2; the Mo—Mo pairs are outlined.

References

* Present address: Laboratoire des Matériaux et du Génie Physique, Ecole Nationale Supérieure de Physique de Grenoble, 38402 Saint-Martin d'Héres, France.

1. F. Wöhler, *Ann. Chim. Phys.* (2) **43**, 29 (1823).
2. P. Hagenmuller, *Progress in Solid State Chem.*, Vol. 5 (H. Reiss Ed.), p. 71, Pergamon (1971).
3. P. G. Dickens and P. J. Wiseman, *MTP International Review of Science*, Series 2, Inorg. Chem., Vol. 10 (L. E. Roberts Ed.), p. 211, Butterworths (1975).
4. P. G. Dickens, *Advances in Chemistry Series*, Vol. 163 (J. B. Goodenough and M. S. Wittingham Eds.), p. 165, American Chemical Society, Washington (1977).
5. G. Hägg and A. Magnéli, *Arkiv Kemi, Mineral. Geol.* **A19**, 2, (1944).
6. A. Magnéli, *Acta Cryst.* **6**,495 (1953).
7. C. Gleitzer, 'Molybdène', *Compléments au Nouveau Traité de Chimie Minérale* **5**, (P. Pascal Ed.), Masson, Paris (1976).
8. A. Wold, W. Kunnmann, R. J. Arnott, and A. Ferretti, *Inorg. Chem.* **3**, 545 (1964).
9. T. A. Bither, J. L. Gillson, and H. J. Young, *Inorg. Chem.* **5**, 1559 (1966).
10. J. M. Réau, C. Fouassier, and P. Hagenmuller, *J. Solid State Chem.* **1**, 326 (1970).
11. A. F. Reid and J. A. Watts, *J. Solid State Chem.* **1**, 310 (1970).
12. P. Strobel and M. Greenblatt, *J. Solid State Chem.* **36**, 331 (1981).
13. L. F. Schneemeyer, S. E. Spengler, F. J. di Salvo, J. V. Waszczak, and C. E. Rice, *J. Solid State Chem.* **55**, 158 (1984).
14. W. H. Mc Caroll and M. Greenblatt, *J. Solid State Chem.* **54**, 282 (1984).
15. O. Glemser and G. Lutz, *Z. Anorg. Allg. Chem.* **264**, 17 (1951).
16. J. J. Birtill and P. G. Dickens, *Mat. Res. Bull.* **13**, 311 (1978).
17. M. Ganne, A. Boumaza, and M. Dion, *Mat. Res. Bull.* **20**, 1297 (1985).
18. J. Marcus, C. Escribe-Filippini, R. Chevalier, and R. Buder, *Solid State Comm.* **62**, 221 (1987).
19. R. Schöllhorn, R. Kühlmann, and J. O. Besenhard, *Mat. Res. Bull.* **11**, 83 (1976).
20. T. Iwamoto, Y. Itoh, K. Ohwaka, and M. Takashi, *Nippon Kogaku Kaishi* **2**, 273 (1983).
21. D. M. Thomas and E. G. McCarron, *Mat. Res. Bull.* **21**(8), 945 (1986).
22. I. D. Brown and K. K. Wu, *Acta Cryst.* **B32**, 1957 (1976).
23. W. H. Zachariasen, *J. Less-Com. Metals* **62**, 1 (1978).
24. L. Pauling, *The Nature of the Chemical Bond*, 3rd ed., Cornell University Press, Ithaca N.Y. (1960).

25. A. F. Wells, *Structural Inorganic Chemistry*, 3rd ed., Clarendon Press, Oxford (1962).
26. A. Magnéli, *Arkiv Kemi* **32**, 269 (1949).
27. A. Magnéli, *Arkiv Kemi* **24**, 213 (1949).
28. A. Magnéli and B. Blomberg, *Acta Chem. Scand.* **5**, 372 (1951).
29. A. Magnéli, *Acta Chem. Scand.* **7**, 315 (1953).
30. M. Onoda, K. Toriumi, Y. Matsuda, and M. Sato, *J. Solid State Chem.* **66**, 163 (1987).
31. N. C. Stephenson, *Acta Cryst.* **20**, 59 (1966).
32. M. Onoda, Y. Matsuda, and M. Sato, *J. Solid State Chem.* **69**, 67 (1987).
33. H. Vincent, M. Ghedira, J. Marcus, J. Mercier, and C. Schlenker, *J. Solid. State Chem.* **47**, 113 (1983).
34. S. C. Abrahams, P. Marsh, L. F. Schneemeyer, C. E. Rice, and S. E. Spengler, *Acta Cryst.* (forthcoming).
35. W. G. Mumme and J. A. Watts, *J. Solid State Chem.* **2**, 16 (1970).
36. P. G. Dickens, J. J. Birtill, and C. J. Wright, *J. Solid State Chem.* **28**, 185 (1979).
37. P. P. Tsai, J. A. Potenza, M. Greenblatt, and H. J. Schugar, *J. Solid State Chem.* **64**, 47 (1986).
38. N. C. Stephenson and A. D. Wadsley, *Acta Cryst.* **18**, 241 (1965).
39. J. Graham and A. D. Wadsley, *Acta Cryst.* **20**, 93 (1966).
40. M. Ghedira, J. Chenavas, M. Marezio, and J. Marcus, *J. Solid State Chem.* **57**, 300 (1985).
41. A. Magnéli, *Acta Chem. Scand.* **2**, 501 (1948).
42. A. Magnéli, *Acta Chem. Scand.* **2**, 861 (1948).
43. A. Magnéli, *Nova Acta Regiae Soc. Sci. Upsaliensis* **4**, 14, No. 8 (1950).
44. A. Magnéli, B. Blomberg-Hansson, L. Kihlborg, and G. Sundkvist, *Acta Chem. Scand.* **9**, 1382 (1955).
45. G. Hägg and A. Magnéli, *Rev. Pure and Appl. Chem.* (*Australia*) **4**, 235 (1954).
46. O. Glemser and G. Z. Lutz, *Z. anorg. allgem. Chem.* **263**, 2 (1950).
47. L. Kihlborg, *Acta Chem. Scand.* **14**, 1612 (1960).
48. L. Kihlborg, *Acta Chem. Scand.* **17**, 1485 (1963).
49. L. Kihlborg, *Arkiv Kemi* **21**, 427 (1963).
50. L. Kihlborg, *Arkiv Kemi* **21**, 443 (1963).
51. L. Kihlborg, *Arkiv Kemi* **21**, 461 (1963).
52. L. Kihlborg, *Arkiv Kemi* **21**, 471 (1963).
53. L. Kihlborg, *Arkiv Kemi* **21**, 365 (1963).
54. S. Asbring and L. Kihlborg, *Acta Chem. Scand.* **18**, 1571 (1964).
55. M. Ghedira, H. Vincent, M. Marezio, J. Marcus, and G. Fourcaudot, *J. Solid State Chem.* **56**, 66 (1986).
56. H. Fujishita, M. Sato, S. Sato, and S. Hoschino, *J. Solid State Chem.* **66**, 40 (1987).
57. H. Bräkken, *Z. Krist.* **78**, 484 (1931).
58. N. Wooster, *Z. Krist.* **80**, 504 (1931).
59. L. Kihlborg, *Arkiv Kemi* **21**, 357 (1963).
60. A. Magnéli, *Arkiv Kemi, Mineral. Geol.* **A24**, No. 2 (1946).
61. B. G. Brandt and A. C. Skapski, *Acta Chem. Scand.* **21**, 661 (1967).
62. M. Ghedira, C. Do-Dinh, M. Marezio, and J. Mercier, *J. Solid State Chem.* **59**, 159 (1985).
63. D. B. Rogers, R. D. Shannon, A. W. Sleight, and J. L. Gillson, *Inorg. Chem.* **8**, 841 (1969).
64. M. Ganne, M. Dion, and A. Boumaza, *C. R. Acad. Sc.*, Sect. II **302**, 635 (1986).
65. M. Ganne, M. Dion, A. Boumaza, and M. Tournoux, *C. R. Acad. Sc.*, Sect. II **302**, 561 (1986).
66. P. P. Tsai, A. Potenza, and M. Greenblatt, *J. Solid State Chem.* **69**, 329 (1987).
67. M. Ghedira, H. Vincent, and M. Marezio, submitted to J. Solid State Chem.
68. L. Kihlborg, G. Hägerström, and A. Rönnquist, *Acta Chem. Scand.* **15**, 1187 (1961).
69. K. A. Wilhelmi, *Acta Chem. Scand.* **23**, 419 (1969).
70. M. Lundberg, M. Sundberg, and A. Magnéli, *J. Solid State Chem.* **44**, 32 (1982).
71. N. Yamazoe and L. Kihlborg, *Acta Cryst.* **B31**, 1666 (1975).
72. L. Kihlborg, *Acta Chem. Scand.* **13**, 954 (1959).

STRUCTURAL INSTABILITIES IN THE LOW DIMENSIONAL MOLYBDENUM BRONZES AND OXIDES

JEAN-PAUL POUGET

Laboratoire de Physique des Solides, Associé au CNRS (L.A. 2), Bât, 510, Univeristé Paris-Sud, 91405 Orsay Cedex, France

1. Introduction

This chapter covers the structural instabilities of the molybdenum bronzes $A_xMo_yO_z$, where A is a monovalent metal, and of the molybdenum oxides Mo_nO_{3n-1}. Most of them have been known for many years, but their low dimensional conducting properties were only recognized recently, and found to be related to their structural anisotropy. As a consequence of such anisotropic properties, it was observed that most of these Mo bronzes and oxides exhibit a low temperature charge density wave (CDW) instability.

In this review, we shall consider only two families of Mo bronzes:

— the blue bronzes: $A_{0.3}MoO_3$ with A = K, Rb or Tl.
— the purple bronzes: $A_{0.9}Mo_6O_{17}$ with A = K, Na, Tl or Li.

Although both families show a layer type crystal structure, in which slabs of MoO_6 octahedra are separated by sheets of monovalent metals, the different geometries of the linkage between octahedra (infinite chains with respect to 2D networks of corner sharing octahedra of the MoO_3 type) make the blue bronzes quasi one-dimensional (1D) conductors and the purple bronzes two-dimensional (2D) conductors, respectively.

The Mo_nO_{3n-1} ($n = 4, 8, 9, 10$) oxides which also contain, in common with the purple bronzes, slabs of MoO_3 type linkages, are 2D conductors. However, in contrast to the bronzes, the linkage between two slabs of MoO_6 octahedra occurs either through a layer of MoO_4 tetrahedra ($n = 4$) or by the formation of a ($1\bar{2}0$) crystallographic shear plane in the MoO_3 prototype structure ($n = 8-10$).

The structure of these Mo bronzes and oxides is reviewed in a previous chapter of this book by Vincent and Marezio.

The metallic behaviour of the Mo bronzes and oxides can easily be understood by reference to the prototype material MoO_3, which is insulating because the d conducting states are unoccupied in a Mo^{6+} ($4d^0$) ion. In the case of the blue bronzes the monovalent metal donates its outer electron to the conduction band which thus becomes partially filled with 0.3 electron per Mo atom. In the case of the Mo_nO_{3n-1} oxides, the lack of oxygen, with respect to the stoichiometry MoO_3, provides $2/n$ electron per Mo atom in the conduction band. In the case of the purple bronzes the conduction electrons are provided both by the monovalent metals (0.15 electron per Mo) and from the lack of oxygen atom (1/3 electron per Mo).

C. Schlenker (Ed.), Low-Dimensional Electronic Properties of Molybdenum Bronzes and Oxides,
87–157.
© 1989 *Kluwer Academic Publishers.*

The low dimensional electronic properties of these materials are revealed by their transport and optical properties which will be reviewed in a subsequent chapter. Such properties can be associated with the presence of a highly anisotropic Fermi surface. In general, an anisotropic Fermi surface contains a large area which can be nested by translation of a wave vector $\mathbf{q}_c$. A maximum in the electron—hole response function occurs generally for this wave vector. This leads to a CDW (electronic) instability which tends to open an energy gap in the nested regions of the Fermi surface. If the electron—phonon coupling is large enough, such a CDW ground state can be stabilized below T_c. It is thus accompanied by a periodic lattice distortion (PLD) of wave vector $\mathbf{q}_c$. If the nesting involves all of the Fermi surface, as in most 1D conductors, an electrical gap is opened and a metal—insulator (Peierls) transition occurs at T_c. If only partial gaps are opened, as in most of the 2D conductors, a metal—semimetal transition occurs at T_c.

Prior to the study of Mo bronzes and oxides, the CDW instability was observed in several families of 1D and 2D conductors: the Krogmann salts $(K_2Pt(CN)_4 \cdot 0.3$ $Br \cdot x\ H_2O\ (KCP):1D)$, the charge-transfer organic salts (TTF—TCNQ:1D), the transition metal tri- and tetrachalcogenides $(NbSe_3:$ quasi 1D) and the transition metal dichalcogenides $(NbSe_2:2D)$. The structural instabilities exhibited by these materials have been the subject of several reviews [1—7].

This review contains three main parts. In the first, we shall recall some basic features of the CDW instability and PLD of low dimensional conductors. Then their manifestation in the 1D case (blue bronzes) and 2D case (purple bronzes and Mo oxides) will be detailed in parts two and three, respectively.

2. The Charge Density Wave Instability

2.1. BASIC FEATURES

2.1.1. *Response Functions*

A microscopic description of the CDW instability requires the study of the $\mathbf{q}$-dependent electron—hole polarizability $\chi_e(\mathbf{q})$. In the case of a noninteracting electron gas, $\chi_e(\mathbf{q})$ is given in second order perturbation theory by the well known expression (where matrix elements between Bloch states have been neglected) [8]:

$$\chi_e^0(\mathbf{q}) = \sum_{\mathbf{k}} \frac{f(E_{\mathbf{k}+\mathbf{q}}) - f(E_{\mathbf{k}})}{E_{\mathbf{k}} - E_{\mathbf{k}+\mathbf{q}}}, \tag{1}$$

where $E(\mathbf{k})$ is the energy of an electron of wave vector $\mathbf{k}$ and $f(E)$ is the temperature dependent Fermi—Dirac distribution function. If $k_B T \ll E_F$, the electronic states close to the Fermi energy, E_F, mainly contribute to $\chi_e^0(\mathbf{q})$. Among them, the electronic states separated by the wave vector $\mathbf{q}$, and for which the denominator of Equation (1) nearly vanishes, have a dominant contribution. Thus, $\chi_e^0(\mathbf{q})$ strongly depends on the geometry of the Fermi surface. Singular behaviour is expected for the wave vectors $\mathbf{q}_c$, leading to a nesting condition of the Fermi surface. The nature of the singularity will depend on the local curvature of the portions of the Fermi surface connected by the nesting wave vector [9]. Typical cases of the dependence of $\chi_e^0(\mathbf{q})$ on $\mathbf{q}$, relevant for 1D and 2D conductors, are shown in Figure 1.

(a) 1D conductors: a logarithmic divergence occurs for the wave vector component $\mathbf{q}_\parallel = 2k_F$ in the chain direction, placing in perfect coincidence the two planes forming the Fermi surface. $\chi_e^0(\mathbf{q})$ is independent of the wave vector components perpendicular to the chain direction.

(b) 2D conductors: a cusp anomaly occurs for the wave vector $\mathbf{q}_0$ (belonging to

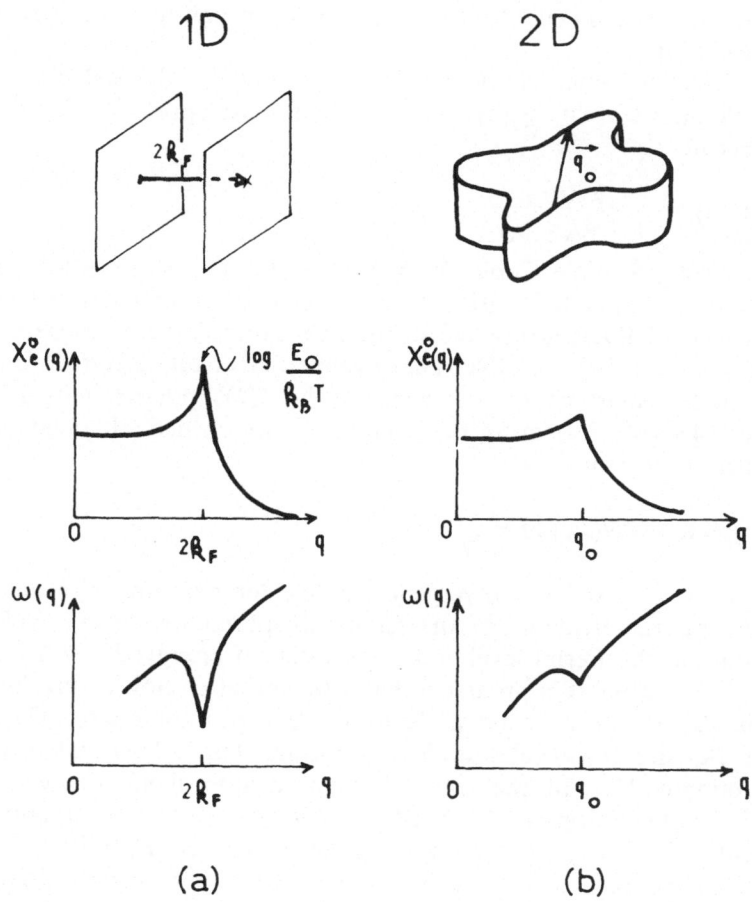

Fig. 1. Schematic representation of the anomaly shown by the electron–hole response function $|\chi_e^0(\mathbf{q})|$ and by the phonon branch coupled to the electron gas (1st order coupling) in the case of a good nesting of the Fermi surface. In the case of 1D conductors, shown in (a), there is a logarithmic divergence for $\mathbf{q} = 2k_F$. In the case of 2D conductors shown in (b), there is a cusp anomaly for $\mathbf{q} = \mathbf{q}_0$.

the conducting plane), which nests portions of the cylindrical Fermi surface having a similar curvature. $\chi_e^0(\mathbf{q})$ is independent of the wave vector component perpendicular to the conducting plane.

The critical wave vector $\mathbf{q}_c$ (i.e. $\mathbf{q}_{\parallel}$ or $\mathbf{q}_0$), which is related in direction to the geometry of the Fermi surface and in modulus to the band filling, has no reason to be in commensurate relation with the reciprocal lattice wave vectors of the atomic array.

In the case of materials containing transition metal elements, such as the Mo oxides and bronzes, several d bands generally cross the Fermi level, leading to a complex Fermi surface composed of several sheets [10—12]. Thus, interband as well as intraband contributions have to be included in $\chi_e^0(\mathbf{q})$. Also nesting processes between the sheets of different symmetry which form the Fermi surface must be considered.

Electron—electron interactions must be included in the calculation of the electron—hole polarizability $\chi_e(\mathbf{q})$. In the random phase approximation (RPA) (see for example [13]):

$$\chi_e(\mathbf{q}) = \frac{\chi_e^0(\mathbf{q})}{1 + U_\mathbf{q}\,\chi_e^0(\mathbf{q})}, \tag{2}$$

where $U_\mathbf{q}$ is an appropriate Coulomb interaction. Deviations from this mean field behaviour must be explicitly considered in 1D and 2D conductors, because of the drastic influence of fluctuations due to the low dimensionality. In particular, 1D conductors experiencing sizeable (short-range) electron—electron interactions show a net enhancement of the divergence of the CDW polarizability $\chi_e(2k_F)$ (see for example [14—16]). Instead of the logarithmic divergence, obtained in the case of noninteracting electrons:

$$\chi_e^0(2\mathbf{k}_F) = N(E_F) \ln \frac{E_0}{k_B T}, \tag{3}$$

$\chi_e(2\mathbf{k}_F)$ presents a power law divergence at low temperature with an exponent involving the matrix elements of the Coulomb interaction [in(3), $N(E_F)$ is the density of state at the Fermi level and E_0 is a cut off energy: $E_0 \approx 2\gamma E_F/\pi$ with $\gamma = 1.781$]. In addition, it is predicted that $\chi_e(\mathbf{q})$ can also show a singularity at the wave vector $4k_F$ for strong enough electron—electron interactions. The interplay between the $2k_F$ and $4k_F$ instabilities is well documented in 1D organic conductors (see, for example, [7]). In the case of $4d$ transition metal oxides and chalcogenides, it has been argued [17, 18] that electron—electron interactions have a less pertinent effect on $\chi_e(\mathbf{q})$ than for organic materials. Thus in the inorganic materials the divergence of the CDW response function is mainly driven by the electron—phonon coupling which will be considered below.

The CDW response function $\chi_\rho(\mathbf{q})$, which leads to a structural instability and to a PLD, must explicitly include the electron—phonon coupling. In the RPA approximation of this coupling, it takes the form:

$$\chi_\rho(\mathbf{q}) = \frac{\chi_e(\mathbf{q})}{1 - \lambda_\mathbf{q}\,\chi_e(\mathbf{q})}, \tag{4}$$

where λ_q is the effective electron—phonon coupling constant:

$$\lambda_q = \frac{|g_q|^2}{N\hbar\Omega(q)}.$$

In λ_q, g_q is the true electron—phonon coupling constant,[1] $\Omega(q)$ is the frequency of the bare phonon mode and N is the number of unit cells.

In the case of weak electron—electron interactions, and smooth dependence of λ_q and U_q in q, $\chi_\rho(q)$ shows a maximum at about the wave vector q_c, at which $\chi_e^0(q)$ is maximum. In the Peierls mean field scenario, where $\chi_e(q)$ has a pronounced peak at q_c and is sizeably temperature dependent, a spontaneous PLD develops when the denominator of (4) vanishes. This occurs at T_c^{MF} given by [13, 21]:

$$1 = \lambda_{q_c} \chi_e(T_c^{MF}, q_c). \tag{4b}$$

In this picture the presence of a well defined singularity at q_c in $\chi_e(q)$ assures pretransitional CDW fluctuations of long coherence length ($\xi_0 q_c \gg 1$) and makes negligible the contribution of phonon entropy terms to the thermodynamics of the phase transition (the opposite situation, where $\xi_0 q_c \sim 1$, is considered in [22]). The instability condition (4b) also assumes a weak electron—phonon coupling. The case of a strong electron—phonon coupling has been analyzed in [23, 24] and by Aubry and Quemerais in this book. The Peierls scenario seems to be qualitatively followed by 1D conductors such as KCP [25]. It has been questioned in 2D conductors like the transition metal dichalcogenides [26].

In addition the mean field approximation (MFA) does not correctly treat the structural fluctuations which strongly depress the temperature at which $\chi_\rho(q)$ diverges. In 1D systems, it is well known that because of their effect no phase transition can occur at a finite temperature ($T_c^{1D} = 0$ K). Depending on the dimensionality, n, of their order parameter, 2D systems may present a true phase transition at a finite temperature ($n = 1$) or no long range order at all for T finite ($n \geqslant 2$). The case $n = 2$ corresponds to a single q incommensurate structural order parameter. In fact, in real materials a small coupling exists between chains or planes which leads to a 3D phase transition at a finite temperature: $T_c < T_c^{MF}$. 3D pretransitional fluctuations are only relevant below a crossover temperature T^x. Above this temperature a true regime of 1D or 2D structural fluctuations (i.e. having the dimensionality of the electron gas of the material) is observed.

It is easy to show that $\chi_\rho(q)$ can be obtained from X-ray diffuse scattering experiments [7], [27]. The CDW lattice instability gives rise to an extra diffuse scattering above the background, $\delta I_d(Q)$, which, in the classical limit ($k_B T > \hbar\omega_{q_c}$, where ω_{q_c} is the characteristic frequency of the unstable phonon mode) is related to $\chi_\rho(q)$ by:

$$\delta I_d(Q) = |F_d(Q)|^2 \frac{\lambda_q}{\hbar\Omega(q)} \cdot k_B T \cdot \chi_\rho(q), \tag{5}$$

where $F_d(Q)$ is the structure factor of the unstable lattice mode and $Q = G + q$, with G being a reciprocal lattice wave vector.

2.1.2. *Spatial Correlations above T_c*

From (5), it is easy to see that the dimensionality and **q** dependence of the CDW structural fluctuations can best be visualized by X-ray diffuse scattering experiments. Above T^x they are manifested in the reciprocal space in the form of:

— diffuse sheets perpendicular to the chain direction, at $\pm\ 2\mathbf{k}_F$ from main Bragg reflections, in the case of 1D conductors (Figure 2b),
— diffuse rods perpendicular to the plane direction, at $\pm\ \mathbf{q}_0$ from main Bragg reflections, in the case of 2D conductors (Figure 2a).

$\chi_\rho(\mathbf{q})$ generally has a Lorentzian **q** dependence. For example in the 1D case, the profile of the X-ray scattering in a direction perpendicular to the diffuse sheet can be approximated by:

$$\chi_\rho(\mathbf{q}_\parallel) = \frac{\chi_\rho(2\mathbf{k}_F)}{1 + \xi_\parallel^2\, \delta\mathbf{q}_\parallel^2}\,, \tag{6}$$

with $\delta\mathbf{q}_\parallel = \mathbf{q}_\parallel - 2\mathbf{k}_F$. The half width at half maximum (HWHM) of the Lorentzian gives the inverse correlation length of the CDW, ξ^{-1}, along the chain direction.

Coupling between chains or planes leads to transverse correlations between CDW, which are manifested in the form of a modulation of the intensity within the diffuse sheets or rods (see Figure 3 for the 1D case). The HWHM of these

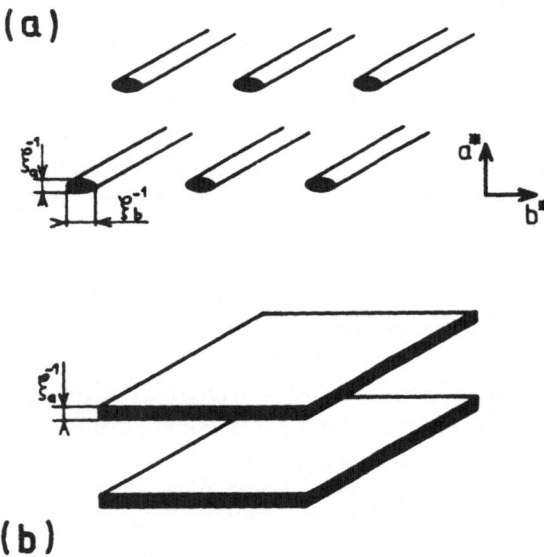

Fig. 2. Anisotropy of the CDW response function in the reciprocal space in the cases: (a) of a 2D conductor ((a, b) is the conducting plane), and (b) of a 1D conductor (a is the conducting direction). The width of the associated scattering in a given q direction is related to the inverse CDW correlation length, ξ^{-1}, in the same (direct space) direction, as schematically indicated.

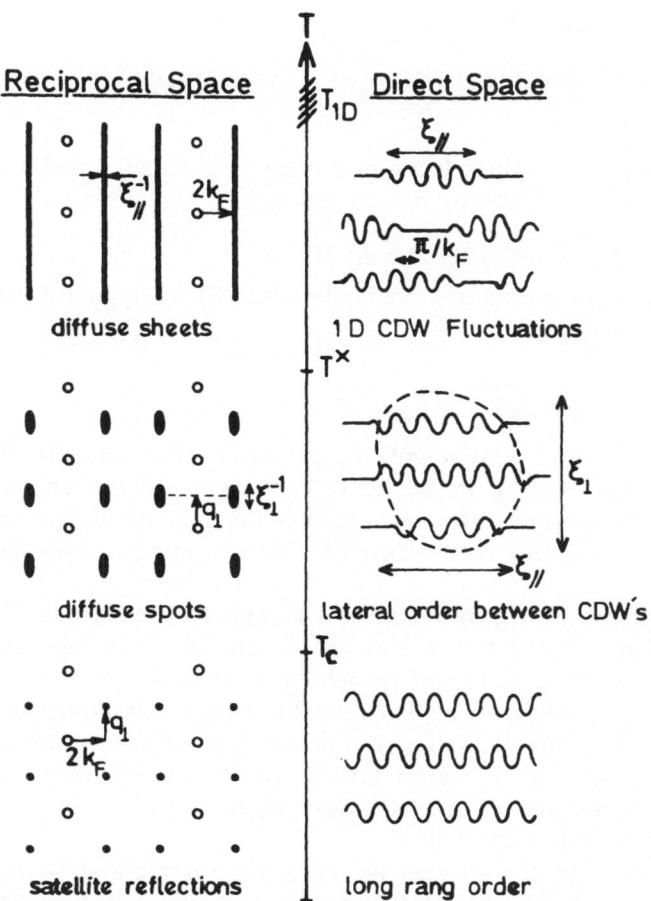

Fig. 3. Schematic representation of the regimes of 1D CDW fluctuations, (below T_{1D}), of short range lateral ordering between CDWs, (below T^x), and of long range order between them, (below T_c), in the direct space (right side). Such informations are deduced from the shape in the reciprocal space (left side) of the X-ray diffuse scattering, giving the Fourier transform of the instantaneous position—position correlation function.

modulations in a given reciprocal direction gives the transverse correlation length $\xi_\perp$ in the associated real space direction. At the crossover temperature, T^x one has $\xi_\perp \sim d_\perp$, where $d_\perp$ is the interchain or interplane distance. In the case of 1D conductors with very anisotropic interchain couplings, it is possible to observe, well decoupled in temperature, firstly (below T^x_{2D}) a regime of 2D fluctuations, then (below T^x_{3D}) a regime of 3D fluctuations.

At the critical temperature T_c, the 3D fluctuations diverge, leading to a long range CDW order. More precisely, as usual for 3D critical systems, χ_ρ and ξ

behaves in the vicinity of T_c as:

$$\chi_\rho(\mathbf{q}_c) \sim (T - T_c)^{-\gamma}$$

$$\xi \sim (T - T_c)^{-v}$$

(7)

The exponents γ and v differ from their mean field value, 1 and 1/2 respectively, below $T_G(< T^x)$ in the Ginzburg critical region.

2.1.3. *Coupling between Charge Density Waves*

The three main mechanisms of coupling between CDW appear clearly in the RPA generalized form of $\chi_\rho(\mathbf{q})$:

$$\chi_\rho(\mathbf{q}) = \frac{\chi_e(\mathbf{q})}{1 - [\lambda_q - U_q] \chi_e(\mathbf{q})},$$

(8)

where $\mathbf{q}$ also contains components in directions transverse to the conducting one(s) (i.e. $\mathbf{q} = \mathbf{q}_c + \mathbf{q}_\perp$). In the RPA approximation the critical wave vector component $\mathbf{q}_\perp$ corresponds to a maximum of the denominator of (8). Each of the three terms $\chi_e(\mathbf{q})$, U_q and $\lambda_q = |g(\mathbf{q})|^2/N\hbar\Omega(\mathbf{q})$ entering in this denominator can depend on $\mathbf{q}_\perp$.

(i) $\chi_e(\mathbf{q})$, defined by (1) in the case of independant electrons, depends on $\mathbf{q}_\perp$, through the shape of the Fermi surface. A warped Fermi surface in transverse direction(s) is the consequence of tunnelling of electrons between adjacent chains or planes in these directions. In that case, the critical component $\mathbf{q}_\perp^c$ corresponds to a maximum of $\chi_e^0(\mathbf{q})$. Its value, as for the $\mathbf{q}_c$ component, is fixed by the best nesting condition of the 3D Fermi surface. In the orthorhombic symmetry, for a Fermi surface composed of a single warped sheet, this critical transverse wave vector component is $\mathbf{q}_\perp^c = \pi/d_\perp$ [28].

(ii) U_q depends on $\mathbf{q}_\perp$ through the Coulomb interactions between CDWs on adjacent chains or planes (the intrachain or intraplane CDW Coulomb interactions are included in $\chi_e(\mathbf{q})$). Because of the repulsive nature of this interaction, maxima of the electronic charge repell each other, so that at equilibrium the CDWs on adjacent chains or planes are out of phase [29, 30]. For an orthorhombic array of chains or planes this corresponds to the wave vector component $\mathbf{q}_\perp^c = \pi/d_\perp$ (as for the tunnelling coupling considered in (i)).

(iii) The electron—phonon coupling constant g_q, as well as the bare phonon branch (of frequency $\Omega(\mathbf{q})$), the softening of which drives the structural instability, usually depends on $\mathbf{q}_\perp$. This is always the case for $\Omega(\mathbf{q})$, because real materials are 3D from the elastic point of view. However there is a relevant elastic coupling between CDWs only if $\Omega(\mathbf{q})$ or $g(\mathbf{q})$ have such an unusual $\mathbf{q}$ dependence that λ_q shows a well-defined maximum for a given value of $\mathbf{q}_\perp$. This could be the case if a well-defined minimum of frequency would already be present in the bare phonon branch in the absence of a coupling with the electron gas.

2.1.4. *The Periodic Lattice Distortion*

At T_c and below, the 3D CDW long range order corresponds to the formation of sharp satellite reflections (Figure 3). Their position, with respect to the main lattice of Bragg reflections, gives the modulation wave vector. Its components are as follows:

— $\mathbf{q}_c$ is given by the critical wave vector of the CDW instability of the low-D electronic gas,
— $\mathbf{q}_\perp^c$ is given by the lateral coupling between CDWs.

Their intensity is related to the amplitude of the normal mode describing the structural distortion, $|A_\mathbf{q}|$, through:

$$I_s(\mathbf{Q}) = |F_S(\mathbf{Q})|^2 |A_\mathbf{q}|^2, \tag{9}$$

where $F_S(\mathbf{Q})$ is the structure factor of the satellite reflections, the $\mathbf{Q}$ dependence of which gives the polarization of the atomic displacements at the transition (for a distortion of weak amplitude: $F_S(\mathbf{Q}) \sim F_d(\mathbf{Q})$ which enters in (5)). $A_\mathbf{q}$ is related to the (half) gap $\Delta_\mathbf{q}$ created in the band structure below T_c by [31]:

$$|\Delta_\mathbf{q}| = g_\mathbf{q} \sqrt{\frac{2\Omega\mathbf{q}}{\hbar N}} |A_\mathbf{q}|, \tag{10}'$$

and to the $\mathbf{q}$ Fourier component of the amplitude of the CDW $\rho_\mathbf{q}$ by:

$$\Delta_\mathbf{q} = -\lambda_\mathbf{q}\rho_\mathbf{q}. \tag{10}''$$

In (10) $A_\mathbf{q}$ is such that the displacement of the jth atom (of mass m_j, position r_j and polarization $\mathbf{e}_j$) in the nth unit cell is:

$$\mathbf{u}_j^n = \frac{1}{\sqrt{N}} \sum_\mathbf{q} \frac{A_\mathbf{q}}{\sqrt{m_j}} \cdot \mathbf{e}_j(\mathbf{q}) \exp[i\mathbf{q}(\mathbf{R}_n + \mathbf{r}_j)]. \tag{11}$$

In the mean field approximation of the electron–phonon coupling, $A_\mathbf{q}$ behaves in temperature as predicted by the BCS weak coupling theory [13, 21].

A more rapid rate of increase of the order parameter in temperature occurs when T_c is sizeably depressed with respect to T_c^{MF} as the result of structural fluctuations involving a large number of phonon degrees of freedom [22, 31]. In addition, the temperature dependence of the order parameter in the form of a power law:

$$|A_\mathbf{q}| \sim (T_c - T)^\beta, \tag{12}$$

with an exponent β differing from the mean field value $1/2$, is expected in the Ginzburg critical region near T_c.

2.1.5. *Temporal Fluctuations*

(a) $T > T_c$: *the Kohn anomaly.* As in any phase transition described by an order parameter, pretransitional fluctuations possess a dynamics above T_c. Because of the electron–phonon coupling, the lattice dynamics is also controlled by the fluctuations of the low-D electron gas. In the limit of weak electron–phonon coupling and in the strong nesting case (for which a well pronounced maximum occurs at $\mathbf{q}_c$ in $\chi_e(\mathbf{q})$) temporal fluctuations are described by the softening of a particular phonon mode at the critical wave vector $\mathbf{q}_c$. More precisely, the bare phonon frequencies $\Omega(\mathbf{q})$ are reduced in the vicinity of $\mathbf{q}_c$ by an efficient screening from the low D conduction electrons. By treating the electron—phonon coupling in perturbation, the renormalized phonon frequencies, $\omega(\mathbf{q})$, become, with notation already introduced:

$$\omega^2(\mathbf{q}) = \Omega^2(\mathbf{q}) \left[1 - \lambda_q \chi_e(\mathbf{q})\right]. \tag{13}$$

If $\chi_e(\mathbf{q})$ shows a well defined maximum at $\mathbf{q}_c$, a sharp dip occurs at $\mathbf{q}_c$ in the phonon spectrum, forming the so-called Kohn anomaly [32, 33] — see Figure 1. Above T^x, in the absence of lateral coupling, softening occurs:

— in the planes of the Brillouin zone perpendicular to $\pm \mathbf{q}_\parallel$ (i.e. $\pm 2\mathbf{k}_F$), in the case of 1D conductors,
— in the rods of the Brillouin zone perpendicular to $\pm \mathbf{q}_0$, in the case of 2D conductors.

In the vicinity of its minimum $\mathbf{q}_c$, (13) can be developed in the form:

$$\omega^2(\mathbf{q}) = \omega^2(\mathbf{q}_c) + |\mathbf{q} - \mathbf{q}_c| \, \overline{\overline{\Lambda}} \, |\mathbf{q} - \mathbf{q}_c|. \tag{13b}$$

A sharp Kohn anomaly has been observed at $2\mathbf{k}_F$ in the acoustic branch polarized in chain direction of the 1D conductor KCP [34]. Only a smooth anomaly occurs at $\mathbf{q}_0$ in longitudinal acoustic phonon dispersion curves for the 2D conductors 2H-TaSe$_2$ and 2H-NbSe$_2$ [35]. These last compounds show the features of the weak nesting situation, where a large number of phonons (i.e. with critical wave vectors filling a substantial part of the volume of the Brillouin zone) are involved in the structural instability. It has been argued [23] that this situation can be described in the framework of a strong electron—phonon coupling theory where the efficient processes of phonon softening are the electronically induced anharmonicity and mode—mode coupling.

Returning to the weak electron—phonon coupling case, it is easy to see in (13) that the progressive divergence in temperature of $\chi_e(\mathbf{q}_c)$ leads to a progressive softening of the frequency $\omega(\mathbf{q}_c)$ of the phonon mode which has the strongest first order coupling with the low D electron gas. In the mean field approximation $\omega(\mathbf{q}_c)$ vanishes at the T_c^{MF} given by the condition (4b). In the same approximation $\omega^2(\mathbf{q}_c)$ behaves, for temperatures close to T_c^{MF}, as (Figure 4a):

$$\omega^2(\mathbf{q}_c) = a(T - T_c^{MF}). \tag{14}$$

It is easy to show [36] that, in the soft mode picture, the softening of a phonon mode at $\mathbf{q}_c$ and the divergence of the thermodynamical susceptibility, $\chi_u(\mathbf{q}_c)$, associated with the order parameter $A_{\mathbf{q}_c}$ are intimately connected; (4) and (13) show that (see also (A4) in the Appendix A):

$$\omega^2(\mathbf{q}_c) \sim \chi_u^{-1}(\mathbf{q}_c) = \chi_e(\mathbf{q}_c)/\chi_\rho(\mathbf{q}_c).$$

As a result of its coupling with the electron gas, the critical phonon decays into electron—hole pairs. It thus has a finite life time, given in the RPA approximation

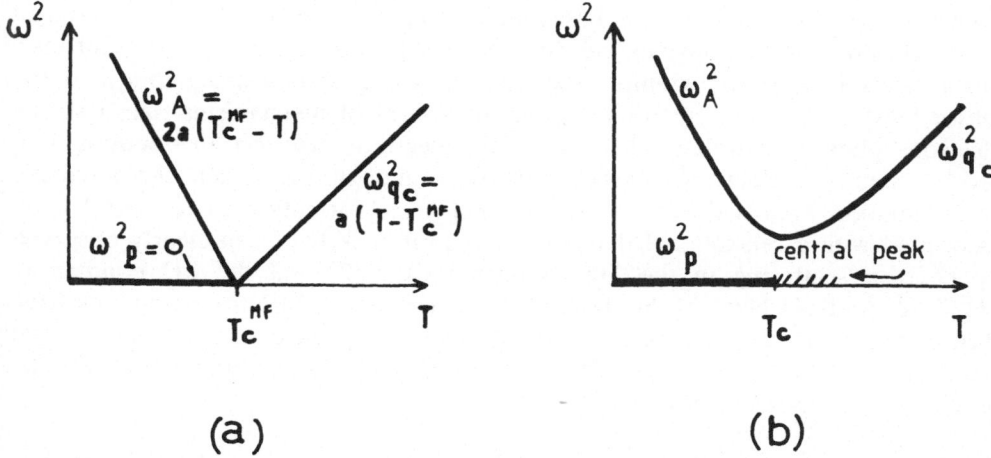

(a)   (b)

Fig. 4. Temperature dependence, at the critical wave vector, of the square of the frequency of the bottom of the Kohn anomaly ($\omega^2(\mathbf{q_c})$) above T_c and of the amplitude (ω_A^2) and phase (ω_P^2) modes below T_c, in the case of a single $\mathbf{q}$ incommensurate modulation. (a) represents the mean field behaviour and (b) the behaviour expected in the case of an incomplete softening associated with the formation of a central peak in energy.

[37] by:[7]

$$\Gamma^{-1}(\mathbf{q_c}) \sim T/\lambda(\mathbf{q_c}) \, \Omega^2(\mathbf{q_c}). \tag{15}$$

The dynamics of the pretransitional fluctuations are thus best described by a damped harmonic oscillator response function:

$$\chi_u(\mathbf{q_c}, \omega) \sim \frac{1}{\omega^2(\mathbf{q_c}) - 2i\Gamma(\mathbf{q_c})\,\omega - \omega^2}. \tag{16}$$

$\omega(\mathbf{q_c})$ and $\Gamma(\mathbf{q_c})$ can be extracted from the neutron scattering differential cross section, which gives a quantity proportional to $S(\mathbf{q}, \omega)$, the Fourier transform of the correlation function $\langle A_\mathbf{q}(0) \, A_{-\mathbf{q}}(t) \rangle$. In the classical limit ($k_B T < \hbar\omega$), $S(\mathbf{q}, \omega)$ is related to the generalized susceptibility $\chi_u(\mathbf{q}, \omega)$ by:

$$S(\mathbf{q}, \omega) = \frac{k_B T}{\hbar\omega} \, \text{Im} \, \chi_u(\mathbf{q}, \omega). \tag{17}$$

In low D conductors, and more generally for displacive phase trnasitions, the dynamics of the structural phase transition is more complex than that predicted by the soft mode picture: the critical mode does not soften completely at T_c. The divergence of $\chi_u(\mathbf{q_c})$ is achieved by the critical growth of another response that is quasi-elastic in frequency and is thus called the central peak [36] (see Figure 4b). The central peak can be viewed as an extra scattering originating from domains of the low temperature phase which are formed above T_c with a finite lifetime. In low D conductors several mechanisms have been considered for the formation of such domains. Existing theories are based either on extrinsic mechanisms [induced CDW (Friedel oscillations) in the electron gas by the defects [37]] or intrinsic mechanisms (this is for example the case of strong anharmonic motion of the lattice [38]). Let us also mention in the latter class of mechanisms that a strong electron—phonon coupling [24] or strong electron—electron interactions [16] favor, when the electronic response is slow enough (i.e. when the adiabatic approximation breaks down for $\omega_{el} < \Omega(\mathbf{q}_c)$) the formation of a central peak associated with a relaxational dynamic. A central peak has been clearly observed by neutron scattering in the 1D conductor KCP [3] and the 2D conductors $2H\text{-}TaSe_2$ and $2H\text{-}NbSe_2$ [35] above T_c. In $2H\text{-}NbSe_2$, NMR measurements have shown that a substantial contribution to this central peak is inhomogeneous and static or the NMR time scale (10^{-3} s). It is certainly due to anomalous Friedel oscillations around impurities [39].

(b) $T < T_c$: *phason and amplitudon*. Below T_c, let us now consider the temporal fluctuations of the long range P.L.D. In the case of a single $\mathbf{q}$ incommensurate modulation, the charge varies spatially according to:

$$\rho(\mathbf{r}) = \rho_0 + |\rho_\mathbf{q}| \cos(\mathbf{q}\mathbf{r} + \varphi), \qquad (18)$$

where the first term represents the average charge and the second term its spatial modulation. In incommensurate systems, one has to distinguish between the fluctuations of the amplitude, $|\rho_\mathbf{q}|$, and of the phase φ of the (two-component) order parameter [29, 36, 38]. A specific phonon mode is associated with each kind of fluctuations, called the amplitudon and phason, respectively. Figure 5a gives a schematic representation of the dispersion of the corresponding phonon branches.

The mode associated with the fluctuations of the amplitude of the order parameter, $|\rho_\mathbf{q}|$ behaves in function of $\mathbf{q}$ as follows:

$$\omega_A^2(\mathbf{q}) = \omega_A^2(\mathbf{q}_c) + |\mathbf{q} - \mathbf{q}_c| \overline{\overline{\Lambda}} |\mathbf{q} - \mathbf{q}_c|, \qquad (19)$$

with, for the first contribution, a temperature dependence given in the MFA by:

$$\omega_A^2(\mathbf{q}_c) = 2a(T_c^{MF} - T),$$

and in the second contribution $\overline{\overline{\Lambda}}$ defined by (13b).

The temperature dependence of ω_A^2 is that expected below T_c for the soft mode of a displacive phase transition.

A singular dispersion is shown by the mode associated with the fluctuations of the phase φ of the order parameter:

$$\omega_p(\mathbf{q}) = \sqrt{\Lambda} |\mathbf{q} - \mathbf{q}_c|. \qquad (20)$$

In (20), $\omega_p(\mathbf{q})$ vanishes when $\mathbf{q} = \mathbf{q}_c$. This corresponds, by means of the Goldstone theorem, to a new symmetry of free sliding of the phase of the incommensurate CDW with respect to the underlying lattice. In that case it is easy to see that such a displacement of the modulation costs no energy. This is not true for a commensurate modulation: a gap occurs at $\mathbf{q}_c$ in the "phason" spectrum. In addition, the phase of the incommensurate modulation of real systems is generally pinned by

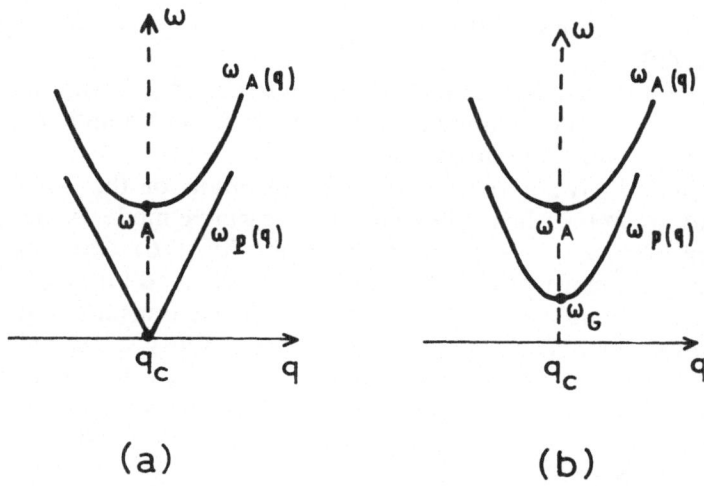

Fig. 5. $\mathbf{q}$ dependence below T_c of the amplitude and the phase modes (a) in the case of free sliding of the incommensurate modulation and (b) in the case of a pinning (at a frequency ω_G) of the phase of the modulation.

impurities or by interchain interactions [29, 40]. In these cases a pinning energy $\hbar\omega_G$ (see Figure 5b) must be overcome in order to obtain the phase sliding. As the CDW is electrically charged, the depinning is achieved above an electric field threshold. This gives rise to the nonlinear conductivity effects described in the following chapter of this book by Schlenker *et al.*

The pinning processes have been mostly considered for 1D conductors. In the case where impurities or defects can locally pin the phase of the CDW, a distribution of pinning energies is expected in the crystal. In the case of a commensurate modulation, where there is lock-in of the phase of the CDW on the underlying (high temperature) lattice, the gap ω_G depends on the strength of the lock-in interaction potential between the prototype lattice and the CDW sublattice (if the wave vector of modulation is $\mathbf{q} = m\mathbf{G}/p$, this coupling decreases when the order of commensurability p increases). Pinning due to the effects of lattice discretness on a primarily incommensurate modulation is also considered in a following chapter by Aubry and Quemerais. In materials composed of segregated chains of positively and negatively charged CDW (the case of TTF—TCNQ), interchain Coulomb interactions can pin the phase of the CDW with respect to the action of a uniform electric field.

The amplitude mode is fully symmetric (Ag symmetry below T_c) and is thus Raman active. The phase mode, which involves the motion of CDW against a background of positive charge, carries a dipole moment and is thus expected to be optically active in the infrared. If the long range Coulomb forces involved in the phase oscillations are not screened, which occurs at low temperature in the insulating Peierls state of 1D conductors, Coulomb interactions raise at finite frequencies the excitations of the phase for wave vectors propagating along the chain direction [29, 41].

In the case of a modulation of multiple **q** wave vectors (i.e. with more than two components for the order parameter) several phase and amplitude modes are expected below T_c in the phonon spectrum [36].

Figures 4 and 5 give a schematical representation of the dynamics in the displacive limit, when the phason branch is a propagative mode. Assuming that the ω dependence of $\chi_p(\mathbf{q}, \omega)$ is given by the expression (16),[2] the phase mode, is overdamped in the vicinity of $\mathbf{q}_c$ and becomes underdamped for larger values of $|\mathbf{q} - \mathbf{q}_c|$. In the strongly anharmonic case, when the phase transition more likely has an order—disorder character, the phase mode has a diffusive dynamic. The cases intermediate between the two limits can only be studied by numerical calculations (see for example [24] and the chapter by Aubry and Quemerais in this book).

In the case of a two component order parameter, $S(\mathbf{q}, \omega)$, defined by (17), can be written in the form [38]:

$$S(\mathbf{q}, \omega) = S^P(\mathbf{q}, \omega) + S^P(\mathbf{q}, \omega) \otimes S^A(\mathbf{q}, \omega). \qquad (21)$$

When the characteristic frequencies of the phase and amplitude fluctuations are well decoupled, as in the displacive limit at low temperature or in the order—disorder limit, (21) reduces to:

$$S(\mathbf{q}, \omega) \simeq S^P(\mathbf{q}, \omega) + S^A(\mathbf{q}, \omega).$$

The response function is the sum of two damped harmonic oscillator response functions, the first one describing the dynamics of the phase of the order parameter and the second one the dynamics of its amplitude. Such a decoupling procedure naturally occurs in the MFA approximation. However in the vicinity of T_c, this decoupling procedure is questionable, and thus a more complex ω dependence of $S(\mathbf{q}, \omega)$ is expected, especially from the contribution of the amplitude fluctuations, through the convolution product in (21). The dynamics of these fluctuations also crucially depend on the degree of anharmonicity of the system.

Only few experimental studies of the CDW dynamics have been performed below T_c. In 2H-TaSe$_2$ [35], the dynamics are more closely of the order—disorder type. In TTF—TCNQ, a similar dynamic is expected from the weakness of the dip of the Kohn anomaly observed in the phonon spectrum just above T_c [42, 43]. The dynamics of KCP are clearly of the displacive type [3], but the sharpness of the Kohn anomaly and the incompleteness of the Peierls transition prevent an unambiguous analysis of the neutron cross section. However, the data can be interpreted [44, 45] in terms of amplitude and (pinned) phase excitations.

2.1.6. *The Electron—Phonon Coupling*

In the CDW instabilities considered in this section the electron—phonon coupling occurs at first order. It leads to anomalies in the phonon modes which are linearly coupled to the electrons. In a tight binding description of the electron gas, such modes must modulate, to first order, the transfer integrals (t) or the site energies (ε). In the case of linear conductors containing one repeat unit (as the $Pt(CN)_4$ unit in the 1D conductor KCP), there is clearly 1st order coupling with the acoustic phonon polarized in the conducting direction. In the case of chains composed of equispaced M and O atoms, as in the oxides, the phonon mode modulating the MO distances is coupled at first order both at the t_{MO} transfer integral and at the site energy through the crystalline field (see for example [17]). With a linear repeat unit composed of two atoms (M and O), two such modes are important to consider: the acoustic mode of low frequency, and the optical mode ('breathing' mode of the bonds) of higher frequency but of stronger electron—phonon coupling constant, because the M and O atoms are displaced in opposite directions on each bond. In real materials the unit cell is more complex and generally gives rise to several conduction bands at the Fermi level. The bands affected to first order by the electron—phonon coupling depend on their symmetry and on that of the phonon mode which couples them. An important class of materials ($K_{0.3}MoO_3$, $NbSe_3$) contains pairs of identical chains from which, by bonding (B) and anti-bonding (AB) combinations of the highest occupied intrachain orbital, originate two conduction bands. Within this band structure it is easy to calculate the electron—phonon matrix elements for first order coupling with the phonons [46].

The symmetry of the phonon mode linearity coupled to the conduction electrons can be found as follows. As the B (AB) orbitals are even (odd) by the mirror symmetry which relates the pair of chains, a structural modulation of the atomic positions:

— in phase between the two chains (i.e. even by mirror symmetry) couples only similar (B—B or AB—AB) bands.
— out of phase between the two chains (i.e. odd by mirror symmetry) couples the dissimilar (B—AB) bands.

A modulation of the interchain distance keeping the mirror symmetry couples only the similar bands. The same symmetry argument holds for materials composed of pairs of identical planes.

2.2. THE ONE-DIMENSIONAL CASE

The purpose of this section is to add some complements, specific to 1D conductors, to the general introduction given in Section 2.1.

2.2.1. Intrachain Thermal Fluctuations

Because of fluctuations, due to the entropy term, no symmetry breaking phase transitions can occur at a finite temperature in 1D systems [47]. Theoretical treatments going beyond the mean field approximation are thus required.

In the case of one dimensional systems, exact calculations can be performed [38] with the Landau—Ginzburg free energy:

$$F^{1D}(\Delta_q) = a|\Delta_q|^2 + b|\Delta_q|^4 + c\,\delta q_{\parallel}^2|\Delta_q|^2. \tag{22}$$

The coefficients a, b, c, can be calculated microscopically for an electron—phonon coupled system. In that case the gap Δ_q (related to A_q by (10)') is usually chosen as the order parameter. In the simplest treatment of the Peierls instability which neglects the phonon entropy, the elastic energy term and the development of the free energy of a gas of independant electrons up to the fourth order gives [48]:

$$
\begin{aligned}
a &= \frac{1}{2}\left[\frac{1}{\lambda_{2k_F}} - \chi_e^0(2k_F)\right] = \frac{N(E_F)}{2}\log\frac{T}{T_c^{MF}} \\[2mm]
b &= -\frac{1}{4}\left.\frac{\partial\chi_e}{\partial\Delta^2}\right|_0 = \frac{N(E_F)}{2}\frac{7\zeta(3)}{16\pi^2(k_BT)^2} \approx N(E_F)\left(\frac{1}{2\pi k_BT}\right)^2 \\[2mm]
c &= b(\hbar v_F)^2 = \frac{N(E_F)}{2}\xi_{0\|}^2.
\end{aligned}
\qquad (23)
$$

$$
\xi_{0\|}^2 = -\left.\frac{\partial\log\chi_e^0(q)}{\partial q_\|^2}\right|_{2k_F} \approx 2\left(\frac{\hbar v_F}{2\pi k_BT}\right)^2,
$$

fixes the spatial scale of the thermally induced electronic fluctuations in chain direction (i.e. $\xi_{0\|}^{-1}$ is the fluctuation of k_F due to the thermal broadening of the Fermi surface). In a 1D electron gas the density of state at the Fermi level, including the two spin directions, is:

$$
N(E_F) = 2d_\|/\pi\hbar v_F,
$$

where $\hbar v_F = 2d_\| t_\| \sin k_F d_\|$, and $t_\|$ is the transfer integral in chain direction.[3]

The Ginzburg criterion applied to (22) [49] indicates that deviations from the mean field behaviour become important in the reduced temperature range:

$$
\Delta t = \frac{T - T_c^{MF}}{T_c^{MF}} = 2\left(\frac{bk_B T_c^{MF}}{a'^{3/2}c^{1/2}}\right)^{2/3},
\qquad (24)
$$

where we have assumed that $a = a'(T/T_c^{MF} - 1)$. With the coefficients given by (23) the reduced critical temperature range is $\Delta t \approx 1.6$.

From (22) it is possible to calculate $\chi_\Delta^{1D}(2k_F)$ and ξ_Δ^{1D} in the presence of the fluctuations of the phase and of the amplitude of the order parameter Δ_q (see Appendix A for the definition of χ_Δ and ξ_Δ). The temperature dependence of these quantities depends crucially on Δt. If $\Delta t < 1$, there is a net decoupling in temperature between the regimes of amplitude and phase fluctuations (if Δt is small, the crossover temperature between the two regimes is close to T_c^{MF} — see Figure 6a). At high temperature ($T \geqslant T_c^{MF}$) the amplitude fluctuations dominate. The corrections to the mean-field behaviour, taking into account the amplitude

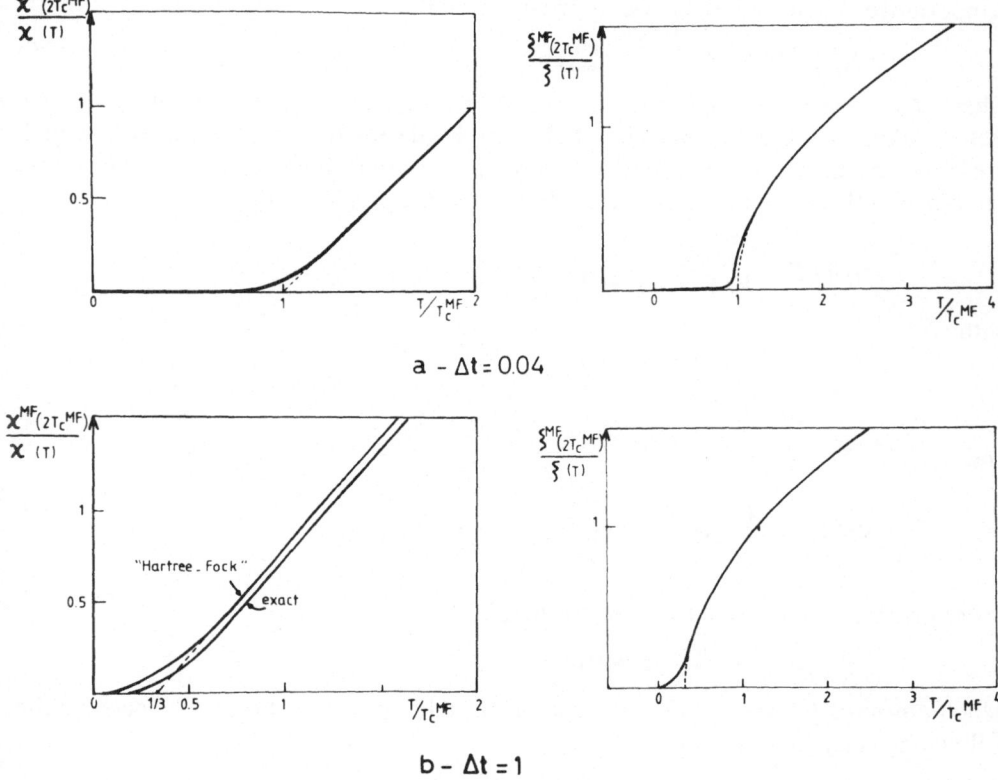

Fig. 6. Temperature dependence of the inverse susceptibility $\chi^{-1}(T)$ and the CDW correlation length $\xi^{-1}(T)$ associated to the order parameter Δ_q for a 1D electron—phonon coupled system in the case $\Delta t = 0.04$ (a) and $\Delta t = 0.1$ (b) (ΔT is defined by relationship (24) in the text). In the left bottom side the exact and 'Hartee—Fock' calculations of $\chi^{-1}(T)$ are compared. The data are normalized to the mean field values taken at $2T_c^{MF}$ ($\chi^{MF}(T) = [2a(T)]^{-1}$ and $\xi^{MF}(T) = \sqrt{c/a(T)}$, with $a(T)$ and c defined by (23)).

fluctuations in the 'Hartree—Fock' approximation, give [50]:

$$\begin{cases} 2\chi_A(2k_F) = \tilde{a}^{-1} = (a + \Sigma(T))^{-1} \\ \xi_A = \sqrt{c/\tilde{a}}, \end{cases} \qquad (25)$$

where $\Sigma(T) = 4b \sum_q \langle|\Delta_q|^2\rangle_T$ is determined by the following self-consistent equation based on the classical fluctuation dissipation theorem:

$$\Sigma(T) = 4bk_B T \sum_q \chi_A(q) = \frac{2bk_B T}{\tilde{a}\xi_A}.$$

This equation can easily be solved. Assuming that b and c do not depend on the temperature, it is found to a good approximation that:

$$(2\chi_A)^{-1} = \tilde{a} \approx a_0(T - T_{eff}), \tag{26}$$

where T_{eff} is depressed with respect to T_c^{MF} by a quantity which increases with Δt. Below about T_{eff} the fluctuations of the amplitude of the order parameter, which are too energetic, are ineffective. Only the thermal fluctuations of the phase remain, and these are of lower energy. In this regime one has [38]:

$$\chi_p(q) = \frac{S(2k_F)}{k_B T} \cdot \frac{1}{1 + \delta_{q\parallel}^2 \xi_p^2}, \tag{27}$$

with

$$\xi_p = \frac{2|a|c}{bk_B T},$$

and

$$S(2k_F) = \int \langle \Delta(x) \Delta(0) \rangle_T \, dx \approx \langle |\Delta_{2k_F}|^2 \rangle_T \, 2\xi_p.$$

More precisely, one has at low temperature:

$$\langle |\Delta_{2k_F}|^2 \rangle_T \approx \Delta_{2k_F}^2(T=0),$$

which amounts to $\Delta_0^2 \approx |a|/2b$ in the mean field approximation. (27) leads to the following divergences in temperature:

$$\begin{cases} \chi_p(2k_F) \sim T^{-2} \\ \xi_p \sim T^{-1} \end{cases}. \tag{28}$$

When $\Delta t \sim 1$, the amplitude fluctuations remain important well below T_c^{MF}. The phase fluctuation regime is realized only near 0 K. The transition between the two regimes is smooth in that case. Figure 6b gives $\chi_A^{-1}(2k_F)$ and ξ_A^{-1} in the case $\Delta t = 1$. It shows, with respect to the exact results of Scalapino et al. [49], that the 'Hartree—Fock' treatment is a good approximation down to $T_c^{MF}/2$, with χ_A^{-1} given by (26) where $T_{eff} \approx T_c^{MF}/3$ [50].

2.2.2. Interchain Coupling

(a) General features. In the above section it has been shown that the consideration of thermal fluctuations, treated even under the form of the lowest order term Δ_q^4 in the free energy, suppress the mean field Peierls transition of isolated chains. However χ and $\xi_\parallel$ diverge as T approaches 0 K, with power laws given by (28) in the classical limit (the exponents are different when quantum fluctuations are considered). Thus only a very weak interchain coupling in necessary to drive the 3D Peierls transition. In the limit of very weak interchain coupling, treated in the mean-field approximation, T_c is given by [51, 38]:

$$1 - zJ_\perp \chi(T_c) = 0, \tag{29}$$

where $J_\perp$ is the interchain coupling constant between the order parameters Δ (defined by (33) below) and z is the number of neighbouring chains. If the Peierls transition occurs in the phase fluctuation regime: $\chi = \chi_p$ given by (27). Thus (29) can rewritten under the form:

$$2zJ_\perp \Delta_0^2\, \xi_p(T_c) = k_B\, T_c, \tag{30}$$

which shows that $k_B\, T_c$ corresponds to the coupling energy between neighbouring 1D correlated fluctuations. With the expressions of ξ_p and Δ_0^2, one gets:

$$k_B\, T_c \approx \frac{|a|}{b}\, \sqrt{2zcJ_\perp}\, . \tag{31}$$

The square-root dependence of T_c with $J_\perp$ agrees with general scaling arguments [52].

For stronger interchain coupling T_c is more weakly depressed with respect to T_c^{MF}. In the Hartree—Fock approximation of the amplitude fluctuations it is found [31, 50] that, if $zJ_\perp \ll a'$:

$$T_c \simeq T_c^{MF}\left(1 + \frac{2bk_B\, T_c^{MF}}{a'\, \sqrt{czJ_\perp}}\right)^{-1}. \tag{32}$$

The Ginzburg—Landau model used in Section 1 can be generalized to include the interchain coupling:

$$F[(\Delta_i(q_{\parallel})] = \sum_i F_i^{1D}(\Delta_i(q_{\parallel})) + \sum_{\langle i,j\rangle} J_{ij}\mathrm{Re}[\Delta_i(q_{\parallel})\,\Delta_j^*(q_{\parallel})], \tag{33}$$

where i and j are chain indices, and F^{1D} is given by (22). By introducing the Fourier transform of the order parameter in transverse directions:

$$\Delta(\mathbf{q}) = \sum_i \Delta_i(q_{\parallel})\, e^{i\mathbf{q}_\perp \cdot \mathbf{R}_i},$$

(33) can be rewritten in the form:

$$F(\Delta(\mathbf{q})) = [a + c\,\delta q_{\parallel}^2 + 2J_\perp(\cos q_\perp^{(1)}d_\perp +$$
$$+ \cos q_\perp^{(2)}d_\perp)]\,|\Delta(\mathbf{q})|^2 + b|\Delta(\mathbf{q})|^4, \tag{34}$$

where $\mathbf{q} = (q_{\parallel}, q_\perp^{(1)}, q_\perp^{(2)})$ and where only the nearest neighbour interactions on a square lattice of parameter $d_\perp$ have been considered ($z = 4$). The coupling considered in (34) favours an out of phase ordering of CDW if $J_\perp$ is positive. By setting $q_\perp = \pi/d_\perp + \delta q_\perp$, (34) becomes, assuming $\delta q_\perp$ is small:

$$F(\Delta(\mathbf{q})) = \{a - 4J_\perp + c\delta q_{\parallel}^2 + J_\perp\, d_\perp^2[(\delta q_\perp^{(1)})^2 +$$
$$+ (\delta q_\perp^{(2)})^2]\}\,|\Delta(\mathbf{q})|^2 + b|\Delta(\mathbf{q})|^4. \tag{35}$$

In the Hartree—Fock approximation of the amplitude fluctuations $\chi \simeq \chi_A$, given by (25), enters in (29), and T_c occurs for:

$$\tilde{a}(T_c) - zJ_\perp = 0.$$

This leads to expression (32). From (35), the transverse correlations lengths are defined by:

$$\xi_\perp^{1,2} = d_\perp \sqrt{\frac{J_\perp}{\tilde{a} - zJ_\perp}}. \tag{36}$$

The crossover temperature T^x from the regime of 1D fluctuations to the regime of 3D fluctuations occurs for $\xi_\perp \approx d_\perp$, i.e. for $\tilde{a}(T^x) \approx (z + 1)J_\perp$. Below T^x, 3D critical fluctuations become very important between T_G and T_c in the 'Ginzburg critical width' which is [31]:

$$\Delta t_G = \frac{T_G - T_c}{T_c} = \frac{4T_c^2 b^2 d_\perp^2}{\pi^4 c^3 a'} \left(\frac{\xi_\parallel}{\xi_\perp}\right)^4. \tag{37}$$

(b) *Tunneling coupling.* A very important mechanism of interchain coupling is due to tunneling of electrons between adjacent chains. This mechanism introduces complications in the above calculation which deserve further attention. Assuming a tunneling coupling in a single transverse direction[6] (which leads to the 2D warped Fermi surfaces shown in Figure 7) $\chi_e^0(\mathbf{q})$, defined by (1), can be rewritten in the compact form [53]:

$$\chi_e^0(\mathbf{q}) = N(E_F) \int_{x_0}^\infty \frac{\left(\dfrac{\mathrm{d}x}{x_T}\right)}{\mathrm{Sh}\left(\dfrac{x}{x_T}\right)} \int_0^{2\pi} \frac{\mathrm{d}p}{2\pi} \cos[\delta q_\parallel(p)x], \tag{38}$$

where: $x_0^{-1} = 2\gamma k_F$ is a cut off; $x_T = \hbar v_F/2\pi k_B T$ is the thermal length previously introduced ($x_T = \xi_{0\parallel}/\sqrt{2}$); and $\mathbf{q} = [2k_F + \delta q_\parallel, q_\perp]$.

If $\mathbf{q}$ connects the sheets I and II of the Fermi surface, built on the following dispersion relation ($i = \mathrm{I}$ or II):

$$\varepsilon^i(k) = \hbar v_F^i(|k_\parallel| - k_F) + \varepsilon_\perp^i(k_\perp), \tag{39}$$

in (38) one has:

$$\delta q_\parallel(p) = \delta q_\parallel + \frac{\varepsilon_\perp^{\mathrm{I}}(p)}{\hbar v_F^{\mathrm{I}}} + \frac{\varepsilon_\perp^{\mathrm{II}}(p - q_\perp d_\perp)}{\hbar v_F^{\mathrm{II}}}. \tag{40}$$

In the case of a tight binding relationship $\delta q_\parallel(p)$ varies sinusoidally with p. Two contributions are included in $\varepsilon_\perp(k_\perp)$ [54]:

— the transverse tight binding dispersion itself: $2t_\perp \cos k_\perp d_\perp$, in the case of a 2D rectangular lattice,

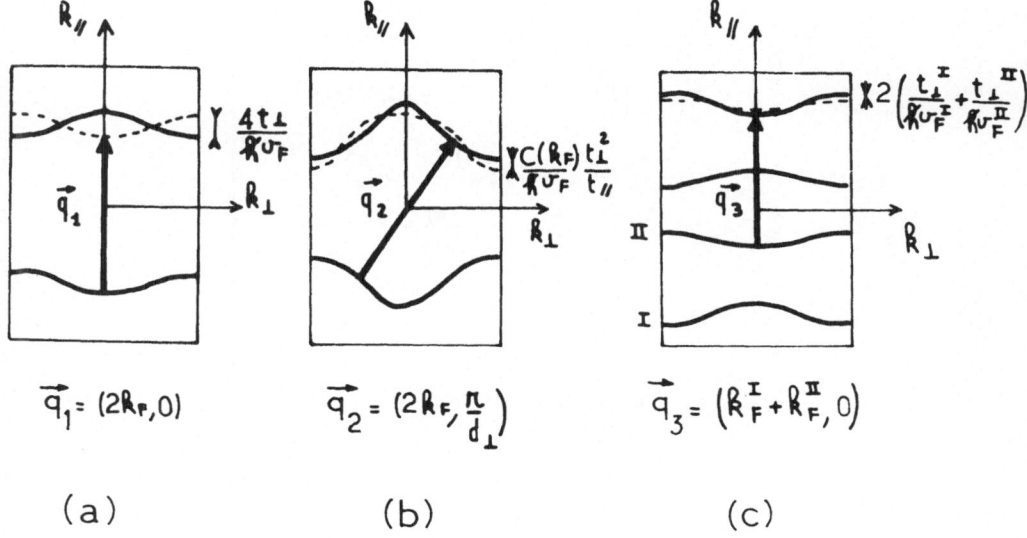

Fig. 7. Pockets left by a longitudinal (a) and a transverse (b) nesting of a single band and (c) by a longitudinal nesting of 2 bands onto each other in the case of 2D open Fermi surfaces. The maximum size of the remaining pockets in the chain direction, $2\Delta q_\|$, is indicated.

— a correction term due to the linearisation of the tight binding relation $2t_\|$ cos $k_\| d_\|$ by $\hbar v_F(|k_\|| - k_F)$ in (39). In the vicinity of E_F this correction is at the lowest order given by:

$$\frac{C(k_F)}{2} \frac{t_\perp^2}{t_\|} \cos 2k_\perp d_\perp, \tag{41}$$

where:

$$C(k_F) = \frac{\cos k_F d_\|}{\sin^2 k_F d_\|}.$$

The maximum of variation of $\delta q_\|(p)$, given by (40), $2\Delta q_\|$, is indicated in Figure 7 in the case of: (a) a longitudinal $\mathbf{q}_1 = (2k_F, O)$, (b) a transverse $\mathbf{q}_2 = (2k_F, \pi/d_\perp)$ nesting of a Fermi surface composed of one set of sheets and (c) a longitudinal $\mathbf{q}_3 = (k_F^I + k_F^{II}, O)$ nesting of a Fermi surface composed of two sets of sheets. The effect of the correction term (41) is relevant only in the case of nearly perfect nesting (case b of Figure 7).

The first consequence of the transverse tunnelling is that $\chi_e^0(\mathbf{q})$ depends on $q_\perp$. More precisely Figure 8 compares the temperature dependence of $\chi_e^0(\mathbf{q})$ (a) for a planar Fermi surface (where $\chi_e^0(q)$ does not depend on $q_\perp$), and (b) for a warped Fermi surface, where the imperfect nesting leaves pockets of size $2\Delta q_\|$. This figure shows clearly that, with respect to the case of a planar Fermi surface (where $\chi_e^0(q)$ shows a logarithmic divergence given by (3) on all the temperature range), the effect of such pockets is [123]:

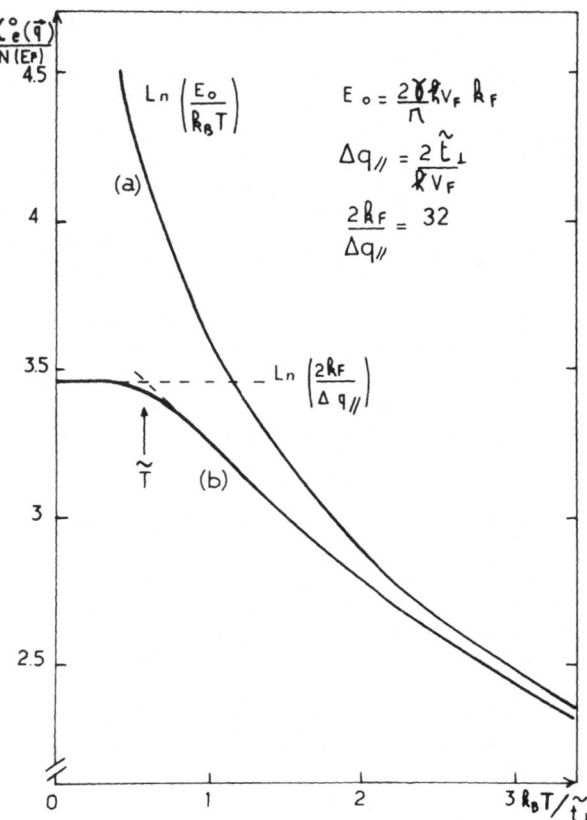

Fig. 8. Temperature behaviour of the reduced electron–hole response function $\chi_e^0(q_c)/N(E_F)$ of a 1D electron gas in the case: (a) of a planar Fermi surface and (b) of an imperfectly nested Fermi surface, leaving pockets of half width $\Delta q_\parallel$. The parameters of the calculation are indicated on the upper right side of the figure (by courtesy of G. Montambaux).

— to diminish $\chi_e^0(2k_F)$ and to smooth its temperature dependence at high temperature. This can be accounted for by a reduction of the cut off energy E_0 by a quantity ΔE_0 which depends on the temperature (with the parameters of Figure 8: $\Delta E_0/E_0$ increases from 5% to 30% when $k_B T/t_\perp$ decreases from 3 to 1).

— to cut the divergence of $\chi_e^0(2k_F)$ when $2\pi k_B T$ corresponds to the electronic energy contained in a pocket: $\hbar v_F \, 2\Delta q_\parallel$. In Figure 8b, this cutoff temperature is defined by:

$$k_B \tilde{T} \approx \frac{\gamma \hbar v_F}{2\pi} \Delta q_\parallel. \tag{42}$$

This cutoff is higher, by a factor 2, than that given in [55]. The expression (42) shows that the smaller the size of the pocket, the smaller is the thermal cutoff $\tilde{T}$, and the stronger is the divergence of $\chi_e^0(2k_F)$ at a given temperature. Thus the strongest instability will occur for the transverse component $q_\perp$ leading to the best nesting condition of the Fermi surface. For the single and double band warped Fermi surfaces shown in Figure 8, these conditions are realized in (b) and (c), respectively. For a Fermi surface composed of one band (case (b)) the best nesting wave vector connects its inflection points [54]. If $t_\perp/t_\parallel$ is not very small, this wave vector may have components which deviate appreciably from $2k_F$ and $\pi/d_\perp$ [53]. Moreover these deviations will increase for decreasing temperature [56].

From (38) it is easy to calculate $\chi_e^0(q)$ with $\mathbf{q}$ in the vicinity of the critical wave vector $\mathbf{q}_c$. For example for $\mathbf{q} = (2k_F + \delta q_\parallel, \pi/d_\perp)$, one gets [28]:

$$\chi_e^0(\mathbf{q}) = \chi_e^0(\mathbf{q}_c) - 2N(E_F)\left(\frac{\hbar v_F}{2\pi k_B T}\right)^2 \left[\delta q_\parallel^2 + \right.$$

$$\left. + 2\left(\frac{2t_\perp}{\hbar v_F}\right)^2 \sin^2\left(\frac{\delta q_\perp d_\perp}{2}\right) \right]. \tag{43}$$

$\chi_e^0(\mathbf{q})$ enters in the free energy through the term: $-(1/2)\chi_e^0(q)|\Delta_q|^2$. (43) again corresponds to the expression (34) with $J_\perp = N(E_F)\xi_{0\parallel}^2(t_\perp/\hbar v_F)^2$. If $q_\perp d_\perp$ is small, the sinus can be developed, leading to the following relationship between the transverse and longitudinal correlation lengths:

$$\frac{\xi_\perp}{\xi_\parallel} = \sqrt{2}\,\frac{t_\perp d_\perp}{\hbar v_F}. \tag{44}$$

This ratio is of the order of $v_\perp/v_F$, where $v_\perp$ is the tunnelling velocity between chains, $(t_\perp d_\perp/\hbar)$. It is independent of the temperature.

2.2.3. Amplitude and Phase Excitations

In Section 2.1.5.b. we have already introduced the excitations of the phase and the amplitude of the CDW modulation. The amplitude mode at $2k_F$ is the soft phonon describing the displacive instability below T_c. Its frequency has been calculated within the mean field theory. In the low temperature limit it is given by the simple expression [29, 57]:

$$\omega_A^0(2k_F) = \sqrt{\tilde{\lambda}}\,\Omega_{2k_F} \tag{45}$$

Let us now consider in greater detail the phase excitations in the case of a 1D conductor.[4] It is important to distinghuish between the transverse and longitudinal deformations of the phase of the CDW sublattice (see for example [18]). As shown by Figure 9: (a) The longitudinal mode corresponds to a modulation of the phase of the CDW in the chain direction. It locally changes the value of $2k_F$, which leads to a charge redistribution $\delta\rho$. This deformation mode involves long-range Coulomb interactions which drastically modify the $\mathbf{q}$ dependence of the phason branch if these interactions are not screened efficiently by the free carriers (see below).

(b) The transverse mode corresponds to a shear deformation of the CDW

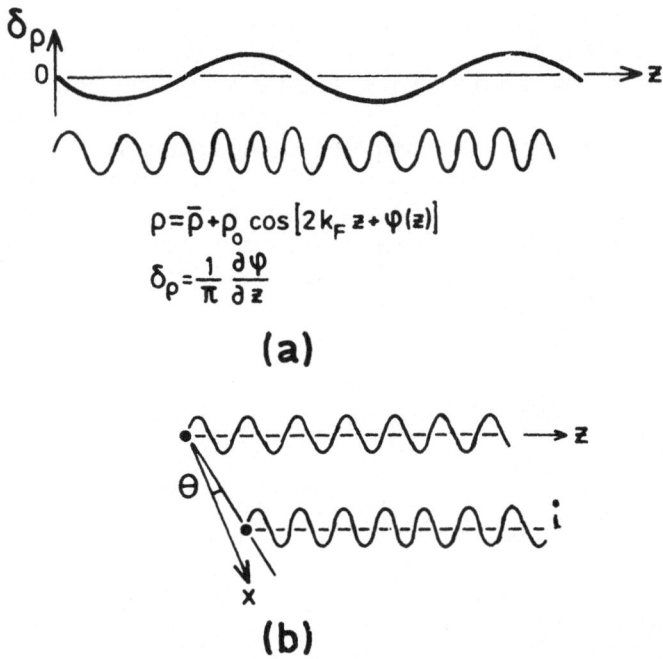

Fig. 9. (a) longitudinal deformation of the phase of the CDW, $\varphi(z)$, leading to a redistribution of the charge density $\delta\rho$ in chain direction. (b) transverse shear of the CDW sublattice leading to a global shift by θ_i of the phase of the CDW situated on chain i, and to no charge redistribution in chain direction.

sublattice. It changes the phase shift between CDW without affecting the charge distribution within each CDW. The acoustic-like dispersion of the phason branch given by (20) thus remains unaffected for transverse wave vectors.

We shall now discuss qualitatively the screening effects mentioned in (a) (a quantitative analysis is given in [59]). There is a (dynamical) screening if the velocity of the free carriers V_c is larger than that of the phase mode V_Φ which controls the charge redistribution within the CDW. It occurs in the vicinity of the Peierls transition when the Peierls gap is not too highly developed (if $k_B T > \Delta$ the free carriers are still quantum particles with $V_c \simeq v_F \gg V_\Phi$). This situation is also relevant whatever the temperature when the Fermi surface does not disappear at the Peierls transition (case of NbSe$_3$). The phase velocity V_Φ is given by the square root of the curvature of the Kohn anomaly in the vicinity of $2k_F$. With (13), (13b) and (23) one has:

$$V_\Phi = \sqrt{\bar{\Lambda}_\parallel} = \sqrt{\bar{\lambda}}\, \Omega_{2k_F} \frac{\hbar v_F}{\sqrt{2\pi k_B T_c}}, \tag{46}$$

where $\bar{\lambda} = \lambda N(E_F)$ is the dimensionless electron–phonon coupling constant. In the mean-field approximation, where $3.5\, k_B T_c^{MF} \approx 2\Delta_0$, (46) can be written in the form:

$$V_\Phi = \sqrt{\tilde{\lambda}}\ \Omega_{2k_F}\ \tilde{\xi}_0. \tag{47}$$

In this expression $\tilde{\xi}_0 = \hbar v_F / \pi \Delta_0$ is the Pipard coherence length associated with of the Peierls gap at 0 K: Δ_0. The velocity given by (47) is very close to that obtained in the RPA approximation limit $T \to 0$ K (with the long range Coulomb interactions also being neglected) [29, 59]

$$V_\Phi = v_F \sqrt{m_e/m^*}. \tag{48}'$$

In (48)' the reduced CDW mass m^*/m_e amounts to:

$$\frac{m^*}{m_e} = 1 + \frac{4\Delta_0^2}{\tilde{\lambda} \hbar^2 \Omega_{2k_F}^2}, \tag{48}''$$

or

$$\frac{m^*}{m_e} = 1 + \left(\frac{2\Delta_0}{\hbar \omega_A^0} \right)^2 .$$

The screening is spatially realized beyond a Thomas—Fermi length ξ_{TF}, which depends on the temperature through the number of free carriers created by thermal activation through the Peierls gap. For decreasing temperature one expects, together with an increase of the Thomas—Fermi screening length, an enhancement of the phason velocity from its value given by (46). Qualitatively one should have (Figure 10a):

$$\tilde{V}_\Phi \simeq \sqrt{\tilde{\lambda}}\ \Omega_{2k_F}\ \xi_{TF}. \tag{49}$$

This would result in an increase of the 'elastic stiffness' of the CDW, making it more rigid as the temperature diminishes. This could have important consequences for the sliding of the CDW at low temperature.

Together with the above mentioned increase of the phason velocity, the quasi particle velocity decreases when the temperature decreases (at low temperature, when $k_B T < \Delta$, the free carriers are classical particles with a Maxwell—Boltzmann velocity:

$$V_c \simeq \sqrt{\frac{k_B T}{m_{eff}}} \simeq v_F \sqrt{\frac{k_B T}{\Delta_0}},$$

which is slower than the Fermi velocity). When V_c becomes smaller than V_Φ, the dynamical screening by the quasi particles is inefficient. The longitudinal phason becomes an optical mode (Figure 10b) of frequency, for $\delta q_\parallel (> \delta q_\perp) \to 0$, given by [29]:

$$\omega_p^2 = \sqrt{\frac{3}{2}\ \tilde{\lambda}}\ \Omega_{2k_F}. \tag{50}$$

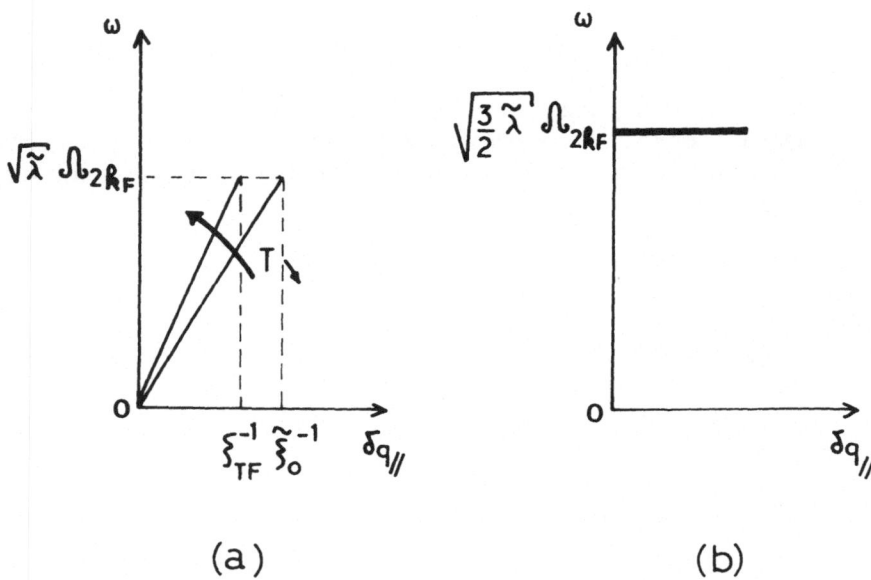

Fig. 10. Dispersion of the phase mode in chain direction: (a) in the vicinity of T_c, showing its acoustic nature due to the dynamical screening by the normal carriers. The stiffening of the phase mode velocity, due to the increase of the Thomas—Fermi length when T decreases, is schematically indicated. (b) at low temperature, showing its optical character due to the absence of dynamical screening.

In the mean field theory, there is a decoupling at T_c^{MF} between the excitations of the phase and the amplitude of the CDW (Figure 4). In real materials, because of the large regime of 1D fluctuations, the 3D critical temperature occurs at a T_c which can be much lower than the T_c^{MF}. In that case the phase and amplitude degrees of freedom could already be partly decoupled above T_c inside domains of the size of the CDW correlation length $\xi_{\parallel}$. Generally one expects to observe decoupled fluctuations when $\xi_{\parallel}$ is large. With a local Peierls distortion there is a net reduction of the one electron density state leading to the formation of a pseudo gap [60]. In the 1D fluctuation regime, these effects probably occur around the temperature at which the crossover between the phase and amplitude fluctuation regimes is observed on the thermodynamical quantities χ and ξ considered in Section 2.2.1.[5] Numerical calculations of $S(\mathbf{q}, \omega)$ have shown in the regime of 1D fluctuations of a Peierls chains, that this decoupling occurs around $T_c^{MF}/3$ [61], in agreement with the results of Figure 6b.

3. The Blue Bronzes $A_{0.3}MoO_3(A = K, Rb, Tl)$

3.1. ROOM TEMPERATURE STRUCTURE

The structure of the blue bronzes $A_{0.3}MoO_3$ is detailed by Vincent and Marezio in this book. It belongs to the monoclinic $C2/m$ space group with 20 formula units per unit cell [62]. A projection of the structure along the monoclinic b axis is shown in Figure 11. The structure is essentially of a layer type with infinite $(\bar{2}01)$ slabs of MoO_6 octahedra, separated by the monovalent ions A^+. The layer is built with clusters of 10 distorted MoO_6 octahedra. These clusters are linked together via corner sharing along the b and [102] directions. The octahedra built on the Mo(2) and Mo(3) atoms form infinite chains in the b (high conducting) direction. Within each cluster, the Mo(2) and Mo(3) octahedra are edge sharing in the transverse directions. They share a face with the isolated octahedron built on the Mo(1). If the stoichiometry corresponds exactly to the formula $A_{0.3}MoO_3$, all crystallographic sites are occupied and therefore no crystallographic disorder exists.

The structural anisotropy appears clearly on the phonon dispersion curves. This is particularly the case for the acoustic branches where neutron measurements of the sound velocity [68, 72] show a decreasing stiffness from the b to the $\mathbf{a} + 2\mathbf{c}$, then to the $2\mathbf{a}^* - \mathbf{c}^*$ (perpendicular to the $Mo_{10}O_{30}$ layers) directions (directions 1, 2, 3 in Figure 12 respectively). This anisotropy is also manifested by electronic properties (see the following chapter of this book) and by the critical fluctuations driving the Peierls instability (Figure 18).

In the following, we shall use the face centered monoclinic setting of [62], shown in Figure 11, to express the wave vector of the low temperature structural modulation (another setting exists in the literature, see [63]).

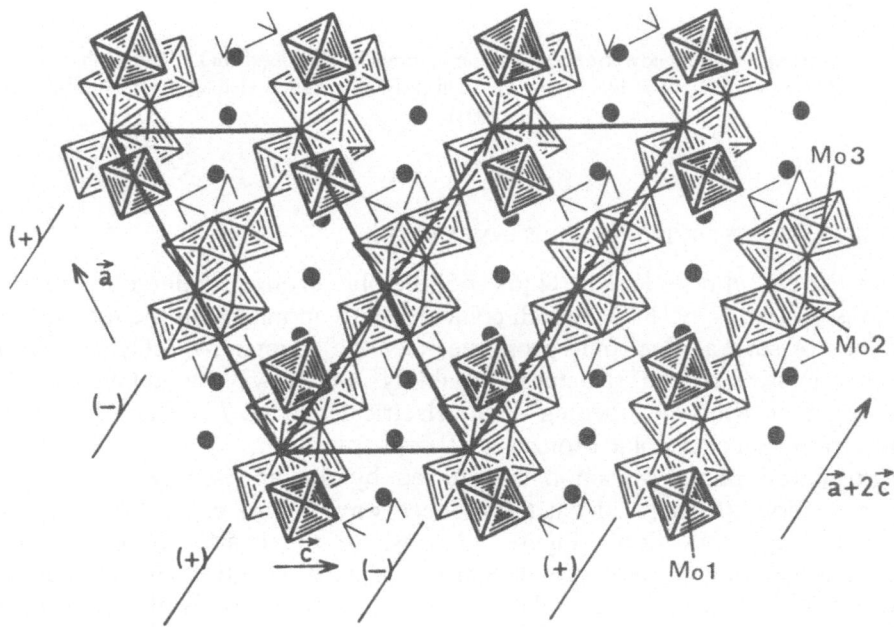

Fig. 11. Structure of the blue bronzes projected along the b direction. The 3 kinds of Mo atoms are indicated as well as the monovalent metals (dots). +(0) and −(π) give the phase of layers of CDW below T_c.

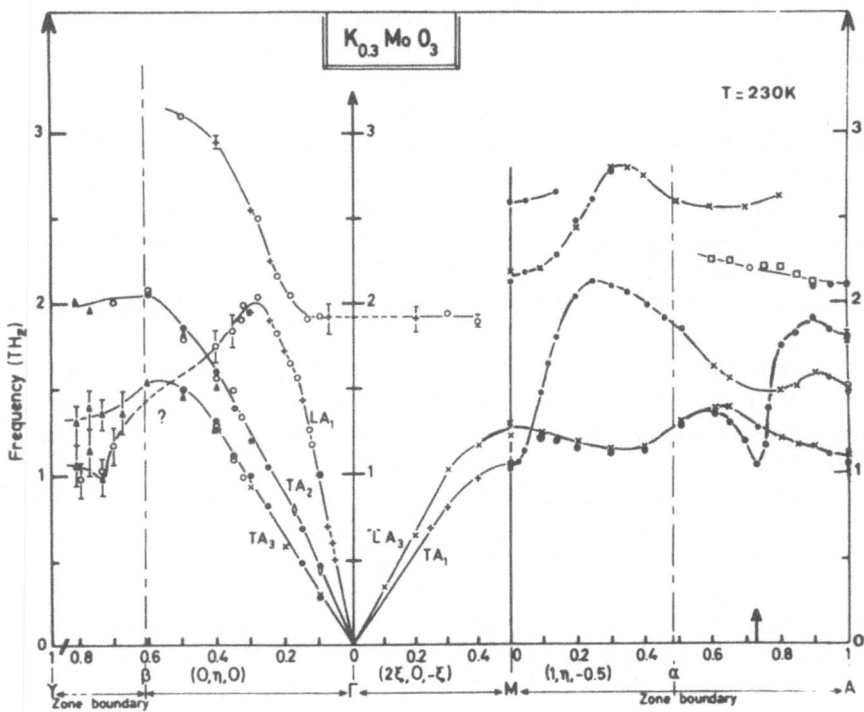

Fig. 12. Dispersion of some low frequency phonon branches of $K_{0.3}MoO_3$. The Kohn anomaly measured at 230 K in an acoustic-like phonon branch (arrow) is clearly visible. (From Hennion *et al.* [91]).

3.2. BASIC ASPECTS OF THE PEIERLS INSTABILITY

Whatever the cation A = K, Rb, Tl [64, 65] the blue bronzes undergo a structural phase transition at $T_c = 183$ K, which coincides with anomalies in the temperature dependence of several electronic properties (see the chapter by Schlenker *et al.*). The characteristics of the structural instability, which will be detailed in this section, together with the opening of an electrical gap at T_c, establish that the structural transition of the blue bronzes is a Peierls transition.

The structural phase transition is detected by the appearance below T_c of satellite reflections (Figure 13b) at the reduced wave vector $\mathbf{q}_R = (0, \pm \mathbf{q}_b, 1/2)$ from main Bragg reflections (Figure 13d), where the chain component q_b is incommensurate and temperature dependent (Figure 14 gives the temperature variation of $1 - q_b$, more closely related to $2k_F$). The transition is of second order, as seen by the continuous vanishing of the satellite intensity at T_c (Figure 15). $2\mathbf{q}_R$ satellite reflections are also observed below T_c [27] — see the insert of Figure 15. However, because of their weak intensity, they are likely to be a diffraction harmonic.

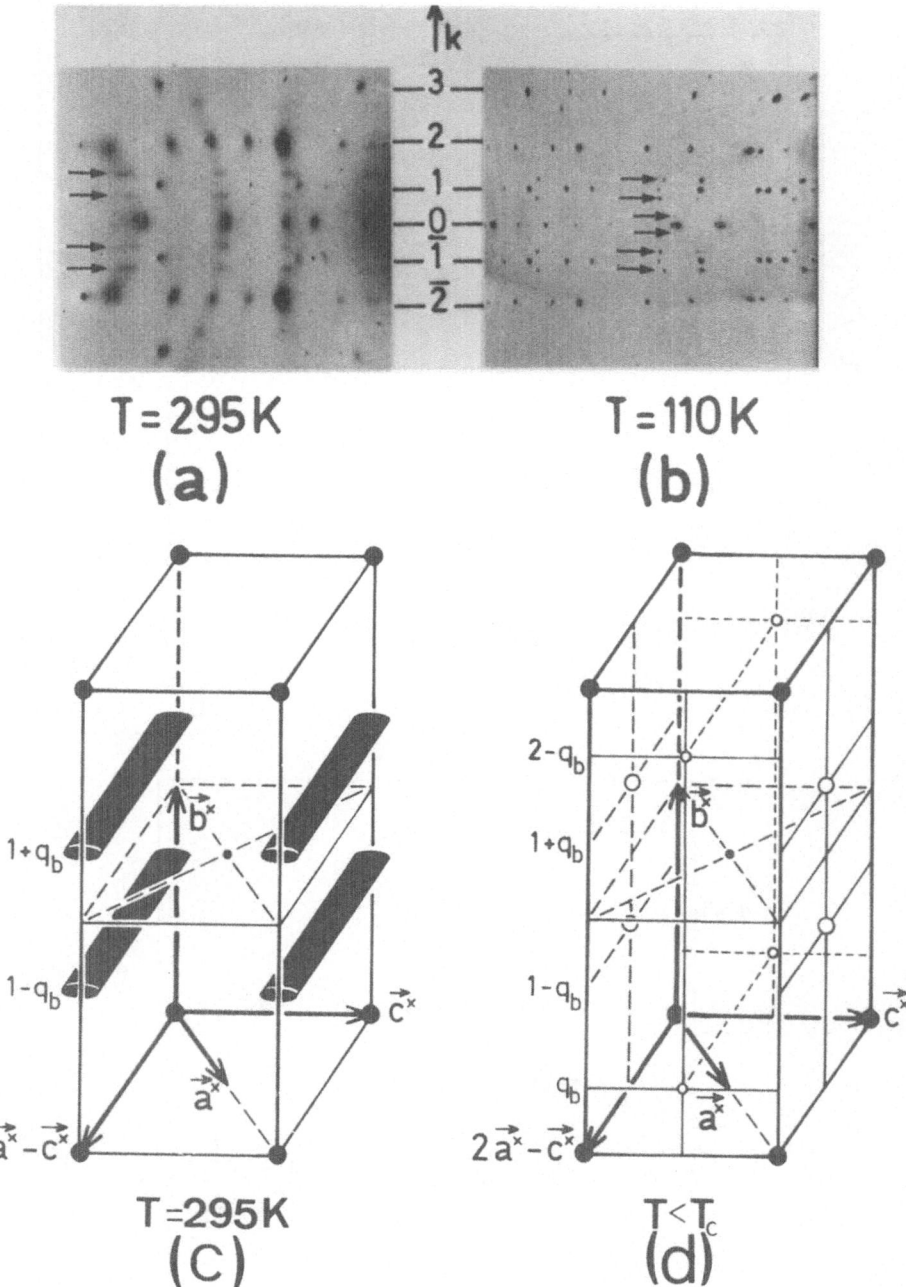

Fig. 13. X-ray diffuse scattering from the blue bronzes at (a) 295 K and (b) at 110 K showing respectively the 2D elongated nature of the diffuse scattering above T_c and its 'condensation' into satellite reflections below T_c. (c) shows more precisely the anisotropy and position with respect to the main Bragg reflections of the X-ray diffuse scattering observed at room temperature, while (d) gives the wave vector and the extinction rules of the satellite reflections in the reciprocal space (From Pouget *et al.* [64]).

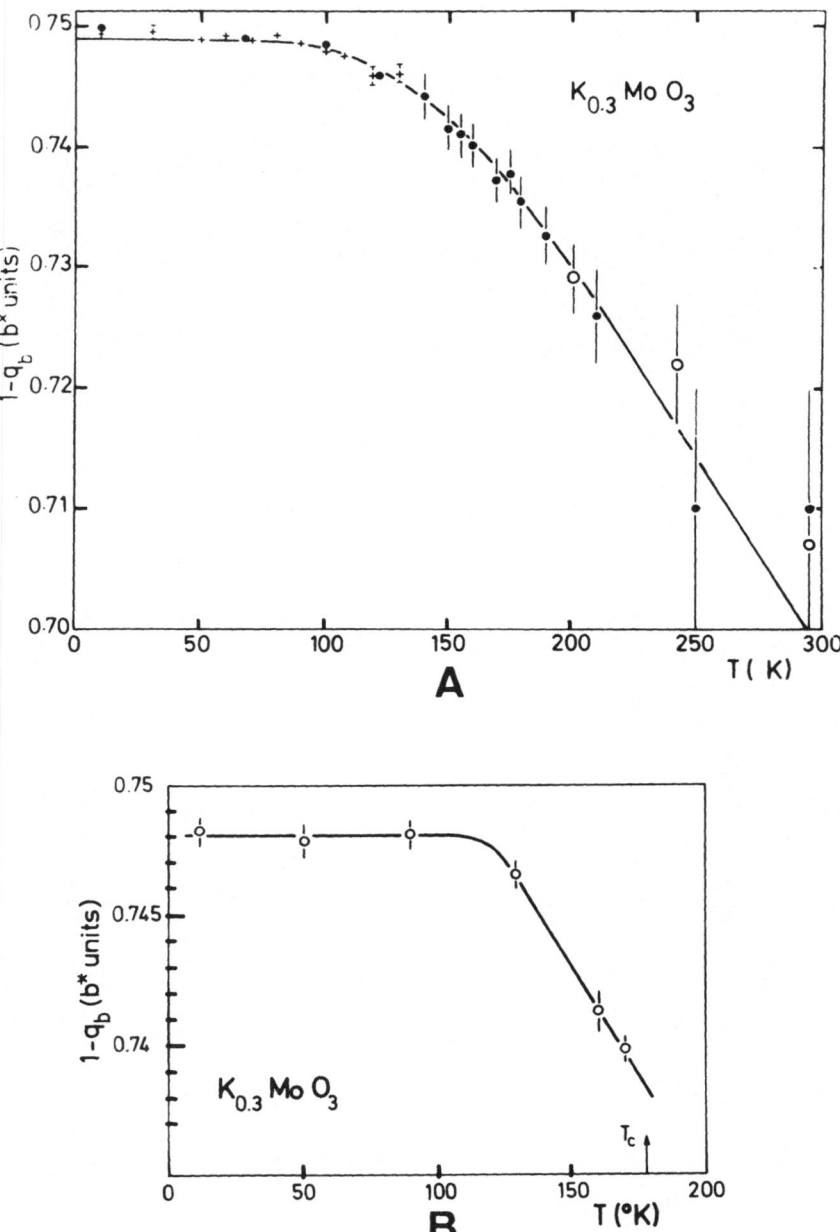

Fig. 14. Temperature dependence of the chain wave vector component $1 - q_b (\equiv k_F^{\perp} + k_F^{\parallel})$ of $K_{0.3}MoO_3$ measured on two different samples: (a) on all the temperature range by X-ray diffuse scattering (from Pouget *et al.* [27]). (b) below T_c by neutron scattering (from Escribe-Filippini *et al.* [72]). Note the net saturation in the temperature dependence of $1 - q_b$ below about 100 K (the measurements have been performed upon heating in (b)).

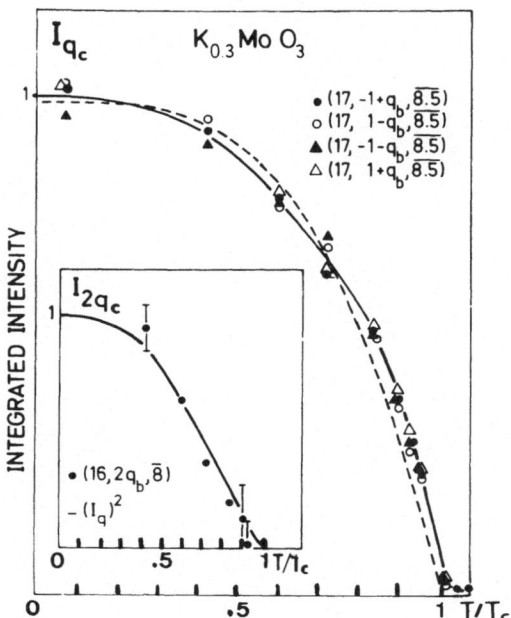

Fig. 15. Temperature dependence of the primary I_{q_c} and secondary I_{2q_c} (insert) satellite reflections in the Peierls insulating state of $K_{0.3}MoO_3$ (from Pouget *et al.* [27]).

In a large temperature range above T_c the phase transition is announced by important structural fluctuations which appear in the form of:

— an important X-ray diffuse scattering (arrows in Figure 13a), having at ambient temperature the shape of elongated cigars (Figure 13c),
— a (Kohn) anomaly in the frequency dependence of a particular phonon branch (Figure 12).

Several neutron investigations [66—68] have shown the displacive nature of the structural instability of the blue bronze by the progressive softening of the frequency $\omega(\mathbf{q}_R)$ of the bottom of the Kohn anomaly, as T_c is approached from above. However, as in many displacive structural instabilities, together with the phonon softening, a quasi-elastic (central) peak is found to develop in the vicinity of T_c (Figure 16).

The critical nature of the structural fluctuations at $\mathbf{q}_R$ was best proved by X-ray diffuse scattering experiments [27, 64, 69] which gave the temperature dependence of the CDW response function $\chi_\rho(\mathbf{q}_R)$ (Figure 17) and the correlation length ξ in several directions (Figure 18). It is important to remark that a true regime of 1D structural fluctuations has not been observed in the blue bronze up to 1.5 T_c, in contrst to other 1D conductors such as KCP of TTF—TCNQ. There is already a well established 2D (anisotropic) regime of fluctuations at ambient temperature (giving rise to the diffuse cigars shown in Figure 13 a and c). A crossover from a

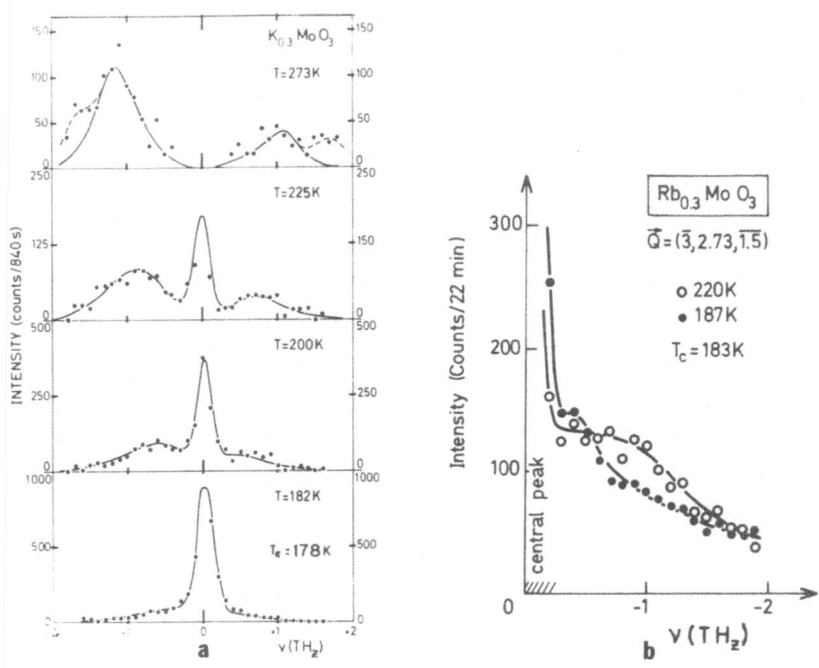

Fig. 16. Temperature dependence above T_c of the neutron scattering cross section measured at q_c (bottom of the Kohn anomaly) as a function of the transferred frequency $v = \omega/2\pi$. The softening of ω_{q_c} and the growth of a central peak in frequency is clearly shown. (a) Thermal neutron data obtained on $K_{0.3}MoO_3$ (from Pouget *et al.* [68]). (b) Cold neutron data obtained on $Rb_{0.3}MoO_3$ (from [107]).

critical regime of 2D fluctuations to a critical regime of 3D fluctuations occurs around $T^x \sim 200$ K. Figure 18 gives the temperature dependence of ξ in the 3 main directions of anisotropy b, $2\mathbf{a}^* + \mathbf{c}^*$ ($\approx \mathbf{a} + 2\mathbf{c}$) and $2\mathbf{a}^* - \mathbf{c}^*$ defined in Section 3.1.

The incommensurate distortion considered above, whose associated fluctuations diverge at T_c, is the primary order parameter of the Peierls transition. Additional anomalies in structural quantities, like the lattice parameters, are observed around T_c. The associated fluctuations, measured by the elastic constants, do not show a power law singularity at T_c, but only an anomaly in their temperature dependence. Such a behaviour is typical of a secondary order parameter. In the blue bronze, a sizeable anomaly is shown at T_c by the Young modulus [70] measured in the [102] direction, i.e. the direction of the polarization (at small $\mathbf{q}$) of the acoustic branch which bears the Kohn anomaly. Associated with this jump one may expect an anomaly in the temperature dependence of the $\mathbf{a} + 2\mathbf{c}$ lattice parameter at T_c. A weaker anomaly is also observed at T_c in the Young modulus measured in the b direction. Such elastic behaviours are discussed on general grounds in Appendix B.

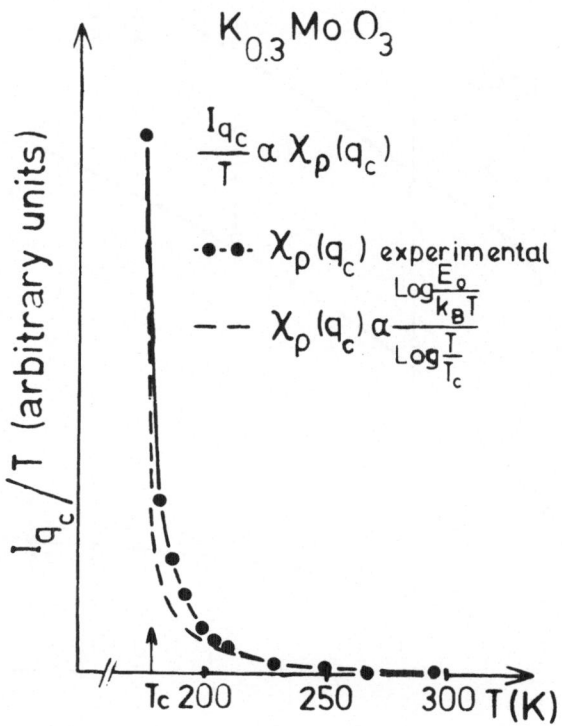

Fig. 17. Temperature dependence of the CDW response function $\chi_\rho(T) \sim I_{qc}(T)/T$ of $K_{0.3}MoO_3$. $I_{qc}(T)$ has been corrected by the experimental resolution (From Girault [65]). The dotted line is a fit by the mean field CDW susceptibility given by (54), and adjusted on the high temperature data.

3.3. CRITICAL WAVE VECTOR

Within the first Brillouin zone the modulation wave vector is defined by $\mathbf{q}_R = (0, \pm q_b, 1/2)$. However, as the physics of the Peierls transition corresponds to a 3D coupling between CDW of wave vector $2k_F$, it is better to define the critical wave vector by $\mathbf{q}_c = (1, 1 - q_b, 1/2)$, where $2k_F = 1 - q_b$ appears directly (this definition of $\mathbf{q}_c$ includes the extinction conditions for satellite reflections shown in Figure 13d).

3.3.1. *The Chain Component*

The chain component of the wave vector of modulation of the K bronze has been measured by several authors [27, 66—68, 71—73]. All these determinations give a low temperature value of $1 - q_b$ close to $0.75b^*$ (however there is always a systematic deviation by less than 0.5% towards lower values). The same low temperature value is obtained in the Rb and Tl bronzes [65]. $1 - q_b$ decreases significantly when the temperature increases (Figure 14).

In first we shall relate the low temperature value of $1 - q_b$ to the filling of

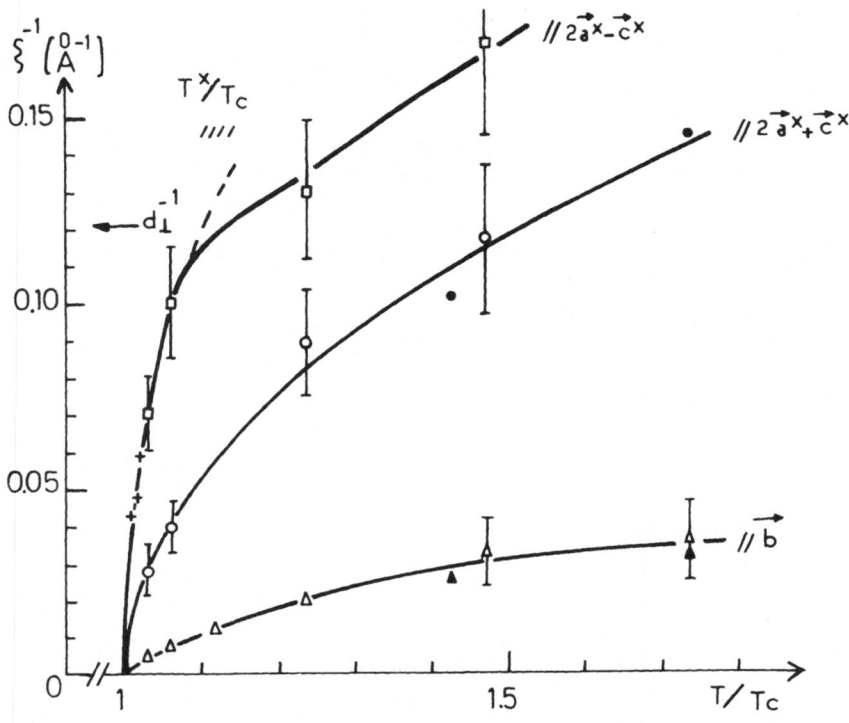

Fig. 18. Temperature dependence of the inverse CDW correlation length of $K_{0.3}MoO_3$ measured along the 3 principal directions **b**, $2a^* + c^*$ ($\sim a + 2c$) and $2a^* - c^*$. The temperature range where there is crossover from 2D to 3D fluctuations is shown. At this crossover, occurring for $d_\perp = \xi 2a^x - c^x$ corresponds a break in the temperature dependence of $\xi^{-1}_{2a^* - c^*}$ (data of Pouget *et al.* [27] corrected by a (Gaussian) experimental resolution).

pseudo 1D conduction bands by the electrons provided by the monovalent metal. As these atoms donate 3 electrons to each $Mo_{10}O_{30}$ cluster, which form the basic unit to consider in a band description, the experimental wave vector of $\sim 0.75b^*$ ($\equiv 3/4b^*$) corresponds either to the $2k_F$ value in the case of two degenerate filled conduction bands [69] or to the $k_F^I + k_F^{II}$ value in the case of two distinct ones [27].

As there is no symmetry argument in favour of two degenerate conduction bands, the second possibility seems to be more likely. Actually a recent LCAO calculation of the band structure of a $Mo_{10}O_{30}$ slab [11] provides evidence of two quasi 1D conduction bands crossing the Fermi level at about $k_F^{II} \sim 0.33b^*$ and $k_F^I \sim 0.42b^*$ (see Figure 19a). The average value of k_F ($\sim 0.375b^*$) was directly determined by means of angle-resolved ultraviolet photoemission spectroscopy [74]. These two bands are due to bonding and antibonding combinations of orbitals $|\phi_a\rangle$ and $|\phi_b\rangle$ of the pseudo cluster built on one half of the Mo(2) and Mo(3) octahedra of the cluster and schematically shown in Figure 19b (for the combination of the d orbitals of the Mo atoms entering in each $|\phi\rangle$, see [11]).

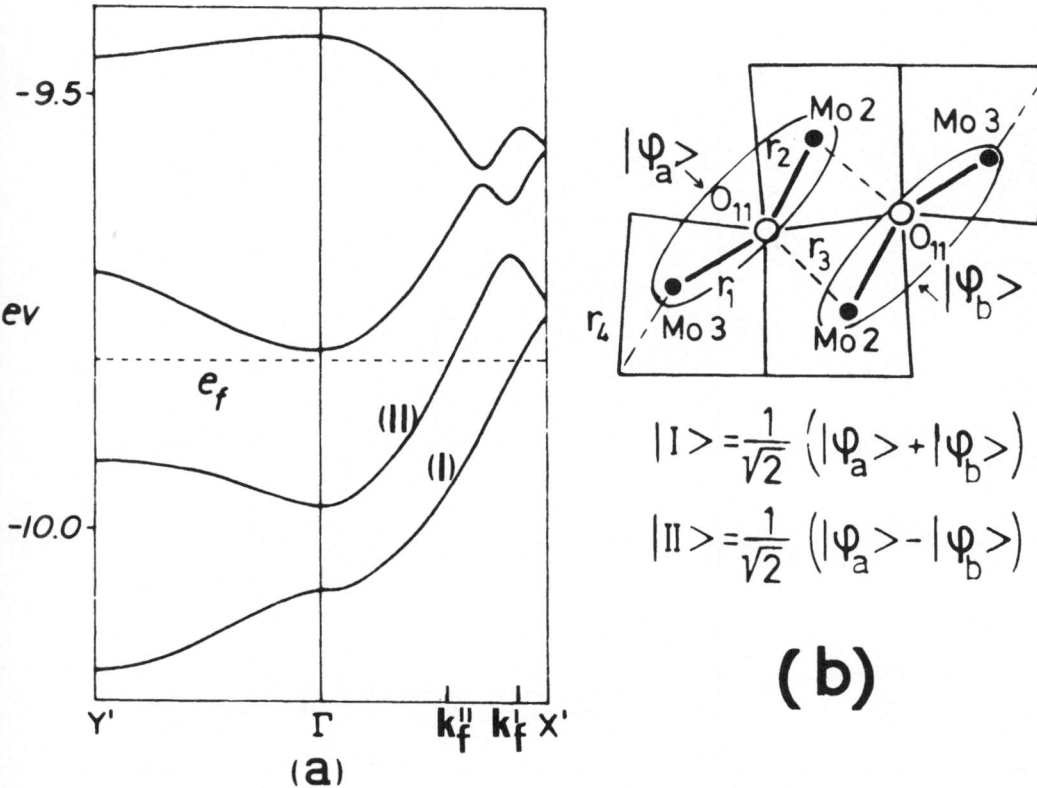

$$|I> = \frac{1}{\sqrt{2}} \left(|\varphi_a> + |\varphi_b> \right)$$

$$|II> = \frac{1}{\sqrt{2}} \left(|\varphi_a> - |\varphi_b> \right)$$

(b)

(a)

Fig. 19. (a) Dispersion of the energy bands in the vicinity of the Fermi level for the wave vector (nearly) in chain direction ($\Gamma X'$) and in transverse direction ($\Gamma Y'$). I and II correspond respectively to the bonding and antibonding states defined in (b). (Adapted from Whangbo and Schneemeyer [11]). (b) Pseudo-cluster unit based on $2Mo_2$ and $2Mo_3$ octahedra which form the infinite chains along b, and which generate the conduction bands shown in (a). The r_1, r_2, r_3 intracluster Mo—O distances ($r_2 \leqslant r_1 \ll r_3$) as well as the r_4 intercluster Mo—O distance ($r_4 \approx r_2$) are indicated. The a and b 'Mo$_2$—O$_{11}$—Mo$_3$' entities are shown with their H.O.M.O.: $|\phi_a\rangle$ and $|\phi_b\rangle$. The intracluster transfer integral t_{ab} leads to the bonding $|I\rangle$ and antibonding $|II\rangle$ combinations, as indicated.

With two conduction bands crossing the Fermi level, a Peierls semiconducting state is achieved at T_c only by the nesting of one band into the other, as is schematically shown by Figure 7c.

If $1 - q_b$ is really given by the band filling, its value should change if the total number of conduction electrons per cluster can be varied by chemical modification. The simplest way is to substitute a few Mo atoms by a non-isoelectronic element such as V. Figure 20 gives $1 - q_b$ as a function of the temperature for $K_{0.3}V_xMo_{1-x}O_3$ with $x = 0.024$ [75]. With a defect of one valence electron per V atom, there is on the average $3 - 10x$ conduction electrons per cluster. This gives for $x = 0.024 : 1 - q_b = 0.69b^*$, which agrees very well with the low temperature experimental finding of $1 - q_b = 0.685 \pm 0.008b^*$. Consistently, the substitution

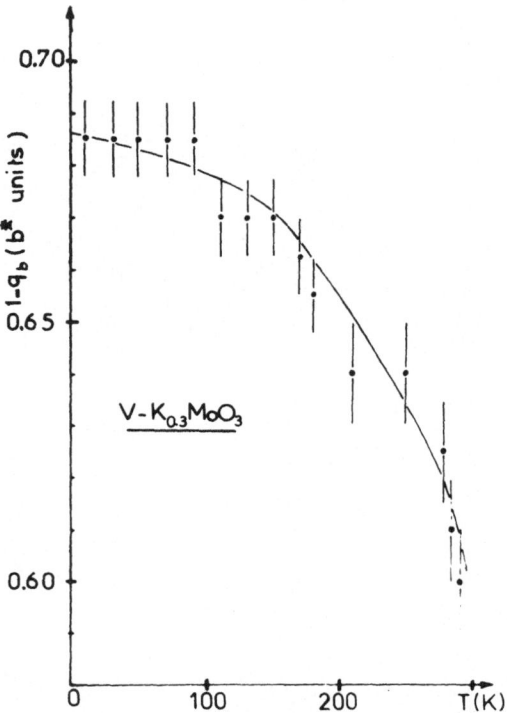

Fig. 20. Temperature dependence of the chain wave vector component, $1 - q_b$, for the substitutional alloy $K_{0.3}Mo_{1-x}V_xO_3$ with $x = 0.024$ (from Girault *et al.* [75]).

of Mo by isoelectronic impurities such as W does not significantly change the value of $1 - q_b$ [76]. In a similar way, substitution among the monovalent metals, A, does not change $1 - q_b$. (The non-observation of a shift of $1 - q_b$ in the '$K_{0.28}MoO_3$' sample studied in [77] can be understood by the presence of additional Na impurities leading to the chemical formula $K_{0.28}MoO_{2.99}(Na_2O)_{0.01}$, where the total number of conduction electron per cluster remains unchanged [65]). The difference found in the literature among the various determination of $1 - q_b$ could be due to non-stoichiometry in the alkali content or to substitution of a small amount of Mo atoms by non-isoelectronic impurities during the preparation of the samples.

Surprisingly, $1 - q_b$ decreases when the temperature increases with a deviation from the low temperature value which could be accounted for, over the complete temperature range, by an exponential law:

$$\delta q_b \approx 0.5 \exp - (\Delta E/k_B T),$$

where $\Delta E \sim 60$ meV in $K_{0.3}MoO_3$ (Figure 14) and $\Delta E \sim 50$ meV in $K_{0.3}V_{0.024}Mo_{0.976}O_3$ (Figure 20). The physical origin of such a thermally activated process is still unclear. It has been proposed [27] that this could be due to a progressive decrease of the conduction electron concentration by thermal activation of these electrons towards electronic levels situated just above the Fermi level by about ΔE (in that case δq_b is proportional to the Fermi–Dirac function). Band calculation [11] indicates that there is an additional band just above the Fermi level (Figure 19a). However other mechanisms, such as the thermal breaking of a bipolaron sublattice (obtained in the strong electron–phonon coupling limit) have also been proposed [78] to account for the exponential dependence of $1 - q_b$ on temperature. It has been also proposed [117] that the temperature dependence of q_b could be accounted for by the formation of a kink lattice in the relative phase of CDW's of wave vectors $2k_F^I$ and $2k_F^{II}$. However this leads to the formation of a semimetallic state at T_c, instead of the semiconducting one experimentally observed.

For completness let us mention that it is easy to find mechanisms giving a T^2 dependence of δq_b in the metallic phase. In the case of the blue bronzes the prediction of such variations (see below) are not in quantitative agreement with the experimental results. These mechanisms rely on:

(a) The temperature dependence of the Fermi energy. It is easy to show that in a metal:

$$E_F(T) - E_F(0) = - \frac{\pi}{6} (k_B T)^2 \frac{\partial \log N(E)}{\partial E}\bigg|_{E_F},$$

which in the 1D case becomes:

$$\delta k_F = k_F(T) - k_F(0) = \frac{\pi}{6} \frac{(k_B T)^2}{\hbar v_F^3 m_{eff}}.$$

In the blue bronze, transport and optical studies [79] show that the carriers are hole-like with an effective mass $m_{eff} \approx -m_e$. With $v_F = 2.10^5$ m/s (see Section 3.8.1.) this gives $\delta k_F \approx -1.5 \times 10^{-4} b^*$ at room temperature. Such a deviation is much smaller than the one that is experimentally observed: $\delta(k_F^I + k_F^{II}) \approx -5 \times 10^{-2} b^*$ (the larger shift occurs for band I which has the lighter effective mass — see Figure 19a).

(b) The temperature dependence of the position of the Kohn anomaly, when the bare phonon frequency varies with q: $\Omega(q) = \Omega(q_c) - s\delta q_{\parallel}$. Starting from the expression (13), the frequency of the minimum of the Kohn anomaly occurs for:

$$\frac{\partial \omega(q)}{\partial q} = 0 \quad \text{or}$$

$$\frac{\partial \log \Omega(q)}{\partial q} [\omega_{(q)}^2 + \Omega_{(q)}^2] = \left[\frac{2\partial \log g_q}{\partial q} + \frac{\partial \ln \chi_e(q)}{\partial q} \right] [\Omega_{(q)}^2 - \omega_{(q)}^2].$$

If (i) g_q does not vary sizeably with q, and if (ii) the q dependence of $\chi_e^0(q)$ is given by (23), the minimum of the Kohn anomaly is shifted from $2k_F$ by:

$$\delta q_{\parallel} = - \frac{s}{2\Omega_{(q_c)} \xi_{0\parallel}^2} \frac{\omega_{(q_c)}^2 + \Omega_{(q_c)}^2}{\Omega_{(q_c)}^2 - \omega_{(q_c)}^2},$$

with $s = 2 \times 10^2$ m/s (Figure 14a) and the room temperature values of $\xi_{0\parallel}$, $\omega_{(q_c)}$

and $\Omega_{(q_c)}$ (see Section 3.8.1.) one gets $\delta q_{\parallel} = 6 \times 10^{-3} b^*$, which is one order of magnitude lower than the experimental value.

3.3.2. *The Transverse Components*

Figure 12d shows that the extinction rule of the satellite intensities is identical to that of the main Bragg reflections. This means that the CDW modulation keeps the face-centered symmetry of the lattice. More physically the Peierls transition corresponds to a coupling between the 1D CDW of wave vector $1 - q_b$ [27, 64]:

— in phase between clusters forming the layer of MoO_3 octahedra: $\mathbf{q}_c(\mathbf{a}/2 + \mathbf{c}) = 0$ (this phase relation can be obtained from the position, in reciprocal space, of the diffuse rods at room temperature — see Figure 12c),
— out of phase between neighbouring layers: $\mathbf{q}_c \cdot \mathbf{c} = \pi$.

The resulting CDW phase pattern is schematically represented in Figure 11.

Within the slab of MoO_3 octahedra the modest electrical anisotropy ratio measured, $\sigma_{|010|}/\sigma_{|102|} \approx 30$ [80] suggests that there is a sizeable electronic tunneling $(t_{\perp})$ between the clusters. This certainly provides the dominant coupling mechanism between the CDW along the [102] direction. In the case of an open Fermi surface, it is easy to show (see for example [50]) that:

$$\sqrt{\frac{\sigma_{\perp}}{\sigma_{\parallel}}} \sim \frac{t_{\perp} d_{\perp}}{t_{\parallel} d_{\parallel}} = 1.3 \frac{t_{\perp}}{t_{\parallel}} \ (\approx 0.2),$$

(with the cluster repeat periodicities of $d_{\parallel} = 7.6$ Å and $d_{\perp} = 9.8$ Å along $\mathbf{b}$ and $\mathbf{a} + 2\mathbf{c}$ respectively). More precisely the band structure calculation of Whangho and Schneemeyer [11] allows us to estimate, for the two conduction bands shown in Figure 19a (see also Figure 25 for a schematic representation):

— the intrachain tunneling integrals $t'^{I}_{\parallel} = 0.08$ eV and $t'^{II}_{\parallel} = 0.1$ eV, from the bandwidth $W_{\parallel} = 4t'_{\parallel} (= 2t_{\parallel}$ in Figure 25) in the $\Gamma X'$ direction,
— the Fermi velocities: $v^{I}_{F} \approx v^{II}_{F} \approx 2 \times 10^5$ m/s,
— the tunneling integral between the a and b 'Mo(2)—Mo(3)' entities defined in Figure 19b: $t_{ab} \sim 50$—75 meV, from the $2t_{ab}$ splitting between the two conduction bands (t_{ab} increases slightly with $k_{\parallel}$),
— the transverse tunneling integrals: $t^{I}_{\perp} \sim -25$ meV and $t^{II}_{\perp} \sim +10$ meV, from the bandwidth $W_{\perp} = 4t_{\perp}$ in the $\Gamma Y'$ direction.

This gives a ratio: $t_{\perp}/t'_{\parallel} \sim 0.19$, in good agreement with that deduced from the electrical anisotropy.

The opposite sign of $t^{I}_{\perp}$ and $t^{II}_{\perp}$ gives a Fermi surface composed of sheets of opposite curvature. In that case, shown by Figure 7c, it is easy to understand that the best nesting of the whole Fermi surface leads to a zero transverse component for the CDW wave vector. However, as $|t^{I}_{\perp}|$ differs from $|t^{II}_{\perp}|$, the nesting is not perfect. It leaves pockets of size $2\Delta q_{\parallel} = 2 \times 10^{-2}$ Å corresponding to a ratio $2k_F/2\Delta q_{\parallel} \sim 30$. Using the expression (42), such pockets cut the divergence of $\chi^0_e(k^I_F + k^{II}_F)$ at about $T \sim 80$ K. As this value is lower than the Peierls transition temperature $T_c = 183$ K, the pockets are too small to inhibit the growth of the CDW instability above T_c.

In the direction perpendicular to the layers of MoO_3 octahedra, where the slabs are separated by sheets of monovalent metals, the coupling between the CDWs is more likely of electrostatic origin. Minimization of Coulomb coupling between the CDWs leads to the observed out-of-phase ordering between slabs of CDW (see Section 2.1.3.).

3.4. SPATIAL FLUCTUATIONS $(T > T_c)$

The pretransitional spatial fluctuations are characterized by the measurement of the correlation length in the three principal directions of anisotropy of the crystal [27, 69, 81]. Figure 18 gives the temperature dependence of the ξ^{-1}. At 250 K one has for example $\xi_b \approx 30$ Å, $\xi_{2a^*+c^*} \approx 8$ Å and $\xi_{2a^*-c^*} \approx 5$ Å. These values follow the structural anisotropy.

The weakest coupling is along the direction perpendicular to the $Mo_{10}O_{30}$ slabs. The crossover from a regime of 2D to 3D fluctuations occurs at about $T^x = 200$ K, when $\xi_{2a^*-c^*}$ reaches 8 Å, the interslab distance. This crossover corresponds to a break in the temperature dependence of $\xi_{2a^*-c^*}$. Within experimental error the break is not observed in the temperature dependence of the correlation length measured in the other directions.

The direction of intermediate coupling is $\mathbf{a} + 2\mathbf{c}$. Along this direction, there is clearly a coupling between CDW belonging to neighbouring clusters until room temperature at least. If one assumes that the CDW will be decoupled when the correlation length ξ_{a+2c} is comparable to the size of an octahedron (~ 4 Å), the crossover temperature towards a regime of 1D CDW fluctuations will occur well above room temperature.

Finally, as expected for a quasi 1D conductor, the strongest coupling (i.e. the longest correlation length) is observed in the chain direction, b. The anisotropy of the CDW correlation lengths will be more quantitatively discussed in Section 3.8.2.

Using Equation (23), and the value of v_F quoted in the last section, it is easy to calculate the electronic thermal length $\xi_{0\parallel}$ in the chain direction. At 250 K one gets $\xi_{0\parallel} \sim 13$ Å, which is about half the observed value ξ_b. In fact because of the electron—phonon coupling, one expects a CDW correlation length ξ_b somewhat enhanced from the electronic one $\xi_{0\parallel}$ (see Appendix A). The enhancement increases as T_c is approached from above, leading to the divergence of ξ at T_c.

The divergence of ξ and of the CDW response function at the critical wave vector $\chi_\rho(q_c)$ has been followed by means of X-ray diffuse scattering (Figures 18 and 17, respectively). As expected below T^x, in the 3D coupling regime, the same power law of divergence of ξ is observed in all directions. A fit of the divergences of ξ and χ_ρ, according to the power laws given by the expressions (7) leads [69], for $(T - T_c)/T_c < 0.1$, to the following exponents:

$$\gamma = 1.33 \pm 0.04$$

$$\nu = 0.68 \pm 0.05.$$

These exponents are those expected (1.32 and 0.67 respectively) for a 3D phase transition with a two components ($n = 2$) order parameter. This regime of 3D critical fluctuations with non classical exponents should be contrasted with the mean field one observed in KCP [122]. In this last compound the "harmonic" behaviour of the fluctuations is enhanced by the larger value of the in chain thermal length $\xi_{0\parallel}$ and by the pertinent effect of the disorder.

In fact, high resolution X-ray measurements at T_c and below show that there is no true long-range order in the direction of weakest coupling between CDW: $2\mathbf{a}^* - \mathbf{c}^*$ [76, 82]. The broadening in excess of the experimental resolution gives a coherence of about 10^3 Å at T_c. As the temperature is lowered the coherence between layers of CDW increases, reaching nearly the experimental resolution by 90 K ($\xi_{2\mathbf{a}^*-\mathbf{c}^*} > 5 \times 10^3$ Å).

3.5. THE MODULATED STRUCTURE ($T < T_c$)

3.5.1. *The CDW Order*

With the restriction mentioned at the end of the previous section, for the $2\mathbf{a}^* - \mathbf{c}^*$ direction, the CDW order is characterized by the presence of sharp satellite reflections. The wave vector of ordering has been analyzed in Section 3.3. The integrated satellite intensity, proportional to the square of the order parameter, has been measured several times [27, 66, 69, 73] with an overall agreement of its temperature dependence (Figure 15). In the determination of reference [69] a fit of the satellite intensity in the vicinity of T_c, with the expression (12), allowed a determination of the critical exponent of the order parameter:

$$\beta = 0.31 \pm 0.05.$$

The same value of β (0.32 $\pm$ 0.03) has also been obtained from an NMR study of the ^{87}Rb in $Rb_{0.3}MoO_3$ [83]. This value agrees also with the critical exponent expected (0.35) for a 3D transition with a two components order parameter.

It has been suggested [73] that the blue bronze could exhibit an incommensurate—commensurate phase transition around 100 K with a lock-in of q_b at the commensurate value $q_b = 0.25\ b^*$. All the measurements really show a saturation in the temperature dependence of q_b below about 100 K, but its value significantly deviates from 0.25 b^* (see for example Figure 14b) by a quantity which seems to depends on the sample [27, 66—68, 71, 72]. Such a deviation can be due to nonstoichiometry and/or to the presence of impurities with various concentrations for different samples. It is also conceivable that the modulation wave vector measured below 100 K is not that of the thermodynamical equilibrium because the thermal adjustment of the wave length could require a long time (such a possibility is suggested by the high resolution data of Figure 14b, showing a net change in the rate of variation of $1 - q_b$ around 100 K). The slow kinetics involved in the change of wave length could be due to the pinning of the nearly commensurate modulation wave by lattice defects. The metastability of the wave vector could have a link with hysteresis and memory effects observed below about 100 K [84].

Effects of metastability in the CDW sublattice, disordered by the application of an electric field, are also observed below 100 K in the blue bronze (see Section 3.6.2.).

NMR [87]Rb line shape measurements have clearly demonstrated that the modulation is incommensurate below 100 K [85, 86]. A careful simulation of the NMR line shape at 100 K has shown that if the CDW lattice contains commensurate domains separated by walls (discommensurations), the walls must have a width comparable to their spacing [83], which means that the modulation is nearly sinusoidal at this temperature.

It has been shown [27] that, because of the C centering symmetry of the lattice, the lock-in potential provided by the lattice discretness is, at the lowest order, of the 8th power of the order parameter (i.e. order of commensurability $p = 8/3$ — see Figure 21 for a simplified presentation of the group theory argument in a centered sheet of MoO_6 octahedra modulated along b). The effect of such a high-order potential on the incommensurate modulation is expected to be particularly weak, as it seems to be experimentally the case.

3.5.2. Atomic Displacements

The atomic displacements involved in the structural modulation have been obtained by structural refinement from a collection of low temperature satellite intensities [65, 67, 87]. The best determination [87] performed at 100 K on 7961 unique reflections, with a reliability factor of 3.3% gives:

(i) the largest amplitude of modulation (of about 0.06 Å) on the Mo(2) and Mo(3) atoms which belong to the infinite chains running along b; their atomic

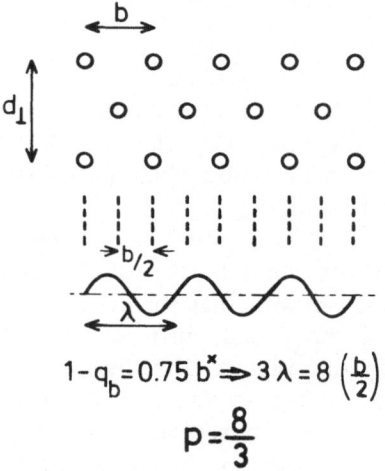

$$1 - q_b = 0.75\, b^x \Rightarrow 3\lambda = 8\left(\frac{b}{2}\right)$$

$$p = \frac{8}{3}$$

Fig. 21. Schematic representation of a "C" centered layer of clusters of MoO_6 octahedra (upper part) showing that a wave propagating in chain direction experiences a lattice periodicity of $b/2$. If such a modulation has a wave length close to $4b/3$, the lattice discreteness tends to impose phase relationship every 3 wave lengths corresponding at $8(b/2)$ lattice periodicities and at a commensurability order $p = 8/3$.

displacement is mainly transverse ($\sim$[101], i.e. along a direction close to the long distance of the Mo(3)—O(11)—Mo(2) entities defined in Figure 19b).

(ii) Weak displacements for the coordinated oxygens in distorted octahedra.

(iii) Displacement of the alkali atoms along the chain direction (0.03 Å), in phase with their surrounding oxygens and the Mo(1).

The displacement of the Mo(2) and Mo(3) leads to a modulation of the Mo—O distance within each octahedra by about 0.1 Å. It is easy to see in Figure 19b, that the modulation of the distances:

(a) r_1 and r_2 ($r_1 > r_2$) changes, to 1st order, the site energy of the orbitals $|\phi_a\rangle$ and $|\phi_b\rangle$ of the a and b entities respectively,
(b) r_3 changes, to 1st order, the tunneling integral t_{ab} within the cluster,
(c) Mo—O in all the directions changes, to 1st order, the in-chain overlap integrals $t_{\parallel}$ between Mo octahedra,
(d) r_4 changes, to 1st order, the transverse overlap integrals $t_{\perp}$ between $Mo_{10}O_{30}$ clusters.

The electron—phonon coupling in a band structure composed of bonding (B) and antibonding (AB) conduction bands has been already considered in Section 2.1.6. The nature of the bands coupled by the phonon field depends on the symmetry of the phonon mode. In the present case, the acoustic-like critical phonon mode leads to a modulation:

— of the distance r_3,
— of the distances r_1 and r_2 (out of phase between the a and b entities because of the inversion symmetry relating a and b),
— of the distances Mo—O along b (also out of phase between the a and b entities).

The modulation of r_3 leads to a coupling between bands of the same symmetry (B—B or AB—AB), while the other modulations, because of their antiphasing, couple bands of different symmetry (B—AB). This last coupling is the only one that leads to the opening of a gap on the whole Fermi surface and to the stabilization of a semiconducting state below T_c.

The key role played by the Mo(2) and Mo(3) atoms has already been pointed out by an earlier analysis of the intensity of the X-ray diffuse scattering in reciprocal space [64] which showed that the pretransitional fluctuations involve mainly the deformation of the pseudoclusters built on the Mo(2) and Mo(3) octahedra (Figure 19b). In this picture, the displacement of the alkali atoms is more likely a consequence of the structural instability than its cause. Such an interpretation agrees also with the observation that the value of T_c and that the pretransitional fluctuations (correlation lengths, dynamics) do not depend significantly on the nature of the alkali atom.

3.6. CDW DISORDER

In this part, we shall discuss the CDW disorder in the blue bronzes (1) produced by substitutional alloying and (2) induced below T_c by an external electric field.

The observations have been performed by X-ray diffraction techniques. They are essential to build a microscopic picture of the CDW transport.

3.6.1. Substitutional Disorder

Structural studies of K/Rb substitutions on the alkaline sublattice [76] as well as the substitutions of Mo atoms by isoelectronic impurities such as W [76] or non-isoelectronic impurities such as V [75] have been performed. In the case of the V doped samples, structural refinements [65] show a preferential substitution on the Mo(2) sites.

These two kinds of substitution have different effects on the CDW transport: the alkaline substitution gives rise to weak pinning centers and the Mo substitution to strong pinning ones (see the following chapter). The perturbation they induce on the CDW sublattice is also quantitatively different.

The study [76] of the solid solution $(K_{1-x}Rb_x)_{0.3}MoO_3$ shows that the alkaline metal substitution induces a weak disorder on the CDW sublattice:

— the coherence length remains several hundred Å below 'T_c' along all directions,
— the Peierls transition temperature 'T_c', determined by the maximum of the resistivity derivative (and the onset of well defined satellite spots) shifts only slightly with x, reaching a minimum of about 171 K for $x = 0.5$.

Studies of the solid solutions $K_{0.3}Mo_{1-x}W_xO_3$ [76] and $K_{0.3}Mo_{1-x}V_xO_3$ [75] show that, as expected, the Mo substituent induces a strong disorder on the CDW sublattice. Very small amount of substituant reduces significantly the coherence lengths, especially in the direction $2a^* - c^*$ of weakest coupling between layers of CDW. For $x \approx 1\%$ the interlayer correlation is lost. With the dramatic loss of interlayer coupling, the structural transition is suppressed.

The most likely interpretation is that the CDW order is destroyed by random phase pinning effects, formation of Friedel oscillations and local $2k_F$ distortions around impurities. In particular, the formation of Friedel oscillations simply explains the enhancement of the high temperature diffuse scattering and the profile asymmetry of the scattering in the b direction (case of the alloy $K_{0.3}Mo_{1-x}V_xO_3$ [75]).

3.6.2. Electric Field Induced Disorder

An electric field is directly coupled to the CDW. As shown by electrical studies (following chapter) such a field is expected to move locally the CDW sublattice or to deform it. It is only recently that these field induced motional deformations have been detected directly by X-ray diffraction in the blue bronzes [71, 82, 88, 89].

The key point is the observation of a dramatic transverse broadening of the satellite reflections (mainly along $2a^* - c^*$) when an electric field is applied in the chain direction. A much weaker broadening is also observed in the chain direction when a very large electric field is applied ($E \sim 24$ V/cm at 10 K) [89]. The broadening appears as soon as the dc field is applied, (i.e. below the threshold field), increasing with the amplitude of the field. This shows that the induced disorder is due either to a local deformation of the CDW sublattice or to its break into domains, due to the relative motion of pieces of differently pinned CDW. The broadening is obtained in field cooled sample [82], but it can also develop when the field is set at low temperature with a time scale which depends strongly of the temperature: ~500 s at 10 K [89], ≤ 1 s at 70 K [82] (probably ~1 ms from the time resolved data of [71] and [88]). The broadening remains unchanged on a longer time scale (several hours at 10 K) when the electric field is removed. The initial spatially coherent state can be restored by the thermal cycling of the sample above about 100 K [82]. This implies that the field perturbed CDW sublattice populates metastable states from which the relaxation time is very long at low temperature.

If asymmetric electric contacts are applied at the sample, the peak broadening is asymmetric or even shifted in the transverse direction [71, 88]. The shear deformation of the CDW sublattice is ascribed to its coupling with an inhomogeneous electric field distribution within the sample. Time resolved X-ray experiments have shown that the shift is reversed when an electric field of opposite direction is applied. This process occurs with a switching time of about 1 ms around 70 K, a value comparable to the electrical polarization transient time (see the following chapter). This shows directly the contribution of the CDW sublattice to the transient response. Also a large improvement of the CDW transverse order was noticed during the transient period between the 2 steady positions of the CDW sublattice (at 70 K the field perturbed CDW sublattice is still in a metastable state, because the suppression of the electric field does not affect, on the switching time scale, the satellite reflections in position and width, as noticed in the last paragraph).

The most fundamental question concerns the microscopic nature of the field induced CDW disorder. In principle the satellite broadening can be due either to an inhomogeneous field or pinning resulting in a continuous variation of the CDW phase (amplitude) across the sample, or from induced finite range order effects. In the first case, the inhomogeneous broadening will be a function of the sample and of the part of the sample probed by the X-ray beam. If the disorder is described by microdomains with a gaussian or a random distribution of domain sizes, the X-ray diffraction peak will have a Gaussian or a Lorentzian squared line-shape (Good statistics are necessary to make the difference between these two profiles). In the crudest analysis, the average domain size is given by the Scherrer formula $L = 0.89 \ \pi \Delta Q_{1/2}^{-1}$, where $\Delta Q_{1/2}$ is the half width at half maximum (H.W.H.M.) of the Gaussian. In the second case, where the loss of correlation is more progressive and described by an Ornstein—Zernike law, the line shape of the peak is Lorentzian and its H.W.H.M. gives the inverse of the real-space correlation length, ξ^{-1}, as for the pretransitional fluctuations (see Section 2.1.2.).

It has been shown that the field induced transverse displacement (along $2\mathbf{a}^* - \mathbf{c}^*$) of the superlattice spots is due to inhomogeneous effects [88]. In addition, the high resolution profile obtained for transverse wave vectors is closer to a Gaussian than to a Lorentzian [82], which suggests that the transverse disorder is more appropriately described by microdomains (i.e. with a rapid loss of the CDW transverse order at the domain boundaries) rather than by short range order effects. The CDW domain average size obtained for a sample cooled at 14 K in a field of ~1 V/cm is about L ~2000 Å [82].

The broadening of the satellite reflections in the direction of the applied electric field directly shows that polarization effects break the long range spatial coherence of the CDW in chain direction. The high resolution profile obtained for longitudinal wave vectors, under strong electric field ($E \sim 24$ V/cm), can be reasonably well fitted by a Lorentzian [89]. Its H.W.H.M., corrected by the Lorentzian resolution, gives an in-chain correlation length $\xi_b \sim 5000$ Å, which is larger by a factor 20 than the effective transverse correlation length ξ_{c*}, deduced from the H.W.H.M. of the profile recorded in the $c*$ direction in the same conditions. This anisotropy ratio is larger by a factor 3 than the one of the 3D pretransitional fluctuations ($\xi_b/\xi_{c*} \sim 6$ deduced from the data of Figure 18). It shows that the field induced disorder, at least for large electric field and low temperature, cannot be described by the CDW elasticity tensor determined above T_c. The rigidity of the CDW sublattice along the chain direction is increased below the Peierls transition, as a consequence of the non screening of the long range Coulomb forces for longitudinal deformation of the phase of the CDW. One should also consider below T_c the plasticity of the transverse deformation of the CDW sublattice under a shear constraint. Also, the difference of profile in the broadening of the satellite reflections which is observed in the longitudinal and transverse scans, suggests that the CDW sublattice could experience deformations of different nature in the directions parallel and perpendicular to the applied electric field. The elastic and plastic deformations of CDW are considered by Feinberg and Friedel in this book.

3.7. DYNAMICS

3.7.1. *Pretransitional Temporal Fluctuations* ($T > T_c$)

Because of the great number of low frequency modes, the phonon spectrum of the blue bronzes is imperfectly known. Only a few number of branches have been followed by inelastic neutron scattering measurements [66—68, 72, 90]. One of them shows a well formed Kohn anomaly above T_c at the critical wave vector ($\mathbf{q}_c$), on the Brillouin zone boundary (see Figure 12). However the crossing of the critical phonon branch with other phonon branches of low frequency makes its identification very delicate. By continuity, it seems to belong to a branch which, near the zone center (Γ point), could be the transverse acoustic branch polarized along $\mathbf{a} + 2\mathbf{c}$. Such a polarization is consistent with that, below T_c, of the (dominant) atomic displacement of the Mo(2) and Mo(3) (see Section 3.5.2.).

In the vicinity of $\mathbf{q}_c$, the $\mathbf{q}$ dependence of the Kohn anomaly can be accounted for by the expression (13b). Along the chain direction one gets, at 190 K: $\sqrt{\Lambda_{\parallel}} = 3.1 \times 10^3$ m/s. The bottom of the Kohn anomaly, $\omega(\mathbf{q}_c)$, clearly softens when T_c is approached from above (Figure 16). Together with the phonon softening, a quasielastic (central) peak grows in intensity. Within experimental error, $K_{0.3}MoO_3$ and $Rb_{0.3}MoO_3$ show the same pretransitional dynamics.

The phonon softening has been analyzed [68, 81] with a simple damped harmonic oscillator (Equation 16). The deconvolution of the experimental data from the experimental resolution gives an increase of Γ and roughly a linear decrease of $\omega^2(\mathbf{q}_c)$ when the temperature decreases. However, $\omega(\mathbf{q}_c)$ does not soften completely at T_c where $\omega(\mathbf{q}_c) \approx 0.4$ THz. The divergence of the critical fluctuations seems to be achieved by the critical growth of the central peak in the vicinity of T_c. One should note that the continuity in the temperature behaviour of $S(q, \omega)$ at a 2nd order phase transition suggests, from the observation of an amplitude mode at finite frequency at T_c (see Section 3.7.2. below), that the critical fluctuations, especially in the vicinity of T_c, cannot be simply modelized by the sum of a single damped harmonic oscillator and of a central peak. This point will be further discussed in the next section.

Finally, a strongly temperature dependent anomalous broadening of the [87]Rb NMR lines at temperatures as high as 200 K in $Rb_{0.3}MoO_3$ has been observed [85]. This broadening is interpreted as being due to an inhomogeneous and static (on the NMR time scale of 10^{-3} s) distribution of CDW, more likely $2k_F$ Friedel oscillations induced in the vicinity of impurities or defects. These features may contribute to the central peak observed in neutron scattering experiments.

3.7.2. Phase and Amplitude Excitations of the Incommensurate Structure $(T < T_c)$

Below T_c the modes corresponding to the excitation of the amplitude and the phase of the incommensurate structural modulation have been observed by neutron scattering [67, 81, 90, 91].

Figure 22 shows scans through a phonon branch which exhibits, within experimental uncertainties, a linear dispersion starting from a satellite reflection [90]. This branch very likely corresponds to the phase mode. Its slope is very anisotropic, leading to the following phason velocities at 175 K [81, 91]:

$$V_\phi = 3.3 \pm 0.5 \times 10^3 \text{ m/s} \quad \text{for } \mathbf{q} \| b^*$$

$$V_\phi \approx 1.8 \times 10^3 \text{ m/s} \quad \text{for } \mathbf{q} \| 2\mathbf{a}^* + \mathbf{c}^*.$$

The longitudinal velocity is twice as high as that of the steepest acoustic branch (L A_1 longitudinal branch for $\mathbf{q} \| \mathbf{b}^*$ — see Figure 12). As expected by the relation (45) V_ϕ is equal to the square root of the curvature of the Kohn anomaly measured just above $T_c (\sqrt{\Lambda_\|} \approx 3.1 \times 10^3$ m/s — see Section 3.7.1.). The longitudinal slope seems to increase sizeably when the temperature decreases (this increase may be due to the reduction of the screening effects of long range Coulomb forces by the reduction of electron—hole carriers thermally created through the Peierls gap — see Section 2.2.3.). The transverse velocity seems to be nearly independent of the temperature, as expected.

Between 130 K and 175 K, a high resolution neutron study of the phase mode in the vicinity of a satellite reflection ($\delta q_\| \sim 0$) has been performed [91]. A fit of this mode with a damped harmonic oscillator response function (Equation 16) shows its overdamped character ($\Gamma_p = 0.8 \pm 0.2$ THz). The quasi harmonic frequency obtained from the fit, $\omega_p = 0.2 \pm 0.1$ THz, is very small with respect to Γ. Its damping constant Γ_p is comparable to that found in the same temperature range for the amplitude mode (see below and Figure 23). Far infrared reflectance measurements have also been performed at low temperature in the blue bronze

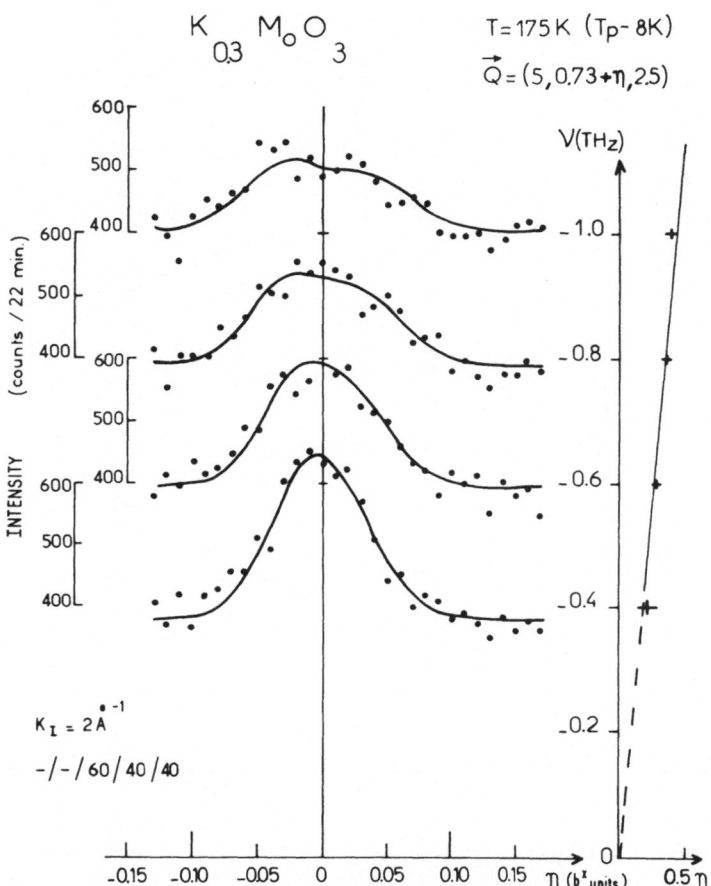

Fig. 22. Constant energy scans (left side) and dispersion (right side) of the phase mode branch of $K_{0.3}MoO_3$ just below $T_c(T = 175$ K) (Adapted from Escribe-Filippini *et al.* [90]). The slope gives a phason velocity $V_\phi = 3.3 \pm 0.5 \times 10^3$ m/s in the chain direction.

[92, 93]. They reveal several infrared active responses at $\omega_0 = 85$ GHz [93] and $\omega_0 = 0.45$ [92]/0.66 [93] THz which have been attributed to the oscillation of the pinned (transverse) CDW phase mode. The characteristics of the 1st infrared response ($\omega_0 \sim 0.1$ THz and $\Gamma_0 \sim 0.5$ THz), measured at 2 K, compare favourably (within a factor two) with that found in the neutron study of the phase mode response.

First observed by Raman scattering [94], the amplitude mode was also measured by inelastic neutron scattering [67, 81, 91]. The analysis of this mode in term of a damped oscillator gives the same characteristics for the neutron and Raman data, as shown by the Figure 23. It is interesting to remark that its quasi harmonic frequency ω_A does not soften to zero at T_c, as predicted by the mean field

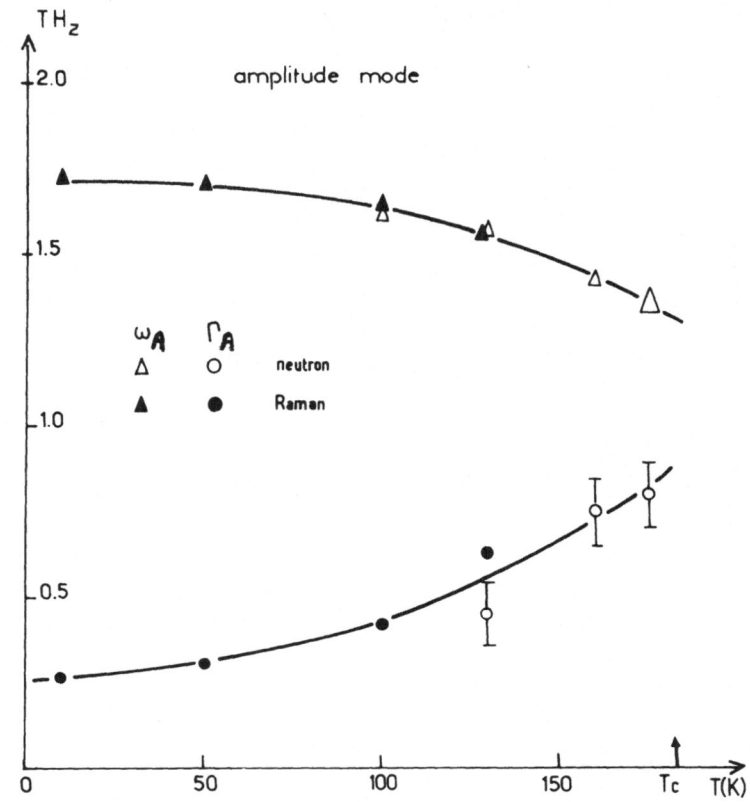

Fig. 23. Quasi harmonic frequency (ω_A) and damping frequency (Γ_A), defined by (16), for the amplitude mode of $K_{0.3}MoO_3$. The full symbols correspond to the Raman data (Travaglini *et al.* [94]) and the empty ones to the neutron data (Pouget *et al* [81]).

treatment (see Figure 4a). A fit of its temperature dependence according to Equation (19) gives a T_c^{MF} around 300 K.

The amplitude mode is clearly observed at T_c and below. It seems to emerge around T_c from a structuration at finite frequency of the high temperature overdamped soft mode response. This interpretation qualitatively agrees with the result of numerical simulations of $S(q, \omega)$ for a Peierls chain [61]. In the case of the blue bronze, the pertinent effect of the 3D coupling below T^x, very close to T_c, may help in the decoupling between the fluctuations of the phase and of the amplitude of the order parameter.

Knowing the Peierls gap $2\Delta_0$ and the frequency of the amplitude mode at 0 K, it is possible to obtain from (48)″ the CDW mass enhancement. With $\omega_A^0 = 1.7$ THz [94] (see Figure 23) and $2\Delta_0 = 0.1–0.15$ eV [95, 79], one gets the ratio $m^*/m_e \sim 200–450$, which is somewhat lower than the ones 900 [92]–10^4 [93] deduced from far infrared measurements. From the ratio of longitudinal phason velocity at 175 K (3.3×10^3 m/s) to the Fermi velocity (2×10^5 m/s — see

Section 3.8.1.) it is possible from (48)′ to estimate the CDW mass enhancement m^*/m_e. The value of 3.6×10^3 found is in reasonable agreement with that obtained from far infrared measurements, but is one order of magnitude larger than that previously deduced from (48)″.

3.8. MICROSCOPIC PARAMETERS

3.8.1. *In-Chain Parameters*

With the X-ray and neutron data of Sections 3.4. and 3.7. it is possible to extract the microscopic parameters of the electron—phonon coupled entity driving the Peierls instability of the blue bronzes [81]. With the notation already introduced, one has:

(i) from (45), for the frequency of the amplitude mode at $T = 0$ K:

$$\omega_A^0(2k_F) = \sqrt{\bar{\lambda}}\, \Omega_{(2k_F)} = 1.7 \text{ THz},$$

(ii) from (46), for the longitudinal phase mode velocity at 175 K ($\approx T_c$):

$$V_\phi = \omega_A^0(2k_F)\, \xi_{0\parallel}(T_c) = 3.3 \pm 0.5 \; 10^3 \text{ m/s},$$

(iii) from (46), for the frequency at $2k_F$ of the Kohn anomaly above T_c:

$$\omega^2(2k_F) = \Omega_{(2k_F)}^2[1 - \lambda\, \chi_e(2k_F)],$$

(iv) from (A.9), for the in-chain CDW correlation length above T_c:

$$\xi_\parallel^{-2} = \xi_{0\parallel}^{-2}\, \frac{\chi_e(2k_F)}{N(E_F)}\, [1 - \lambda\, \chi_e(2k_F)].$$

In the last two expressions, the electron—phonon coupling is treated in the RPA approximation. Although the validity of this approximation can be questioned in the blue bronzes, because of the presence of a sizeable electron—phonon coupling constant (see below), we shall only use these expressions with the room temperature data where the corrections due to the electron—phonon coupling are the smallest. These expressions are also mean-field-like. This behaviour is still acceptable in the temperature range near T_c^{MF} where the fluctuations of the amplitude of the order parameter dominate.

At room temperature ($\approx T_c^{MF}$):

— the frequency of the Kohn anomaly is: $\omega(2k_F) = 1.2 \pm 0.05$ THz,
— the in-chain CDW correlation length measured on $\chi_\rho(q)$ is: $\xi_\parallel = 25 \pm 3$ Å.

From (i) and (ii) one deduces $\xi_{0\parallel}(T_c) \approx 19 \pm 3$ Å. As, according to (23), $\xi_{0\parallel}$ behaves in temperature as T^{-1}, one gets at room temperature:

$$\xi_{0\parallel}(295 \text{ K}) = 12 \pm 2 \text{ Å}.$$

With the expression (i), (iii) and (iv) it is easy to show that:

$$\tilde{\lambda} \frac{\chi_e(2k_F)}{N(E_F)} = \frac{\omega_A^{02}}{\omega^2(2k_F)} \frac{\xi_{0\parallel}^2}{\xi_\parallel^2}. \tag{51}$$

At room temperature, (51) gives:

$$\tilde{\lambda} \frac{\chi_e(2k_F)}{N(E_F)} = 0.46 \begin{pmatrix} +0.42 \\ -0.23 \end{pmatrix}, \tag{52}$$

and (iii) leads to:

$$\Omega(2k_F) = 1.65 \begin{pmatrix} -0.2 \\ +1.7 \end{pmatrix} \text{THz}.$$

1.65 THz is the frequency, extrapolated to the wave vector $2k_F$, of the phonon branch which bears the Kohn anomaly (see Figure 12). However, one should expect at this wave vector a bare phonon mode of slightly higher frequency because the phonon branch is already screened by the non singular part of $\chi_e(q)$. More precisely such screening should be at $q = 2k_F$ such that:

$$1.65 \text{ THz} \simeq \Omega(2k_F)(1 - \tilde{\lambda}/2).$$

A reasonable compromise between all the data is to take $\Omega(2k_F) = 2.3$ THz and $\tilde{\lambda} = 0.55$. (Within uncertainties, these values agree with that of ref. [81] obtained from a more restricted set of data). Although the uncertainty on $\tilde{\lambda}$ is quite large $(+0.5/-0.3)$, its average value indicates that the blue bronzes experience a sizeable electron—phonon coupling. A similar value of $\tilde{\lambda}$ has been estimated from NbSe$_3$ [17].

With the expression (23) it is easy to calculate the Fermi velocity v_F from the experimental value: $\xi_{0\parallel} = 19$ Å at 175 K. The Fermi velocity thus obtained, 2×10^5 m/s, is identical with that deduced in Section 3.3. from the band calculation of Whangbo and Schneemeyer [11]. With the two conduction bands, the density of states at the Fermi level (including the two spin directions) is thus:

$$N(E_F) = \frac{4d_\parallel}{\pi \hbar v_F} \simeq 7.7 \text{ states/eV cluster}.$$

As a cluster contains 10 formula units, this value is consistent with that deduced from the Pauli paramagnetic susceptibility: 0.75 state/eV molecule [95, 96].

Knowing $N(E_F)$, it is easy to obtain the electron—phonon coupling constant g_{2k_F}:

$$\frac{g_{2k_F}^2}{N} = \tilde{\lambda} \frac{\pi \hbar^2 v_F \Omega_{(2k_F)}}{4d_\parallel}.$$

With $d_\parallel (\equiv b) = 7.65$ Å, the intercluster distance, v_F, $\tilde{\lambda}$ and $\Omega_{(2k_F)}$, previously determined, one obtains:

$$\frac{g_{2k_F}}{\sqrt{N}} = 26 \begin{pmatrix} +5 \\ -3 \end{pmatrix} \text{meV}.$$

By the relationship (10), g_{2k_F} links the (half) Peierls gap, Δ_{2k_F}, to the amplitude of the unstable normal mode $|A_{2k_F}|$, which itself is related through the expression (11) at the amplitude of the atomic displacements $\{u_j\}$. If in (11) one assumes that only the 4 Mo(2) and 4 Mo(3) atoms of each cluster are displaced of the same amplitude u_{2k_F}(Mo), one gets with $|e_{Mo}| \approx 1/\sqrt{8}$:

$$u_{2k_F}(\text{Mo}) = \frac{\Delta_{2k_F}}{g_{2k_F}} \sqrt{\frac{\hbar N}{4 m_{\text{Mo}} \Omega_{(2k_F)}}}. \tag{53}$$

At 0 K, $\Delta_{2k_F} \equiv \Delta_0$ ranges between 50 meV (magnetic susceptibility measurements) [95] and 75 meV (optical measurements) [79]. Taking for Δ_0 the average between these two determinations, one gets from (53) a displacement of the Mo atoms, u_{2k_F}(Mo), of about 7×10^{-2} Å at $T = 0$ K; a value which compares favorably with that obtained in the structural refinements: $5-6 \times 10^{-2}$ Å [65, 67, 87].

From the Peierls gap at 0 K, $2\Delta_0$, it is easy to deduce the mean field Peierls temperature T_c^{MF} using the well known BCS relationship:

$$3.5 \, k_B \, T_c^{MF} \approx 2\Delta_0.$$

With the above quoted values of Δ_0, one gets $T_c^{MF} \sim 320-500$ K. The lower limit agrees with the mean field Peierls critical temperature deduced from the temperature dependence of the amplitude mode (see Section 3.7.2.). Typically, the true transition temperature of the blue bronze occurs at about $T_c^{MF}/2$.

From (51) it is possible, using the above determined quantities, to deduce $\chi_e(2k_F)$. If one assumes that there is no enhancement of $\chi_e(2k_F)$ due to the electron–electron interactions, and if one takes for $\chi_e^0(2k_F)$ the logarithmic temperature dependence given by (3), one gets a cutoff energy $E_0 \sim 0.1$ eV, comparable to $t'_\parallel$. If one uses the relationship (4), this gives an effective critical temperature:

$$T_{\text{eff}} \approx E_0 \, e^{-1/\tilde{\lambda}}$$

which amounts to about 180 K (with $\tilde{\lambda} \approx 0.55$), a temperature comparable to the T_c of the blue bronzes, but lower than the estimated 1D mean field temperature T_c^{MF}. This means probably that the cutoff frequency of 0.1 eV has been already renormalized by the effect of fluctuations; probably the amplitude fluctuations (see Section 2.2.1.) corrected eventually by the 2D coupling already effective at room temperature (see Section 2.2.2.b, and especially Figure 8). With the constants $\tilde{\lambda}$ and E_0 previously determined, the temperature dependence of the CDW susceptibility (Figure 17) can be roughly accounted for by the mean field expression A(5) (with $U_\perp = 0$):

$$\chi_\rho(T) = \frac{\chi_e^0(T)}{1 - \lambda \, \chi_e^0(T)} = \frac{N(E_F)}{\tilde{\lambda}} \frac{\log(E_0/k_B T)}{\log(T/T_c)}, \tag{54}$$

at high temperature (i.e. for $T \geqslant 200$ K). Significant deviations from (54) are expected below T^x, because the coupling in the 3rd direction and the 3D criticality of the fluctuations have to be included explicitly in $\chi_\rho(T)$.

3.8.2. *Transverse Coupling*

Let us now consider more explicitly the transverse coupling energies. These quantities can be obtained from the measurement of the correlation lengths and of their anisotropy [81] (see Section 3.4 and Figure 18). The anisotropy ratio inside the layer of MoO_6 octahedra amounts to about:

$$\frac{\xi_b}{\xi_{a^* + 2c^*}} \sim 3.7{-}4.2,$$

at room temperature [27, 69]. With relationship (44), this gives an effective transverse tunneling integral of about $|t_\perp^{\text{eff}}| \sim 21{-}24$ meV, in good agreement with the average value $\frac{1}{2}(|t_\perp^{\text{I}}| + |t_\perp^{\text{II}}|) \sim 18$ meV deduced from band calculation [11]. This tunneling energy is comparable to $k_B T_c^{\text{MF}}$.

The coupling energy between layers in the $2\mathbf{a}^* - \mathbf{c}^*$ direction:

$$W^c = U_\perp |\rho_q|^2 \cos q_\perp d_\perp = 2J_\perp |\Delta_q|^2 \cos q_\perp d_\perp,$$

can be estimated from two measurements:

(i) from the anisotropy ratio just below $T^x \sim 200$ K [27, 69]:

$$\frac{\xi_b}{\xi_{2a^* - 2c^*}} \sim 11{-}13.$$

With expressions (A.9) and (A.10) defined in Appendix A, this ratio is:

$$\frac{\xi_{0\|}}{d_\perp} \left| \frac{\chi_e(q_c)}{N(E_F)} \right. \sqrt{U_\perp N(E_F)}.$$

It leads to $U_\perp \sim 1.5$ meV, using $\xi_{0\|}(T^x) = 17$ Å, $N(E_F) \sim 7.7$ states/eV cluster and

$$\frac{\chi_e^0(q_c)}{N(E_F)} = \log \frac{E_0}{k_B T^x} \sim 1.7.$$

(ii) From the definition of the crossover temperature T^x, at which $\xi_{2a^* - c^*} = d_\perp$. With $\xi_{2a^* - c^*}$ expressed by (A.10), it gives:

$$U_\perp = \frac{1}{\chi_\rho(T^x)}.$$

With $\chi_\rho(T^x) \sim 7{-}14$ times $\chi_\rho(295$ K$)$ [27, 65] (see also Figure 17) and $\chi_\rho(295$ K$)/N(E_F) \sim 4.9$, according to the previous determinations, one gets $U_\perp \sim 2.5$ meV.

Both determinations lead to a very weak interlayer coupling energy of about $U_\perp \sim t_\perp^{\text{eff}}/10 \, (= V_\perp/2$ in ref. [81]).

4. Other Oxides

In this part we shall briefly consider first the titanium bronze $Na_{0.25}TiO_2$ which bears some ressemblance with the blue bronze. Then we shall present the structural instabilities of the Mo oxides having 2D electrical properties: the purple bronzes $A_{0.9}Mo_6O_{17}$ and the series Mo_nO_{3n-1}.

4.1. THE TITANIUM BRONZE $Na_{0.25}TiO_2$

The titanium bronze and the blue bronze belong to the same space group: $C2/m$ [97]. Figure 24 shows that the Ti bronze is composed of (a, b) layers of clusters, each cluster containing 4 TiO_6 octahedra, as the pseudocluster built on the of Mo_2 and Mo_3 octahedra of the blue bronze and shown by Figure 19b. However, in contrast to the blue bronze, the octahedra belonging to neighbouring clusters are linked via edge sharing in the a and b directions. This forms zigzag (Ti_4O_{14}) chains running along the b direction. In the other direction, c, the clusters are corners sharing, delimiting an array of channels filled by the Na^+ ions. The main structural differences with the blue bronze are summarized in Table I. The linkage shows that the Ti bronze is certainly more isotropic than the blue bronze. The physical properties of the Ti bronzes and their anisotropy are not known.

The similarity with the blue bronze resides in the stabilization of an incommensurate modulation below 430 K, with a critical wave vector $\mathbf{q}_c = (1, q_b, 0)$, where q_b is incommensurate and varies slightly in temperature: q_b tends towards $1/4$ with decreasing temperature [97]. By analogy with the blue bronze, the component q_b

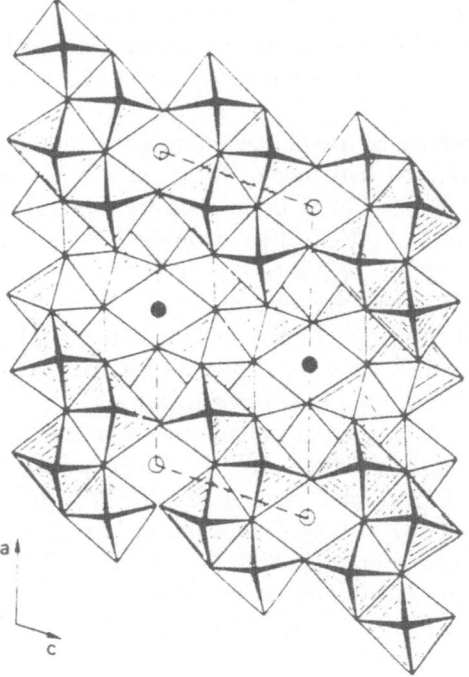

Fig. 24. Structure of $Na_{0.25}TiO_2$ projected along the b direction (from Brohan [98]). The dots represent the Na atoms. Compare this projection with that of the blue bronze shown by Figure 11.

TABLE I.

Comparison between the structural data of the blue Mo bronze $A_{0.3}MoO_3$ and of the Ti bronze $Na_{0.25}TiO_2$.

	$A_{0.3}MoO_3$	$Na_{0.25}TiO_2$
space group	$C2/m$	$C2/m$
cluster	$Mo_{10}O_{36}$	Ti_4O_{18}
	— corner sharing within the $(\mathbf{b}, \mathbf{a} + 2\mathbf{c})$ layer	— edge sharing within the $(\mathbf{b}, \mathbf{a})$ layer
	— separated along c by a layer of A^+	— corner sharing along c
T_c	183 K	430 K
$\mathbf{q}_c$	$(1, 1 - q_b, \frac{1}{2})$	$(1, q_b, 0)$
	$1 - q_b \nearrow$ when $T \searrow$	$q_b \nearrow$ when $T \searrow$
	$1 - q_b \approx \frac{1}{4} b^*$ at low T	$q_b \approx \frac{1}{4} b^*$ at low T

~1/4 could correspond to the $2k_F$ wave vector of a quasi 1D band structure if the electron provided by the Na^+ ion to each cluster of 4 octahedra is shared by two conduction bands. Such an interpretation agrees with the decrease of q_b for x increasing in $Na_{1-x}Ti_4O_8$ [118]. As for the blue bronze, the decrease of the q_b component with increasing temperatures can be accounted for by an exponential law (defined in Section 3.3.1.) but having a $\Delta E \sim 0.12$ eV, twice as great as in the blue bronze.

However, by the presence of a sizeable intercluster tunnelling ($t_{ab}^{(2)}$ in Figure 25) in the zigzag chains, the band structure of the Ti bronze certainly differs from that of the Mo bronze. Its dispersion is controlled by the ratio: $t_{ab}^{(1)}/2t_{ab}^{(2)}$. Figure 25 compares the band structure of the two bronzes. The situation depicted in this figure corresponds to $t_{ab}^{(1)} < 2t_{ab}^{(2)}$ which, according to the band calculation of Brohan [98], seems to be the case of $Na_{0.25}TiO_2$. With this ratio, the two-sheet Fermi surface is closed (an open Fermi surface can be obtained in the opposite case: $t_{ab}^{(1)} > 2t_{ab}^{(2)}$). The band structure shows also a sizeable dispersion in the 3rd direction, c^*, along which clusters share oxygen atoms.

At the present stage of the band calculation, it is not clear if the transverse components of the modulation wave vector of the Ti bronze can be understood by the nesting properties of its double sheet Fermi surface. It is not even clear from the physical properties of the Ti bronze that the structural transition results from a CDW instability. But the incommensurability of $Na_{1-x}Ti_4O_8$ ($0 < x < 0.2$) is certainly not the result of an order-disorder transformation on the occupancy of Na sites, as suggested in ref. [118], because the modulation is already present in the stoichiometric ($x = 0$) compound.

Here let us just mention that in the Ti bronze these transverse components differ from that of the blue bronze:

— within the layer, by an out-of-phase ordering between clusters: $\mathbf{q}_c \cdot \mathbf{a}/2 = \pi$.

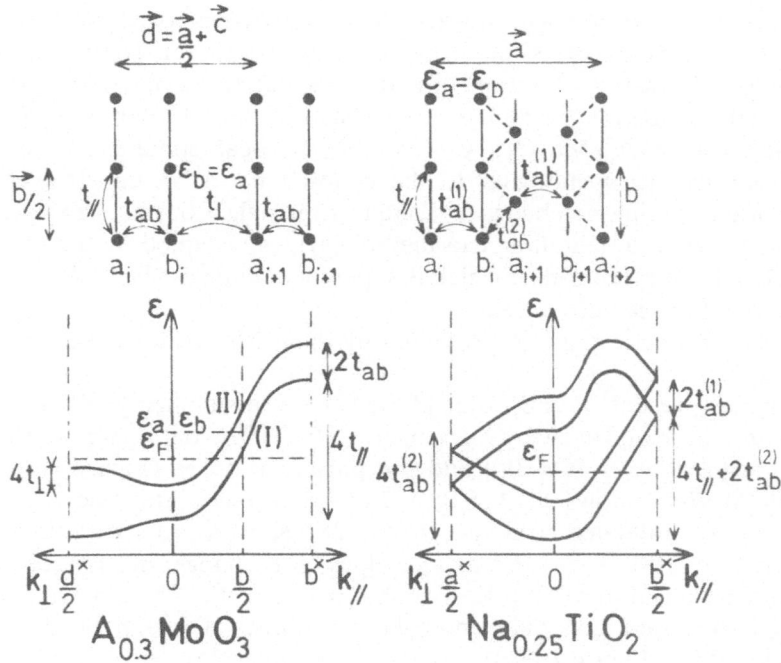

Fig. 25. Schematic representation of the array of a and b entities (built on two Mo or Ti octahedra, defined by Figure 19b) within the $(\mathbf{b}, \mathbf{a} + 2\mathbf{c})$ and $(\mathbf{b}, \mathbf{a})$ layers of $A_{0.3}MoO_3$ and $Na_{0.25}TiO_2$ respectively. The main tunneling directions are shown together with a schematic representation of the band structure of these layers.

(To be compared with an in-phase ordering, $\mathbf{q}_c(\mathbf{a}/2 + \mathbf{c}) = 0$, in the blue bronze).

— between the layers, by an in-phase ordering: $\mathbf{q}_c \cdot \mathbf{c} = 0$. (To be compared with an out-of-phase ordering, $\mathbf{q}_c \cdot \mathbf{c} = \pi$, in the blue bronze).

In addition, the higher value of $T_c \approx 430$ K in the Ti bronze, compared to 183 K in the Mo bronze, can be explained by a strengthening of the transverse coupling between clusters. Also the observation, at room temperature, of $2q_c$ satellite reflections of sizeable intensity [97] shows that the modulation is certainly not sinusoidal.

4.2. THE LAYER TYPE Mo OXIDES AND BRONZES

4.2.1. *Phase Diagrams*

The structure of these oxides, which are 2D metals, is detailed in the previous chapter. They have in common highly conducting slabs of the ReO_3 type where the Mo octahedra are linked by the corners. In the Magneli phases, $Mo_nO_{3n-1}(n \geqslant 8)$, the $(\mathbf{b}, \mathbf{c})$ slabs are connected by edge sharing octahedra, along the $(\bar{1}20)$ crystallographic shear deformation plane of the prototype ReO_3 structure. In the series with $n = 4$, the linkage between (**b**, **c**) slabs (composed with 3 layers of corner sharing octahedra) is done by a layer of corner sharing tetrahedra. Depending on the array of tetrahedra, the orientation between two successive ReO$_3$-like slabs is identical (η phase) or related by mirror symmetry (γ phase). The purple bronzes A$_{0.9}$Mo$_6$O$_{17}$(A = Na, K, Tl) bear some resemblance with Mo$_4$O$_{11}$: each slab (perpendicular to the c hexagonal axis), constructed with 4 layers of corner sharing octahedra, is linked to the neighbouring slab by a double layer of Mo tetrahedra delimiting icosahedral cavities occupied by the monovalent metals A. The Li purple bronze, which is superconducting at about $T_s \sim 2$ K, has a rather distorted linkage between slabs.

The structural phase diagram presented by these Mo oxides is shown in Table II.

These oxides presents a 2nd order phase transition stabilizing either an incommensurate modulation in Mo$_4$O$_{11}$ (Figure 26) [102] or a superstructure in Mo$_n$O$_{3n-1}$ with $n = 9$, 10 [100] and the purples bronzes (Figures 27 and 28) [103, 104]. Mo$_8$O$_{23}$ shows at 315 K a 2nd order phase transition towards an incommensurate modulated structure, then at 285 K a 1st order transition towards a superstructure [99]. However as $\mathbf{q}_{inc}$ does not change in temperature, the commensurate transition at 285 K is probably not of the lock-in type. At these transitions correspond generally anomalies in transport, magnetic and thermal properties [105] (see Figure 28 in the case of the K purple bronze).

4.2.2. Mo$_4$O$_{11}$

The critical temperature T_c below which the incommensurate modulation is stabilized, and its wave vector in the (**b***, **c***) plane is nearly the same for the η and γ phases [102], which means that the interslab linkage has no influence on the

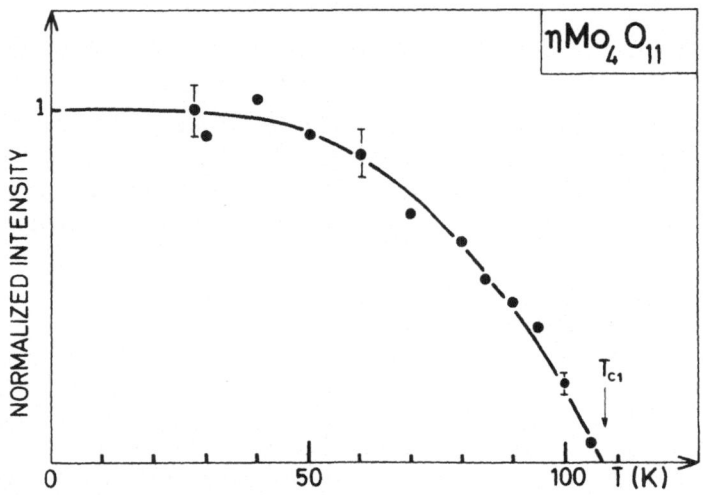

Fig. 26. Temperature dependence of the CDW satellite intensities showing the second order nature of the incommensurate transition of η Mo$_4$O$_{11}$ (from Guyot *et al.* [102]).

TABLE II

Structural phase diagrams of the quasi 2D metals Mo_nO_{3n-1} and $A_xMo_6O_{17}$. The relevant references are indicated.

Compound				
$(Mo_{0.88}W_{0.12})_{10}O_{29}$	$\mathbf{q}_1 = (0, \frac{1}{2}, 0)$	ref. [100]	$\geqslant 700$ K	
Mo_9O_{26}	$\mathbf{q}_2 = (\frac{1}{2}, \frac{1}{2}, 0)$	ref. [100]	500 K	$P2/c$
Mo_8O_{23}	$\mathbf{q}_1 = (0, \frac{1}{2}, 0)$ Pc	ref. [101] 285 K	$q_{inc} = (0.195, 0.5, \overline{0.12})$ 315 K ref. [99]	$P2/c$
Mo_4O_{11} (η phase)	30 K? $\mathbf{q}_{inc} = (0, 0.23, 0)$	ref. [102]	109 K	$P2_1a$
Mo_4O_{11} (γ phase)	$\mathbf{q}_{inc} = (?, 0.23, 0)$	ref. [102]	100 K	$Pn2_1a$
$Na_{0.9}Mo_6O_{17}$	$\mathbf{q}'_1 = (\frac{1}{2}, 0, 0)$	ref. [104]	90 K	$A2$
$K_{0.9}Mo_6O_{17}$	$\mathbf{q}'_1 = (\frac{1}{2}, 0, 0) +$	equivalent by trigonal symmetry ref. [103]	105 K	$P\bar{3}$
$Tl_1Mo_6O_{17}$	$\mathbf{q}'_1?$		~ 115 K	$P\bar{3}m1$

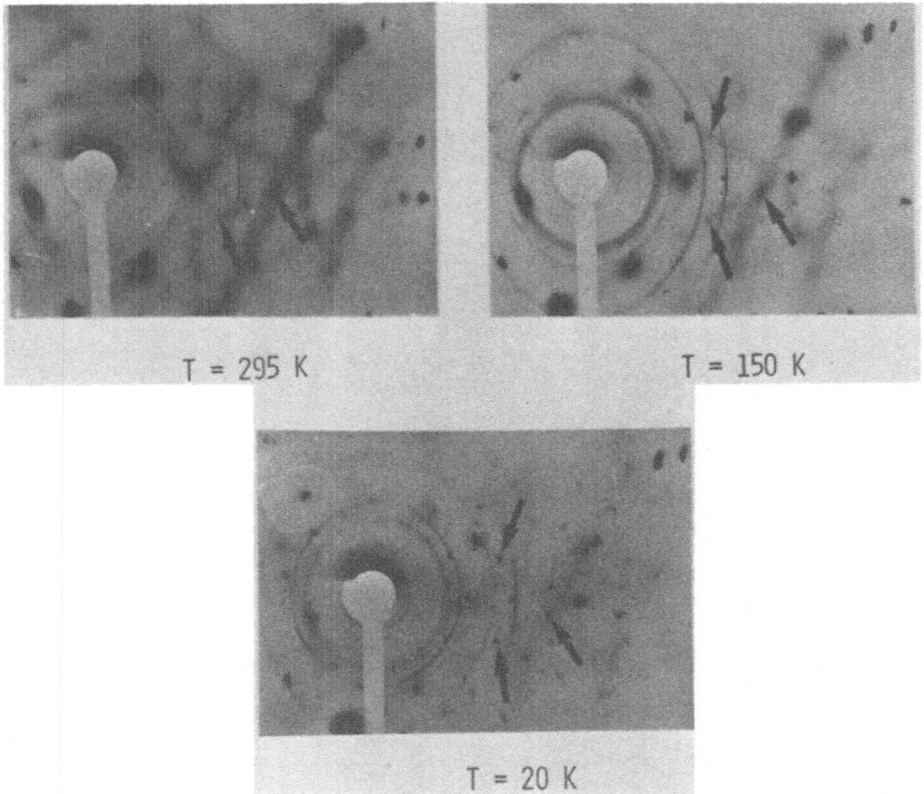

Fig. 27. Monochromatic X-ray patterns from $Na_{0.9}Mo_6O_{17}$ above T_c (295 K and 150 K) and below T_c (20 K), showing at high temperature 3 kinds of anisotropic diffuse scattering. Only the diffuse scattering shown by arrows is critical at T_c. It leads below T_c to a pseudo hexagonal array of satellite reflections at $\mathbf{q}'_1 = (1/2, 0, 0)$ and symmetry related reduced wave vectors. The incoming X-ray beam is close to the pseudo hexagonal axis. The X-ray patterns are free from $\lambda/2$ contamination ($\lambda = 1.542$ Å:CuK_a). (from S. Kagoshima and J. P. Pouget [104]).

intraplanar structural instability. Structural refinements [106] show that the greatest part of the d conduction electron density belongs to the Mo(4) atoms (a little less density is on the Mo(3) sites) situated at the inner part of the slab. The octahedra built on the Mo(4) atoms from zigzag Mo_2O_{10} chains along the b direction (the direction of higher conductivity). These chains are linked together by the Mo(3) octahedra (Figure 29a).

   Tight binding band calculations [10] show that, depending on the symmetry of the $t_{2g}\,d$ orbitals, the electronic structure in the vicinity of the Fermi level possesses both the 1D feature of the Mo(4) zig zag chains (for one kind of dt_{2g} orbitals) and the 2D feature of the Mo(3)—Mo(4) linkage (for the two others kinds of dt_{2g} orbitals). This leads, in the ($\mathbf{b}^*$, $\mathbf{c}^*$) plane, to a complex Fermi surface composed of planes (1D feature) and of 2 shifted squares (2D feature) filled by the 4 conduction electrons provided to each inner slab (Mo_4O_{16}) repeat unit. The

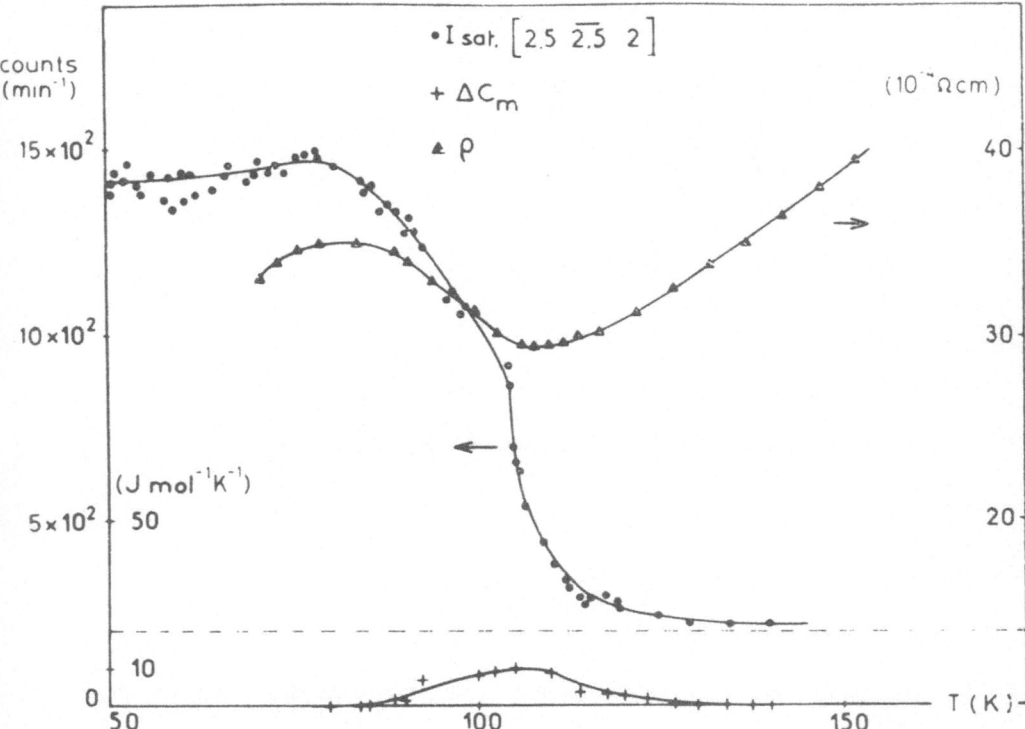

Fig. 28. Temperature dependence of the q'_i superstructure intensity, I sat. in plane resistivity ρ and excess specific heat ΔC_m in the vicinity of T_c for the purple bronze $K_{0.9}Mo_6O_{17}$ (from Schlenker *et al.* [105]).

incommensurate modulation ($q_c = 0.23b^*$) nests nearly perfectly the two shifted squares of the Fermi surface. This nesting property is certainly at the origine of the CDW instability of Mo_4O_{11} both in its η and γ phases. The nesting, which is not perfect, especially for the 1D portions of the Fermi surface, leaves in the (b^*, c^*) plane, small pockets of area less than 1% of the 2D Brillouin zone, which have been detected in η Mo_4O_{11} by magneto transport studies [102, 105].

It is interesting to remark that the tungsten bronze $P_4W_{12}O_{44}$, isostructural with γ Mo_4O_{11}, shows two resistivity anomalies at 114 K and 60 K [119]. They can be temptatively associated to CDW instabilities of a Fermi surface very similar to that of Mo_4O_{11}.

4.2.3. *Purple Bronzes and Magneli Phases*

In the purple bronzes, the slab contains one more layer of octahedra than in Mo_9O_{11}. Thus, as shown on Figure 29b, the inner part of the slab built on $Mo(2)$ octahedra, is composed of two Mo_2O_{15} chains sharing the axial oxygen atoms. The Mo_4O_{18} layer thus formed has an hexagonal structure. The electronic band structure is now more clearly 2D [12].

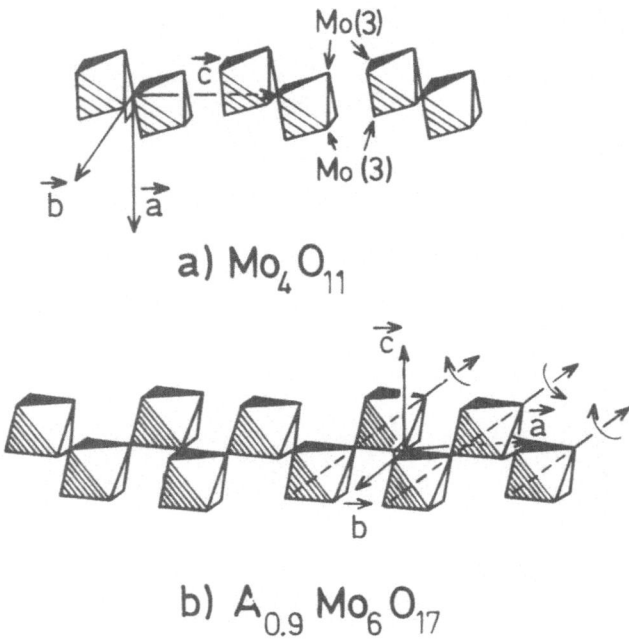

Fig. 29. Inner part of the slab of Mo_4O_{11} (a) and $A_{0.9}Mo_6O_{17}$ (b) showing respectively the zig-zag array of chains of Mo(4) octahedra running along b and the (a, b) hexagonal layer of Mo(2) octahedra. In (b) a possible rotation pattern of the Mo(2) octahedra, which can explain the doubling of the a periodicity, is indicated.

The stabilization of the $q'_1 = (1/2, 0, 0)$ modulation in the purple bronzes has been interpreted as resulting from a CDW nesting instability of the Fermi surface [12, 103]. Although the band calculation quoted above shows that the Fermi surface of the purple bronzes can admit nesting wave vectors close to q'_1, the finding of similar commensurate wave vectors of modulation in the Mo_nO_{3n-1} series, where the number of conduction electrons per Mo atom differs from that of the purple bronzes, suggests a different mechanism (see below).

The structural refinements of the q_1 superstructure of Mo_8O_{23} [101] and of the q_2 superstructure of Mo_9O_{26} [100] have shown that the distortion mainly consists in rotation of MoO_6 octahedra within the ReO_3 type cubic slabs (in the Magneli phase the slab is perpendicular to a, instead of c for the purple bronzes). This deformation strongly resembles the M3 instability, of wave vector $(1/2, 1/2, 0)$, of the ReO_3 prototype cubic structure. It consists of an in-phase rotation of chains of octahedra along a direction of oxygen sharing and in an antiphase rotation in the perpendicular directions. In the Magneli phases, Mo_nO_{3n-1}, the rotation axis is in the (a, c) plane, which leads to the doubling of the b periodicity ($q_b = 1/2$). The pattern of rotation ([100], [101]) shows that the periodicity c is retained, while the periodicity a is doubled when n is odd.

The q'_1 superstructure of the purple bronzes could be explained in a similar way by assuming a rotation of octahedra around the axis shown in Figure 29b. The pattern keeps the b periodicity (i.e. that of the Mo_2O_{10} chains), but doubles the periodicity along the direction a which makes an angle of $2\pi/3$ with b (in the prototype ReO_3 cubic structure, this corresponds to a rotation around [001] with a and b being respectively the [101] and [110] cubic directions).

It should be remarked that these rotations mainly occur in the inner part of the slab (i.e. for the corner sharing MoO_6 octahedra, where there is a maximum of the conduction electron density). The outer part of the layer composed of edge sharing octahedra, in the case of the Magneli phases, cannot rotate without deformation of these octahedra. The physical origin of such a lattice instability is not clear. In general, rotations of octahedra are not coupled at 1st order with the electron gas

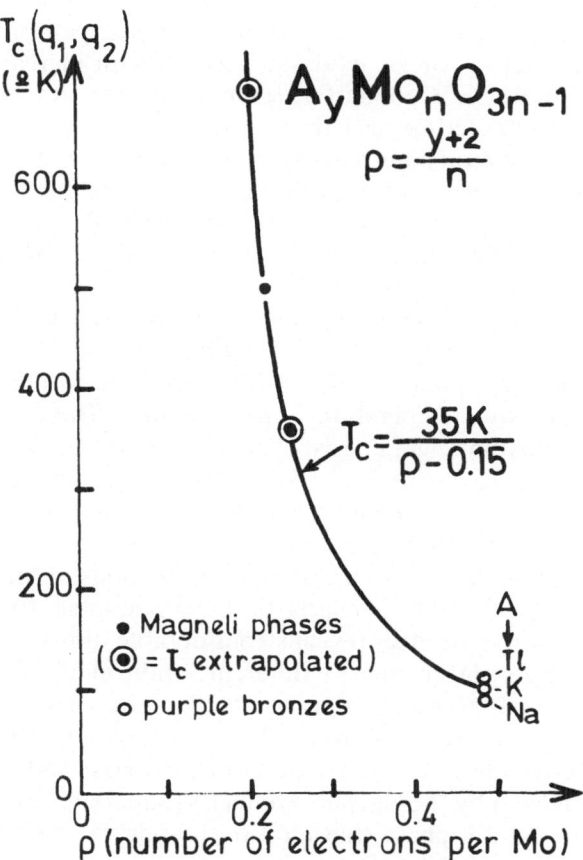

Fig. 30. 2nd order critical temperature, towards the q_1, q_2 and q'_1 superstructures, as a function of the average number of electrons per Mo(ρ) for 2D oxides and bronzes. The decrease of T_c can be fitted by the empirical law: $T_c = 35$ K$/(\rho - 0.15)$. In Mo_4O_{11}, corresponding to $\rho = 0.5$, $T_c \sim$ 100 K, but below this temperature an incommensurate CDW modulation is stabilized instead of a superstructure.

(rotations of $\pm \theta$ change equally the transfer integrals), thus at the lowest order in θ, the transfer integrals "t" varie as θ^2, which correspond to a 2nd order electron—phonon coupling. However, rotations of octahedra can change sizeably the crystal field on the ionic sites in the conducting layers [120]. It has been noticed [100] that the transition temperature, below which these superlattices are stabilized, depends strongly on the average number of conduction electron per Mo atoms, ρ. Figure 30 shows that T_c decreases strongly when ρ increases, which seems to indicate that the filling of the conduction bands destabilizes the structural transition. A similar observation has been done for the CuO_6 tilt instability of the $La_{2-x}Sr_xCuO_{4-y}$ layered oxides, with deviations from half band filling, when $x - 2y$ increases [121]. Usually in ionic materials such a structural instability results from a competition between long range Coulomb (Madelung) forces and short range forces, and generally the pressure evolution of T_c allows one to decide which force drives the structural transition [108]. If the rotational pattern of octahedra is due to long range attractive forces, its destabilization with increasing ρ could be achieved by screening of the Coulomb interactions by the intralayer conduction electrons. Such an effect could be also accounted for by a ρ dependent (covalent) part of the short range interactions. However, as shown by the unusual temperature dependence of magnetic and electrical properties below T_c [99, 105] (see Figure 28), there is also a very subtle coupling between the structural and electronic degrees of freedom in these oxides. A detailed analysis of this coupling has not been performed.

The origin of the incommensurate modulation of Mo_8O_{23} between 315 K and 285 K is not known. However it has been remarked [99] that the projection of its wave vector, in the $(\mathbf{a}^*, \mathbf{c}^*)$ plane defines a direction close to the rotation axis of the octahedra in the low temperature superstructure. This suggests that the incommensurate modulation might consist also in a rotation wave of the octahedra. Preliminary inelastic neutron scattering measurements [109] have shown that the pretransitional dynamics of the incommensurate modulation is of the displacive type: phonon softening and growth of a central peak in energy.

Only a brief X-ray study of the pretransitional fluctuations of the purple bronzes have been performed [103, 104]. Similarly to the perovskites, room temperature X-ray patterns of the purple bronzes reveal a considerable diffuse scattering related to anisotropic atomic displacements (3 different kinds of diffuse scattering are observed on the X-ray patterns shown in Figure 27 — see also Figures 2 and 3 in [103]). Only the more isotropic scattering (in the $(\mathbf{a}^*, \mathbf{b}^*)$ plane), shown by arrows on Figure 27, behaves critically when the temperature decreases and condenses below T_c in the $\mathbf{q}'_1$ (and equivalent by "hexagonal" symmetry) superlattice reflections. These data show that the structural phase transition of the purple bronzes is announced above T_c by a large regime of (2D ?) fluctuations extending over more than 200 K (2 T_c).

5. Conclusion

This review has covered the structural instabilities exhibited by some compounds belonging to the large class of the molybdenum bronzes and oxides. Among them, only the blue bronzes have been thoroughly investigated, and now the origin of the structural phase transition is relatively well understood, providing links with the CDW collective transport. In this respect the structural features of the blue bronzes might be representative of the behaviour of the class of quasi 1D inorganic materials (trichalcogenides of transition metals, Ti bronzes?), where the CDW instability is basically due to the electron—phonon coupling at 1st-order. However the blue bronzes differ from the quasi 1D organic conductors where electron—electron interactions contribute crucially at the CDW instability.

Two classes of structural instabilities have been discovered in these molybdenum compounds:

— CDW instabilities related to nesting of a Fermi surface composed of several sheets (blue bronzes $A_{0.3}MoO_3$, Mo_4O_{11}), through a 1st-order electron-phonon coupling
— rotation of MoO_6 octahedra in the case of materials containing 'ReO$_3$' type cubic slabs (Magneli phase Mo_nO_{3n-1}, $n \geqslant 8$, $A_{0.9}Mo_6O_{17}$?). The detailed mechanism of this instability is not understood, neither is its relationship with the electronic properties, but these features could be associated, through a 2nd order electron-phonon coupling, to a modulation of the crystal field in the conducting layers.

These instabilities are also found in other families of bronzes and oxides exhibiting low dimensional electronic properties.

A great amount of the structural data reported in this chapter have been obtained in collaboration with R. Currat, C. Escribe-Filippini, S. Girault, B. Hennion, S. Kagoshima, J. Marcus, R. Moret and A. H. Moudden. We have also benefited from discussions with S. Barisic, J. L. de Boer, E. Canadell, J. Friedel, H. Guyot, G. Montambaux, C. Noguera, M. Sato, C. Schlenker and M. H. Whangbo.

Appendix A

ORDER PARAMETERS, SUSCEPTIBILITIES AND CORRELATION LENGTHS

Three order parameters (η_q) can be used to describe the Peierls CDW state:

— the amplitude of the Peierls gap: Δ_q,
— the amplitude of the CDW: ρ_q,
— the amplitude of the normal phonon mode: A_q.

Δ_q and ρ_q are quantities which are thermodynamically conjugated:

$$\Delta_q = \frac{\partial F}{\partial \rho_{-q}}. \tag{A.1}$$

Δ_q is a very convenient order parameter for microscopic calculations, because $\chi_e(q)$ is an implicit function of Δ_q. A_q is directly measured by structural studies below T_c and its fluctuations above T_c are measured by neutron scattering. The fluctuations of ρ_q above T_c are directly probed by X-ray diffuse scattering.

The correlation function involving the fluctuations of each kind of amplitude defines a susceptibility which, in the classical limit, is expressed as:

$$S_\eta(q) = \langle |\eta_q|^2 \rangle = k_B T \, \chi_\eta(q), \tag{A.2}$$

where:

$$\chi_\eta^{-1}(q) = \frac{\partial^2 F}{\partial^2 |\eta_q|}.$$

We shall consider below only the susceptibilities given in the RPA approximation.

The susceptibility associated to $|\Delta_q|$ has been introduced in Sections 2.2.1. and 2.2.2. From (22, 23 and 33) one has:

$$\chi_\Delta(q) = \frac{\lambda_q}{1 - \lambda_q \chi_e(q) + 2\lambda_q J(q)} \tag{A.3}$$

($J(q)$ is the fourier transform of J_{ij}).

The susceptibility associated to $|A_q|$ is related to the inverse of the square of the frequency of the soft phonon mode. Thus, a straightforward generalization of (13) gives:

$$\chi_u(q) \sim \frac{1}{\omega^2(q)} = \frac{1}{\Omega^2(q) \, [1 - \lambda_q \chi_e(q) + 2\lambda_q J(q)]}. \tag{A.4}$$

The susceptibility associated with $|\rho_q|$ is given by (8):

$$\chi_\rho(q) = \frac{\chi_e(q)}{1 - (\lambda_q - U(q)) \, \chi_e(q)}. \tag{A.5}$$

As all the susceptibilities diverge at the same temperature, one should have:

$$U(q)\chi_e(q) = 2\lambda_q \, J(q). \tag{A.6}$$

It is also interesting to remark that the electron—phonon coupling constant is the cross susceptibility:

$$\frac{g(q)}{\sqrt{\hbar\Omega(q)} \, N} = \frac{\partial^2 F}{\partial A_q \partial \rho_{-q}}. \tag{A.7}$$

The correlation lengths are defined by the q dependent part of the susceptibility according to:

$$\xi_i^2 = -\frac{\partial \log \chi(q)}{\partial q_i^2} \bigg|_{q_c}. \tag{A.8}$$

$\chi_e^0(q)$ depends strongly on $q_{||}$. Its derivative (A.8) leads to to correlation length $\xi_{0||}$ given by (23). $\chi_e^0(q)$ can also depend on $q_\perp$. In that case, the derivative gives $\xi_{0\perp}$, which is related to $\xi_{0||}$ by the ratio (44).

The correlation lengths defined on χ_Δ are given by (25) for $\xi_{||}$, and by (36) or (44) for $\xi_\perp$, when $J_\perp$ or $t_\perp$ are considered, respectively.

The correlation lengths defined on χ_ρ, more suitable to interpret the X-ray diffuse scattering experiments, are given in the RPA approximation of $\chi_\rho(q)$ by:

$$\xi_\| = \xi_{0\|} \bigg/ \sqrt{\frac{\chi_e(q_c)}{N(E_F)} \left[1 - (\lambda_{q_c} - U(q_c)) \chi_e(q_c)\right]} \tag{A.9}$$

$\xi_\perp = \xi_\| \sqrt{2} \, t_\perp \, d_\perp / \hbar v_F$ (44), when the transverse tunneling is considered, and by:

$$\xi_\perp = d_\perp \sqrt{\frac{U_\perp \chi_e(q_c)}{1 - (\lambda_{q_c} - U(q_c)) \chi_e(q_c)}}, \tag{A.10}$$

when the Coulomb coupling is considered ($U_\perp$ is related at $J_\perp$ through (A.6), and thus (A.10) is identical to (36)).

Appendix B

ELASTIC ANOMALIES AT THE PEIERLS TRANSITION

Elastic anomalies can be simply accounted for by a very simple Landau theory, using two order parameters η_q and e describing respectively the CDW incommensurate modulation (of wave vector $\mathbf{q}$) and a uniform elastic deformation. The free energy contains 3 terms:

$$F = F_{CDW}(\eta_q) + F_{el}(e) + F_c(e, \eta_q) \tag{B.1}$$

where:

(i) $F_{CDW}(\eta_q) = a|\eta_q|^2 + b|\eta_q|^4 + f|n_q|^6.$ (B.2)

Such a development has already been considered with, in (22) and (35), $\eta_q = \Delta_q$ (the Peierls gap). (23) gives the microscopic expression of the 1D part of the coefficients a and b.

(ii) $F_{el}(e) = \dfrac{E}{2} e^2.$ (B.3)

The elastic constant E is proportional to the square of a sound velocity s. In an electron–phonon coupled system, the elastic vibrations are generally screened by the conduction electrons. By analogy with (13), one has for a $q \to 0$ acoustic mode coupled to first order with the electron gas:

$$s^2 = s_0^2 (1 - \lambda_0 \chi_e(0)) \text{ or}$$
$$E = E_0 (1 - \lambda_0 \chi_e(0)) \tag{B.4}$$

(iii) $F_c(e, \eta_q) = h|\eta_q|^2 e^r.$ (B.5)

At the lowest order the coupling term involves $|\eta_q|^2$ because of the translational symmetry. The power of e is $r = 1$ if the symmetry is such that the $|e|$ and $-|e|$

elastic deformations are inequivalent or $r = 2$ otherwise (the coupling term with $r = 2$ is always allowed whatever the symmetry). The coupling (B.5) is due to a change of the microscopic parameters like λ_{2k_F}, $N(E_F)$, $t_\parallel$, $t_\perp$, $J_\perp$ or the charge transfer with the elastic deformation (linear change if $r = 1$, quadratic change if $r = 2$). In addition, the development of a Peierls gap below T_c reduces the screening of the elastic constant below T_c, through the reduction of $\chi_e(0)$. As $\chi_e(0)$ has an even dependence on Δ_q, one gets from (B.4):

$$\tilde{E} = E_0 \left[1 - \lambda_0 \chi_e(0) + \lambda_0 \left| \frac{\partial \chi_e(0)}{\partial |\Delta_q|^2} \right| |\Delta_q|^2 \right],$$

which as the general form:

$$\tilde{E} = E + 2h|\Delta_q|^2. \tag{B.6}$$

(B.6) contributes to the coupling term with $r = 2$.

Generally the biquadratic coupling term ($r = 2$) leads to a stiffening of the elastic constant E below T_c (if $h > 0$), by a quantity which is proportional to the square of the primary order parameter η_q, as shown in Figure 31b. Such an increase of the sound velocity has been observed below the 52 K Peierls transition

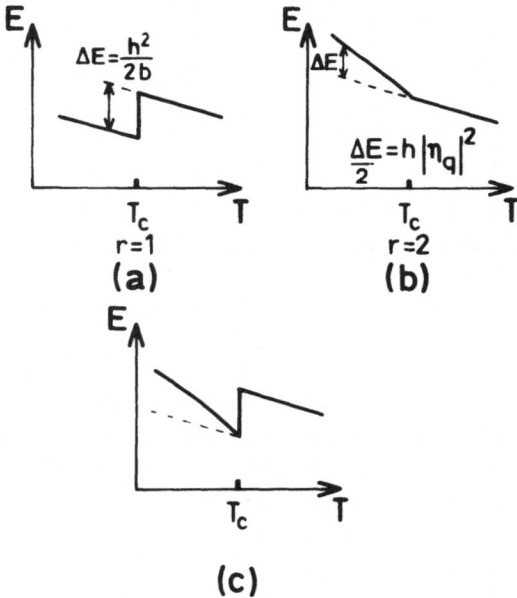

Fig. 31. Schematical temperature dependence of the elastic constant in the vicinity the Peierls transition for either a linear coupling with the square of the CDW order parameter (a — case $r = 1$) or a quadratic coupling (b — case $r = 2$). (c) represents either the superposition of cases $r = 1$ and $r = 2$ or the case $r = 1$ with the effect of a 6th order term in the CDW free energy.

of TTF—TCNQ [110], the 12 K Slater transition of $(TMTSF)_2PF_6$ [111], and the 86 K $(\frac{1}{2}, \frac{1}{2}, \frac{1}{2})$ anion ordering transition of $(TMTSF)_2FSO_3$ [112].

The case of a linear-quadratic coupling term $(r = 1)$ is more delicate. The minimization of the free energy with respect to e gives:

$$e = -\frac{h}{E}|\eta_q|^2. \tag{B.7}$$

This leads to an elastic deformation which appears either under the form of a spontaneous distortion e_s which breaks the high temperature symmetry of the Bravais lattice (case of a true secondary order parameter) or by an extra variation Δe in the temperature dependence of a lattice parameter of the high temperature Bravais lattice (case of no symmetry breaking). This last feature can be best detected by an anomaly, at T_c, in the thermal expansion coefficient ($\alpha = \partial \log e/\partial T$). e_s or Δe behaves like $|\eta_q|^2$ below T_c, according to (B.7). The direction of variation, however, depends on the sign of h.

If one ignores the coefficient f in (B.2), it is easy to show that the elastic constant, given by $\partial^2 F/\partial^2 e$, is:

$$\left.\begin{aligned} \tilde{E} &= E \quad \text{for} \quad T > T_c \\ \tilde{E} &= E - \frac{h^2}{2b} \quad \text{for} \quad T < T_c \end{aligned}\right\}. \tag{B.8}$$

There is (in the mean-field approximation) a sudden decrease of the elastic constant and thus an elastic softening of the lattice below T_c, as shown by the Figure 31a. If the coefficient f is now considered, (B.8) becomes for $T < T_c$:

$$\tilde{E} = E - \frac{h^2}{2(b + 3f|\eta_q|^2)}. \tag{B.9}$$

As $|\eta_q|^2$ increases when T decreases, the softening diminishes as schematically represented on the Figure 31c. The same behaviour can be accounted for, if f is very weak, by the superposition of the effects of (B.8) and (B.6) due to the simultaneous presence of the $r = 1$ and $r = 2$ coupling terms in the free energy.

Such a softening of the lattice is observed at the 145 K 'Peierls' transition of $NbSe_3$ [113], the 183 K Peierls transition of $K_{0.3}MoO_3$ [70] and the 139 K $(\frac{1}{2}, \frac{1}{4}, 0)$ anion ordering transition of $(TMTSF)_2PO_2F_2$ [112].

The anomaly in thermal expansion at T_c, $\Delta\alpha$, can be related to anomalies in other physical quantities (2nd derivatives of the free energy) through the Ehrenfest relations for a 2nd order transition:

$$\Delta\alpha = -\frac{\partial T_c}{\partial \sigma}\frac{\Delta c_p}{T_c e} \quad \text{or}$$

$$\Delta\alpha = +\frac{\Delta E}{E\sigma \dfrac{\partial T_c}{\partial \sigma}}, \tag{B.10}$$

where σ is the stress conjugated to e ($\sigma = Ee$ obtained from $\partial(F_{el} + F_{ext})/\partial e = 0$ with $F_{ext} = -\sigma e$). In the case $r = 2$, where $\Delta E = 0$ at T_c (see Figure 31b), no anomaly in the thermal dependence of the lattice parameters is expected. This is the case of TTF—TCNQ [114] and (TMTSF)$_2$PF$_6$ [115]. By contrast, in the blue bronzes, which correspond to the case $r = 1$ and shows a jump ΔE of elastic constant at T_c, one expects, from (B.10), to detect by dilatometry an anomaly in the thermal contraction of the $\mathbf{a} + 2\mathbf{c}$ direction below T_c; anomaly similar to the one reported below the 24 K $(0, \frac{1}{2}, 0)$ anion ordering transition of (TMTSF)$_2$ClO$_4$ [115].

Notes

[1] More generally the electron—phonon coupling constant g is a function of the wave vectors $\mathbf{k}$ and $\mathbf{k} + \mathbf{q}$ [19], involved in the electron—hole excitations of the denominator of (1). It has been argued [20] that the $\mathbf{k}$ dependence of the electron—phonon matrix element must be explicitly considered in the weak nesting case, i.e. when $\chi_e^0(\mathbf{q})$ does not show a pronounced peak at $\mathbf{q}_c$.

[2] In the case of 1D fluctuations, the ω dependence of $S^P(\mathbf{q}, \omega)$ or $\chi^P(\mathbf{q}, \omega)$ is more complex than that of a damped harmonic oscillator [38].

[3] The case of a 1D electron gas with two conduction bands (of Fermi velocities v_F' and v_F''), crossing the Fermi level has been considered in [27]: in $\xi_{0\|}$, v_F must be replaced by $2v_F' v_F'' (v_F' + v_F'')^{-1}$, in the density of states v_F^{-1} must be replaced by $4(v_F' + v_F'')^{-1}$.

[4] The linear dispersion of the phason branch considered in this section holds for $\delta q_\| \xi_{0\|} < 1$. Deviations from linearity for larger values of $\delta q_\|$, and due to the presence of non local 'elastic forces', has been considered in the literature [58].

[5] These quantities 'integrate' the dynamical fluctuations as can be seen in the classical approximation from the relationship:

$$k_B T \chi(\mathbf{q}) = S(\mathbf{q}, t = 0) = \int S(\mathbf{q}, \omega) \, d\omega$$

[6] In this subsection it is still assumed that the interchain coupling is small ($t_\perp \ll t_\|$) so that its effect on the primarily 1D fluctuations can be still treated in the R.P.A. approximation, as in subsection (a). If this is not the case, the 2D fluctuation pattern (i.e. inside the layer considered in part (b)) must be calculated explicitly. For a single q incommensurate modulation (xy like order parameter) fluctuations lead to the formation of vortex lines piercing the layer and to a singular temperature dependence of χ and ξ near the Kosterlitz-Thouless temperature, T_{KT}, defined by the intraplanar interactions. The 3D ordering, at T_c, is due to the interplanar coupling in the 3rd direction, $U_\perp$. If the interplanar coupling does not modify the 2D vortex pattern, T_c occur slightly above T_{KT}. In the R.P.A. approximation of the interlayer coupling, T_c is given by:

$$T_c^0 = T_{KT} \left(1 + \pi^2 \left/ \log^2 \left| \frac{k_B T_{KT}}{U_\perp} \right| \right. \right).$$

In the case where interplanar fluctuations lead to the formation of vortex loops parallel to the layers, T_c is found to be intermediate between T_c^0 and T_{KT} [116].

[7] A more effective contribution to Γ is the strong anharmonicity of the Peierls chain below T_c^{MF}, due to the $|\Delta_q|^4$ term in the free energy (22) [38, 61].

References

1. J. A. Wilson, F. J. Di Salvo, and S. Mahajan, *Adv. in Physics* **24**, 117 (1975).

2. P. M. Williams, in *Physics and Chemistry of Materials with Layered Structure* Vol 2, Edited by F. Levy (Dordrecht, Reidel) p. 51 (1976).

3. R. Comès and G. Shirane, in *Highly Conducting One Dimensional Solids* Edited by J. T. Devreese, R. P. Evrard and V. E. Van Doren (New York, Plenum) p. 17 (1979).

4. S. Kagoshima, in *Extended Linear Chain Compounds II*, Edited by J. S. Miller, (New York Plenum) p. 303 (1982).

5. P. Monceau, in *Electronic Properties of Inorganic Quasi One Dimensional Materials II* Edited by P. Monceau (Dordrecht, Reidel) p. 139 (1985).

6. R. Moret and J. P. Pouget, in *Crystal Chemistry and Properties of Materials with Quasi One Dimensional Structure* Edited by J. Rouxel (Dordrecht, Reidel) p. 87 (1986).

7. J. P. Pouget, in 'Highly conducting quasi one dimensional organic crystals' Edited by E. M. Conwell (Pergamon Press) — *Semiconductors and Semimetals*, Vol. 27, p. 87 (1988).

8. See for example J. M. Ziman, *Principles of the Theory of Solids*, 2nd. Edition (Cambridge University Press), Chapter 5 (1972).

9. See for example J. Friedel, in *Electron—Phonon Interactions and Phase Transitions* Edited by T. Riste (New York, Plenum) (1977).

10. E. Canadell, M. H. Whangbo, C. Schlenker, and C. Escribe-Filippini, *Inorg. Chem.* **28**, 1466 (1989).

11. M. H. Whangbo and L. F. Schneemeyer, *Inorg. Chem.* **25**, 2424 (1986).

12. M. H. Whangbo, E. Canadell, and C. Schlenker, *JACS* **109**, 6308 (1987).

13. S. K. Chan and V. Heine, *J. Phys. F.: Metal Phys.* **3**, 795 (1973).

14. V. J. Emery,*Highly Conducting One Dimensional Solids*, Edited by J. T. Devreese, R. P. Evrard and V. E. Van Doren, (New York Plenum) p. 247 (1979).

15. J. Solyom, *Adv. Physics* **28**, 101 (1979).

16. S. Barišić and A. Bjelis, in *Theoretical Aspects of Band Structures and Electronic Properties of Pseudo One Dimensional Solids*, Edited by H. Kanimura (Dordrecht, Reidel) p. 49 (1985).

17. S. Barišić, in *Electronic Properties of Inorganic Quasi One Dimensional Materials I* Edited by P. Monceau (Dordrecht Reidel) p. 1 (1985).

18. S. Barišić, in *Low Dimensional Conductors and Superconductors*, Edited by D. Jérome and L. G. Caron, NATO ASI series (New York Plenum) **B155**, p. 395 (1987).

19. S. Barisic, J. Labbé, and J. Friedel, *Phys. Rev. Lett.* **25**, 919 (1970).

20. N. J. Doran, *J. Phys. C: Solid St. Phys.* **11**, L959 (1978).

21. M. J. Rice and S. Strassler, *Solid State Comm.* **13**, 125 (1973).

22. W. L. MacMillan, *Phys. Rev.* **B16**, 643 (1977).

23. C. M. Varma and A. L. Simons, *Phys. Rev. Lett.* **51**, 138 (1983).

24. P. Y. Le Daeron and S. Aubry, *J. de Physique Colloq.* **44**, C3-1573 (1983).

25. See for example G. A. Toombs, *Phys. Rep.* **C40**, 181 (1978).

26. See for example F. J. Di Salvo, in *Electron Phonon Interactions and Phase Transitions* Edited by T. Riste (New York Plenum) p. 107 (1977).

27. J. P. Pouget, C. Noguera, A. H. Moudden, and R. Moret, *J. de Physique* **46**, 1731 (1985) (erratum *J. de Physique* **47**, 147 (1986)).

28. B. Horowitz, H. Gutfreund, and M. Weger, *Phys. Rev.* **B16**, 1468 (1975).

29. P. A. Lee, T. M. Rice, and P. W. Anderson, *Sol. State Comm.* **14**, 703 (1974).

30. K. Saub, S. Barišić, and J. Friedel, *Phys. Lett.* **A56**, 302 (1976).

31. See for example H. J. Schulz in *Low Dimensional Conductors and Superconductors* Edited by D. Jérome and L. G. Caron (New York Plenum) NATO ASI **B155**, p. 95 (1987).

32. W. Kohn, *Phys. Rev. Lett.* **2**, 393 (1959).

33. A. M. Afanasev and Yu. Kagan, *Soviet. Phys. JETP* **16**, 1030 (1963).

34. B. Renker, L. Pintschovius, W. Glaser, H. Rietschel, R. Comès, L. Liebert, and W. Drexel, *Phys. Rev. Lett.* **32**, 836 (1974).

35. D. E. Moncton, J. D. Axe, and F. J. Di Salvo, *Phys. Rev.* **B16**, 801 (1977).

36. See for example R. A. Cowley, *Adv. Physics* **29**, 1 (1980).

37. See for example L. J. Sham, in *Highly Conducting One Dimensional Solids* Edited by J. T. Devreese, R. P. Evrard and V. E. Van Doren (New York Plenum) p. 227 (1979).

38. See for example W. Dieterich, *Adv. Physics* **25**, 615 (1976).

39. C. Berthier, D. Jérome, and P. Molinie, *J. Phys. C.: Solid State Phys.* **11**, 797 (1978).
40. A. J. Berlinsky, *Rep. Prog. Phys.* **42**, 1243 (1979).
41. P. A. Lee and H. Fukuyama, *Phys. Rev.* **B17**, 542 (1978).
42. G. Shirane, S. H. Shapiro, R. Comès, A. F. Garito, and A. J. Heeger, *Phys. Rev.* **B14**, 2325 (1976).
43. J. P. Pouget, in *Low Dimensional Conductors and Superconductors* edited by D. Jérome and L. G. Caron (New York Plenum), NATO ASI **B155**, p. 17 (1987).
44. K. Carneiro, G. Shirane, S. A. Werner, and S. Keiser, *Phys. Rev.* **B13**, 4258 (1976).
45. L. K. Hansen and K. Carneiro, *Physica* **143B**, 216 (1986).
46. C. Noguera, *J. Phys. C: Solid State Phys.* **19**, 2161 (1986).
47. See for example L. D. Landau and E. M. Lifshitz, *Statistical Physics* (Pergamon Press, London) p. 482 (1959).
48. D. Allender, J. W. Bray, and J. Bardeen, *Phys. Rev.* **B9**, 119 (1974).
49. D. J. Scalapino, M. Sears, and R. A. Ferrell, *Phys. Rev.* **B6**, 3409 (1972).
50. D. Jerome and H. J. Schulz, *Adv. in Physics* **31**, 299 (1982).
51. D. J. Scalapino, Y. Imry, and P. Pincus, *Phys. Rev.* **B11**, 2042 (1975).
52. S. Barišić and K. Uzelac, *J. de Physique* **36**, 1269 (1975).
53. G. Montambaux, Thesis, Université Paris-Sud (unpublished, 1987).
54. See for example G. Montambaux, in *Low Dimensional Conductors and Superconductors*, Edited by D. Jérome and L. G. Caron, NATO ASI **B155** (Plenum Press) p. 233 (1987).
55. V. J. Emery, *J. de Physique Colloq.* **44**, C3-977 (1983).
56. S. Jafarey, *Phys. Rev.* **B16**, 2584 (1977).
57. M. J. Rice and S. Strassler, *Solid State Comm.* **13**, 1931 (1973).
58. L. K. Hansen and D. Baeriswyl, *J. Phys.* **C19**, 5615 (1986).
59. Y. Nakane and S. Takada, *J. Phys. Soc. Japan* **54**, 977 (1985).
60. P. A. Lee, T. M. Rice, and P. W. Anderson, *Phys. Rev. Lett.* **31**, 462 (1973).
61. E. Tutiš, Thesis, Zagreb (unpublished, 1989).
62. J. Graham and A. D. Wadsley, *Acta Cryst.* **20**, 93 (1966).
63. M. Ghedira, J. Chenavas, and M. Marezio, *J. Solid State Chem.* **57**, 300 (1983).
64. J. P. Pouget, S. Kagoshima, C. Schlenker, and J. Marcus, *J. de Physique Lettres* **44**, L113 (1983). In this paper one should read p. L-117, 2 lines before the end, $\mathbf{q} = [0, \pm q_b, \frac{1}{2}]$ instead of $\mathbf{q} = [0, \pm(1 - q_b), \frac{1}{2}]$.
65. S. Girault, Thesis, Université Paris-Sud (unpublished, 1987).
66. M. Sato, H. Fujishita, and S. Hoshino, *J. Phys. C: Solid State Phys.* **16**, L877 (1983).
67. M. Sato, H. Fujishita, S. Sato, and S. Hoshino, *J. Phys. C: Solid State Phys.* **18**, 2603 (1985).
68. J. P. Pouget, C. Escribe-Filippini, B. Hennion, R. Currat, A. H. Moudden, R. Moret, J. Marcus, and C. Schlenker, *Mol. Cryst. Liq. Cryst.* **121**, 111 (1985).
69. S. Girault, A. H. Moudden, and J. P. Pouget, *Phys. Rev.* **B39**, 4430 (1989).
70. L. C. Bourne and A. Zettl, *Solid State Comm.* **60**, 789 (1986).
71. T. Tamegai, K. Tsutsumi, S. Kagoshima, Y. Kanai, M. Tani, M. Tomozawa, M. Sato, K. Tsuji, J. Harada, M. Sakata, and T. Nakajima, *Solid State Comm.* **51**, 585 (1984).
72. C. Escribe-Filippini, J. P. Pouget, R. Currat, B. Hennion, and J. Marcus, *Lecture Notes in Physics* **217** (Springer Verlag) p. 71 (1985).
73. R. M. Fleming, L. F. Schneemeyer, and D. E. Moncton, *Phys. Rev.* **B31**, 899 (1985).
74. J. Y. Veuillen, R. C. Cinti, and Nemeh E. Al Khoury, *Europhys. Lett.* **3**, 355 (1987).
75. S. Girault, A. H. Moudden, J. P. Pouget, and J. M. Godard, *Phys. Rev.* **B38**, 7980 (1988).
76. T. Tamegai, K. Tsutsumi, and S. Kagoshima, *Synthetic Metal* **19**, 923 (1987).
77. S. Girault, A. H. Moudden, G. Collin, J. P. Pouget, and R. Comès, *Solid State Comm.* **63**, 17 (1987).
78. P. Quemerais, Thesis, Université de Nantes (unpublished, 1987).
79. G. Travaglini, P. Wachter, J. Marcus, and C. Schlenker, *Solid State Comm.* **37**, 599 (1981).
80. E. Bervas, Thesis Grenoble (unpublished, 1984).
81. J. P. Pouget, S. Girault, A. H. Moudden, B. Hennion, C. Escribe-Filippini, and M. Sato, *Physica Scripta* **T25**, 58 (1989).
82. R. M. Fleming, R. G. Dunn, and L. F. Schneemeyer, *Phys. Rev.* **B31**, 4099 (1985).
83. C. Berthier and P. Segransan, in *Low Dimensional Conductors and Superconductors*, Edited

by D. Jérome and L. G. Caron, NATO ASI series B, Vol. 135 (Plenum Press) p. 455 (1987).

84. M. Mutka, F. Rullier-Albenque, and S. Bouffard, *J. de Physique* **48**, 425 (1987).
85. P. Butaud, P. Segransan, C. Berthier, J. Dumas, and C. Schlenker, *Phys. Rev. Lett.* **55**, 253 (1985).
86. K. Nomura, K. Kume, and M. Sato, *Solid State Comm.* **57**, 611 (1986).
87. W. J. Schutte and J. L. de Boer, *Acta Cryst. B* (in press, 1988).
88. T. Tamegai, K. Tsutsumi, S. Kagoshima, Y. Kanai, H. Tomozawa, M. Tani, Y. Nogami, and M. Sato, *Solid State Comm.* **56**, 13 (1985).
89. L. Mihaly, K. B. Lee, and P. W. Stephens, *Phys. Rev.* **B36**, 1793 (1987).
90. C. Escribe-Filippini, J. P. Pouget, B. Hennion, and M. Sato, *Synthetic Metals* **19**, 931 (1987).
91. B. Hennion, J. P. Pouget, C. Escribe-Filippini, and M. Sato, in preparation.
92. G. Travaglini and P. Wachter, *Phys. Rev.* **B30**, 1971 (1984).
93. H. K. Ng., G. A. Thomas, and L. F. Schneemeyer, *Phys. Rev.* **B33**, 8755 (1986).
94. G. Travaglini, I. Morke, and P. Wachter, *Solid State Comm.* **45**, 289 (1983).
95. D. C. Johnston, *Phys. Rev. Lett.* **52**, 2049 (1984).
96. R. Brussetti, B. K. Chakraverty, J. Devenyi, J. Dumas, J. Marcus, and C. Schlenker, in *Recent Developments in Condensed Matter Physics*, vol. 2, Edited by J. T. Devreese, L. F. Lemmens, V. F. van Doren, and J. van Royen (Plenum), p. 181 (1982).
97. L. Brohan, R. Marchand, and M. Tournoux, *J. Solid State Chemistry* **72**, 145 (1988).
98. L. Brohan, Thesis, Université de Nantes (unpublished, 1986).
99. M. Sato, H. Fujishita, S. Sato, and S. Hoshino, *J. Phys. C: Solid State Phys.* **19**, 3059 (1986). According to Figure 3 of this paper the wave vector of the incommensurate modulation is ± (0.195, 0.5, 0.12) and not ± (0.195, 0.5, 0.12) as mentioned in the text.
100. M. Onoda, H. Fujishita, Y. Matsuda, and M. Sato, *Synthetic Metals* **19**, 947 (1987).
101. H. Fujishita, M. Sato, S. Sato, and S. Hoshino, *J. Solid State Chem.* **66**, 40 (1987).
102. H. Guyot, C. Schlenker, J. P. Pouget, R. Ayroles, and C. Roucau, *J. Phys. C: Solid State Phys.* **18**, 4427 (1985).
103. C. Escribe-Filippini, R. Almairac, R. Ayroles, C. Roucau, K. Konate, J. Marcus, and C. Schlenker, *Phil. Mag.* **B50**, 321 (1984).
104. S. Kagoshima and J. P. Pouget, (unpublished results).
105. C. Schlenker, J. Dumas, C. Escribe-Filippini, M. Guyot, J. Marcus, and G. Fourcaudot, *Phil. Mag.* **B52**, 643 (1985).
106. M. Ghedira, H. Vincent, M. Marezio, J. Marcus, and G. Fourcaudot, *J. Solid State Chem.* **56**, 66 (1985).
107. J. P. Pouget, C. Escribe-Filippini, B. Hennion, R. Currat, and J. Marcus, (unpublished data).
108. G. A. Samara, *Comments Sol. State Phys.* **8**, 13 (1977).
109. H. Fujishita, M. Sato, S. M. Shapiro, and S. Hoshino, *Physica* **143B**, 201 (1986).
110. T. Tiedje, R. R. Haering, M. H. Jericho, W. A. Roger, and A. Simpson, *Solid State Comm.* **23**, 713 (1977).
111. P. M. Chaikin, T. Tiedje, and A. N. Bloch, *Solid State Comm.* **41**, 739 (1982).
112. R. C. Lacoe, P. M. Chaikin, F. Wudl, S. D. Cox, and J. Brennan, *Mol. Cryst. Liq. Cryst.* **119**, 155 (1985).
113. J. W. Brill and N. P. Ong, *Solid State Comm.* **25**, 1075 (1978).
114. J. P. Pouget, S. M. Shapiro, G. Shirane, A. F. Garito, and A. J. Heeger, *Phys. Rev.* **B19**, 1992 (1979).
115. C. Gaonach, G. Creuzet, and A. Moradpour, *Mol. Cryst. Liq. Cryst.* **119**, 265. (1985).
116. J. Friedel, 'The high T_c superconductors: a conservative view' (1989), to be published.
117. A. Bjelis and S. Barišić, *J. Phys. C: Solid State Phys.* **19**, 5607 (1986).
118. D. Colaitis, W. Coene, S. Amelinckx, L. Brohan, and R. Marchand, *J. Solid State Chem.* **75**, 156 (1988).
119. E. Wang, M. Greenblatt, I. E. Rachidi, E. Canadell, M. M. Whangbo, and S. Vadlamannati, *Phys. Rev.* **B39**, 12969 (1989).
120. S. Barišić and I. Batistić, *Europhys. Lett.* **8**, 765 (1989).
121. R. Moret, J. P. Pouget, C. Noguera, and G. Collin, *Physica C153-155*, 968 (1988).
122. G. Soda, C. Bourbonnais, and D. Jérome, *J. Phys. Soc. Japan* **56**, 3951 (1987).
123. Y. Hasegawa, and H. Fukuyama, *J. Phys. Soc. Japan* **55**, 3978 (1986).

CHARGE DENSITY WAVE INSTABILITIES AND TRANSPORT PROPERTIES OF THE LOW DIMENSIONAL MOLYBDENUM BRONZES AND OXIDES

CLAIRE SCHLENKER, JEAN DUMAS, CLAUDE ESCRIBE-FILIPPINI
and HERVÉ GUYOT

Laboratoire d'Etudes des Propriétés Electroniques des Solides — C.N.R.S.
Associated with Université Joseph Fourier
B.P. 166—38042 GRENOBLE CEDEX, France

1. Introduction

Low dimensional conductors have been the object of extensive studies during the last fifteen years. While the quasi-two-dimensional (2D) compounds, provided by the layered transition metal dichalcogenides [1], had already been studied long ago, interest in the quasi-one-dimensional (1D) conductors has grown faster, as they could be compared to more simple theoretical models. The organic conductors, such as the $(TMTSF)_2X$ series, have been studied mostly in relation with their superconducting properties [2], while the inorganic transition metal trichalcogenides, such as $NbSe_3$ have attracted considerable interest due to nonlinear transport properties related to the electronic quasi 1D instability [3]. During the last five years, another family of transition metal compounds has become very successful: the low dimensional molybdenum bronzes and oxides now provide a rich variety of both quasi 1D and quasi 2D conductors.

The specific properties of the low-dimensional conductors are due to the instability of the 1D or 2D electron gas at a wave vector $\mathbf{q} = 2\mathbf{k}_F$ ($\mathbf{k}_F$ is the Fermi wave vector). This is related to the presence on the Fermi surface (FS) of large parallel areas separated by a single wave vector $2\mathbf{k}_F$ and usually described as the nesting properties of the FS. The electron gas instability leads, through the electron—phonon coupling to a periodic lattice distortion (PLD), known as the Peierls transition [4]. The corresponding periodic modulation of the electronic density is called the charge density wave (CDW). As the period λ of the CDW and therefore of the superlattice, is determined completely by the $2\mathbf{k}_F$ value, it may be different from a simple multiple of the lattice period a. The CDW is then incommensurate with the underlying lattice. In this case, the PLD and the CDW are not fixed absolutely and have no preferred position with respect to the average lattice: the phase of the CDW in the referential of the lattice is arbitrary. The CDW is in principle free to move and the corresponding sliding will carry a current: this was a mechanism for superconductivity proposed by Fröhlich [5] in 1954.

In real crystals, the incommensurate CDW is pinned by crystal defects, such as impurities. Depinning of the CDW occurs only above a finite threshold electric

159

C. Schlenker (Ed.), Low-Dimensional Electronic Properties of Molybdenum Bronzes and Oxides,
159—257.
© 1989 *Kluwer Academic Publishers.*

field and gives rise to an excess conductivity, the so-called Fröhlich conductivity. This was first reported by Monceau *et al.* for NbSe$_3$ more than ten years ago [6].

We will review in this article the physical properties, mostly transport, of the low dimensional molybdenum bronzes and oxides [7]. The bronzes are ternary oxides, of general formula $A_xM_yO_z$, where A is often an alkali metal. The relevant oxides belong to the Mo$_n$O$_{3n-1}$ series. These compounds are metallic conductors either because of the charge transfer from the A metal into the conduction band or because of the lack of oxygen compared to the semiconducting MoO$_3$ compound. Their crystal structure is layer-type, so that most of them are quasi 2D metals. The so-called blue bronze $A_{0.30}MoO_3$, because of the presence in the structure of infinite chains of MoO$_6$ octahedra, is the only compound of the family to be a quasi 1D metal.

The transition metal bronzes are introduced in the first chapter of this book [8]. The crystal structures and various aspects of the structural instabilities of these materials are reviewed in two other chapters [9, 10]. In the next section of the present chapter, a brief theoretical background will be given. Section 3 will be devoted to the quasi 1D properties, including CDW transport of the blue bronzes $A_{0.30}MoO_3$ and Section 4 to the quasi 2D systems, purple bronzes $A_{0.9}Mo_6O_{17}$ and Mo$_n$O$_{3n-1}$ oxides.

2. Theoretical Background

In the first part of this section we will recall some aspects of the Peierls transition and of the CDW state. The second part will briefly review some of the theoretical models accounting for the mechanisms of CDW transport. Another chapter of this volume considers the transition by breaking of analyticity of CDW systems [11], while the last one describes the CDW state and CDW transport in relation with the CDW defects [12].

2.1. THE PEIERLS TRANSITION AND THE CDW STATE

One of the parameters which governs the low dimensional electronic properties of a conductor is the transverse coupling between chains in the 1D case and between planes in the 2D one. However, in the unrealistic situation of a perfectly 1D solid, the Fermi surface reduces to two infinite parallel planes at $-\mathbf{k}_F$ and $+\mathbf{k}_F$, while in the 2D analogues, it is an infinite cylinder with an axis perpendicular to the plane of the layers. The nesting between the two planes is perfect in the 1D case and only partial between parallel portions of the FS in the 2D one (Figure 1).

The basic origin of the Peierls instability lies in the special properties of the density response function of the electron gas, $\chi(q)$. Figure 2 shows $\chi(q)$ vs q in the case of a free electron gas in the 3D case with a spherical Fermi surface, in the 2D one with a cylindrical FS and in the 1D case with a FS made of two parallel planes. In the 2D and in the 1D cases, $\chi(q)$ shows a pronounced singularity at $q =$

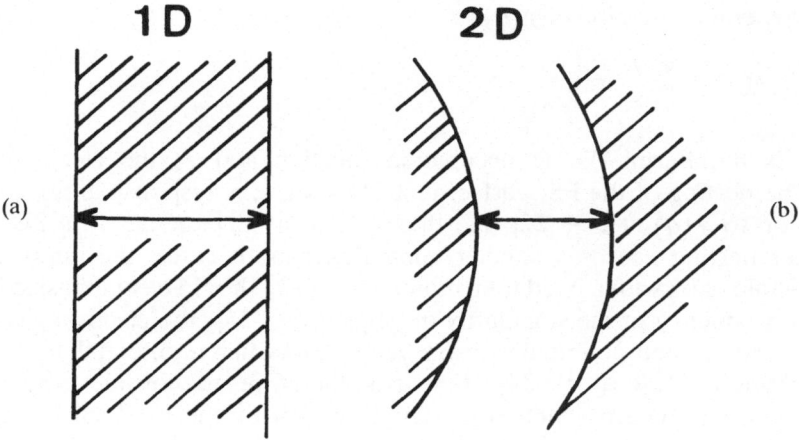

Fig. 1. Fermi surface showing the nesting property in the one dimensional (a) and in the two dimensional (b) case.

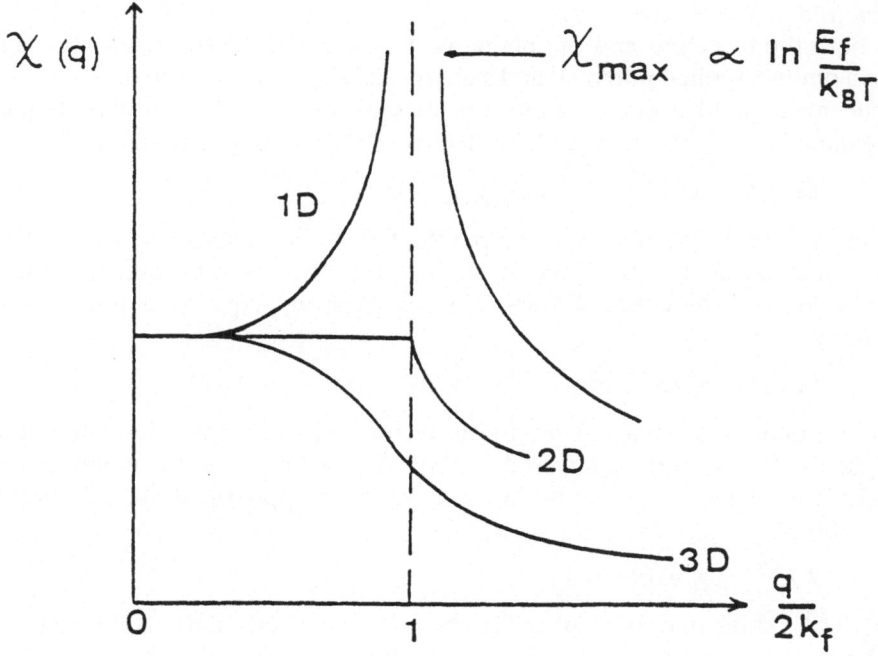

Fig. 2. Electron density response function versus wave vector for a one, two and three dimensional electron gas.

$2k_F$. The general expression of $\chi(q)$ is:

$$\chi(q) = \sum_k \frac{f_k - f_{k+q}}{\varepsilon_k - \varepsilon_{k+q}}$$

where f_k is the Fermi—Dirac occupation function and ε_k the electron kinetic energy. The nesting of the FS, perfect in a 1D system, is responsible for a gigantic contribution to $\chi(q)$ at $q = 2k_F$ and therefore to its divergence. In a 2D system, the weaker nesting causes a smaller anomaly. In both cases, the importance of electron—hole pairs with a fixed total momentum $2\hbar k_F$ should be emphasized [13].

As a consequence of the singularity of $\chi(q)$ at $q = 2k_F$, the electron gas may be unstable to an applied potential of wave vector $2k_F$. Such a potential V_q is created by the phonon mode $q = 2k_F$. One has therefore to consider the coupled electron—phonon system which is usually described by the Fröhlich Hamiltonian [14]:

$$H = \sum_k \varepsilon_k a_k^+ a_k + \sum_q \hbar\omega_q b_q^+ b_q + \frac{1}{\sqrt{N}} \sum_{k,q} g a_{k+q}^+ a_k (b_q + b_{-q}^+)$$

where a_k, a_k^*, b_q, b_q^* are the electron and phonon creation and annihilation operators with momenta k and q, ε_k and $\hbar\omega_q$ are the electron and phonon energies and g the electron—phonon coupling constant. The $2k_F$ instability thus affects both the electrons and the phonons. The electron density is set up in order to screen out the applied potential and a charge density wave is formed.

In the mean field approximation and in the 1D case, the phonon frequency $\Omega(q)$ obtained by taking account of the electron—phonon interaction, is [15]:

$$\Omega^2(q) = \omega_q^2 \{1 - [2g^2/\hbar\omega_q]\chi(q, T)\}.$$

A lattice instability occurs at a temperature T_P^{MF} corresponding to $1/\chi(q) = 2g^2/\hbar\omega_q$ and $\Omega(2K_F) = 0$. There is in this case no restoring force for the $2k_F$ phonon mode and the lattice distorts. The $2k_F$ phonon frequency is found to obey the relation:

$$\Omega^2(2k_F) = \lambda\omega_{2k_F}^2 \ln(T/T_P^{MF}) \quad \text{with} \quad \lambda = g^2 g(\varepsilon_F)/N\hbar\omega_{2k_F}$$

$g(\varepsilon_F)$ is the density of states at the Fermi energy. The corresponding softening of $\Omega(2k_F)$ as the temperature is lowered towards T_P^{MF} is known as the Kohn anomaly. The Peierls transition temperature in the mean field approximation, T_P^{MF} is found on the order of:

$$kT_P^{MF} \sim \varepsilon_F \exp(-1/\lambda).$$

At $T < T_P^{MF}$ a static periodic lattice distortion of wave vector $2k_F$ is stabilized.

In the electron energy spectrum, a gap 2Δ builds up at $k = \pm k_F$ for $T < T_P^{MF}$, as a consequence of the coupling between states k and $k + 2k_F$ (Figure 3). The Peierls gap 2Δ is found to obey the same type of equation as the BCS gap of superconductivity: $\Delta \sim \exp(-1/\lambda)$. At zero temperature: $\Delta(0) = 1.76 k_B T_P^{MF}$. Figure 4 shows the temperature dependence of Δ.

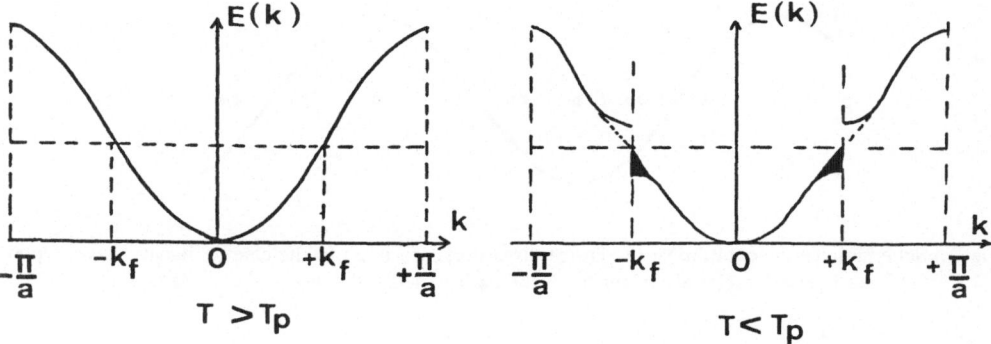

Fig. 3. Band structure of the undistorted (left) and the Peierls modulated (right) one dimensional electron system. States with $|k| < k_F$ gain energy by the distortion.

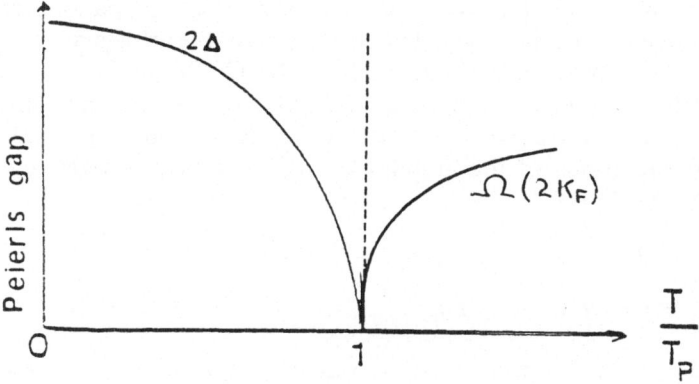

Fig. 4. Peierls gap (left part) and $2k_F$ phonon frequency (right part) vs temperature, in the mean field approximation, in the one dimensional case.

The periodic lattice distortion must be accompanied by a modulation of the electronic density. Figure 5 shows the $2k_F$ lattice distortion and the associated charge density wave:

$$\rho(x) = \rho_0 + \rho_1 \cos(2k_F x + \phi)$$

$$\rho_1 = \rho_0 \Delta/\lambda v_F k_F \quad (v_F \text{ is the Fermi velocity}).$$

The lattice modulation can be written for the nth ion of a chain of parameter a along x:

$$U_n = \sqrt{\frac{2}{\omega_{2k_F}}} \frac{\Delta}{g} \cos(2k_F na + \phi)$$

ϕ is the phase of the CDW and of the lattice modulation with respect to the

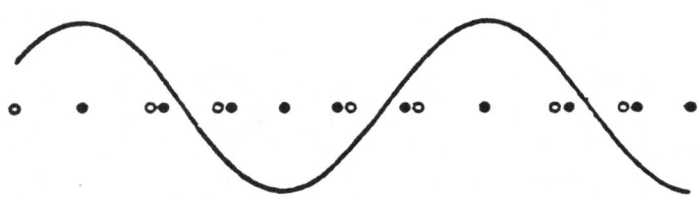

Fig. 5. Schematic representation of the lattice distortion associated to the charge density wave. Open and closed circles are respectively the undistorted and distorted lattice sites.

undistorted lattice. One can therefore define a complex order parameter $\Delta e^{i\varphi}$, with two degrees of freedom, amplitude and phase.

Since the period π/k_F of the modulation is determined by the band filling, it may have no simple relation with the undistorted lattice parameter a. If it is incommensurate with the underlying lattice, the phase ϕ is arbitrary.

The phonon spectrum is also affected by the CDW, in a way similar to the electron spectrum: the degeneracy of the $\pm k_F$ phonons is lifted and the two modes are mixed [16]. For wave vectors close to k_F, two modes ω_+ and ω_-, corresponding to modulations of the amplitude and the phase respectively, are found to be; if $m^* \gg m$ [172]

$$\omega_+^2 = \lambda \omega_{2k_F}^2 + (mv_F^2/3m^*)|q - 2k_F|^2$$
$$\omega_-^2 = (m/m^*)v_F^2|q - 2k_F|^2.$$

The so-called amplitude and phase mode or amplitudon and phason spectra depend on the effective mass m^* for electrons in the CDW: $m^*/m = 1 + 12\Delta^2/\lambda\omega_{2k_F}^2$. The phase mode velocity in this model is found to be:

$$v_\varphi = v_F (m/m^*)^{1/2}.$$

The mean field theory of the Peierls transition summarized above is in principle incorrect since it is well known that one dimensional systems without long range forces cannot undergo phase transitions at finite temperatures. The existence of a phase transition in this model is due to the neglect of phonon modes other than the $2k_F$ one, which turns the short range electron—phonon interaction into a long range term. However, the presence of interchain couplings and three dimensional effects restore the main results of the MF treatment. When these effects are included together with the short range character of the electron—phonon interaction, a pseudo-gap is found in the electronic density of states at $T < T_P^{MF}$. This is a consequence of the fluctuations. The correlation length has also been calculated and found to diverge only far below T_P^{MF}. The critical temperature for 3D ordering is then expected to be $\sim T_P^{MF}/4$ or smaller depending on details of the theory [14].

2.2. CHARGE DENSITY WAVE TRANSPORT

2.2.1. *The Fröhlich Mechanism*

Fröhlich proposed long ago that in an ideal 1D system, in the Peierls distorted case, the system of electrons condensed in the CDW, coupled to the lattice displacements, could move through the lattice without friction, provided the velocity v of the motion would be small enough ($v < \Delta/\hbar k_F$) [5]. Figure 6 shows the band structure corresponding to this case. δk is related to the velocity v by $\delta k = mv/\hbar$, where m is the electron band mass. At $T = 0$, this state carries a Fröhlich electric current $J = nev$, n being the condensed electrons density. At $T \neq 0$, the presence of the Peierls gap prevents in principle scattering of the electrons as long as $kT \ll 2\Delta$ and the system would be a perfect conductor. An infinitely small electric field would excite the Fröhlich mode.

This is not observed in real systems for various reasons. The main one is that the translational invariance of the infinite system is removed by the presence of static defects, impurities, dislocations, etc. The motion of the CDW then requires a finite dc electric field, larger than a threshold value E_t.

2.2.2. *Rigid CDW: Phenomenological Description of the Motion*

Since the discovery of nonlinear conductivity in $NbSe_3$ attributed to the sliding of CDW, several other materials of the families of the transition metal trichalcogenides, halogenated tetrachalcogenides such as $(TaSe_4)_2I$ and of the molybdenum bronzes have proved to be CDW conductors [18]. The studies received a considerable stimulus when it was established that the response to a steady current larger than the threshold value contains periodic components with frequencies F,

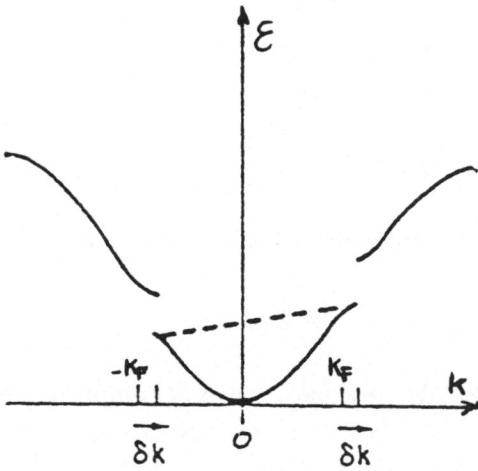

Fig. 6. Band structure of the one dimensional Peierls distorted state when a Fröhlich current is present.

$2F$ ... together with broad band noise. The frequency F is found to be approximately proportional to the excess CDW current J_{CDW} and the ratio $F/J_{CDW} = 1/nel$, such that l is a length of the order of the superlattice period λ. The value $l = \lambda$ can be interpreted simply by assuming that the CDW behaves as a classical rigid object moving with a velocity v and interacting with the fixed impurities, thus leading to a characteristic frequency $F = v/\lambda$ [18, 19].

An oversimplified phenomenological model based on this assumption describes the interaction of the CDW with the impurities by a periodic potential

$$V = \frac{\omega_0^2}{2k_F} (1 - \cos 2k_F x)$$

where x is the center of mass of the rigid CDW and ω_0 is a characteristic pinning frequency. The corresponding equation of motion of the CDW is:

$$\ddot{x} + \Gamma\dot{x} + \frac{\omega_0^2}{2k_F} \sin 2k_F x = \frac{eE}{m^*}$$

where the damping is described by the coefficient Γ. This equation is often written in terms of the phase $\phi = 2k_F x$ as:

$$\ddot{\phi} + \Gamma\dot{\phi} + \sin \phi = \frac{E}{E_c} \quad \text{with} \quad E_c = \frac{m^*\omega_0^2}{2k_F e}.$$

The experimental properties are partly well fitted by the approximation of an overdamped harmonic oscillator, the inertial term being negligible compared to the dissipative one:

$$\Gamma\dot{\phi} + \sin \phi = \frac{E}{E_c}.$$

Non linear conduction occurs for $E > E_c$ and the CDW conductivity is found to be:

$$\sigma(E) = 0 \quad \text{for} \quad E < E_c$$

$$\sigma(E) = \frac{ne^2}{m^*\Gamma} \left[1 - \left(\frac{E_c}{E} \right)^2 \right]^{1/2} \quad \text{for} \quad E > E_c.$$

With this model, the low frequency dielectric constant is found to be related to E_c by:

$$\varepsilon(\omega \to 0) \quad E_c = 2ne\lambda = 2\pi ne/k_F.$$

2.2.3. Deformable CDW: Microscopic Models of Pinning by Impurities

A realistic description of CDW dynamics must take into account the internal degrees of freedom of the CDW. This was first done by Fukuyama and Lee [20], who assumed a slowly varying phase $\phi(x)$ interacting weakly with impurities at random positions R_j in the underlying lattice. The corresponding Hamiltonian is:

$$H = \int dx \left[\frac{K}{2} (\nabla \phi)^2 - V_0 \rho_1 \sum_j \delta(x - R_j) \cos[2k_F x + \phi(x)] \right].$$

The first term corresponds to the elastic energy of the deformable CDW and K represents the phase stiffness. The second term describes the interaction of the CDW with the impurities. V_0 is the impurity pinning potential and ρ_1 the amplitude of the CDW. The relevant dimensionless parameter of the problem is:

$$\varepsilon = \frac{V_0 \rho_1}{v_\varphi n_i}$$

where v_φ is the velocity of the phase mode and n_i the number of impurities per unit length. The strong pinning case corresponds to $\varepsilon \gg 1$ and is associated with a strong impurity potential (or a dilute impurity concentration). In the other limit, $\varepsilon \ll 1$, or in the weak pinning case, the elastic energy dominates. These two extreme cases are illustrated in Figure 7. In the first case ($\varepsilon \gg 1$), the total reduction in energy per unit volume is just proportional to n_i: the so-called pinning energy is $\Delta E = -V_0 \rho_0 n_i$ and the correlation length is the average distance between impurities L_0. In the weak pinning case, the phase coherence is lost over a distance L much larger than L_0 [21]. L_0 is found to depend on n_i^{-1} as:

$$L_0 = \left(\frac{\pi v_\varphi}{V_0 \rho_1} \right)^2 \frac{1}{n_i}$$

L_0 is often considered to be the size of the so-called Lee and Rice domains, although no real domains with defined boundaries are present in this model.

A treatment of the dynamics of the CDW based on the Fukuyama, Lee and Rice model has been performed more recently by D. S. Fisher [22]. It describes the behavior of the sliding of CDW near threshold as a dynamical critical phenomenon. A mean field approximation, valid when the stiffness of the CDW has infinite

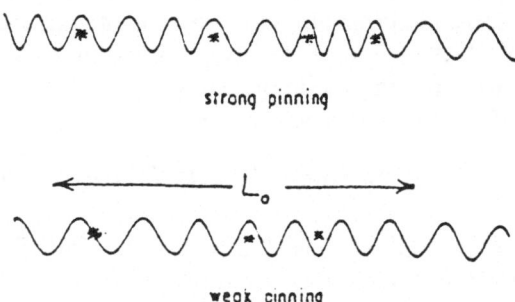

Fig. 7. Strong and weak pinning of a CDW by impurities.

range, has been developed. A critical exponent is derived from the average velocity of the CDW when the field is lowered towards the threshold:

$$v \sim f^{\zeta} \quad \text{with} \quad f = \frac{E - E_t}{E_t} \quad \text{and} \quad \zeta = 3/2.$$

The model predicts two diverging time scales near the threshold. The first one is associated with the average CDW velocity and behaves as $f^{-\zeta}$ while the second one corresponds to rapid motions of a given phase and behaves as $f^{-\mu}$ with $\mu = 1/2$. Below threshold, many metastable stationary states are found: the behavior of the system depends on its past history and shows considerable hysteresis.

Several numerical studies based on the Fukuyama, Lee and Rice model have also been performed [23, 24]. The behaviour at electric fields below threshold has been worked out in a 1D model in [24], with emphasis on the effects of hysteresis and metastability. A large number of metastable states with a distribution of energy barriers between nearby states extending down to very low energies are found. These calculations also provide mechanisms for the hysteresis and memory effects experimentally observed in CDW materials (see Section 3).

2.2.4. *Model Based on the Coupling of CDW with Lattice Phonons*

In a completely different approach, S. Aubry proposes that the Fröhlich conductivity in an incommensurate system is not limited by the pinning by impurities as usually accepted, but by the coupling with the lattice. His theory predicts that the phason excitations are not gapless at $q = 2k_F$ as obtained by Lee, Rice and Anderson [16], but that, due to the strong electron–phonon coupling in CDW materials, this gap is finite. There is thus no zero frequency Fröhlich sliding mode and the threshold field for nonlinear transport is directly related to the phason gap [11, 25, 26].

In a first approximation the model neglects the possible role of impurities and considers the coupling between the CDW and the lattice phonons as essential. In this context, when the electron–phonon coupling increases, a transition from a regime with a zero gap phason to a situation with a finite gap is found. Since this transition corresponds to a nonanalyticity of the physical functions involved in the finite gap case, it has been called transition by breaking of analyticity (TBA).

In this new nonanalytical regime, S. Aubry predicts the existence of low energy configurational excitations for the incommensurate CDW state, such as the CDW structural defects described in the next subsection.

2.2.5. *Models Involving CDW Structural Defects: Discommensurations and Phase Dislocations*

The models sketched in Subsections 2.2.1., 2.2.2., and 2.2.3. assume some deformation of the CDW, but take no account of well-defined CDW phase defects, such as discommensurations or phase dislocations. Only the theory of TBA proposed by S. Aubry leads to an important role for CDW structural defects.

Discommensurations have been proposed and observed long ago in quasi 2D layered transition metal dichalcogenides with a CDW vector close to a commensurate value, $2k_F \sim q_c = G/p$ where G is a reciprocal lattice vector and p and integer [27]. In such systems, a state with commensurate domains separated by narrow walls, called discommensurations, may be more favorable than the uniform incommensurate state (Figure 8). The discommensurations (DC) are the 3D generalization of the soliton phase defect of the 1D chain.

One should note that since the DCs are associated to compression or dilatation of the CDW superlattice, they have to carry an electrical charge which is related to the phase gradient in the DC. For more details, the reader is referred to [17] and references therein.

The existence of phase dislocations and their possible role in CDW transport had already been suggested by Lee and Rice [21]. Such a phase defect of the CDW is sketched in Figure 9. It is analogous to a conventional crystal dislocation, the extra plane of atoms being replaced by an extra equiphase CDW plane in the phase dislocation (DL). The phase DLs and the DCs are not independent phase defects: they must be organized in DC planes delimited by DL loops [21].

The nucleation of phase DLs under the electrical contacts had been proposed independently by Gorkov [28] and by Ong and Maki [29] to account for the voltage oscillations associated with CDW transport. For both groups of authors, the successive creation and destruction of these phase DLs, labelled either phase slip centers or phase vortices, allow the conversion of condensed carriers into normal ones under the contacts.

The role of CDW phase defects in CDW transport may not be limited to the charge conversion under the electrodes. It is very likely that such defects are indeed present in the bulk of the crystals. This has been reconsidered recently by Dumas and Feinberg by analogy between the depinning of a CDW and the onset of plastic deformation of a crystal [30]. In their model, a quasicommensurate CDW

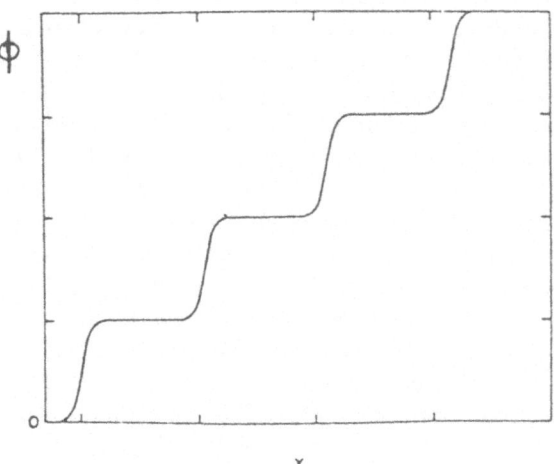

Fig. 8. Model of discommensurations. The phase Φ of the CDW changes by $2\pi/p$ (see text) in the discommensuration. (From [27].)

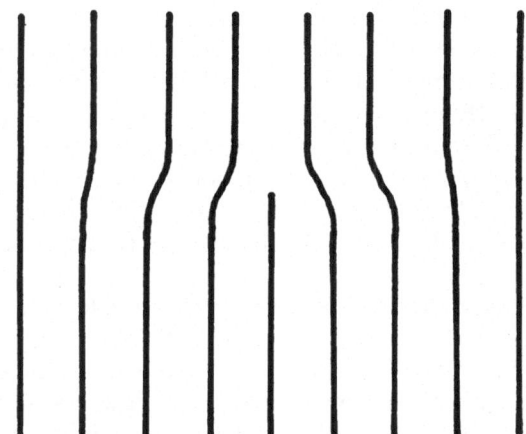

Fig. 9. Phase dislocation: the solid lines represent the contours of constant phase. (From [21].)

is described in terms of commensurate domains separated by DCs limited by phase DLs loops. At low electric fields, below threshold, the CDW phase defects pinned by impurities are distorted and the commensurate CDW is elastically displaced: this accounts for the electric polarization and the large dielectric constants experimentally observed in CDW materials. At electric fields close to threshold, the existing CDW phase defects move, and new ones are nucleated. The CDW excess current is then attributed to a permanent plastic flow of CDW phase defects. Just as the mechanical deformation of a crystal takes place through the motion of common dislocations, CDW transport is made easier, and may be made possible, by the depinning of phase defects. The threshold electric field in this picture is analogous to the yield point in stress—strain curves of a crystal.

More general considerations on CDW or SDW phase defects have been developed by J. Friedel in relation to the problem of friction of weakly commensurate modulations through the underlying lattice [31]. In this context, the motion of weakly commensurate CDW or SDW necessarily involves the development and motion of phase DL. In the case of incommensurate modulations, the friction by various kinds of lattice defects, surfaces and interfaces, conventional dislocations and point defects, is considered. Developments along these lines by Feinberg and Friedel are developed in another chapter of this volume [12].

From another point of view, CDW structural defects may be related to localized states in the Peierls gap and possibly to unpaired spins giving rise to an EPR signal (see Section 3.5). For example, B. K. Chakraverty has considered the effect of the interaction between single particle and CDW phase excitations: the resulting phase solitons are associated to localized states in the Peierls gap [32].

2.2.6. Quantum Models

All the previous models assume that CDW motion can be described classically. In a different approach, Bardeen proposed that a quantum-mechanical tunneling mechanism can account for some of the observed properties [33]. The CDW condensate is considered to be a macroscopic quantum system which undergoes a Zener-type quantum tunneling over large distances, of the order of the Lee and Rice domain size. The relevant energy scale is that of the pinning energy or pinning gap, of the order of 10^{-6} eV. A recent development of this model takes into account the deformation of the CDW over macroscopic lengths [34]. Experimental data involving the response of the CDW to ac fields as well as harmonic mixing may be described by this model combined with the photon-assisted tunneling theory [34]. However, it is not clear at the present time in which situation a quantum description is more appropriate than a classical one.

Barnes and Zawadowsky [35] have proposed a theory of Josephson-type phenomena in which two macroscopic quantum states are the two components of an incommensurate CDW with opposite wave vectors $\pm 2k_F$. These two coherent states are electron—hole pairs with total momentum $2k_F$. Impurities can scatter an electron—hole pair with momentum $2k_F$ to a state $-2k_F$ and can induce quantum oscillations.

We have only summarized above some of the theoretical models currently used to describe CDW transport and the related properties. More details can be found in the Proceedings of recent conferences [36—39]. Clearly, the experimentalist has presently to face a broad choice of theories. It is likely that, depending on the materials and on the physical parameters such as temperature and electric fields, some models are more relevant than others.

3. Quasi One-Dimensional Compounds: The Blue Bronzes $A_{0.30}MoO_3$

3.1. INTRODUCTION

Under the name of blue bronzes, two alkali metal compounds $K_{0.30}MoO_3$ and $Rb_{0.30}MoO_3$ and the thallium bronze $Tl_{0.30}MoO_3$ have been studied up to now. Their properties, especially in relation to the Peierls transition and CDW transport, are analogous. They all show a metal-to-semiconductor Peierls transition due to the onset of an incommensurate CDW with the same wave vector q_b. The Peierls temperature is close to 180 K in these three bronzes.

The possibility of growing comparatively large single crystals, as well as of choosing among three compounds the one best suited for studying a given property, has considerably stimulated the experimental studies on the blue bronze [41, 42]. For example, the direct proof of the sliding of CDW in the nonlinear transport regime has now clearly been established by the motional narrowing of the ^{87}Rb NMR line in $Rb_{0.30}MoO_3$ [43, 44].

The potassium blue bronze was synthesized for the first time in 1964 by Wold et al. by the electrocrystallization technique [45]. The crystal structure was refined by Graham and Wadsley and with a better accuracy more recently by Ghedira et al. [46]. Physical studies by Bouchard et al. established that the blue bronze shows a semiconductor-to-metal transition in the vicinity of 180 K [47]. Later, Perloff et al. noticed a large anisotropy of the electrical conductivity in the plane of the layers [48]. Detailed studies of the transport properties were then performed by Fogle and Perlstein [49]. They especially measured the low-temperature behavior of the conductivity σ and reported a nonohmic behavior at $T < 20$ K, when the sample is insulating ($\sigma = 10^{-14}$ Ω^{-1} cm^{-1}). They proposed a model of an excitonic insulator for the semiconducting state.

It is only recently that the anisotropy of the conductivity was rediscovered and studied in greater details by Brusetti et al. [50]. Optical reflectivity measurements showed indeed that $K_{0.30}MoO_3$ is a quasi-one-dimensional metal in the high-temperature phase [51]. Later, X-ray diffuse scattering studies by Pouget et al. [32] led to the conclusion that the metal-to-semiconductor transition is a Peierls transition towards an incommensurate CDW state. At the same time, in a search for non-ohmic conductivity in the temperature range where the conductivity is not vanishingly small ($T > 50$ K), Dumas et al. [53] established that the blue bronze shows nonlinear transport, due to the sliding of the CDW.

Details concerning the crystal chemistry, the crystal structure and structural instabilities can be found in [8, 9 and 10]. We recall here (Figure 10) only that the structure is monoclinic (space group $C2/m$), with 20 formulae per double unit cell, and is built with clusters of 10 distorted MoO_6 octaheda. It can be described either with a C-centered or an I-centered unit cell. We will use the C-centered cell in the following. The MoO_6 clusters are linked together along the monoclinic b-axis and the [102] direction and form infinite slabs. The A ions lie between these

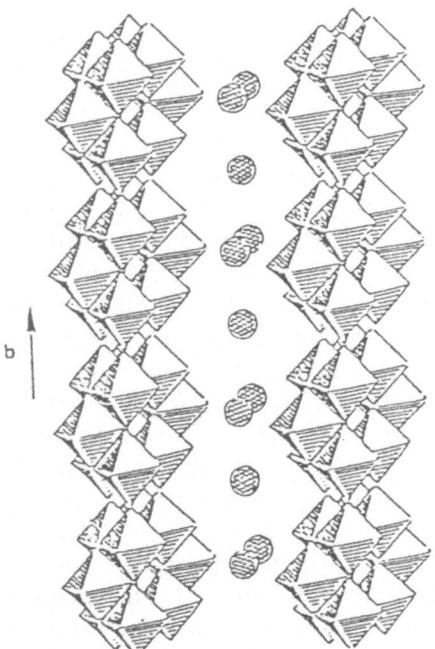

Fig. 10. Crystal structure of $K_{0.30}MoO_3$ showing the infinite slabs separated by the alkali ions (●) and the infinite chains of MoO_6 octahedra parallel to the b axis. (From [46b].)

slabs. The structure is thus primarily layered-type. However, the presence of infinite chains of clusters along b is responsible for quasi-1D electronic properties. One should also note that there are three independent Mo sites, labelled Mo(1), Mo(2) and Mo(3). Crystal structure refinement has established that the $4d$ electrons density is found mostly on sites 2 and 3 [9]. These sites are involved in infinite chains of MoO_6 octahedra.

X-ray diffuse scattering studies have shown that the metal-semiconductor transition is associated with diffuse streaks condensing below 180 K into well-defined satellite reflections with a wave vector $Q = (0, q_b, 0.5)$ [54, 55]. The in-chain component q_b is incommensurate and temperature dependent (Figure 11a). Detailed analysis of the X-ray data show that the wave vector of the Peierls distortion should be taken as $q_0 = (0, 1 - q_b, 0.5)$. $1 - q_b$ is found to increase from ~ 0.70 at room temperature to a value close to 0.75 below 100 K. It is, within the experimental error, constant below ~ 100 K. This shows that the $2k_F$ wave vector is close to the commensurate value of $3/4b^*$ below 100 K and decreases above. However, the presence of a lock-in transition at 100 K has not been established [10]. The temperature dependence of the order parameter of the transition is also obtained from the X-ray data from the square root of the intensity of the satellite (Figure 11b).

In this section, we will first review various aspects of the Peierls transition in the blue bronze, then describe the information presently available, experimentally and theoretically, on the band structure. CDW transport in dc fields, with emphasis on hysteresis and metastability, will be reviewed in subsection 3.4. The response of the CDW in ac fields, including transport and optical properties, are described in the chapter by Fleming and Cava in this volume [50]. Subsection 3.5 is devoted to the description of the hysteresis and metastability phenomena. The local studies of the CDW state and of the CDW dynamics by NMR [43] will be surveyed in subsection 3.6.

3.2. THE PEIERLS TRANSITION

Various physical properties, including transport, magnetic, thermal and elastic properties show drastic changes or anomalies at the Peierls transition in the blue bronzes.

3.2.1. *Ohmic Transport*

The electrical resistivity shows a large anisotropy consistent with a quasi 1D behavior and a metal to semiconductor transition as a function of temperature (Figure 12). This establishes that the Fermi surface is not far from being planar in the metallic state and is completely destroyed by the CDW instability. Both Hall effect [57] (Figure 13) and thermopower [48] data show that the dominant carriers are holes above the Peierls transition and electrons below. At 300 K the carrier concentration is roughly consistent, in a simple Drude model, with a complete charge transfer from the A metal to the conduction band. Since the order parameter and therefore the Peierls gap is increasing with decreasing temperature

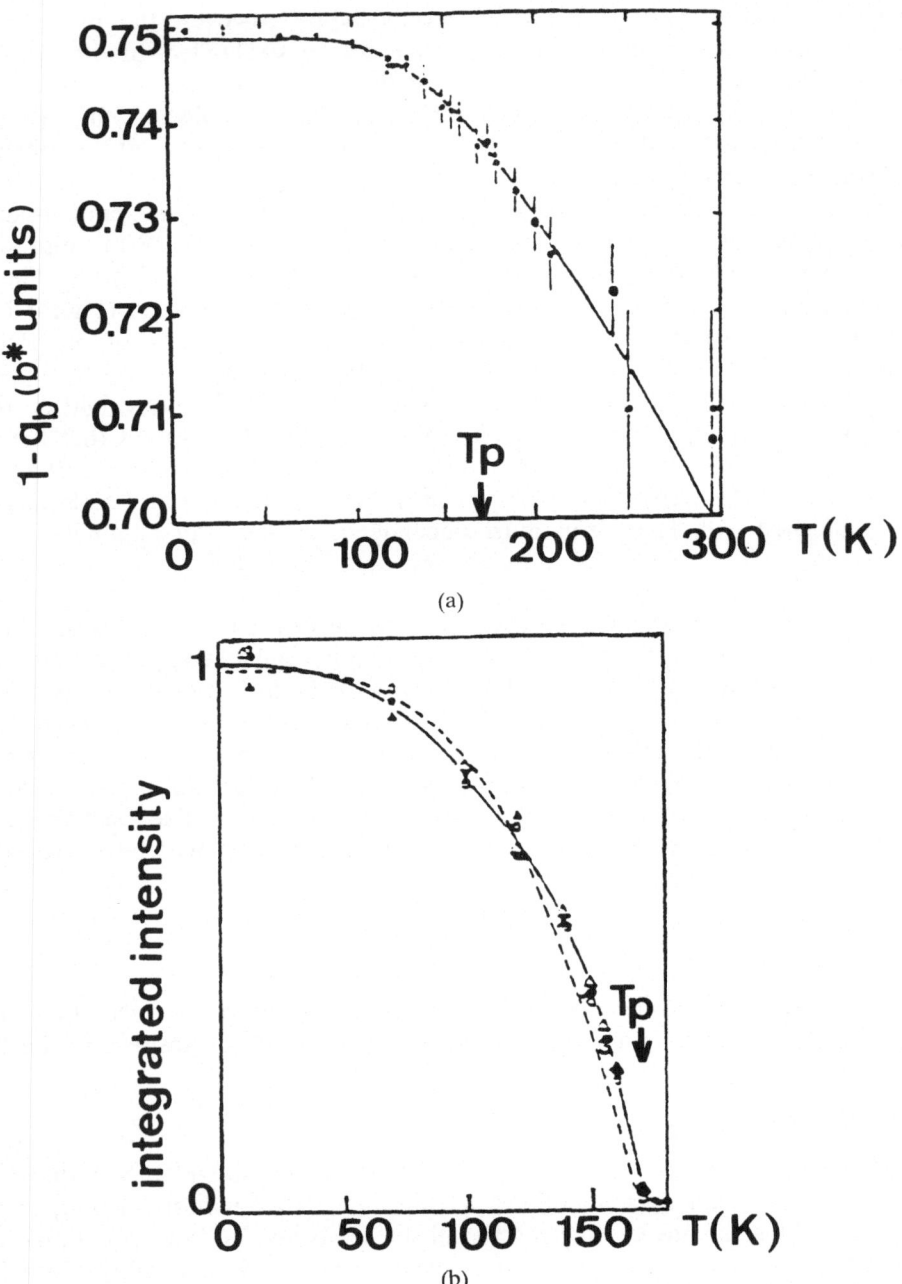

Fig. 11. (a) Temperature dependence of the incommensurate component $1 - q_b$ $(= 2k_F)$ of the satellite reflection (below T_c) and of the diffuse scattering (above T_c). (b) Normalized integrated intensity of the $(17, \mp 1 \pm q_b, \overline{8.5})$ and $(17, \pm 1 \pm q_b, \overline{8.5})$ satellite reflections as a function of the temperature. The dotted line gives the temperature dependence of the square of the B.C.S. order parameter. (From [55].)

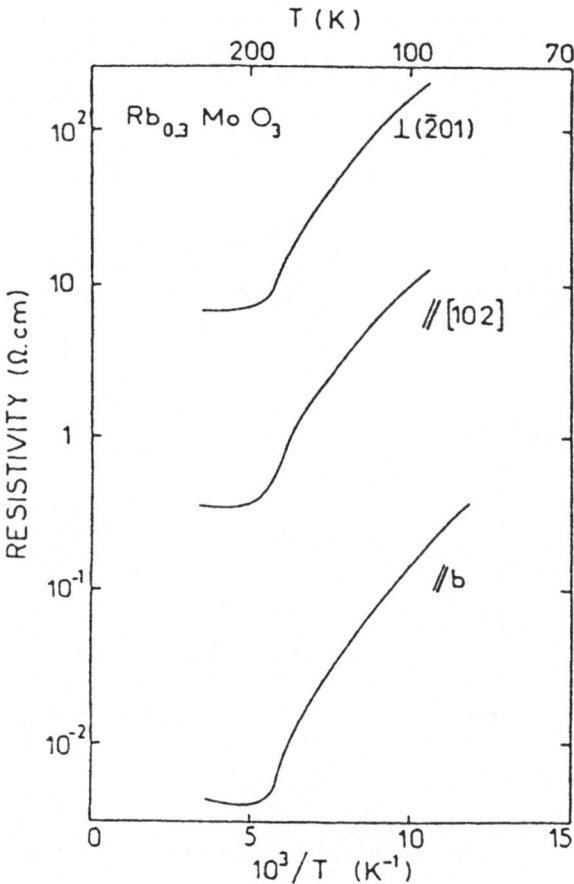

Fig. 12. Electrical resistivity (logarithmic scale) of $Rb_{0.30}MoO_3$ versus inverse temperature along b, [102] and perpendicular to the octahedral layer's ($\bar{2}01$) plane. (From [57].)

down to ~ 50 K (see Figure 11b), one does not expect a simple activated law for the transport properties above 50 K. However, this is probably meaningless because of the narrowness of the temperature interval. Also, the electrical resistivity shows an activation energy of ~ 0.03 eV in the range 30—70 K [50]. As the Peierls gap has been estimated, from optical reflectivity data, to be 0.15 eV [57], this indicates that the blue bronze is not an intrinsic semiconductor and that the transport properties at low temperatures are due to the presence of defects or impurity levels in the Peierls gap. The corresponding localized states may result from a stoichiometry defect giving rise to $Mo^{5+}(4d^1)$ localized states, as will be discussed in Section 2.5. They may also be related to CDW structural defects, such as discommensuration or phase dislocations and to the so-called midgap states predicted by theory [32, 60].

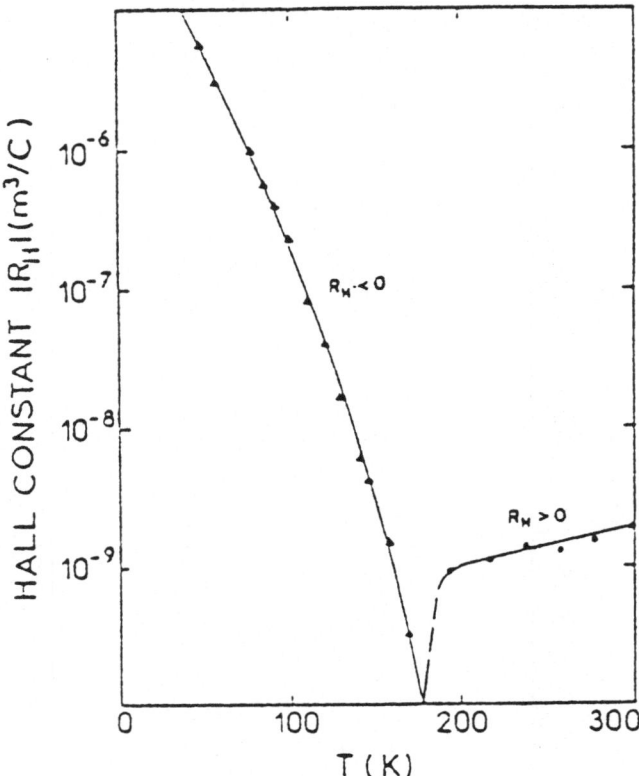

Fig. 13. Hall constant as a function of temperature for $K_{0.30}MoO_3$. (From [57].)

One should note that the ratio of the Peierls gap to kT_P is found to be ~ 9, which is much larger than the mean field value of 3.5. This feature is common to all CDW materials showing a Peierls transition.

3.2.2. *Magnetic Susceptibility*

The magnetic susceptibility of $K_{0.30}MoO_3$ [47, 50, 61, 62] is found to be anisotropic between the ($\bar{2}01$) plane of the layers and the perpendicular orientation (Figure 14). The observed susceptibility is the result of several contributions: $\chi = \chi_{dia} + \chi_{Pauli} + \chi_{Van Vleck}$. The anisotropy is mainly due to the anisotropy of the Van Vleck paramagnetism, related to the anisotropy of the band structure. The decrease of χ with decreasing temperature corresponds to the opening of the Peierls gap and to the vanishing of the Pauli contribution. This result has been analyzed by Johnston by taking the fluctuations into account [62]. A pseudogap of 200 K is found above T_P. The obtained zero temperature Peierls gap is of the order of 1200 K, not too far from the value obtained from optical reflectivity studies.

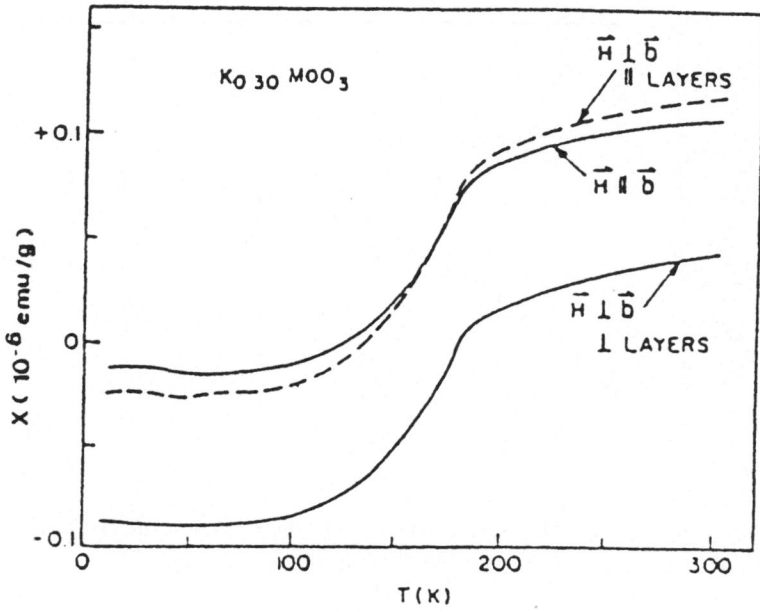

Fig. 14. Magnetic susceptibility of $K_{0.30}MoO_3$ as a function of temperature for different orientations of the magnetic field. (From [61].)

3.2.3. *Thermal and Elastic Properties*

From specific heat studies, the Debye temperature for the potassium and the rubidium bronzes has been found to be close to 320 K (320 ± 10 K for $K_{0.30}MoO_3$ and 310 ± 10 K for $Rb_{0.30}MoO_3$) [63]. This value is similar to what has been obtained either for MoO_3 (340 ± 10 K), for the purple bronze $K_{0.9}Mo_6O_{17}$ and for the oxides η-Mo_4O_{11} and γ-Mo_4O_{11} [63]. This shows that the Debye temperature is mainly determined by the presence of the MoO_6 octahedra in all these materials.

A specific heat anomaly, found in both K and Rb blue bronzes between approximately 150 and 190 K, corresponds to an enthalpy of ~ 25 J mole^{-1} and an entropy change of the order of 150 mJ mole^{-1} K^{-1} (Figure 15). This leads to a value for the density of states in the metallic phase of the order of 0.3 eV^{-1} mole^{-1}, if all the entropy is attributed to the conduction electrons. One also obtains from these data a value of roughly 0.24 R for the discontinuity in the electronic specific heat at the transition. This, compared to the entropy change, suggests that 1D fluctuations are important in this system [62, 63].

At low temperatures (0.1 K < T < 1 K), the specific heat shows an excess contribution that is linear in temperature [64] and is time dependent at the lowest temperatures. This has been attributed to some low energy excitations due to the phase distortions of the randomly pinned CDW. However, the thermal conduc-

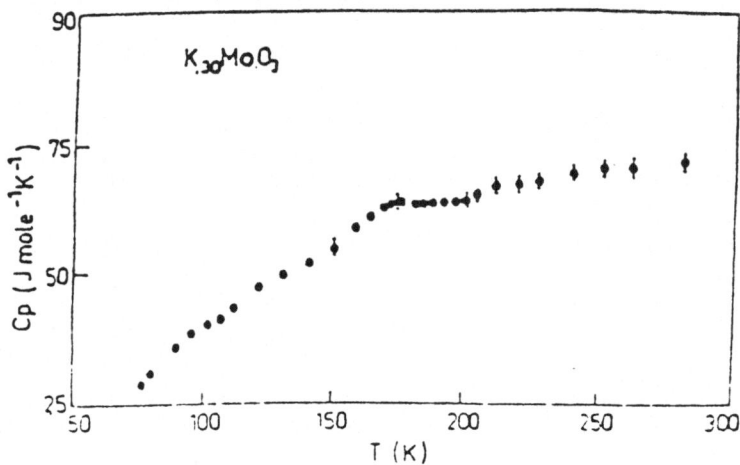

Fig. 15. Specific heat of $K_{0.30}MoO_3$ as a function of temperature in the vicinity of the Peierls transition. (From [63].)

tivity shows no anomalous behavior in the same temperature range [64]. The Young's modulus has been measured in the vicinity of the Peierls transition, both parallel and perpendicular to the *b*-axis [65]. A strong anomaly is found near T_P in the perpendicular case and a weak one in the parallel orientation. These results are consistent with a second order phase transition at T_P, connected with a considerable lattice softening.

3.2.4. *Effect of Impurities and Point Defects*

Doping with isoelectronic impurities such as W substituted onto Mo sites or Rb onto K sites, acts differently on the Peierls transition temperature [50]. Very large effects are found by tungsten substitution whereas alkali substitution produces minor effects. The resistivity in the semiconducting phase is strongly decreased by doping with W and the low temperature activation energy changes from 35 meV down to 15 meV for doping concentrations reaching 12 at. % [66]. This indicates that the W impurities lead to the formation of an impurity band in the Peierls gap.

Similarly, the Peierls temperature seems to be insensitive to a stoichiometry defect for the alkali metal. In order to clarify this point, the b^* satellite wave vector component has been measured in samples with different stoichiometries ($A_{0.28}MoO_3$ and $A_{0.30}MoO_3$): q_b seems to be insensitive to the stoichiometry, thus indicating that the Fermi wave vector does not depend strongly on the A vacancy concentration [67]. This surprising result may be understood by one of the two following pictures: either the stoichiometry defect on the A metal is associated with oxygen vacancies in such a way that the filling of the conduction band is always approximately constant, or there is some warping of the Fermi surface which allows the same nesting vector independently of the exact conduction electron concentration.

On the contrary, defects created by electron irradiation have a strong effect on the Peierls temperature and on the Peierls gap [68, 69] (Figure 16). Rather strong irradiation doses (≈ 0.3 C/cm^2) seem to smear out the Peierls gap, probably because the metallic state is stabilized by disorder.

3.3. BAND STRUCTURE: EXPERIMENT AND THEORY

The band structure was earlier speculated, by analogy with ReO_3, as being built on hybridized Mo and O states leading to filled σ and π bonding bands and empty antibonding bands [51]. A rough LCAO calculation, performed for one chain with two Mo sites only, is consistent with this speculation (Figure 17a).

Tight binding calculations [70] for real chains and for a slab of $Mo_{10}O_{30}$ clusters have also been performed for the bands close to the Fermi level [71] (Figure 17b). They are consistent with the presence of two overlapping bands that are roughly three-quarters filled. They also show that a third band lies above, close to the Fermi level. Figures 17c, d, e show the Fermi surfaces (FS) deduced from this calculation for the two bands and the predicted nesting of the FS along b*: the upper FS of the first band is nested to the lower FS of the second band, and reciprocally, with a wave vector $0.75b^*$. This is in agreement with the existence in the blue bronze of one CDW only.

The temperature dependence of the q_0 wave vector (Figure 11a) has to be attributed to a change of the total number of electrons in the two conduction bands. It has been proposed that this change is due to the presence of the third band lying just above the Fermi level [55]. The experimental data would be consistent with a narrow band located ~650 K above the Fermi level, in agreement with the band structure calculation.

X-ray and ultraviolet photoemission spectroscopy studies corroborate the band

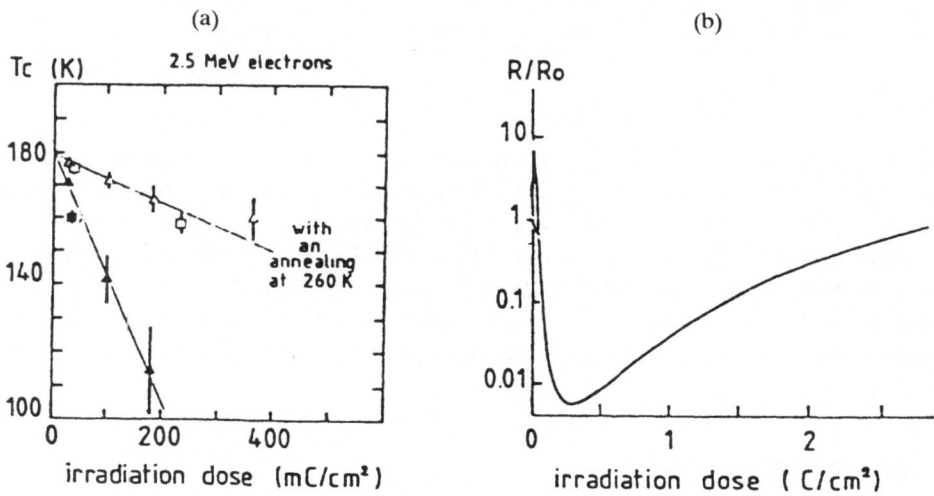

Fig. 16. Effect of electron irradiation on the Peierls transition in $K_{0.30}MoO_3$. (a) Peierls temperature vs irradiation dose. (b) Low field resistivity at 21 K vs irradiation dose ([69]).

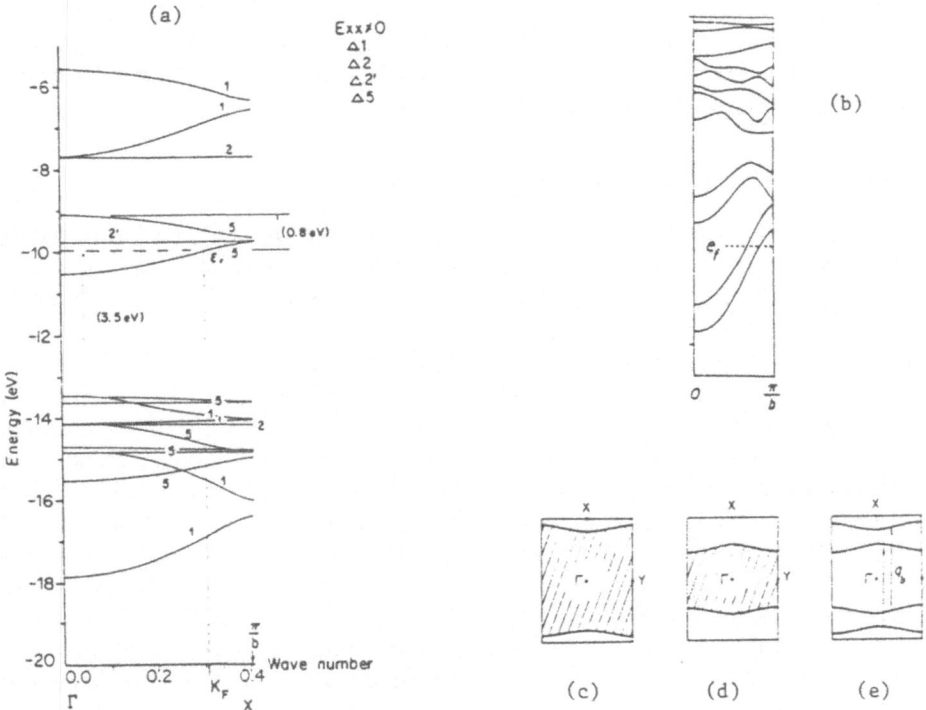

Fig. 17. (a) Band structure calculated with the LCAO method for one chain with 2 Mo sites. (From [70].) (b) Tight binding calculation for a chain of clusters of MoO_6 octahedra showing two bands overlapping at the Fermi level. (c) (d) (e) Fermi surfaces associated with two partially filled d-block bands of a real $Mo_{10}O_{30}$ slab. (c) Fermi surface of the first band, where the wave vectors of the shaded and unshaded regions lead to occupied and unoccupied bands levels respectively. (d) Fermi surface of the second band. (e) Nesting of the Fermi surfaces of the first and second bands. (From [71].)

structure calculations [72, 73]. A remarkable result has also been obtained by angle resolved UV photoemission on ($\overline{2}$ 0 1) cleavage plane: the strong anisotropy of the spectrum obtained with respect to the angle θ between the collected photoelectrons and the normal incident photons allows a direct measurement of the Fermi wave vector (Figure 18). It is found to be 0.31 $\pm$ 0.05 Å^{-1}, in good agreement with the value of 0.312 Å^{-1} obtained from the q_b component of the satellite. This seems to be the first direct measurement of the Fermi wave vector in a low-dimensional solid.

3.4. NONLINEAR TRANSPORT

3.4.1. *Introduction*

The nonlinear transport properties which have been found in certain quasi-one

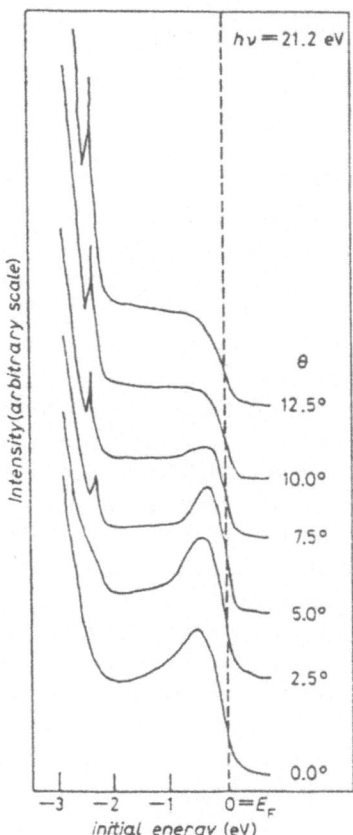

Fig. 18. Angle Resolved Ultraviolet photoemission spectra taken with 21.2 eV photons for $k_{||}$ along b^* showing the conduction band peak just below E_F. (From [73].) $k_{||}$ is the component of the photoelectron wave vector k parallel to the surface, $(\bar{2}\ 0\ 1)$ crystallographic plane. Photoelectrons are collected in the plane defined by the normal to the $(\bar{2}\ 0\ 1)$ plane and the b-axis, at an angle θ from the normal.

dimensional inorganic linear chain compounds such as the trichalcogenides $NbSe_3$, TaS_3, the halogenated tetrachalcogenides $(TaSe_4)_2I$, $(NbSe_4)_{10/3}I$ and the blue bronzes $A_{0.30}MoO_3$ (A = K, Rb, Tl) are a consequence of the motion of a CDW, as described in Section 2. Since the CDW is normally pinned by impurities or defects, the CDW conduction appears above a finite threshold electric field. A rich variety of phenomena associated with this new collective transport have been observed. They include broad band noise voltage, remarkable periodic voltage oscillations, which are viewed as a signature of the collective motion of the CDW, anomalous low frequency dielectric response, ac—dc coupling, metastable phenomena leading to various hysteresis and memory effects [3, 18, 36—39]. At low temperatures, above a second threshold voltage, a dramatic drop of the damping of the CDW occurs. In this regime an average velocity of the CDW as high as

1 m/s is found. This state could approach a Fröhlich superconductor state (see Section 3.4.6).

3.4.2. *Threshold Electric Field*

(a) *Onset of nonlinear conduction.* The threshold electric field E_t above which nonlinear conductivity occurs can be detected on the dc $V-I$ curves, or by measuring the differential resistance dV/dI by lock-in detection as a function of the driving current, or by pulse techniques. Nonlinear conduction occurs only for I along the high conductivity b-axis. E_t depends on the samples and on the contact preparations. For a given sample E_t is strongly temperature dependent. The average CDW excess current $\langle I_{CDW} \rangle = I_{mes} - V/R$ where I_{mes} and V are the measured current and voltage and R the ohmic resistance, respectively, is found to be large in some samples: the ratio of the differential resistance in the non-ohmic state and in the ohmic one can be as large as 6 for electric fields $\sim 2E_t$.

Three different behaviors can be observed on the $V-I$ characteristics: smooth onset of nonlinearity; abrupt switching from the ohmic to the non-ohmic regime with, in some cases, precursor very low frequency voltage spikes; mixed regime with a smooth onset of the nonlinearity followed by a switching. In all cases, large hystereses in $V-I$ curves are found upon elecrical cycling. Figure 19 shows a switching in the dV/dI curve with precursor voltage spikes, the system hopping back and forth between the ohmic and non-ohmic state with a pseudo-frequency of ~ 1 Hz. The amplitude of the spikes increases linearly with the bias current. Switching is observed in a limited number of samples; it may occur in all materials

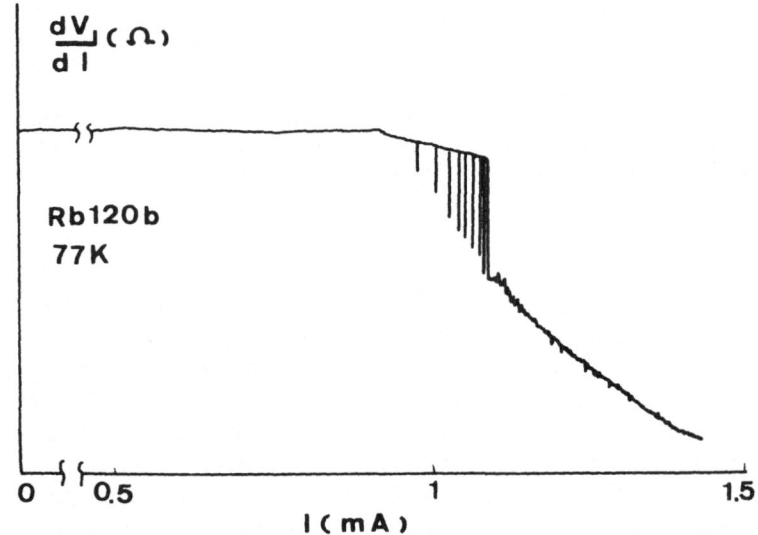

Fig. 19. Differential resistance dV/dI as a function of the dc bias current for $Rb_{0.30}MoO_3$ at 77 K. Precursor voltage spikes are visible (see text). (From [108].)

showing CDW conduction. The switching current I_t found for an increasing bias current is not exactly reproducible from one cycle to the following one. Figure 20 shows a mixed behavior with two thresholds, a smooth one at E_{t_1} followed by a switching at E_{t_2}; the onset of broad band noise voltage is also indicated (see (c) below).

E_t values at 77 K lie in the range 30—500 mV/cm and depend on crystals as well as on contact preparation. The thresholds for switching or for smooth onset of nonlinearity are comparable. When considerable care is taken in the preparation of the contacts, in particular when the ends of the sample are polished before electrodeposition of copper, the threshold is consistently of the order of 50 mV/cm. Threshold values are comparable in $K_{0.30}MoO_3$ and $Rb_{0.30}MoO_3$. However, E_t

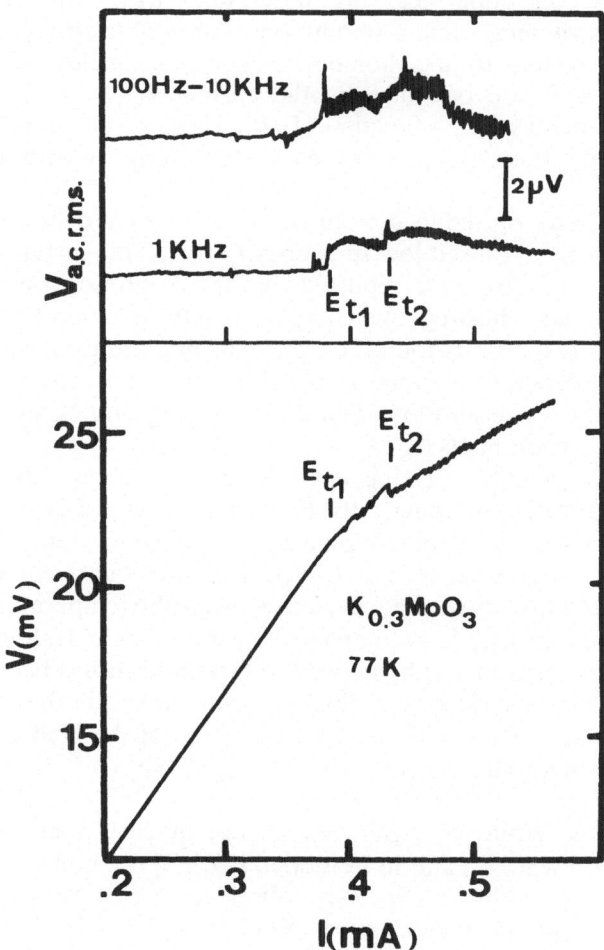

Fig. 20. Bottom curve: $V-I$ characteristic showing two thresholds E_{t_1}, E_{t_2}. Upper curves: onset of broad band noise at E_{t_1}, then at E_{t_2} in the frequency range 100—10 kHz and 1 kHz. (From [36b].)

is found to be of the order of ~ 500 mV/cm in the isotype $Tl_{0.30}MoO_3$ [74]; this is very likely due to a poorer crystal quality.

Crystal defects (impurities, dislocations, nonstoichiometry), and microscopically nonuniform contact resistance are responsible for an inhomogeneous flow of the current throughout the sample [75]. This can explain the dispersion in the absolute values of E_t and of the resistivity, as well as the dispersion in the slopes of the frequency of the periodic voltage oscillations vs CDW current density (see (d) below). There is no evident correlation between E_t values and resistivity values. E_t values measured on cleaved platelets (typical size $2 \times 1 \times 0.1$ mm^3) and on needles (typical size $0.01 \times 0.01 \times 2$ mm^3) are comparable.

The crucial role of contact geometry has been investigated [75]. On a crystal with eight contacts deposited as indicated in Figure 21a, the threshold fields E_{t_1} and E_{t_2} measured on both faces are different. Further, in this example, E_{t_1} corresponds to a switching while a smooth onset of nonlinearity occurs at E_{t_2}. This behavior must be related to an inhomogeneous crystal quality. Figure 21b shows the dependence of E_{t_1} on one face on the current I_2 ($I_2 > I_{t_2}$) passed on the opposite face. E_{t_1} increases as a function of $|I_2|$. These results have been attributed to friction between domains possibly elongated along b with different CDW velocities.

The threshold field E_t deduced from dc $V-I$ characteristics, or from dV/dI versus I curves, or from pulsed low frequency (< 10 KHz) currents have the same values. Tsutsumi et al. [76] have reported that the response to pulsed currents is different for samples showing switching, or only a smooth nonlinearity. A capacitive sluggish response is found for nonswitching samples, while an inductive response with an overshoot is found in switching ones. For nonswitching samples, their results are the same as those found by Fleming and Schneemeyer in their pulse sign memory experiments [77].

In W-doped $K_{0.30}MoO_3$, Fleming et al. [78a] have shown that the threshold depends strongly on the sweeping rate: $E_t \propto \log f$ above a critical frequency, in the range of $10^{-1}-10^2$ Hz, depending on the W concentration. This indicates an extremely sluggish response of the CDW. In pure samples, the response is increasingly time and history dependent as one lowers the temperature [63, 78].

The onset of nonlinearity is accompanied by a nonlinear Hall effect [58a]. The Hall resistance decreases markedly above the threshold. It has been proposed that the motion of the CDW induces a normal electron current in the reverse direction proportional to I_{CDW}. This would lead to a reduction of the Hall voltage above E_t. This situation is similar to that observed in $(TaSe_4)_2I$ [58b].

(b) *Impurity effects.* While no significant change in E_t is found in V-doped (V concentration of a few at. %) and in Fe-doped [79, 80] (Fe concentration of ~ 100 at. ppm) blue bronze, dramatic effects are observed in electron irradiated [68, 82] and in W doped samples (W concentration $\leqslant 12$ at. %) [59, 66].

E_t is found to increase linearly with the dose of electron induced defects and also with W concentration. According to the Lee and Rice theory of pinning of a CDW, this behavior is characteristic of a strong pinning mechanism in which the phase of the CDW adjusts at each impurity site [21].

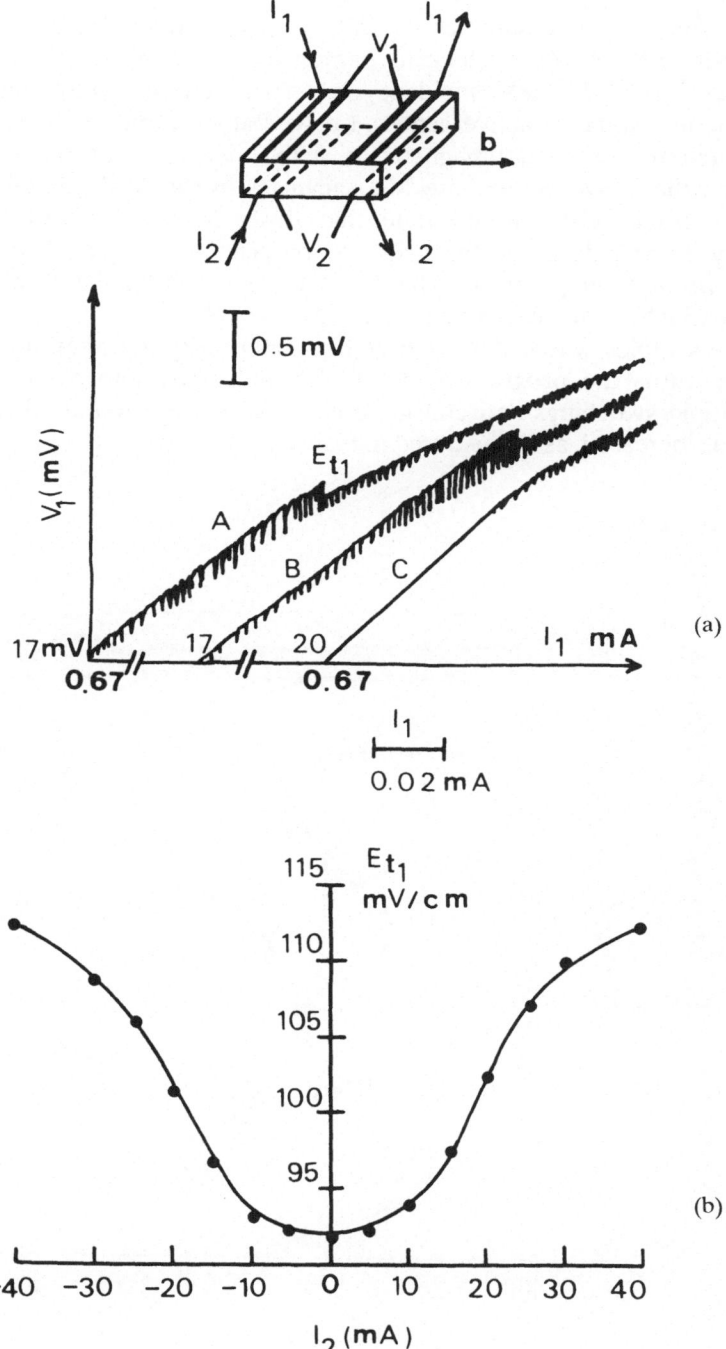

Fig. 21. (a) Upper: lead configuration. Bottom: $V_1(I_1)$ characteristics for different values of dc bias I_2; (A: $I_2 = 0$; B: $I_2 = 10$ mA; C: $I_2 = 25$ mA). $E_{t_1} = 90$ mV/cm; $E_{t_2} = 350$ mV/cm; $I_{t_2} = 2.1$ mA $T = 77$ K. (From [75].) (b) Threshold E_{t_1} as a function of both directions of the dc bias current $\pm I_2$ applied on surface 2. (From [75].)

In electron irradiated samples the defect concentration can be estimated from the intensity of the Mo^{5+} electron paramagnetic resonance signal. A dose of ~ 1 mC/cm^2 of 2.5 MeV incident electrons increases the concentration of paramagnetic resonance active centers by 10^{-5} atomic fraction. It is believed that the production rate of nonmagnetically active defects is of the same order of magnitude; the active defects are Mo vacancies in the MoO_6 octahedra running along the b-axis. One should note that doses as small as μC/cm^2 increase noticeably E_t and doses of the order of mC/cm^2 are required to decrease the Peierls transition temperature. This is in agreement with the fact that the blue bronzes are highly sensitive to electron irradiation [82].

In pure samples where the onset of the nonlinearity is rather smooth, electron irradiation turns the progressive onset of nonohmicity into a switching process [81]. Multiple switchings and considerable hysteresis are found when the irradiation dose is increased, as is illustrated in Figure 22.

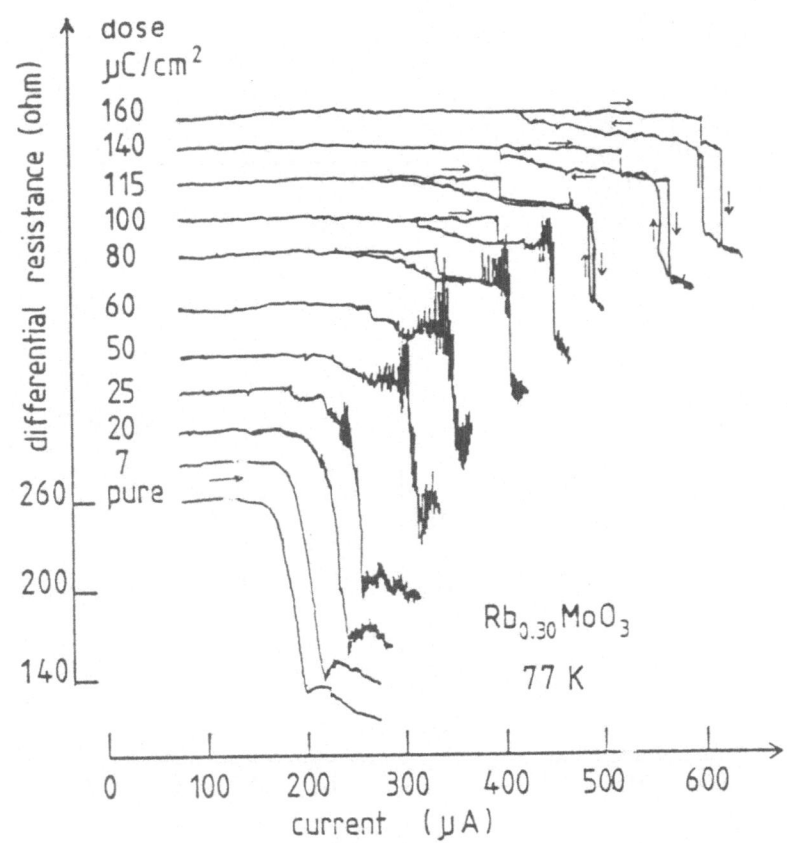

Fig. 22. Differential resistance dV/dI versus dc bias current for different irradiation doses. The curves have been shifted for clarity; irradiation (2.5 MeV electrons) and *in situ* measurements have been performed in liquid nitrogen. (From [36b] p. 449.)

In $(K_{1-x}Rb_x)_{0.3}MoO_3$, E_t seems to increase quadratically with x [59, 66] suggesting a weak pinning process according to the Lee and Rice theory. While in electron irradiated samples as well as in W doped samples the defects are incorporated along the 1D chains and act as strong pinning centers, the alkali substitution plays a minor role in the pinning mechanism.

(c) *Temperature dependence of E_t*. The behavior of E_t is different from that of other CDW materials (NbSe$_3$, TaS$_3$, (TaSe$_4$)$_2$I) and is not yet fully elucidated. Some authors find that E_t increases continuously from 60 K to ~150 K [76]; others find a plateau above 100 K [83] or a small decrease above 100 K [84]; others [41] find a maximum near 100 K in both $K_{0.30}MoO_3$ and $Rb_{0.30}MoO_3$ (Figure 23). The switching, when it is observed, is found over a limited tempera-

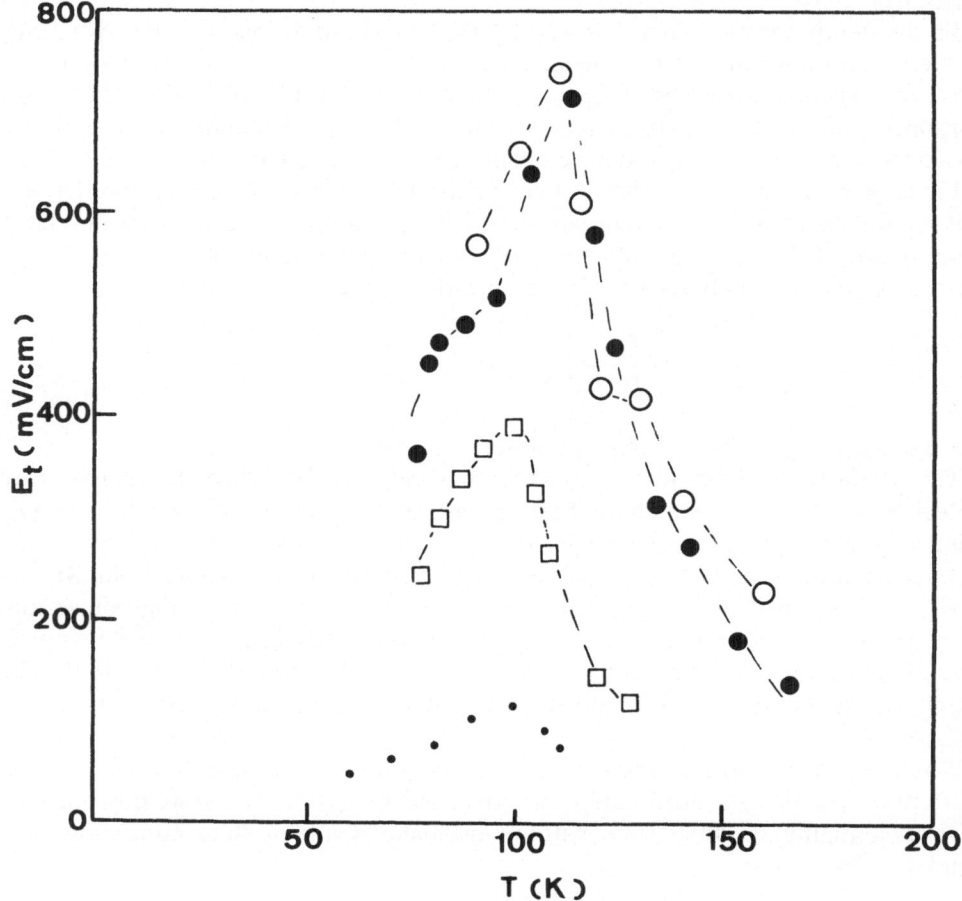

Fig. 23. Temperature dependence of the threshold electric field E_t for nonlinear conductivity in several samples of Rb$_{0.30}$MoO$_3$. (From [75].)

ture range (60—110 K); the amplitude of the voltage drop at E_t decreases with temperature and vanishes at ~ 110 K for samples showing a maximum of E_t.

Below 50 K, the threshold has a complicated temperature dependence [84]. With decreasing temperature, E_t decreases abruptly at ~ 40 K, the temperature at which small anomalies in the resistivity and in the magnetic susceptibility have been reported [59]. Below ~ 20 K, the threshold rises abruptly and exhibits a plateau with values as large as 10—100 V/cm, anomalously large for CDW materials [85]. However, when considerable care is taken in contact preparation E_t drops to ~ 2 V/cm [85d] (Figure 24a). Nonlinearities in $V-I$ curves below ~ 20 K were first reported by Fogle and Perlstein [49]. Below 20 K, the resistivity becomes extremely large ($\sim 10^{12}$ Ω cm at 4.2 K) and a macroscopic polarization of the sample appears [86a]. The sudden rise of E_t is accompanied by a maximum in the dielectric losses [86b] and an increase of the backscattering yield [121] (see also 3.6.1).

In the temperature interval 4—20 K, the threshold is marked by switching, hysteresis and intermittent voltage spikes [85]. At 4.2 K, the switching is such that the differntial resistance above E_t is six orders of magnitude lower than the ohmic resistance [85a] (Figure 24b). It reaches nine orders of magnitude in more recent pulsed measurements [85d] and the current density reaches 10^4 A/cm^2.

Fleming et al. [83] have shown that, in the temperature range explored (24—110 K), the excess CDW current was thermally activated with an activation energy close to that of the ohmic conductivity σ_n. A more quantitative description of I_{CDW} has been given by Mihaly et al. [85f]. These authors find:

$$\sigma_{CDW} \propto \sigma_n \frac{E_t}{E} \left(\frac{E}{E_t} - 1 \right)^\alpha$$

over a broad range of field and temperature.

The nonlinear contribution I_{CDW} to the current vanishes when the temperature is decreased. These results have been interpreted in terms of a CDW strongly influenced by the presence of normal carriers.

At low temperatures ($T \leqslant 30$ K), just above a large switching threshold, Maeda et al. [85] have investigated the transient current response to a voltage pulse and the response to a continuous dc voltage. Quantized steps in the current response to a voltage pulse are observed and are similar to that found in TaS$_3$ [87]. The magnitude of the steps was found to have values that are multiples of a current unit.

When the dc current response to a dc bias voltage is recorded at $T \sim 20$ K, intermittent oscillations are found just above switching [85]. The system is hopping between a highly resistive state without oscillations and a state consisting of a sequence of oscillations.

(d) *Discussion.* The threshold values in the explored temperature ranges are several orders of magnitude smaller than that estimated from optical studies of the phase mode which would be of the order of 100 KV/cm [88, 89]. In the following,

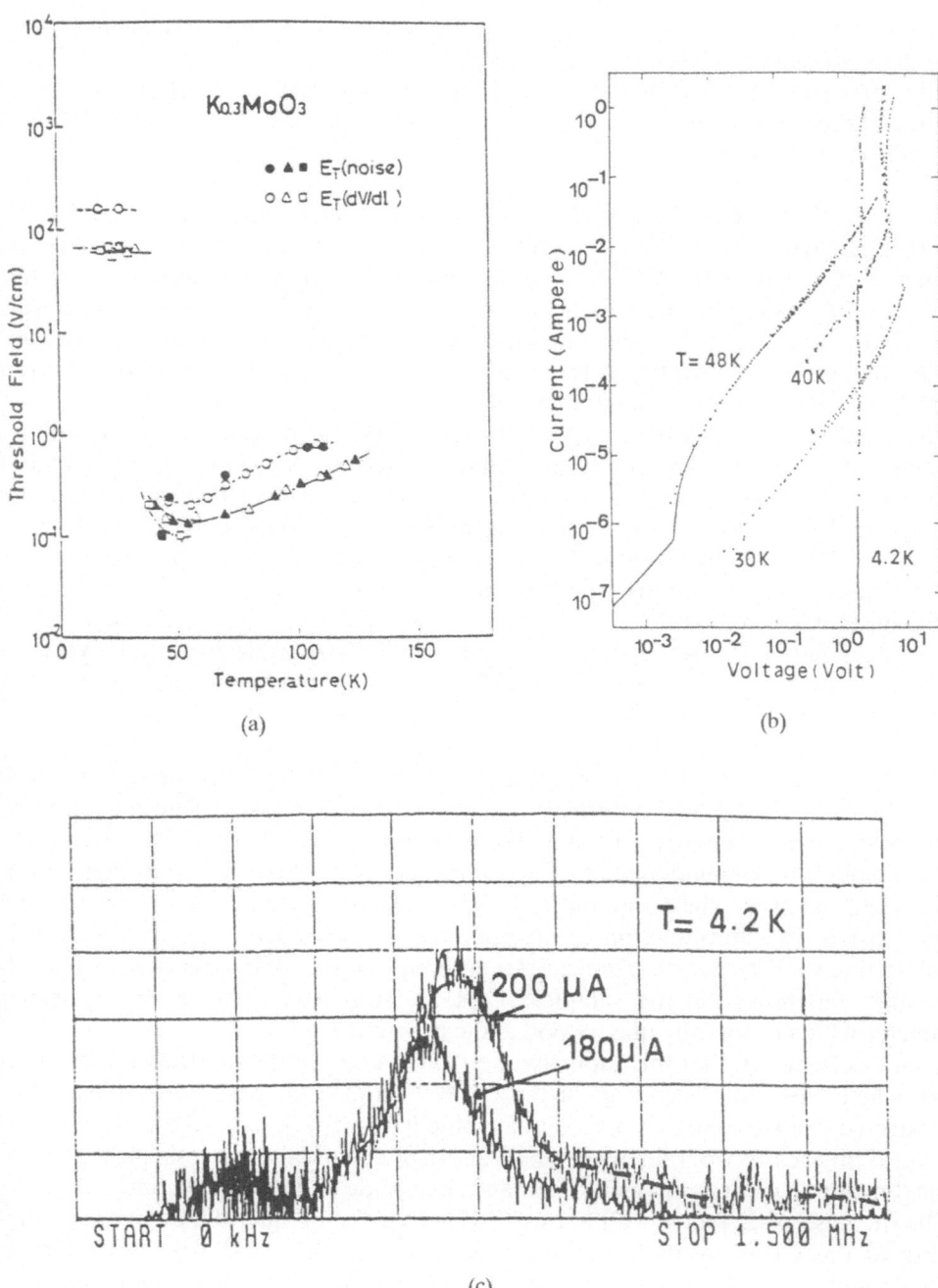

Fig. 24. (a) Temperature dependence of the threshold field of different samples of $K_{0.3}MoO_3$ showing the abrupt rise at low temperatures Logarithmic scale for E_t (From [85a]); (b) $I—V$ characteristic at low temperatures (From [85d]); (c) Noise spectra at 4.2 K. (From [85d].)

we identify tentatively the threshold as the field where CDW defects, instead of the whole deformable CDW, start to move [30].

The temperature dependence of E_t is complicated and unusual for a CDW material. Three temperature ranges have to be distinguished: $50 \leqslant T \leqslant 100$ K; $T \geqslant 100$ K; $T \leqslant 20$ K.

50 K $\leqslant$ T $\leqslant$ 100 K. While E_t decreases with increasing temperature in CDW materials, an increase of E_t is observed. If one assumes that E_t corresponds to depinning of CDW defects [30] (discommensurations or phase dislocations) the increase of E_t would be due to an increase of the viscosity of these structural defects which would be coupled to slowly diffusing lattice point defects. Below 100 K, the motion of CDW defects above E_t would induce a drag of 'atmospheres', deformable clusters of impurities.

The small switching which appears below ~ 100 K in some samples and is sometimes accompanied by precursor voltage pulses would be very similar to the yield drop in stress—strain curves $\sigma(\varepsilon)$ in crystal plasticity. The yield point is accompanied by emission of an acoustic noise which would correspond to the emission of broad band noise at E_t. In crystal plasticity, the yield point is due to the breakaway of dislocations from the 'atmospheres' of solute atoms which they are dragging at low velocity.

The switching at E_t, when it exists, is not fully reproducible from one cycle to the next. This would correspond to different metastable configurations of the CDW defects upon cycling [22].

T $\geqslant$ 100 K. Above 100 K, the impurities would be so mobile that they can always follow the motion of CDW defects. In this picture, E_t decreases when the velocity of impurities is high enough to follow the motion of CDW defects. The maximum in the temperature dependence of E_t, when it exists, would be similar to that found in the yield stress versus temperature in some alloys [12, 31]. The discrepancies which exist in the literature on the temperature dependence of E_t above 100 K could be due to different amounts and to different natures of crystal defects and of impurities depending on the samples. These defects appear to be an important parameter which is not fully understood at the present time.

It should be noted that the anomaly at ~ 100 K corresponds to the temperature below which the wave vector q_b of the CDW becomes temperature independent and extremely close to the commensurate value $0.75b^*$ [67]. In addition, anomalies of several physical properties have been reported: electrical hysteresis [43], proton channeling [90a], nuclear magnetic resonance lineshape and relaxation times [91].

The increase of E_t below 100 K can be viewed as being due to a decrease of the rigidity of the CDW as the temperature is increased. The CDW would interact more strongly with the normal carriers at high temperatures and would become stiffer at low temperatures and therefore would be less sensitive to pinning forces [83b].

T $\leqslant$ 20 K. Below ~ 20 K the threshold is rather large, marked by a switching

[85, 86], and is nearly temperature independent; it is accompanied by periodic low frequency oscillations (Figure 24c). In this temperature range, one should note that anomalies in the photon channeling yield [121] and in the EPR spectra [113] have been reported.

Fogle and Perlstein [49] have suggested that the threshold could be due to a single particle process, namely an electrical breakdown. It is well established that impact ionization of neutral shallow donors in a semiconductor such as n-GaAs at ~ 4 K leads to an electrical breakdown (at $E_t \sim 5$ V/cm) accompanied by quasi-periodic spiky noise [92]. Electrical breakdown occurs inhomogeneously via nucleation of current filaments i.e. conducting paths of low resistances generated by impact ionization.

Another single particle process responsible for nonohmicity at low temperature would be a Zener tunneling; this mechanism is very unlikely since it should not lead to switching with hysteresis and quasi-periodic noise. One can also rule out a Gunn effect since fields as high as kV/cm are involved in this process. An interpretation which seems possible would be a dielectric breakdown; the macroscopic polarization which appears at low temperature ($T \sim 4$ K) is found to diverge near E_t. In this connexion, the blue bronze would be comparable to a disordered dielectric material or to a dipolar glass (see [3, 4, 6]). In the dielectric medium, no screening by normal carriers is available; a configuration of CDW domain walls elongated along the b-axis is likely since they are not electrically charged. Such walls would be analogous to Bloch walls in ferromagnetic materials. They would move perpendicualr to b under an applied electric field $\| b$, thus leading to a CDW excess current for E larger than the threshold field.

It is important to point out that recent narrow band noise measurements by Mihaly and Beauchêne [85d, 85e] at 4.2 K show that the ratio ν/J_{CDW} of the frequency of the narrow band noise to the CDW current density is the same as the ratio ν/J_{CDW} at 77 K, the temperature at which the frequencies have unambiguously been attributed to the sliding motion of the CDW. The estimated average velocity of the CDW is as large as 1 m/s. This therefore establishes that the threshold at $T \sim 4$ K is due to the motion of the CDW. Frequency dependent conductivity studies at 4 K [85g] are consistent with this picture. The gigantic increase of conductivity from a highly insulating state above E_t [85d] would thus be due to a motion with a reduced damping of the CDW. This process could be an approach to the Fröhlich superconductivity. Such a picture seems more likely than the local conversion of condensed electrons to normal electrons which involves very large electric fields [86c].

3.4.3. Broad Band Noise

The low frequency broad band noise which appears at E_t is several orders of magnitude larger than in normal semiconductors; it has been investigated by Maeda et al. [85]. The noise spectrum does not obey a power law. These authors attribute the noise to transitions among metastable CDW states. Tsutsumi et al. [76] have reported an increase of the broad band noise near 100 K in their differential conductivity measurements.

At low temperatures ($T < 20$ K) these authors have investigated the broad band noise in the highly resistive state. In this temperature range, the noise would be related to intermittent oscillations between the pinned and unpinned states.

The analogy between broad band noise and acoustic emission in a tensile test in metallurgy is discussed in [30].

3.4.4. *Periodic Voltage Oscillations*

The high frequency voltage oscillations (also called narrow band noise (NBN)) induced by the sliding motion of the CDW can be visualized by Fourier analysis of the noise voltage [3, 43], or by current pulses experiments in which the NBN appears as oscillations in the transient voltage response [93], or by measuring the real part of the dielectric constant $\varepsilon'(\omega)$ as a function of the dc bias current for given excitation frequencies [94].

Figure 25 shows the Fourier analysis of the noise voltage of $Rb_{0.30}MoO_3$ for a bias current above threshold. One frequency with several harmonics is found. A quality factor at 77 K, $f/\Delta f \sim 30$ has been observed by Janossy *et al.* in needle-like samples [95]. These values remain lower than that found in $NbSe_3$ which can reach 30 000 in some samples.

As in other CDW materials, the frequency F increases linearly with the excess CDW current as illustrated in Figure 26. The slope F/J found in the best samples is ~ 12 kHz/A cm^{-2} [75b, 85f]. In the classical model the scope $F/J_{CDW} = 1/ne\lambda$ where n is the concentration of electrons condensed in the CDW and λ is the superlattice period. In the blue bronze, the theoretical value F/J_{CDW} is 12.5 kHz/A cm^{-2} in assuming a doubly-degenerate three-quarters filled valence band and $\lambda = 4b/3$. The discrepancies observed in some samples can be understood if one assumes that the CDW flows inhomogeneously in the sample; this involves a distribution of CDW velocities. For example, in the case of Figure 26, $\sim 30\%$ of the volume of the sample would participate in the CDW conduction [75]. The CDW current, close to E_t, seems to obey a law $J_{CDW} \sim (E/E_t - 1)^\alpha$ with $\alpha \sim 1$ as

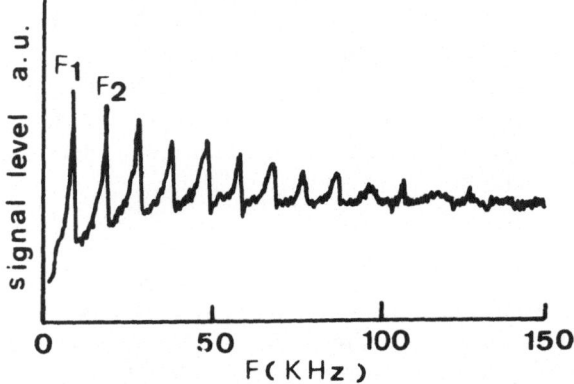

Fig. 25. Fourier transform of the noise voltage at 77 K ($I \sim 2I_t$). (From [75].)

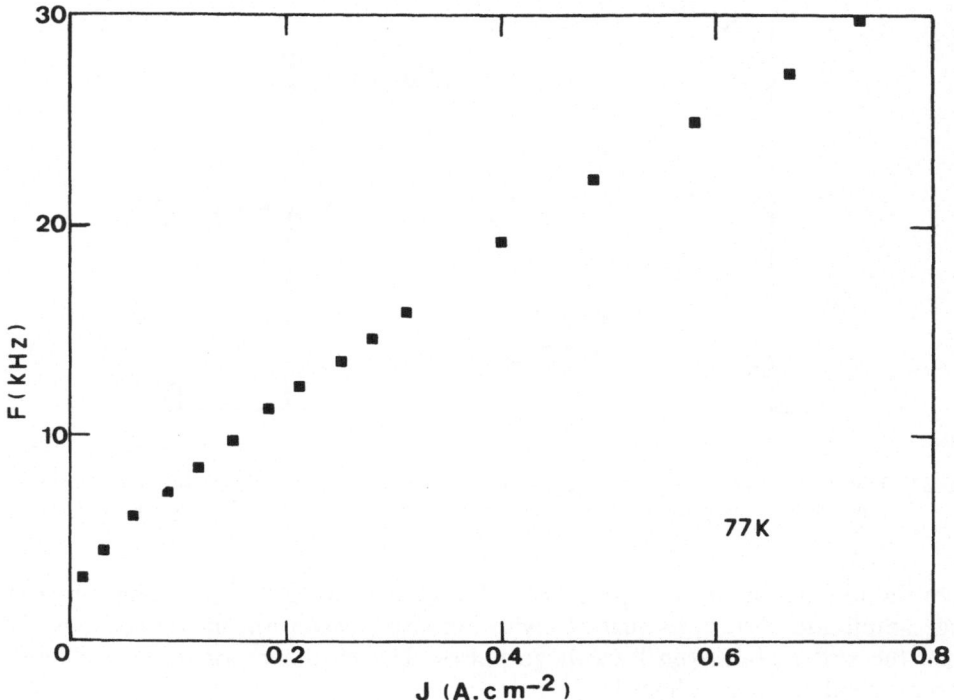

Fig. 26. Noise frequency as a function of the excess CDW current density. (From [75].)

obtained from narrow band noise measurements [95]. These results are different from theoretical predictions $\alpha = 3/2$ for a deformable medium and $\alpha = 1/2$ in the classical rigid CDW model.

Periodic voltage oscillations of large amplitude are observed on some samples; they can be visualized directly on the transient voltage response to a current pulse. Figure 27 shows a sum over 50 unipolar pulses with a pulse repetition rate of 8.5 ms. These oscillations are clearly visible and indicate a phase coherence over large time scales (~ 0.5 s).

Periodic voltage oscillations have been observed at 4.2 K with the same slope F/J_{CDW} as at 77 K [85d]. The origin of the NBN is, as in other CDW materials, still controversial (bulk effect or generated near the electrical contacts). NBN experiments with an applied thermal gradient $\Delta T = T_1 - T_2$ along the high conductivity axis have been reported with conflicting results in NbSe$_3$ (see, for example [36]). Some authors report a splitting of one frequency of the NBN into two frequencies; each of them would correspond to one end of the sample at temperatures T_1, or T_2 respectively. This led to the conclusion that the NBN was a contact effect. Other authors have reported a cascade of frequency splittings. In the case of the blue bronze, no splitting is observed [96]. These latter results have been interpreted in terms of domain structure of the CDW elongated along the high conductivity b-axis.

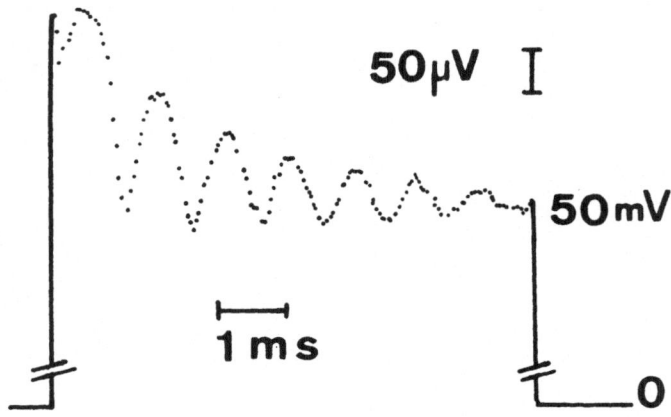

Fig. 27. Transient voltage response averaged over 50 repetitive current pulses of the same polarity. Pulse width 6 ms; pulse repetition rate 8.5 ms; $V_t = 20$ mV; $T = 77$ K. (From [75b].)

In the low temperature regime ($T \sim 20$ K) and above the switching threshold, large amplitude current oscillations with impulsive waveform have been observed when the system is driven by voltage pulses. The frequency of these oscillations increases with the bias voltage [85].

3.4.5. *Very Low Frequency Phenomena*

Just below switching, pulses of transient voltage corresponding to a sudden decrease of the dc voltage value V_{dc}, appear in some samples on the $V-I$ curves for given values of I. The amplitude of the pulses is of the order of 2% of V_{dc}. These pulses vanish as a function of time with no decay of their amplitude [43]. In Figure 28 we show the number of events per unit time as a function of time at 77 K. The average time interval between two pulses increases as a function of time. For this example, extinction of the pulses occurs after several minutes. After a very small increase of the current ($I < I_t$) voltage pulses appear once again.

These pulses do not appear when the current is decreased from above E_t. Similar precursor pulses have been observed by Tessema *et al.* [97]; these authors used silver paste for contacts. When electrodeposited copper electrodes wrapping the sample are employed for contacts, instead of indium evaporation, the amplitude of the pulses is smaller. These pulses appear on the voltage contacts as well as on the current contacts.

This suggests that precursor pulses are related to current distribution inhomogeneities in the sample. The CDW would be depinned in domains with a filamentary shape. These domains would interact with lattice defects and would be unstable.

The regime of pulses bears some analogy with the sudden displacement of a single Bloch wall in a ferromagnetic material [43]. The Barkhausen noise observed when a domain wall is constrained to move with a constant velocity is similar to the regime of voltage pulses.

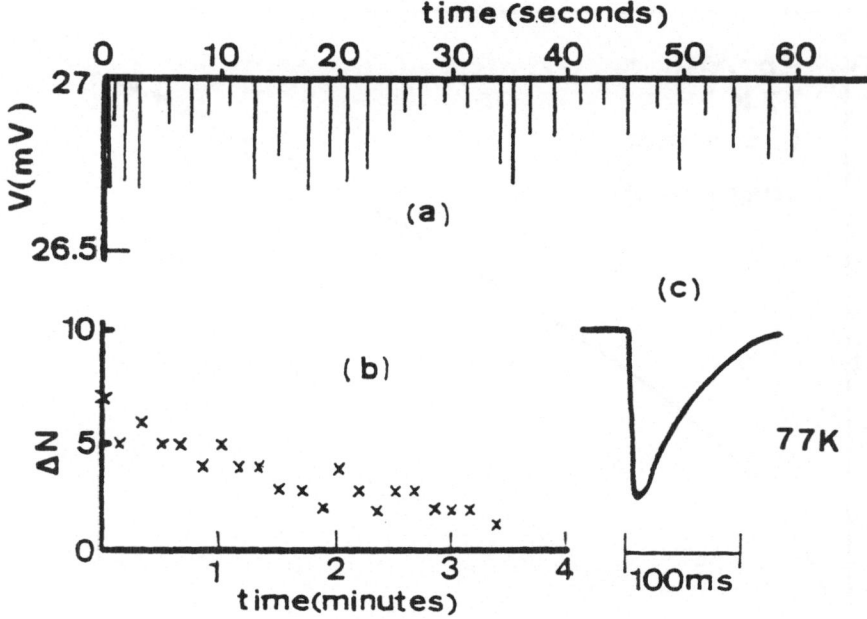

Fig. 28. Intermittent regime found for a current just below I_t. (a) Voltage pulses versus time; (b) number of events ΔN during given a time length (10 s) versus time, showing the vanishing of this regime after a few minutes; (c) Typical pulse shape (oscillogram trace). (From [53].)

In this context, these pulses would be related to the existence of charged CDW lattice defects (CDW phase dislocations or discommensurations). Their sudden displacements in the presence of impurities or other crystal defects could be responsible for the observed phenomena. The metastability of the regime of pulses would be due to trapping of CDW defects in deeper and deeper potential wells as a function of time.

Low frequency voltage pulses have also been found in TaS_3 [98], $NbS_{3+\delta}$ [99] and near the metal insulator transition of the oxide VO_2 [100]. Finite jumps of capacitance have been reported in the insulating incommensurate Rb_2ZnCl_4 [101]. One should also note that transient voltage spikes, at fixed current, have been observed in $NbSe_3$ [102] at high magnetic fields and low temperatures; they appear as a precursor to a magnetically induced switching between well defined low and high resistance states.

Well above the threshold, coherent very low frequency (~ 1 Hz) voltage fluctuations can be generated by quenching a sample from 300 to 77 K with an applied dc bias current I_c [103]. These effects occur especially in $Rb_{0.3}MoO_3$. A typical $V-I$ characteristic for a quenched sample is given in Figure 29. Voltage fluctuations generated well above threshold are superimposed on the dc voltage. Depending on the cooling rate two regimes of voltage fluctuations are found; voltage spikes for fast quenching (~ 5 s) and voltage oscillations for slow quenching (~ 60 s). The frequency of these oscillations increases with the excess CDW

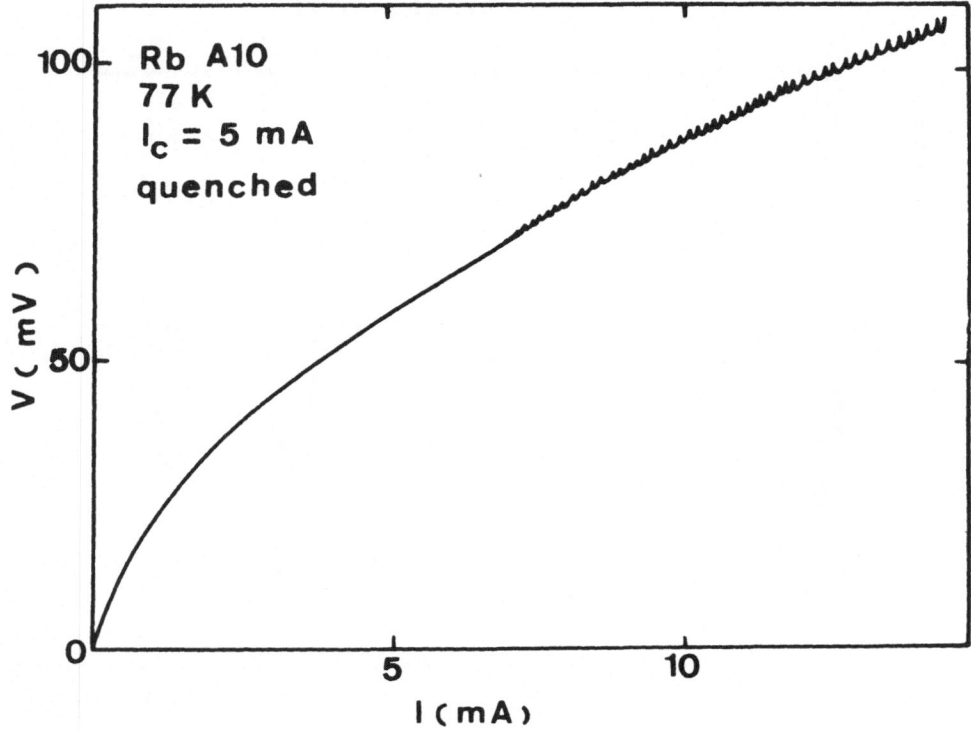

Fig. 29. *V—I* characteristic showing voltage fluctuations well above threshold for a quenched Rb$_{0.30}$MoO$_3$ sample (I_t = 0.4 mA). Current applied during cooling, I_c = 5 mA. (From [36b].)

current; the slope f/J_{CDW} is ~ 0.06 Hz/A cm^{-2}. These results are similar to the onset of 'serrations' in stress—strain curves $\sigma(\varepsilon)$ observed under certain conditions in Al alloys for example (also called Portevin—Le Chatelier effect) [30]. In this situation, irregular alternation of stress increases and decreases are observed in the elastic regime of the $\sigma(\varepsilon)$ curve. In crystal plasticity this effect occurs when the diffusion velocity of solute atoms almost matches the drift velocity of dislocations. In the case of the blue bronzes, the observed oscillations could be related to mobile crystal defects interacting with CDW defects [12].

3.4.6. *High Velocity Sliding of the CDW at Low Temperature*

At low temperature, normal carriers freeze out and at 4.2 K the resistivity is higher than 10^{12} Ω cm. Above a threshold voltage of a few volts, the differential resistivity is unusually low, less than 10^{-4} Ω cm. Narrow band noise, real time oscillations, pulse memory effects and frequency dependent conductivity [85g] studies at 4.2 K indicate that these phenomena arise from sliding of the CDW [85d]. These results demonstrate that a dramatic change in the damping mechanism occurs. In this regime, the velocity of the CDW is unusually high, ~ 1 m/s. This raises the possibility of Fröhlich superconductivity of a rigid CDW.

At higher temperatures up to ~ 40 K, this threshold voltage coexists with a small usual threshold for a strongly damped CDW motion [83b].

3.5. HYSTERESIS AND METASTABILITY PHENOMENA

3.5.1. *Introduction*

It is generally accepted that a pinned CDW is in a state that is out of equilibrium. The metastability of a CDW is related to random pinning by impurities and can be explored only by taking into account the large number of internal degrees of freedom of the CDW [24].

The disorder due to random pinning leads to glassy behavior of the CDW: a linear term in the specific heat at very low temperatures [64], a non-Debye relaxation, a remanent polarization [104], ac conductivity at low frequencies obeying the empirical law [105] $\sigma(\omega) \sim (i\omega)^\alpha$ are the signature of this behavior.

3.5.2. *Low Field Regime*

If one measures the resistivity vs temperature, it is not possible to trace the true equilibrium value of $\rho(T)$ even if the temperature is changed arbitrarily slowly. After the sample has been cooled from 300 K to a given temperature T, an increase of the resistance R as a function of time is observed at the temperature T. When the sample is warmed up, a decrease of the resistance is found at a fixed temperature. The equilibrium value of R lies between the values obtained upon cooling and warming [106].

After a thermal quenching from 300 K to 77 K, the relaxation of the resistance obeys a power law, $R(t) = R_0(1 - at^{-\alpha})$ with $\alpha \sim 0.4$ [107]. The resistance does not drift monotonously as a function of time, it exhibits some plateaus or weak spikes depending on samples [108]. This is comparable to the small discontinuous jumps ('quakes') observed as a function of time by Ong *et al.* [109] in the low field resistance of $NbSe_3$. Similar jumps have also been reported in the case of $(NbSe_4)_{10/3}I$ [110]. The time dependence of the low field resistance is less pronounced in $Tl_{0.30}MoO_3$ [74].

In electron irradiated samples, the phenomena described above are more striking. In particular, a remarkably large thermal memory effect has been reported [106] and is illustrated in Figure 30. After cooling down to a temperature T_A from a virgin state and after relaxation of the resistances during several hours at T_A, the temperature dependences of the resistance below T_A is found to be weaker than that observed upon a continuous cooling. Upon heating a pronounced resistance anomaly is found in the vicinity of T_A which reflects the memory of the temperature at which relaxation was performed. These anomalies are removed by annealing at room temperature.

This effect, similar to that found in incommensurate ferroelectrics such as thiourea, has been attributed to interactions of the CDW with mobile lattice defects diffusing until they reach a spatial distribution more compatible with the modulation, in analogy with the defect density wave model proposed by Lederer *et al.* [111], involving a periodic concentration modulation for the impurities.

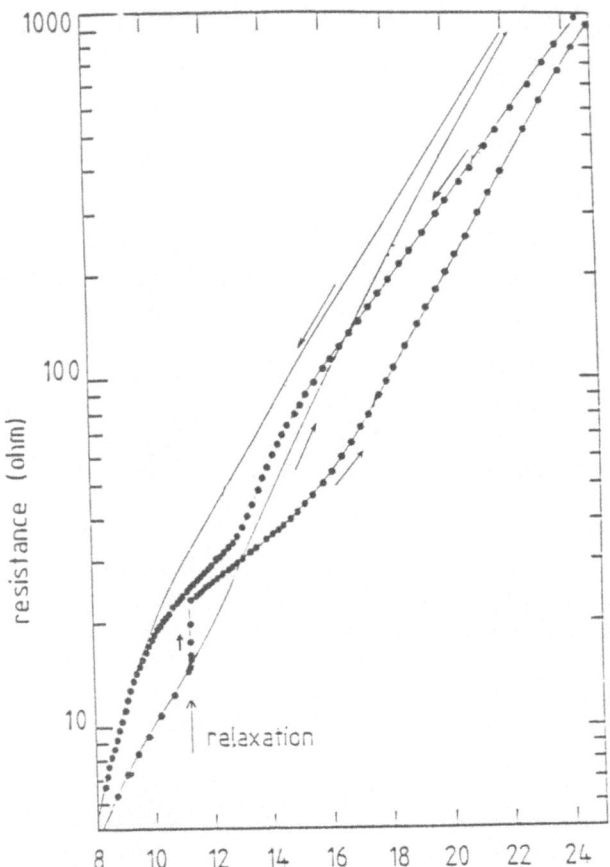

Fig. 30. Logarithm of resistance versus reciprocal temperature for an irradiated sample of
$K_{0.30}MoO_3$ (dose: 12 mC/cm² of 2.5 MeV electrons). Continuous line: cycle with temperature varying
at a constant rate. The arrow shows the relaxation for several hours. (From [106].)

In the vicinity of 180 K, the low field resistance is still time dependent and
reaches an equilibrium value after a few hours [112]. On the other hand, in this
temperature range, a change of the NMR [87]Rb line shape as a function of time has
been observed [91]. These two classes of results suggest that, near 180 K, a
kinetics of the Peierls gap occurs. In the framework of the defect density wave
model, the order parameter is time dependent and leads to a change of the
resistance as well as of the electric field gradient on the Rb ions.

The low field resistance depends not only on the thermal history but also on the
electrical history of the sample at a given temperature. Figure 31 shows the
differential resistance dV/dI as a function of the dc bias current for $Rb_{0.30}MoO_3$.

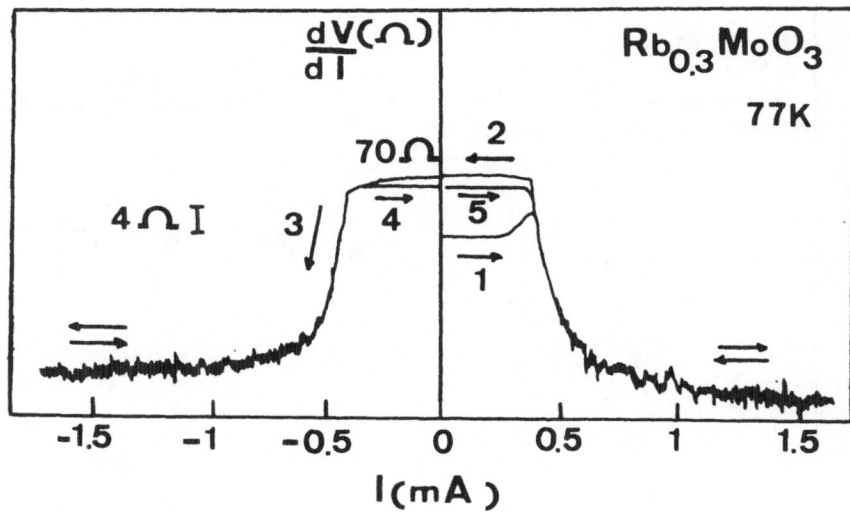

Fig. 31. Differential resistance dV/dI as a function of the dc current bias at 77 K. 1 refers to the virgin state. (From [36b].)

A hysteresis is found when the current has been swept above the threshold I_t up to a given value I_{max}. If one denotes R_1 the low field resistance in the virgin state and R_2 the resistance after the current has been swept above I_t, one can define an isothermal remanent relative resistance (IRR) as $\Delta R/R = (R_2 - R_1)/R_1$. After a full cycle (2—5), the following cycles are reproducible if one keeps I_{max} constant. The resistance value R_2 in the repinned state is more stable as a function of time than R_1.

Figure 32 summarizes the effect of the thermal and electrical history of the sample on the low field resistance. After cooling from 300 K to 77 K with an applied current, the Ohmic resistance R_{th} is found to be larger than the resistance R_1 found under zero current cooling. One can label this increase $\Delta R/R = (R_{th} - R_1)/R_1$, the thermoremanent relative resistance (TRR). The TRR increases noticeably when the current applied during cooling is larger than I_t at 77 K. The TRR becomes vanishingly small above ~ 130 K. These TRR and IRR have some analogies with remanent magnetization of spin glasses. These properties are characteristic of systems with a large number of nearly metastable states.

As reported in Section 3.2 in the low temperature semiconducting state, the low field resistance is extrinsic and arises from impurity levels in the Peierls gap. In the 'pure' samples these levels possibly result from nonstoichiometry and may correspond to paramagnetic Mo^{5+} localized donor levels. These states have been detected by electron paramagnetic resonance at low temperatures [113]. (See Section 3.6.)

The time dependence of the resistivity may be due either to a change of the relative positions of these levels and of the conduction band or/and to a change of

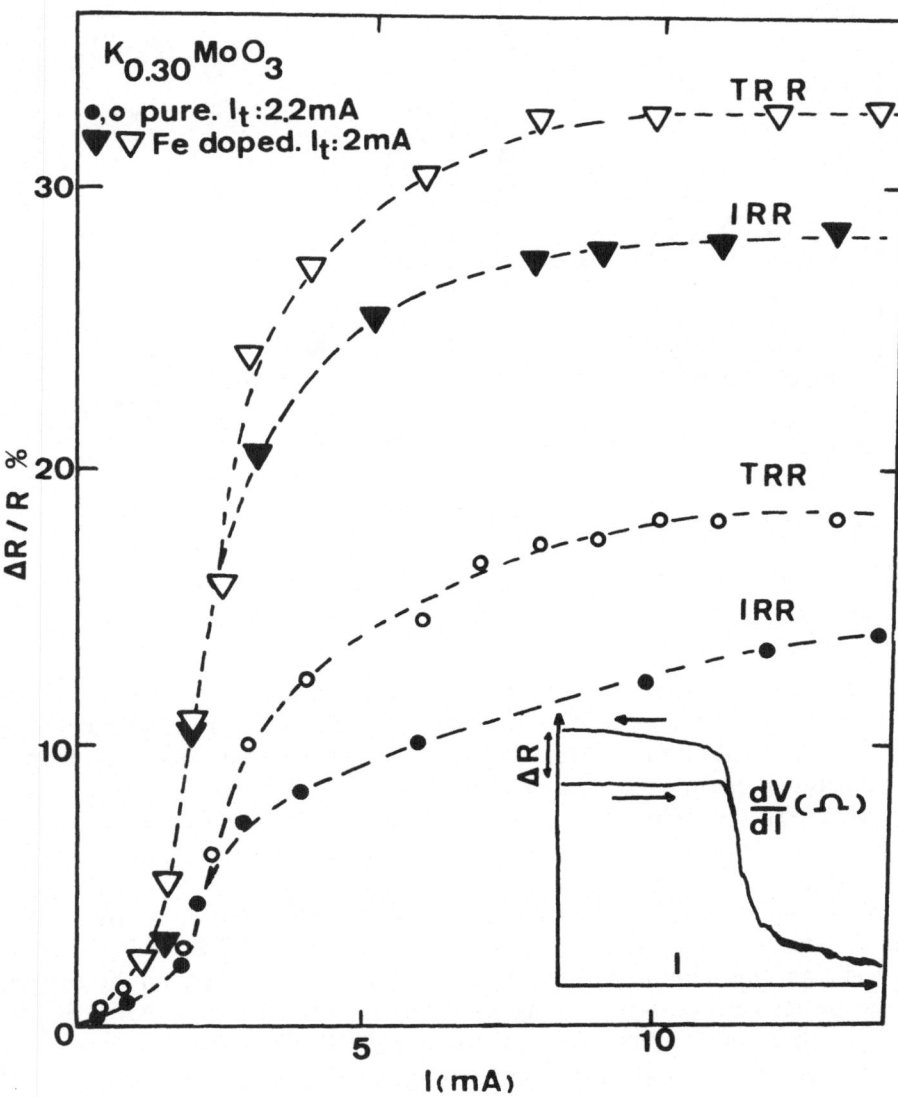

Fig. 32. Thermoremanent relative (TRR) and isothermal relative remanent resistance (IRR) of $K_{0.30}MoO_3$ at 77 K. The horizontal axis corresponds to the current applied during cooling for the TRR and to I_{max} values for the IRR. (From [36b].)

their metastable populations. If the relevant defects are coupled to the CDW, the motion of the CDW may induce both types of phenomena.

3.5.3. Nonlinear Regime

Above thresholds, the $V-I$ curve is not always stable as a function of time. The excess CDW current may decrease with time as indicated in Figure 33; this

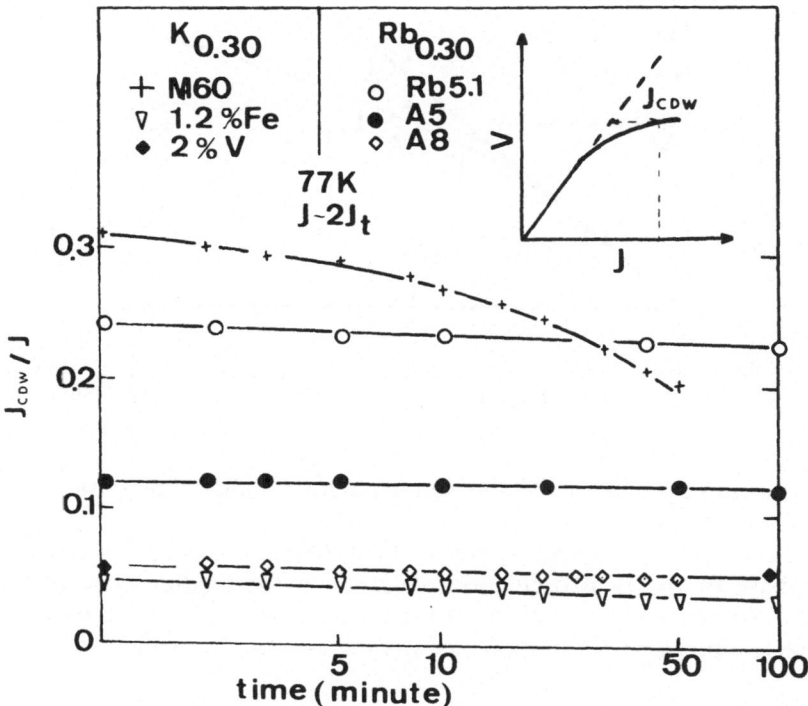

Fig. 33. Decay of the CDW current density J_{CDW}, at fixed current density J_{CDW} as a function of time (logarithm scale). Upper right inset shows how J_{CDW} is obtained. (From [36b].)

decrease is accompanied by either a decrease or an increase of the frequencies of the narrow band noise, or sometimes by the splitting of some frequencies. In $Rb_{0.28}MoO_3$ a more pronounced decay of J_{CDW} is observed. These results indicate the importance of nonstoichiometry in metastability phenomena and suggest that the mobile defects involved may be related to alkali vacancies. These time dependent effects have been attributed to a progressive pinning of CDW defects as a function of time. CDW defects would reach after some time potential barriers that they cannot overcome. This is similar to the dislocation creep in metallurgy.

The time dependence of the frequencies could be related to change in the current density distribution. Near threshold a large hysteresis is observed. In Figure 34 are shown the thresholds $E_{t\uparrow}$ obtained when the current is increased up to I_{max} and $E_{t\downarrow}$ obtained when I is swept back to zero with the same sweep rate. If the sweep rate is much slower when the current is decreased from I_{max}, then $E_{t\downarrow}$ is closer to $E_{t\uparrow}$. After decreasing I above I_{max}, if one fixes the current at a value I_A, the voltage is found to drift slowly from V_A to V'_A.

3.5.4. *Pulse Memory Effects*

As in other CDW materials, the response to a voltage pulse larger than the threshold depends on the polarity of the preceding pulse. A fast response is found

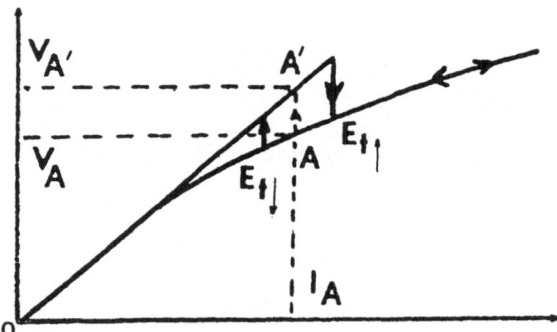

Fig. 34. $V-I$ characteristic showing hysteresis near E_t. $E_{t\uparrow}$ (resp. $E_{t\downarrow}$) are the threshold values found upon increasing (decreasing, respectively) the dc bias current. (From [36b].)

if two consecutive pulses have the same polarity while a sluggish response is observed for bipolar pulses [77].

In addition to this pulse sign memory effect, Fleming *et al.* [114] have reported that the phase of the final oscillation observed in the leading edge of the pulse remained fixed relatively to the end of the pulse as the pulse width was increased. The system exhibits a pulse duration memory that is also observed in NbSe$_3$ [115].

3.5.5. Remanent CDW Polarization

Cava *et al.* [116] have found that $K_{0.30}MoO_3$ can act as a current source upon heating after the sample has been frozen into a metastable state by cooling down to 4.2 K under an applied field. The weak thermally stimulated depolarization current I_{TSD} released by heating depends on the heating rate; I_{TSD} has a maximum near 30 K, a temperature below which E_t is large and temperature independent.

The time dependence of the depolarization current $I(t)$ has been measured by Kriza and Mihaly [104] between 25 and 65 K. Over six decades in time (from 10^{-5} s to 1 s), the time decay of the polarization P obeys a stretched exponential law, $P = P_0 \exp -(t/\tau(T))^n$, well known in glassy materials. The relaxation time τ is thermally activated with an activation energy close to that of the resistivity. This indicates that the collective and one electron excitations are interrelated.

The static polarization aroud 4.2 K shows a large hysteresis when one sweeps the voltage applied to the sample. Nonzero remanent polarization is found for voltages smaller than the threshold value [86]. As the voltage approaches threshold value, the remanent polarization becomes large, close to that of a ferroelectric such as BaTiO$_3$, and seems to diverge at threshold (Figure 35). These results are reminiscent of those observed in ferroelectrics. This remanent polarization has been described by Mihaly *et al.* [86d] using a computer simulation of the Fukuyama—Lee Hamiltonian. Rearrangements of neutral CDW domain walls, possibly elongated along the *b*-axis in order to minimize Coulomb interactions, might be involved in the observed hysteresis. A situation with domain walls perpendicular to the *b*-axis seems unlikely since they would be charged and no

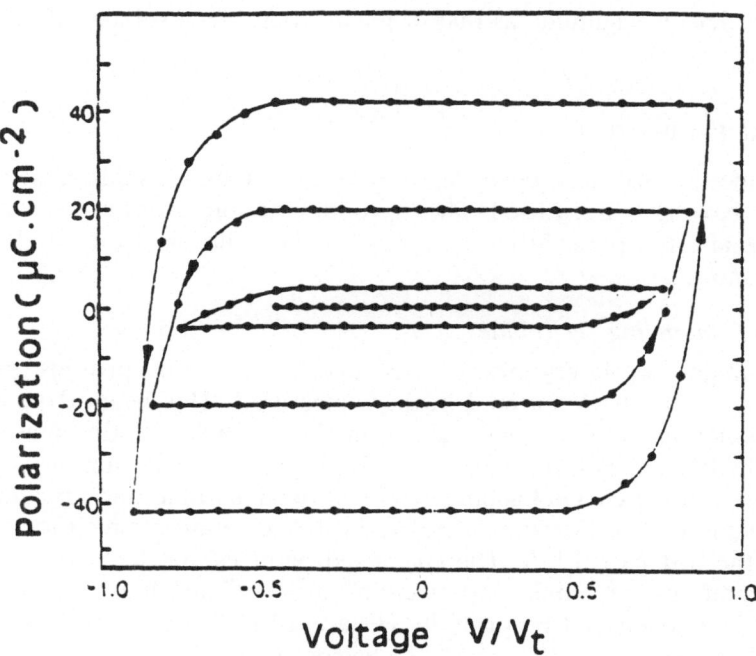

Fig. 35. Hysteresis cycles obtained by sweeping the voltage across the sample showing the remanent polarization $E_t \sim 30$ V/cm. $T = 4.2$ K. (From [86a].)

conduction electrons are available at low temperatures to screen their neutral Coulomb interactions.

3.5.6. *Field Induced Deformation of the CDW*

Structural evidence of metastable CDW states have also been reported. Tamegai *et al.* [117] have found a change of one of the transverse components of the wavevector perpendicular to the *b*-axis when an electric field is applied along *b*. They have proposed that this might be due to an inhomogeneous distribution of the current density in the sample.

Fleming *et al.* [118a] have observed at low temperature an increase of the transverse width of a superlattice peak after the sample has been cooled with a field larger than threshold. The field induced transverse disorder is metastable and can be removed by warming. These authors have proposed that the disorder might be due to domains. One should note that a configuration of domains elongated along the *b*-axis is compatible with this result.

More recently, Mihaly *et al.* [118b] have observed a time dependent variation of the width of a superlattice peak after application at low temperature of an electric field close to threshold. The broadening of the peak occurs with a time scale of ~ 500 s. and remains unchanged after the field is removed. This time scale is of the same order of magntiude as that of polarization [86, 116] and of EPR spectra (see 3.6.3).

3.6. LOCAL PROPERTIES

As the blue bronze has been the object of extensive studies, several local techniques have now been used. They give information both on the nature of the CDW state and the possible mechanism of metastability and on the dynamics of CDW transport.

3.6.1. *Ion Channeling Technique*

In this technique, single crystals are irradiated by small ions, protons or He^+, and the intensity of the backscattered beam is measured. This beam depends on the crystal orientation, since channels giving minimum backscattering may be found. The so-called backscattering yield, χ_{min}, is the ratio of the number of particles backscattered in the channel geometry to that obtained in a random configuration. This technique is therefore very sensitive in the detection of small ionic displacements, of the order of 0.1 Å. The channeled particles are dechanneled either by thermal vibrations, or lattice imperfections such as impurities. Studies of CDW transitions have been first reported by Haga et al. [119a] and by Nunez-Regueiro et al. [119b] in the case of the layered dichalcogenides $TiSe_2$ and $1T-TaS_2$. A first attempt at ion channeling in $K_{0.30}MoO_3$ was performed by Abe et al. with He^+ ions in the temperature range 100—300 K [120]. A minimum of the backscattering yield was observed at 180 K and was attributed to the softening of a phonon mode.

More recent experiments have been performed with 1 MeV protons, since He^+ ions could possibly produce permanent radiation damage [121]. Figure 36a shows the minimum backscattering yield obtained on $A_{0.30}MoO_3$ between 350 K and 50 K by successive decreasing and increasing temperature with heating and cooling rate of 1 K/minute. An anomalous increase of χ_{min}, starting around 100 K, is found with increasing temperatures only. Figure 36b shows the time dependence of χ_{min} measured on the heating cycle at 200 K. These results have been attributed to the interaction of CDW defects or domains interacting with lattice defects. The hysteresis and peculiar temperature dependence of χ_{min} on the heating cycle eliminate a simple relation with the CDW order parameter. They rather correspond to out of equilibrium CDW states, possibly CDW domains, slowly rearranging as a function of temperature and time. They are consistent with both electrical resistivity [121] and NMR time dependence [85]. One should note that the existence of hysteresis above the Peierls transition supports the existence of short range order above T_P. An additional anomaly of χ_{min} with substantial hysteresis has been observed in the vicinity of 20 K [121b]. This effect is rather pronounced when the sample is cooled from a virgin state. A broadening of the anomaly occurs after thermal cyclings.

3.6.2. *Mössbauer Effect*

Mössbauer studies have been performed between 6 K and 300 K on single crystals

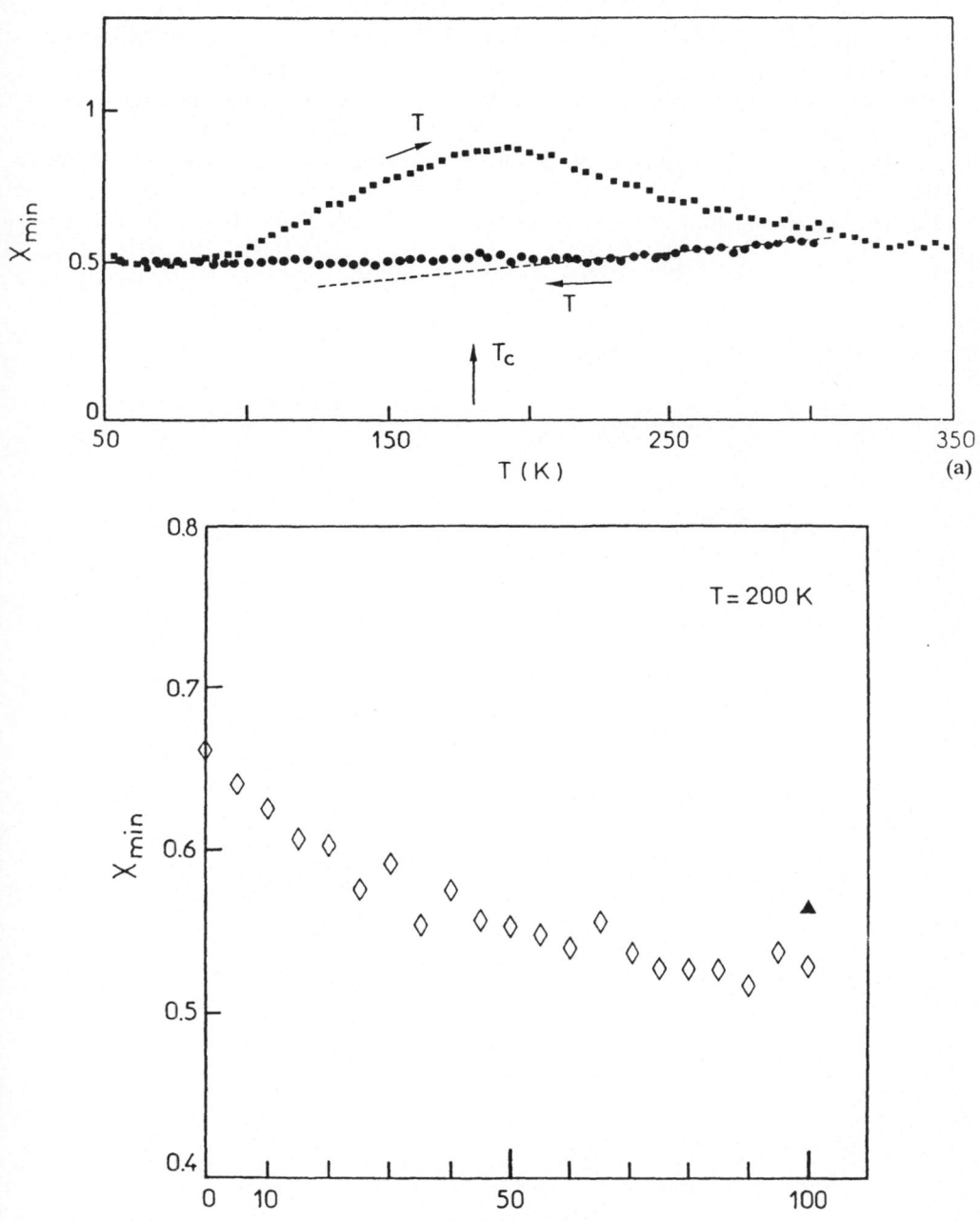

Fig. 36. (a) Minimum backscattering yield χ_{min} obtained by proton channeling experiments on $K_{0.30}MoO_3$ as a function of temperature for decreasing and increasing temperatures. The Peierls temperature (T_c) is indicated. The dotted line is a guide for the eye. (b) Time dependence of χ_{min} at $T = 200$ K (open lozenges). The closed triangle is the value measured at room temperature on a 'virgin' sample. (From [121].)

of ^{57}Fe-doped $K_{0.30}MoO_3$ containing about 300 ppm atomic of iron. Mössbauer spectra can in principle give information on the local environment of the ^{57}Fe ions and on its temperature dependence. In the case of CDW, it may also give some insight on the coupling between the Fe impurities and CDW: this is in fact the main information which has been obtained by Mössbauer studies on the blue bronze [79—80].

Figure 37a shows that the spectra obtained at 300 K and 6 K can be fitted by

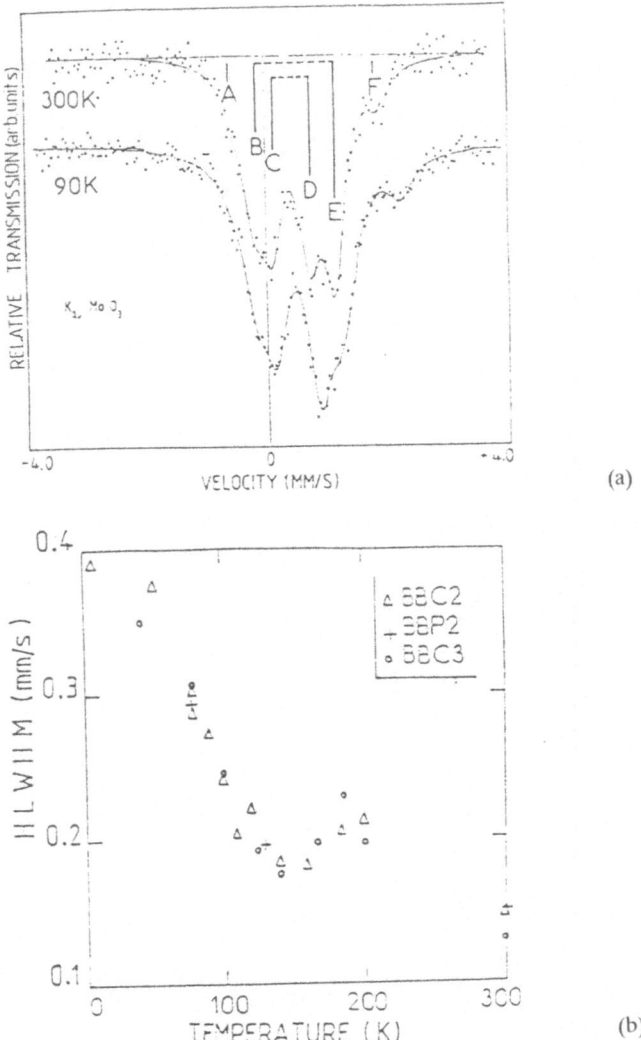

(a)

(b)

Fig. 37. (a) Mössbauer spectra of ^{57}Fe-doped $K_{0.30}MoO_3$ single crystals at 300 K and 90 K. The solid lines are a result of a six-line fit. (b) Half-linewidth at half maximum (HLWHM) of line E vs temperature for three different samples. (BBC2, BBC3 are single crystals and BBP2 is a powder sample obtained by crushing single crystals). (From [122].)

six lines and therefore by three doublets. Due to the complicated crystallographic structure of the blue bronze, a definite assignment of the Fe sites to normal Mo cation sites or to interstitial ones could not be deduced. Also no anomalous temperature variation of the Mössbauer parameters, center shift and quadrupolar splitting was found [122]. Only the linewidth shows an interesting temperature dependence (Figure 36b): it shows little change between 200 K and 120 K, around the Peierls temperature, and increases steeply only below 120 K. This has been attributed to the pinning of CDW by the Fe impurities. The absence of broadening between 180 K and 120 K shows that just below T_P the phase of the CDW at the impurity site is determined by the Fe itself: the Fe ions act as strong pinning centers for the CDW. CDW domains are therefore nucleated around each impurity. Below approximately 120 K, the steep increase of the linewidth suggests that the Fe progressively become weak pinning centers. In the limit of very weak pinning and in the incommensurate phase, the CDW phase would be random at each Fe site.

Such a temperature dependence for the pinning of CDW by impurities, is consistent with the predictions of the Lee and Rice model [21]. In this model, the pinning properties depend on the dimensionless parameter ε which is related to the ratio of the pinning energy to the elastic energy associated to the pinning (see Section 2.2.3). At nonzero temperatures, the strong pinning case corresponds to a first approximation to $\varepsilon > [(T_P - T)/T_P]^{1/2}$ and the weak pinning one to the opposite. The Mössbauer data on the blue bronze indicate that the transition between the two regimes takes place near 100 K and that ε may be of the order of 0.5 to 0.7 for Fe impurities.

3.6.3. *Electron Paramagnetic Resonance Studies*

EPR spectra had been first obtained at low temperature on the blue bronze by Bang and Sperlich [123]. More detailed studies, including the time dependence of the spectra have been performed more recently by Dumas *et al.* between 4 K and 30 K [124]. The spectra consist of anisotropic lines with g-values slightly smaller than 2 [Figure 38a]. At least one line can be identified, from a hyperfine structure characteristic of ^{97}Mo and ^{95}Mo nuclei ($I = 5/2$, total concentration 25%), to Mo^{5+} $(4d^1)$ paramagnetic centers. The total concentration of centers is of the order of 100 ppm. The blue bronze in the CDW state is a Van Vleck paramagnet and the centers are therefore due to crystal defects. These may result from a stoichiometry deviation with vacancies including local distortions and $4d$ electrons localized on some Mo sites.

The spectra at 4 K have been found to be time dependent, on time lengths of the order of one hour, when the crystals are quenched down to low temperature from above the Peierls transition (Figure 38a). The time evolution corresponds to a change of the intensity as well as of the lineshape, with an anomalous asymetric lineshape in some cases. The time variation of the intensity is indicated on Figure 38b.

These results have been attributed to relaxation of the CDW state. The idea of Mo^{5+} centers associated to CDW domain walls or solitons is appealing. Up to now, this has not been proved unambiguously.

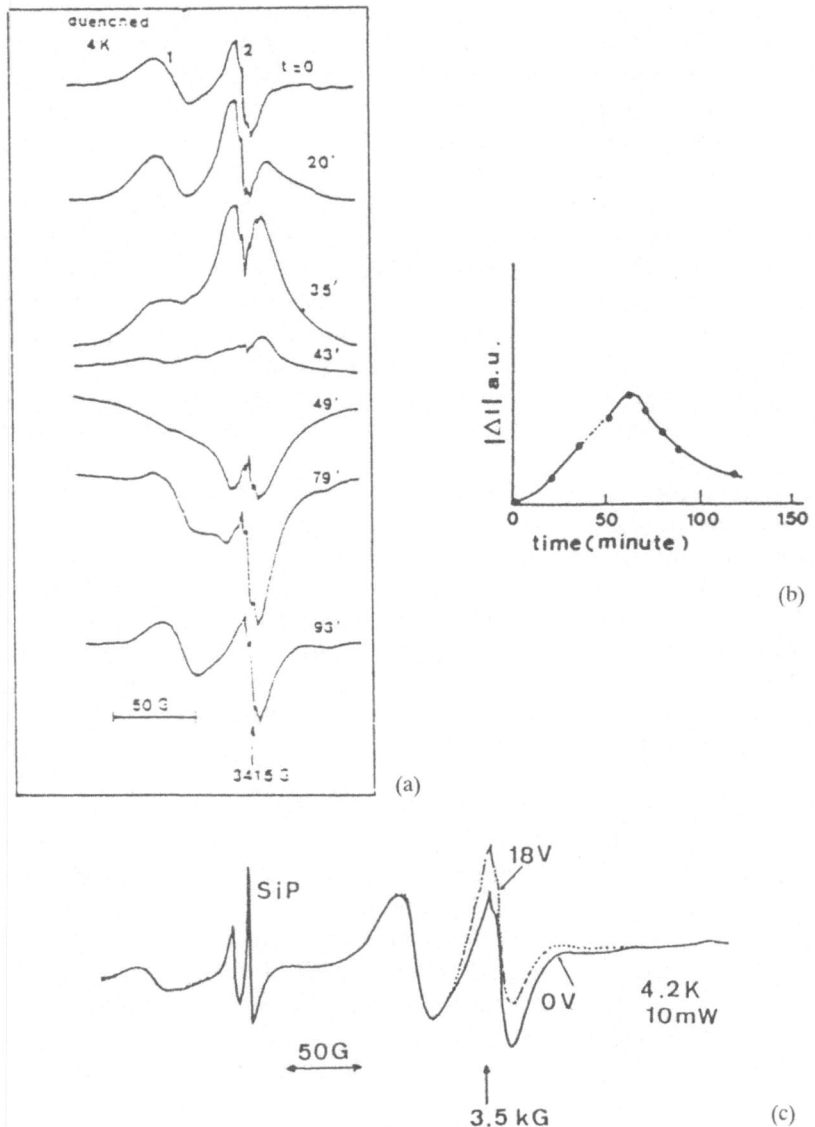

Fig. 38. (a) Evolution of the EPR spectra obtained at 4 K, 9 GHz on $K_{0.30}MoO_3$ single crystals after quenching from room temperature. The spectra have been recorded just after quenching ($t = 0$) and at successive times t. (b) Parameter related to the intensity of the line as a function of time (From [113]). (c) EPR spectra with an applied dc voltage $V_t = 50$ Volts (From [124b]).

Application of an electric field $E \sim E_t$ at low temperature causes a deforma-tion of the EPR spectra (Figure 38c). After the field is removed the EPR spectra relax with a time scale of ~ 600 s.

3.6.4. *Nuclear Magnetic Resonance Studies*

Pulsed NMR, because of its high sensitivity, is a powerful tool to investigate phase transitions involving small atomic displacements. It has now been proved that it allows detailed studies of both the static and dynamic state of a CDW state [44]. This is due to the fact that the NMR linewidth and lineshape are determined by the local hyperfine field distribution, which depends on the CDW distortion.

In the Rb blue bronze, the ^{87}Rb nuclei, with a nuclear spin $I = 3/2$, provide the possibility of studying the CDW state by NMR. Although the Rb ions lie between the 1D chains, remarkable results have now been obtained [43, 91, 125, 126].

Figure 39 shows one of the ^{87}Rb NMR lines recorded at different temperatures, above and below the Peierls transition. The double horned shape observed below T_p is characteristic of an incommensurate state, with a well-defined distribution of electric field gradient tensors at the Rb sites. The CDW order parameter can then be obtained, from the value of the quadrupole splitting. Figure 40 shows the measured order parameter as a function of temperature: this result is consistent with those obtained by X-ray and neutron studies [10].

In the dynamic CDW state, i.e. in the presencec of an electric field larger than the threshold value, the shape of the NMR spectra in principle allows a measurement of the CDW velocity [126]. In this case, the NMR lineshape is modified by the oscillation of the local hyperfine fields as the CDW slides. The frequency of the modulation of the hyperfine field is simply related to the rate of change of the CDW phase or, in other words, to its velocity. Figure 41 shows NMR spectra obtained on $Rb_{0.30}MoO_3$ for a fixed value of the electric field ($E = 15 E_t$), for different values of the CDW velocity v. The change of v is induced by slight changes of temperature, since the CDW current (and velocity) is thermally activated. One observes a striking change of the NMR lineshape, which can be simulated with a CDW velocity distribution. This distribution has been obtained directly from the electrical noise voltage frequency spectrum. Such an analysis

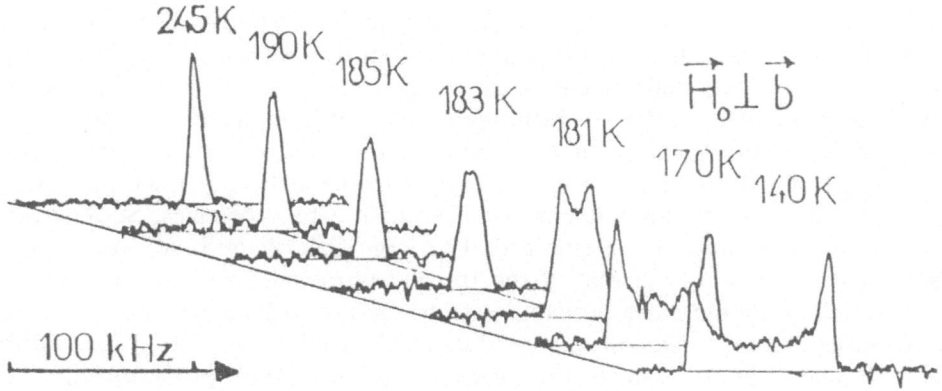

Fig. 39. ^{87}Rb NMR spectra recorded on $Rb_{0.30}MoO_3$ single crystals at different temperatures above and below the Peierls transition. (From [125b].)

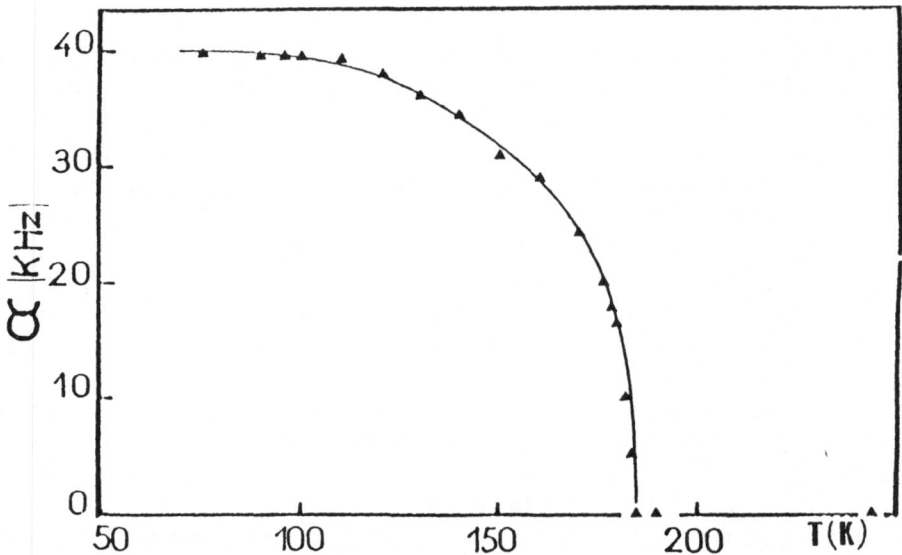

Fig. 40. Temperature dependence of the quadrupole splitting, assumed to be proportional to the CDW order parameter. (From |125b|.)

leads to an estimation of the ratio of the average local field oscillation frequency to J_{CDW} of (11 ± 1)kHz/A cm^{-2}. This value is very close to 12.5 kHz/A cm^{-2} expected from the relations $J_{CDW} = nev$ and $F = v/\lambda$, where λ is the CDW wavelength (see Section 2.2.2.).

3.7. SUMMARY

Since the discovery in NbSe$_3$ ten years ago of an entirely new form of conduction in solids due to the sliding of a CDW, new families where CDW conduction occurs, have emerged, namely the tetrachalcogenides (NbSe$_4$)$_{10/3}$I and (TaSe$_4$)$_2$I, and the blue bronzes. In all these materials this unusual conduction process has led to the discovery of a wealth of phenomena of great complexity. Up to now, no widespread agreement exists on the origin of the coherent voltage oscillations which appear above threshold. The role of interactions between lattice defects and CDW defects seem predominant in the case of the blue bronzes. Sophisticated methods of investigation have emerged. The large size of the blue bronze single crystals in comparison with that of the tri- and tetrachalcogenides have allowed studies which are made more difficult in the tri- and tetrachalcogenides series. The blue bronzes are more anisotropic conductors than NbSe$_3$ or TaS$_3$. The various hysteresis and memory effects described above are more pronounced. Local methods, NMR and EPR for example, seem promising for probing the microscopic nature of the CDW state. In the low temperature (~ 4 K) CDW state where

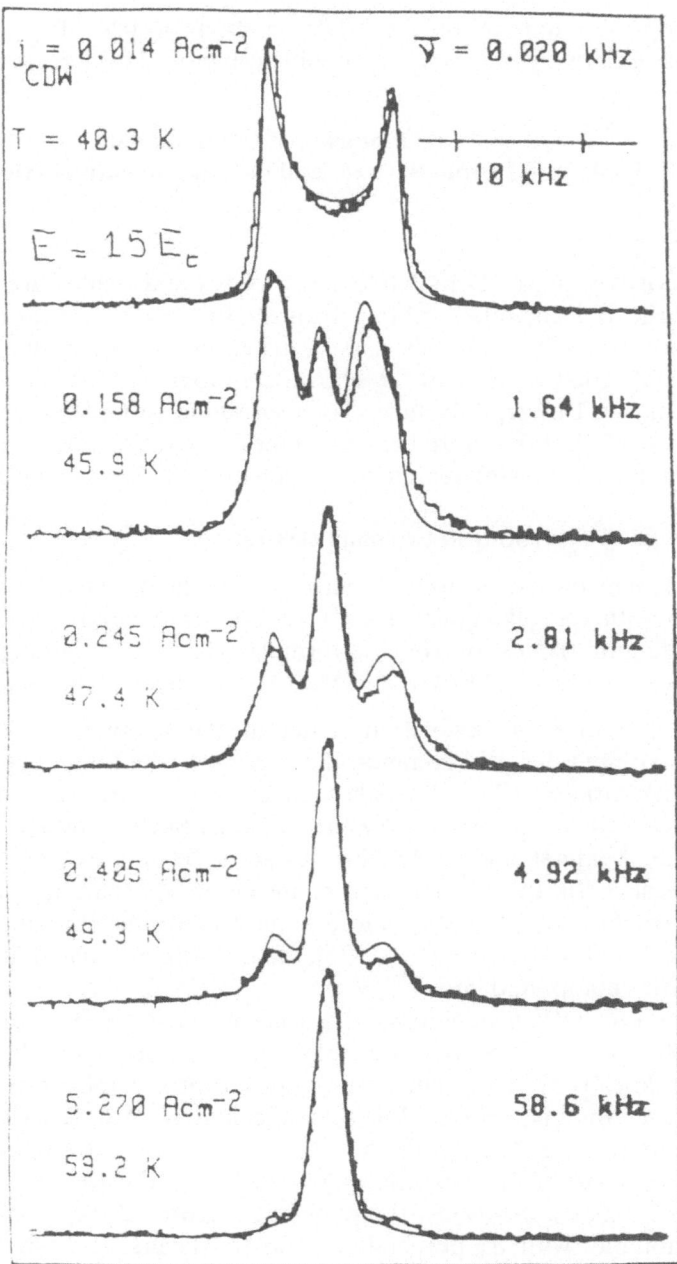

Fig. 41. Variation of the NMR lineshape and computer fitted curves as a function of the CDW current density J_{CDW}, for a value of the electric field fixed at 15 E_t. The change of J_{CDW} is due to a temperature variation (see text). (From [126].)

a dramatic drop of damping of the CDW motion occurs, the possibility of conductivity with extremely low damping should be explored further.

4. Quasi Two-Dimensional Compounds: The Molybdenum Purple Bronzes and the Molybdenum Oxides

4.1. INTRODUCTION

The properties of the quasi 2D molybdenum bronzes and oxides are not so well documented as those of the quasi 1D blue bronzes, because nonlinear conductivity has not up to now been found in this class of materials. However, the existence of CDW is well established in most of the known compounds. One should also note that one of them, the Li purple bronze, has been found to be superconducting at low temperatures [127]. It therefore belongs to the few oxides which, around 1985, have stimulated the successful search for superconductivity in transition metal-based oxides.

The relevant materials belong to two main classes:

(i) the molybdenum purple bronzes of general formula $A_{0.9}Mo_6O_{17}$, synthesized with A being either an alkali metal (A = Li, Na, K) or thallium.
(ii) the molybdenum oxides of general formula Mo_nO_{3n-1}, among which the compounds η-Mo_4O_{11}, γ-Mo_4O_{11} and Mo_8O_{23} have been the most studied.

The crystal structures, described in detail in the chapter by Vincent and Marezio in this volume, have in common the presence of infinite layers of MoO_6 octahedra sharing corners [127—133]. These layers are in most cases separated by MoO_4 tetrahedra and in the bronzes by the A^+ ions. Since conduction electrons provided by the A metal and/or by the deviation from the chemical formula MoO_3, are delocalized in these infinite layers, the compounds are quasi 2D metals. Although the structure of Mo_8O_{23} is more complicated and contains clusters of four edge-linked MoO_6 octahedra [134, 135], the low dimensional character seems to be present in this compound, too.

The existence of CDW instabilities, appearing in most physical properties, as described in the next subsection, are well supported by the observation of structural instabilities [137, 138]. These are reported for several compounds in the chapter by Pouget in this volume. We only recall here that $K_{0.9}Mo_6O_{17}$ has a trigonal structure (Figure 42a), with the c-axis perpendicular to the octahedra layers. Satellites are found in this case below 120 K in the (a^*, b^*) plane, with wave vectors (0.5, 0.5, 0). η-Mo_4O_{11} is monoclinic (Figure 42b), γ-Mo_4O_{11} (Figure 42c) is orthorhombic, with a^* perpendicular to the layers. In both cases, weak incommensurate satellites have been found below T_P (109 K), with wave vector (0, 0.23, 0) along b^*.

4.2. CHARGE DENSITY WAVE INSTABILITIES AND TRANSPORT PROPERTIES

Several transport properties, the electrical resistivity, the thermopower, the Hall

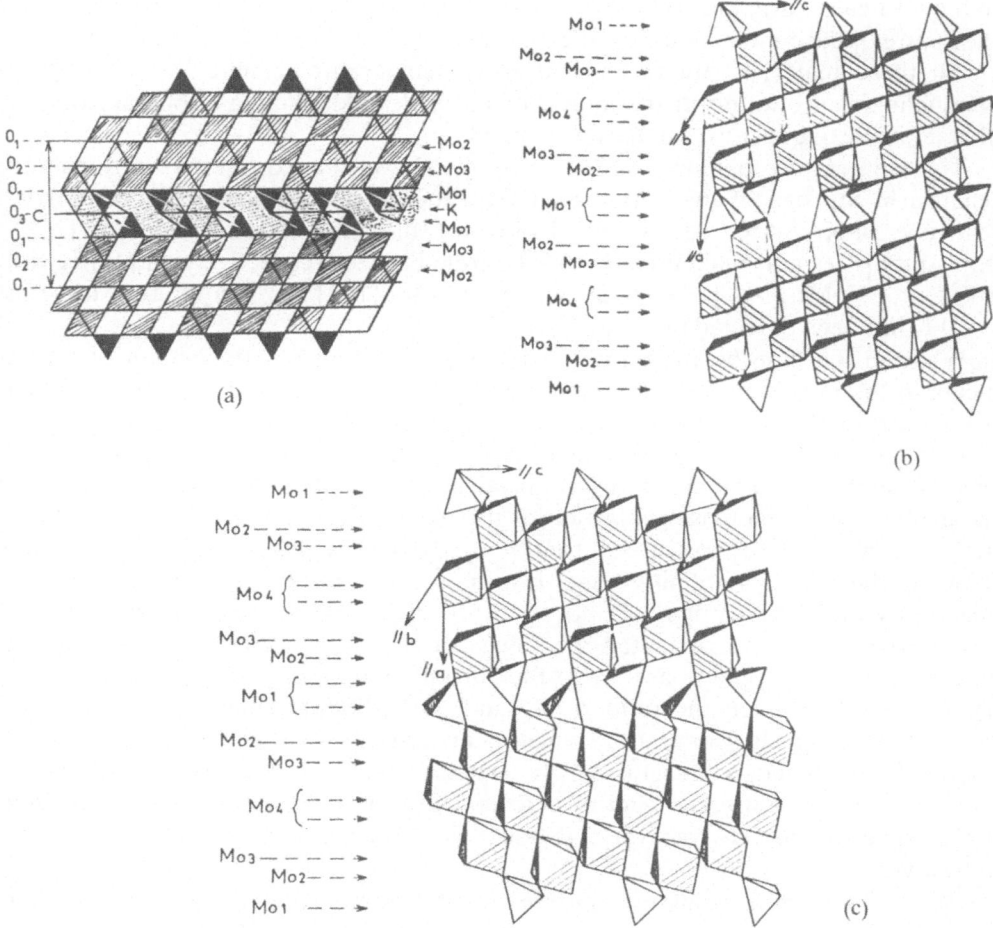

Fig. 42. Idealized crystal structures showing the oxygen polyhedra: (a) $K_{0.9}Mo_6O_{17}$ (viewed along the a-axis). The three Mo sites are shown. The alkali metal lies in the interstices between the oxygen tetrahedra. (From [129].) (b) η-Mo_4O_{11}. (From [132].) (c) γ-Mo_4O_{11}. (From [146].)

effect and the magnetoresistance have been studied in these materials: they all show anomalies or changes of behavior at the Peierls transition.

4.2.1. *Electrical Resistivity*

Electrical resistivity is the basic transport property and one usually defines the Peierls temperature from these data. In all cases, for the molybdenum oxides (γ-Mo_4O_{11} and η-Mo_4O_{11}, Mo_8O_{23}) as well as for the alkali molybdenum purple bronzes (Li, Na, K and Tl), the resistivities have been measured on single crystals with a conventional four-probe configuration. Contacts have been made by various techniques: silver or gold paste, silver paint, evaporation or ultrasonic soldering of indium. In each case, the method giving the lowest contact resistance was used. But the nature of the contact does not affect the general behavior of the conductivity.

All these materials are metallic at room temperature (Table I). The highest conductivities are found in the molybdenum oxides. All show a strong anisotropy characteristic of quasi-two dimensional metals.

The temperature behavior of the resistivity shows the existence of a transition (change in the sign of the temperature dependence of the resistivity) at approximatively 24 K, 80 K, 120 K and 113 K for the lithium [127], sodium [139], potassium [140] and thallium [131, 141] purple bronzes respectively and 98 K and 109 K for γ-Mo_4O_{11} [142, 143] and η-Mo_4O_{11} [144]. A second transition at $\sim$ 30 K has been detected in η-Mo_4O_{11}.

While in the blue bronzes, the transition temperature is independent of the A metal, in the purple bronzes the nature and probably the ionic radius of the alkali metal seem to play a role in the electronic properties.

One can notice that if the crystallographic parameters a and b in a pseudo or real hexagonal system are nearly equivalent for the four purple bronzes, the parameter c is an increasing function of the ionic radius of the A^+ ion. However, the anisotropy of the resistivity cannot easily be correlated with the distance between the Mo_6O_{17} infinite slabs. Except for $Li_{0.9}Mo_6O_{17}$ which presents an anisotropy in the (a, b) plane ($\rho_a/\rho_b \sim 300$), all the purple bronzes are isotropic in this plane. We have reported the values of $\rho_\parallel/\rho_\perp$ in Table I ($\parallel$ refers to the current in the (a, b) plane and $\perp$ to the current along the c axis). No correlation can be found between the ratio $\rho_\parallel/\rho_\perp$ and the lattice parameter c or the ionic radius. One notes also some discrepancies between different results. They might be due to the microscopic structure of the crystals which are not grown in the same conditions and to different kind of defects which affect differently the mean free path, especially in the transverse conductivity. Therefore these values are only indicative.

The thallium and potassium purple molybdenum bronzes have very similar structures with the exception that, in the thallium case, the A site may be fully occupied. These two compounds give similar resistivity temperature dependence, reproducible from one sample to another (Figures 43a, b). This is not the case for the sodium and lithium purple bronzes: the qualitative shape of the temperature variation of the resistivity below T_P is sample dependent as shown on Figure 44 for $Na_{0.9}Mo_6O_{17}$ and on Figure 45 for $Li_{0.9}Mo_6O_{17}$.

Figure 46 shows the temperature dependence of the resistivity for η-Mo_4O_{11} and γ-Mo_4O_{11}. The continuous increase of ρ below $T_P \approx 110$ K could indicate that the CDW gap opening takes place down to the lowest temperature in γ-Mo_4O_{11}. However, one cannot exclude that this is due to a poorer crystal quality and to scattering of electrons by crystal defects.

In the case of Mo_8O_{23}, the resistivity shows anomalies (Figure 47) which have been correlated to the structural transitions: an incommensurate phase appears at $T_{c_1} < 360$ K and an incommensurate—commensurate transition at $T_{c_2} = 285$ K. These transitions have been attributed to charge density waves [149].

One can predict qualitatively the band structure of these compounds. In the

TABLE I

Resistivity along different crystallographic axes for the purple bronzes, the two Mo_4O_{11} oxides and Mo_8O_{23}. (In the case of the purple bronzes, the ionic radius of the A^+ ions and the c parameters are also given.)

	$Li_{0.9}Mo_6O_{17}$	$Na_{0.9}Mo_6O_{17}$	$K_{0.9}Mo_6O_{17}$	$TlMo_6O_{17}$	η-Mo_4O_{11}	γ-Mo_4O_{11}	Mo_8O_{23}
Ionic radius of the A^+ ion (Å)	0.6	0.95	1.33	1.4	—	—	—
c parameter (Å)	12.762	12.983	13.653	14.03	—	—	—
Resistivity at 300 K (Ω cm)							
$\rho \perp c$	2.47	0.21	0.2	—	—	—	—
$\rho \parallel c$	9.5×10^{-3}	3×10^{-3}	7×10^{-4}	5×10^{-2}	2.4×10^{-4}	1.3×10^{-4}	1.8
$\rho \parallel a^*$	—	—	—	6.2×10^{-4}	4.4×10^{-2}	2.7×10^{-3}	1.5×10^{-2}
$\rho \parallel b$	—	—	—	5.6×10^{-4}	4×10^{-6}	7×10^{-5}	1.8
Peierls temperature T_P(K)	24	80	120	113, 120	109	98	—
References	[127]	[139]	[140]	[141, 131]	[142]	[144]	[149]

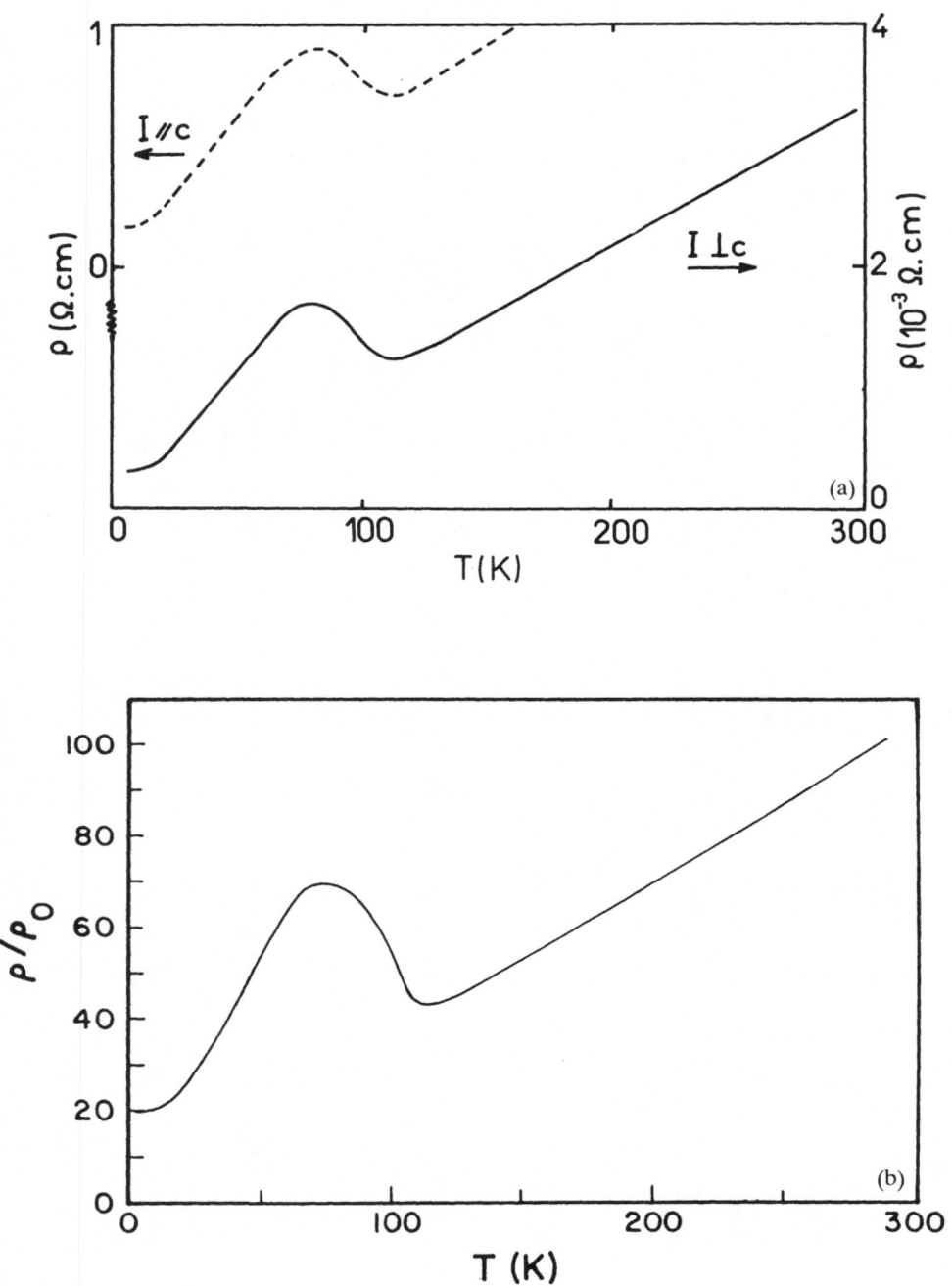

Fig. 43. (a) Electrical resistivity of $K_{0.9}MoO_{17}$ versus temperature measured in the plane $\perp c$ (full curve) and along c (dotted curve). (From [140].) (b) Electrical resistivity of $TlMo_6O_{17}$ versus temperature. (From [141].)

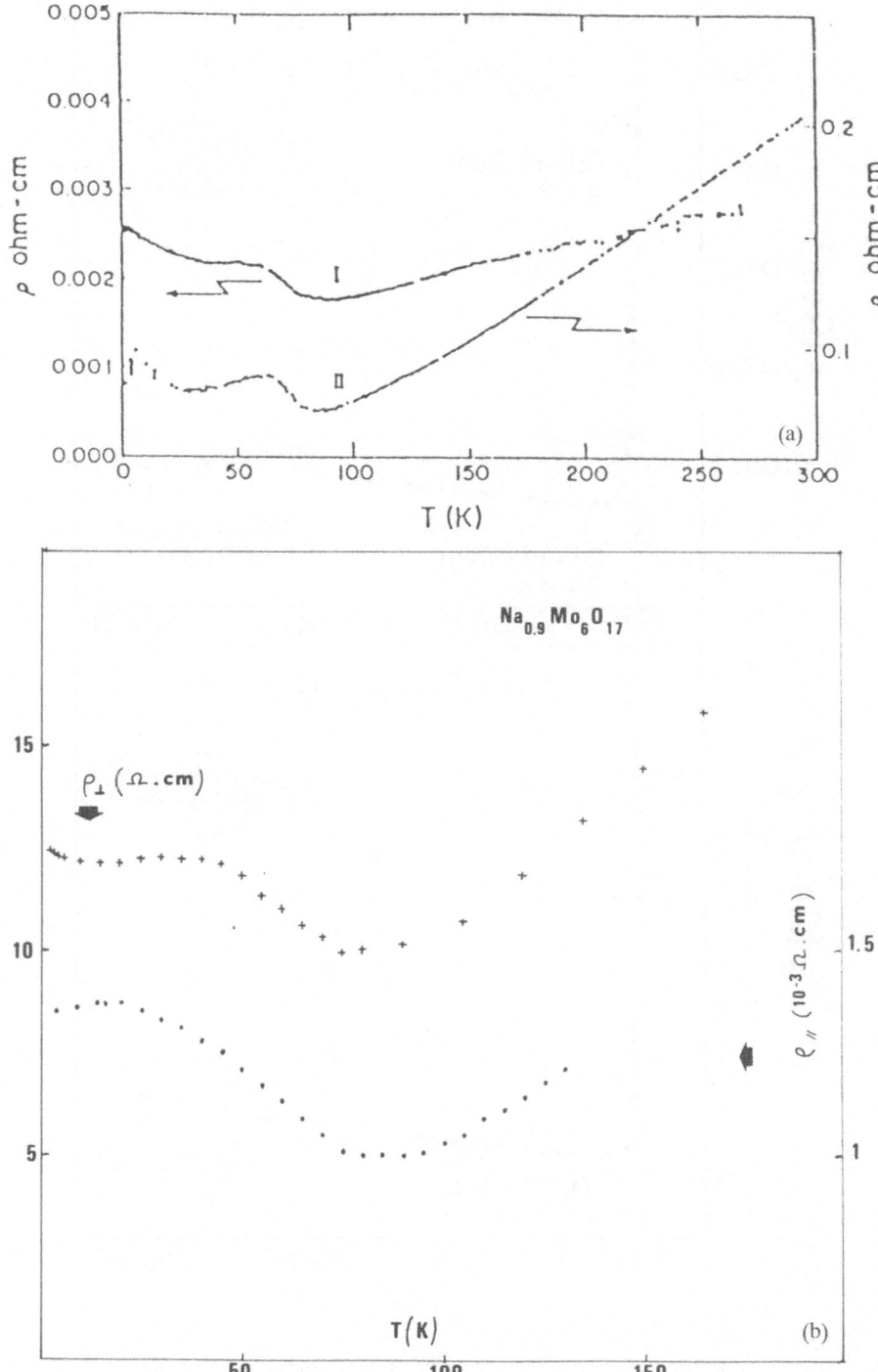

Fig. 44. Electrical resistivity of $Na_{0.9}Mo_6O_{17}$ versus temperature in the plane of the layers (I) and perpendicular to this plane (II). (a) From [139]. (b) Our measurements.

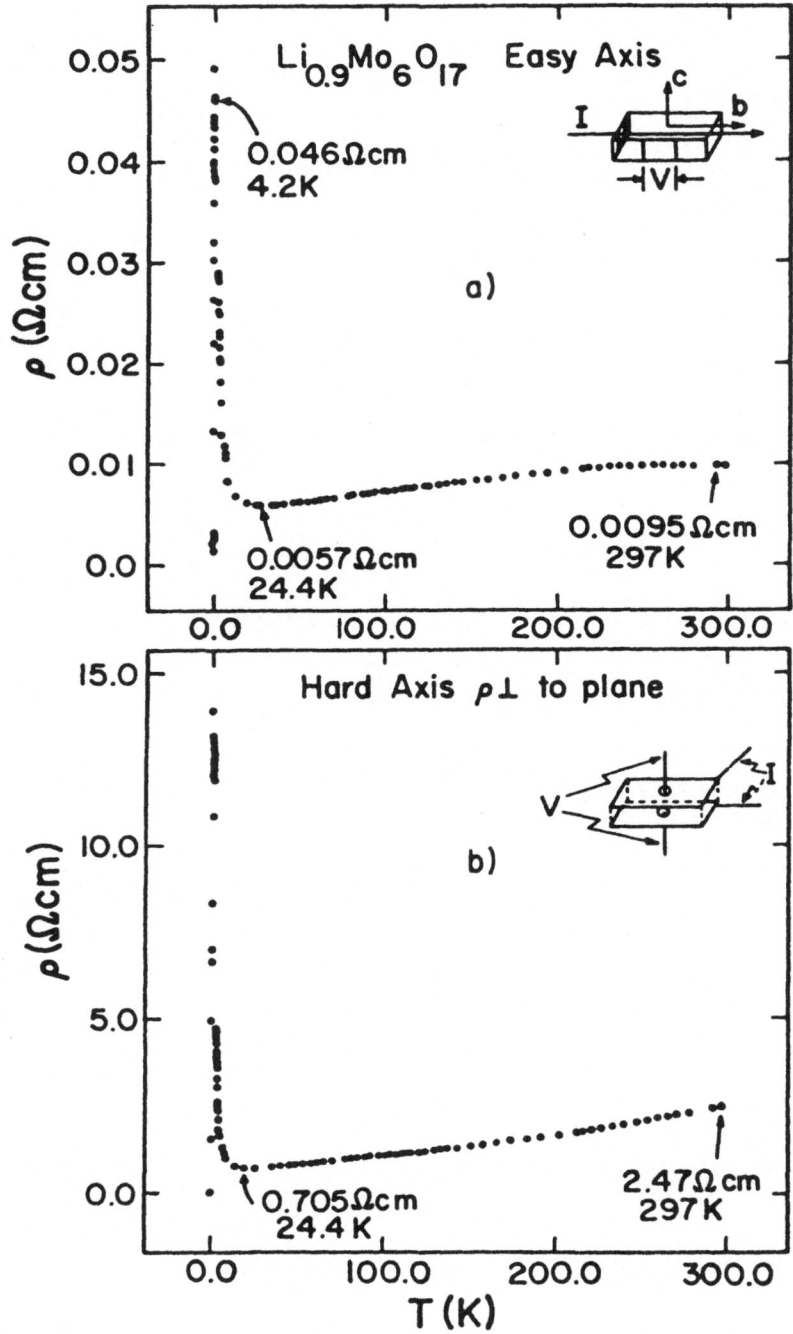

Fig. 45. Electrical resistivity of $Li_{0.9}Mo_6O_{17}$ versus temperature. (From [127].)

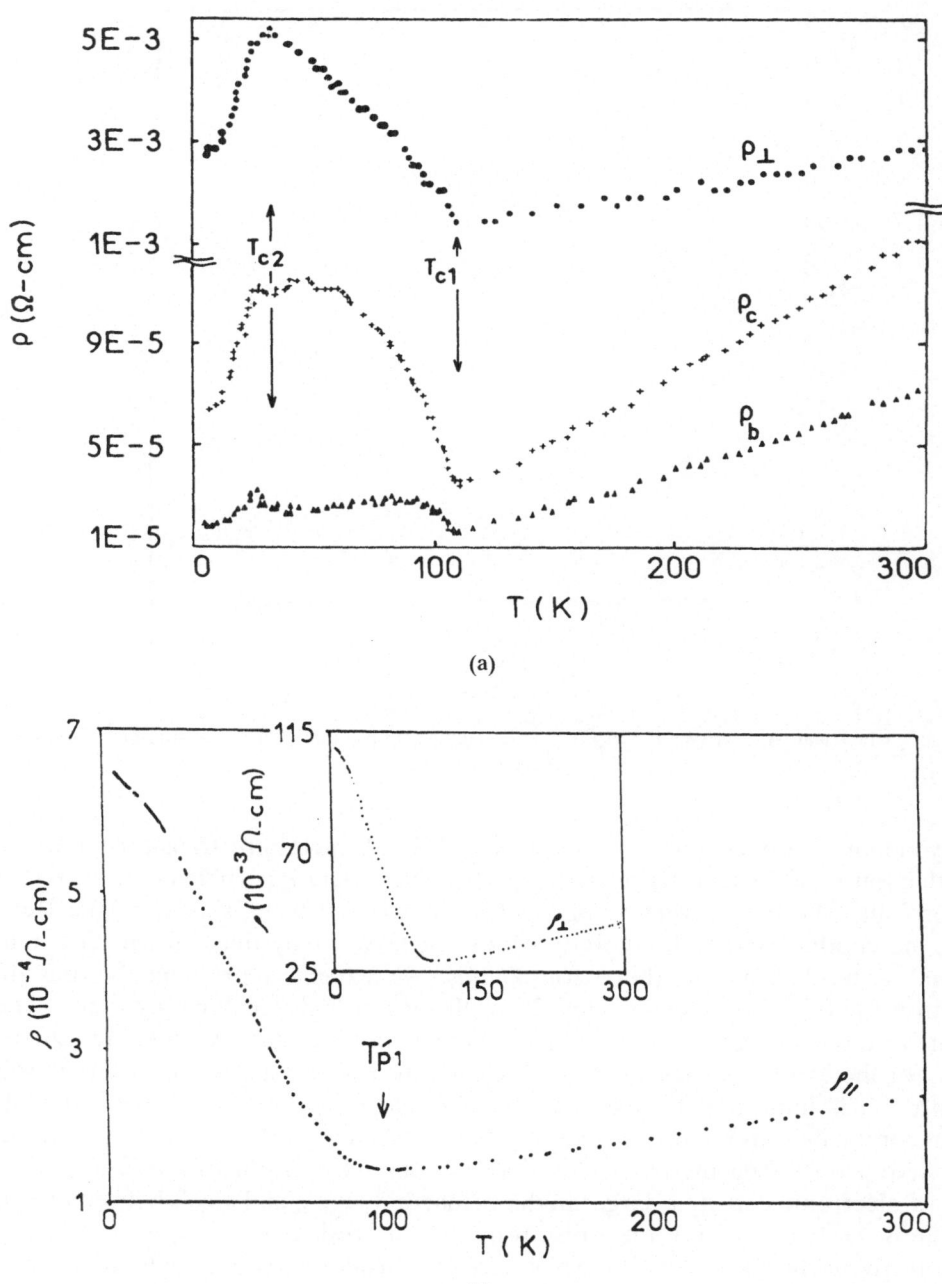

Fig. 46. (a) Electrical resistivity of η-Mo$_4$O$_{11}$ versus temperature. (From [144].) (b) Same for γ-Mo$_4$O$_{11}$. (From [142].) $\rho_{\parallel}$ is measured along c in the (b, c) plane, $\rho_{\perp}$ along a, perpendicular to the layers.

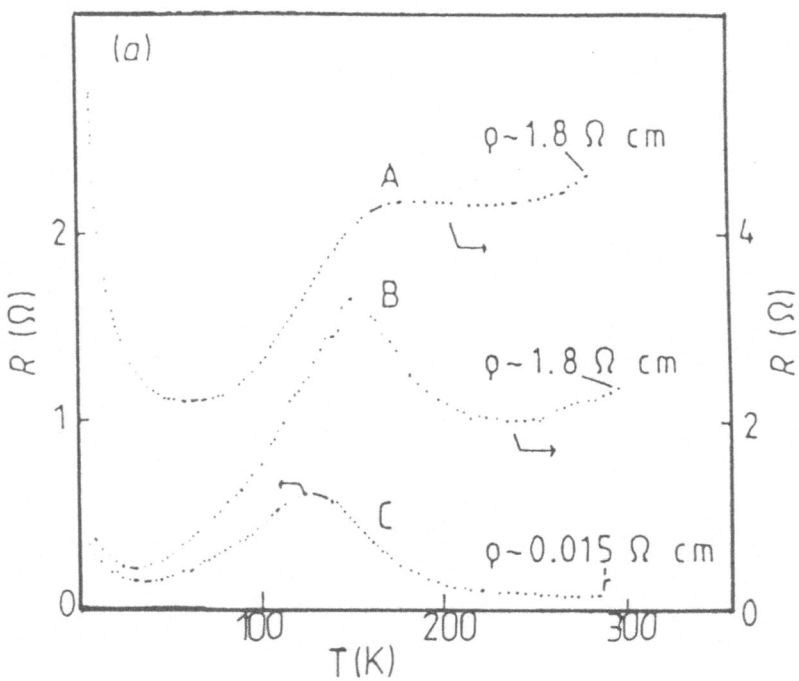

Fig. 47. Electrical resistivity of Mo_8O_{23} versus temperature along three orthogonal directions. (From [149].)

molybdenum bronzes, the outer electrons of the A metal are transfered into the conduction band formed by the overlap of molybdenum t_{2g} orbitals and of oxygen p_π orbitals. The band structure of Mo_4O_{11} is expected to be similar to the ReO_3 one, the conduction band, empty in MoO_3, being partially filled in Mo_4O_{11}. One should note that, in all these compounds, we expect approximately one $4d$ electron for two molybdenum sites. They all show a high conductivity plane. This is due to the presence of layers of MoO_6 octahedra separated by MoO_4 tetrahedra only, for the Mo_4O_{11} oxides, and by MoO_4 tetrahedra plus the A ions in the purple bronzes. It is important to note that the Mo charges are close to $+6$ on the Mo tetrahedral sites and close to $+5$ on the octahedral sites far apart from the tetrahedral ones (see the chapter by Vincent and Marezio in this volume). Then the $4d$ electronic density is larger in the octahedral layers and weak between them, which is consistent with the anisotropy of the resistivity. This results in an anisotropy of the Fermi surface which can be considered as quasicylindrical, with an axis perpendicular to the MoO_6 layers, as in the quasi-two dimensional layered transition metal dichalcogenides [13]. Such a quasicylindrical Fermi surface often allows the possibility of nesting which leads to CDW instabilities. One can then attribute the resistivity anomalies to CDW instabilities. This hypothesis has been confirmed in all the compounds, except $Li_{0.9}Mo_6O_{17}$, by diffraction data (Pouget, this volume).

The increase of the resistivity below T_P suggests that the instabilities are Fermi surface driven and related to the partial openings of gaps. The gaps start to open at T_P and can be fully opened at some lower temperature when the resistivity shows a maximum vs T as in $K_{0.9}Mo_6O_{17}$ or $TlMo_6O_{17}$. The gap opening may continue down to low temperatures when the resistivity further increases. But, due to the two dimensional character of these compounds, the Fermi surface is not completely destroyed and they are all metallic at low temperature.

In the case of η-Mo_4O_{11}, CDW have not been observed at the low temperature transition by diffuse scattering up to now. But electron tunneling experiments (see Subsection 4.5) have established that a gap does indeed open below 30 K. The nature of this gap, CDW or SDW, is not known at present.

The superconducting properties of $Li_{0.9}Mo_6O_{17}$ below 2 K will be discussed in Subsection 4.7. At higher temperatures, the anomaly found in the resistivity at 25 K may be attributed to a CDW instability, the superconducting transition being associated with the remaining Fermi surface below 25 K.

Some resistivity measurements have been performed on doped samples:

— For $K_{0.9}Mo_6O_{17}$ doped with Fe or Eu, one does not notice any variation in the transition temperature and above 150 K the temperature dependence of the resistivities are similar. But below T_P, the thermal variation and the magnitude of the resistivity are very sensitive to the impurities [63]. The same behavior is obtained in the case of η-Mo_4O_{11} doped with Fe or Eu [146].
— The changes as a function of impurity contents have also been investigated for W, Re and V on η-Mo_4O_{11} by Gruber et al. [147]: the electrical resistivity depends strongly on the composition since the conductivity shows a transition from a metallic to a semiconductive behavior in $Mo_{3.96}W_{0.04}O_{11}$.
— Y. Matsuda et al. [148] have measured the temperature dependence of the resistivity of $(Li_{1-x}K_x)_{0.9}Mo_6O_{17}$, $(Li_{1-x}Na_x)_{0.9}Mo_6O_{17}$ and $Li_{0.9}(Mo_{1-x}W_x)_6O_{17}$. The resistivity of the Li-rich ($x \leqslant 0.4$) crystals shows a large upturn at about 25 K while for the K or Na rich side, a CDW transition is obtained. If the transition temperature is not strongly influenced by the value of x ($0.4 < x < 0.8$), the shape of the resistivity below T_P is strongly modified.

The introduction of W substituted onto Mo sites results, as in Mo_4O_{11}, in an effect much larger than in the case of the substitution of the A metal. This is expected since the A ions lie between the high conductivity planes and do not affect too much the path of the conduction electrons.

4.2.2. Thermopower

Thermopower has been measured on $TlMo_6O_{17}$ [138] sintered pressed bars, on $K_{0.9}Mo_6O_{17}$ [140] and Mo_4O_{11} single crystals [142, 144]. The temperature gradient was always applied in the high conductivity plane. In the case of γ-Mo_4O_{11}, no anisotropy could be detected between the b and c axis.

Figure 48 shows the temperature variation of the thermopower in $K_{0.9}Mo_6O_{17}$. Similar data have been obtained on $TlMo_6O_{17}$ for the 77—300 K interval. Above 120 K ($\sim T_P$), the thermopower S is negative and linear with temperature. Below 120 K, S becomes positive and a maximum is found at about 70 K for the

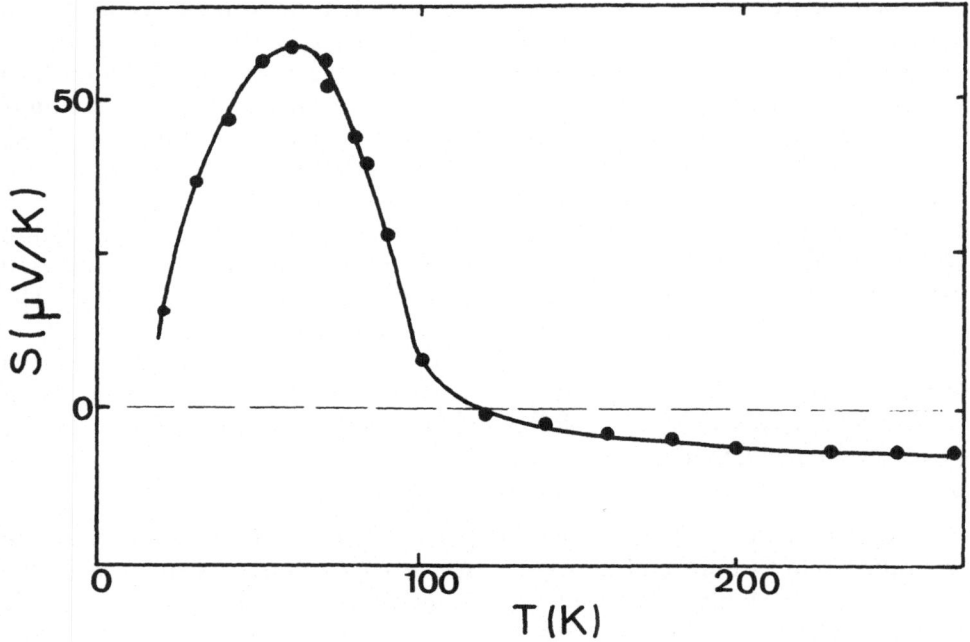

Fig. 48. Thermopower of $K_{0.9}Mo_6O_{17}$ versus temperature measured in the plane of the layers ($\perp c$).
(From [140].)

potassium purple bronze. Figure 49 shows the results obtained on the Mo_4O_{11} oxides: S is always negative. Below T_P, $|S|$ shows a large increase and a maximum in the vicinity of 50 K.

The negative thermopower indicates that at high temperature the dominant carriers are electrons and that the Fermi surface should show mainly electron pockets. Above T_P, all these compounds behave as conventional metal. In a three dimensional free electron model, assuming $S = -2\pi^2 k_B^2 Tg(\varepsilon_F)/9e$ one deduces the density of states at the Fermi level $g(\varepsilon_F)$. The values of $g(\varepsilon_F)$ obtained from the slope of $S(T)$ above T_P are listed in Table II.

The deviation of S from the free electron law may indicate that the band structure is not rigid above T_P and that the effective mass is increasing with decreasing temperatures. One also notes that the room temperature values of S correspond to densities of states at the Fermi level that are slightly different from one compound to the other, while the expected electron concentrations corresponding to 1 electron per 2 Mo sites is roughly 0.9×10^{22} cm^{-3} in all compounds. This leads to a density of states at the Fermi level very close to 1 eV^{-1} per molecule. The experimental deviation from this value may be connected to the in-plane anisotropy of the Fermi surface different from one compound to the other.

Below T_P, one observes a strong deviation of S from linearity. This corresponds to the onset of the CDW metallic state. A third metallic or pseudo-metallic phase

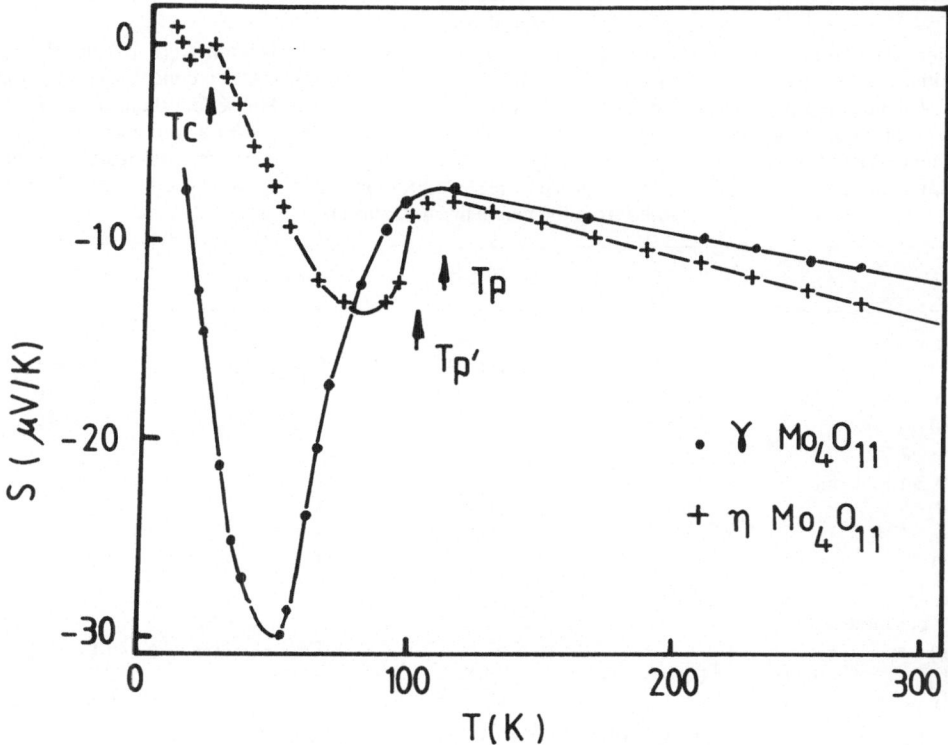

Fig. 49. Thermopower versus temperature for γ-Mo_4O_{11} ($\bullet$) and η-Mo_4O_{11} (+). The temperature gradient is parallel to b.

appears at low temperature in $K_{0.9}Mo_6O_{17}$ and in η-Mo_4O_{11}. In all cases, the thermopower is relatively weak ($|S| < 50 \ \mu V/K$) and does not show a semiconducting behavior.

The thermal variation of the thermopower is correlated to the thermal evolution of the Fermi surface. In spite of a change of sign of S found only in the potassium and thallium bronzes, all the phase transitions are similar. This is consistent with the partial opening of a gap at the Fermi surface which leads to a change in the concentrations of both types of carriers, electrons and holes. In the case of the bronzes, the dominant carriers at $T < T_p$ are holes while for the oxides, they are electrons. In both cases, the anomalies found in the thermopower are the result of a competition between the contributions of carriers of both signs, which are temperature dependent.

In this context and to a first approximation, one can use a two-band model, to account for the behavior of these compounds. This seems necessary for the K purple bronze. In the case of the molybdenum oxides, the existence of both electron and hole carriers has been connected, not to a Fermi surface built with two quasicylindrical sheets with different carriers, but to a Fermi surface which

TABLE II

Carrier concentration n, density of states at the Fermi level $g(\varepsilon_F)$ and variation $\Delta g(\varepsilon_F)$ at the Peierls transition for the purple bronzes $A_{0.9}Mo_6O_{17}$ and the Mo_4O_{11} oxides. Carrier concentration: (a) at $T > T_P$ calculated from the chemical formula, (b) at $T > T_P$ from the Hall effect data, (c) at $T \ll T_P$ from Shubnikov de Haas oscillations. $g(\varepsilon_F)$: (d) calculated at $T > T_P$ with a free electron model (e) deduced from thermopower data at $T > T_P$ (f) obtained from the linear term coefficient γ of the specific heat at $T \ll T_P$. $\Delta g(\varepsilon_F)$: (g) obtained as (e)—(f), (h) obtained from the specific heat anomaly at $T \sim T_P$, (i) obtained from the magnetic susceptibility change at $T \sim T_P$.

	$K_{0.9}Mo_6O_{17}$	$TlMo_6O_{17}$	$\gamma\text{-}Mo_4O_{11}$	$\eta\text{-}Mo_4O_{11}$
$n(cm^{-3})$				
(a)	8.27×10^{21}	8×10^{21}	8.87×10^{21}	8.97×10^{21}
(b)	8.9×10^{21}	—	8.8×10^{21}	9.0×10^{21}
(c)	4.5×10^{18}	—	1.3×10^{20}	8.8×10^{18}
$g(\varepsilon_F)$ (eV^{-1} per molecule)				
(d)	2.99	3.06	1.81	1.87
(e)	1.7	1.22	3.8	2.9
(f)	0.7	—	1.1	1.3
$\Delta g(\varepsilon_F)$ (eV^{-1} per molecule)				
(g)	1	—	2.7	1.6
(h)	1.87	1.18	—	2.2
(i)	—	—	1.1	0.9
References	[140, 150]	[131]	[146, 159b]	[146, 159b]

shows positive and negative curvatures. The variation with temperature of the gap opening is then connected with the evolution of the Fermi surface curvature [146].

It should also be pointed out that if a CDW or SDW instability opens a Peierls gap at the Fermi surface, the partial nesting can be responsible for the existence of small pockets of carriers which can contribute to the transport properties. The anomalies in the thermopower data can be related to the existence of such pockets below each transition. The presence of these pockets are consistent with the magnetic and galvanomagnetic properties.

4.2.3. Galvanomagnetic Properties

We shall restrict this section to the potassium bronze and to the Mo_4O_{11} oxides in the classical regime. The quantum regime will be discussed in Section 4.6.

(1) *Magnetoresistance.* The magnetoresistance results are shown in Figures 50 and 51 [150, 146]. In all cases, the magnetoresistance is positive and strongly anisotropic. It is large, by comparison with usual metals. In the case of Mo_4O_{11}, it rather suggests a semi-metallic behavior, similar to that of bismuth for example [151].

Some measurements had been done on sintered powders of $Mo_{17}O_{47}$ and

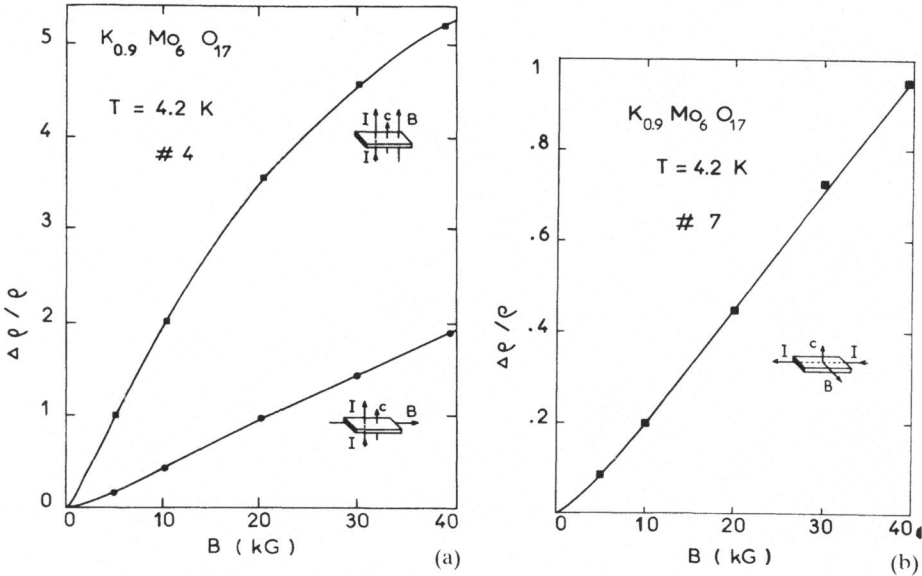

Fig. 50. Magnetoresistance of $K_{0.9}Mo_6O_{17}$ as a function of magnetic field. (a) $I \| c$, $B \| c$ (■), $B \perp c$ (●). (b) I and B in the basal plane $I \perp B$.

Mo_4O_{11} oxides [145]. They indicate for $Mo_{17}O_{47}$ a large magnetoresistance below 103 K, which may indicate a CDW transition in this compound at 103 K. For the Mo_4O_{11} oxides, the results obtained on powders are only roughly consistent with single crystal data reported in [146].

The results obtained on single crystals of $K_{0.9}Mo_6O_{17}$, η-Mo_4O_{11} and γ-Mo_4O_{11} are summarized in Table III. One should add to the data in this table that the magnitude of the magnetoresistance is strongly dependent on impurity concentration. For example, in $K_{0.9}Mo_6O_{17}$ in the case $I \| (a, b)$ plane and $B \| c$, $\Delta\rho/\rho$ changes from 15 in a 'pure' sample to 0.2 in a Fe-doped sample [57]. Similar results have been obtained on W doped $Mo_{17}O_{47}$ [145]: $\Delta\rho/\rho = 2$ in a pure sample and 0.05 in $Mo_{14}W_3O_{47}$.

The longitudinal resistivity is generally weaker than the transverse one, but the variations in the transverse configuration are strongly connected to the orientation of the magnetic field. The highest values of the magnetoresistivity are always obtained when the magnetic field is perpendicular to the high conductivity plane. In view of these results, the magnetic field orientation, in relation to the high conductivity plane, is more important for the variation and the magnitude of the magnetoresistivities than the current direction.

The variation of the magnetoresistance as a function of magnetic field at 4.2 K, therefore in the CDW state, is also strongly dependent on the relative orientation of the magnetic field and the high conductivity plane. One may obtain a linear dependence or a quadratic variation with or without a tendency towards satura-

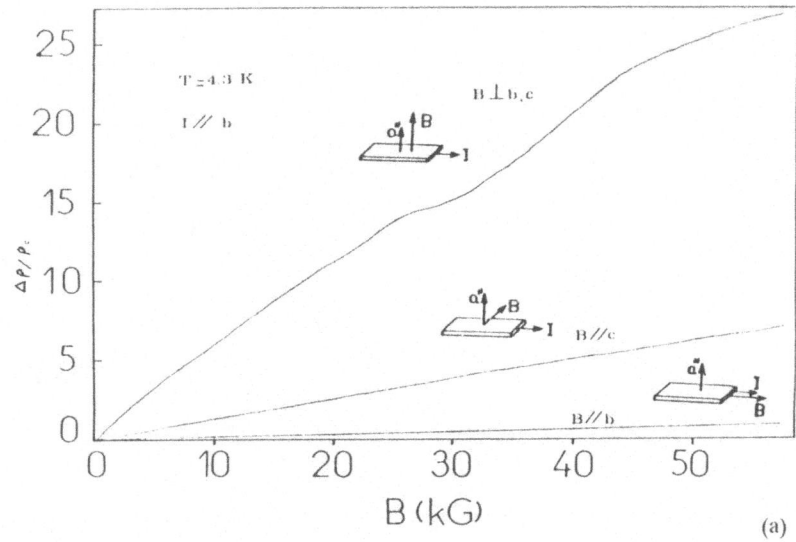

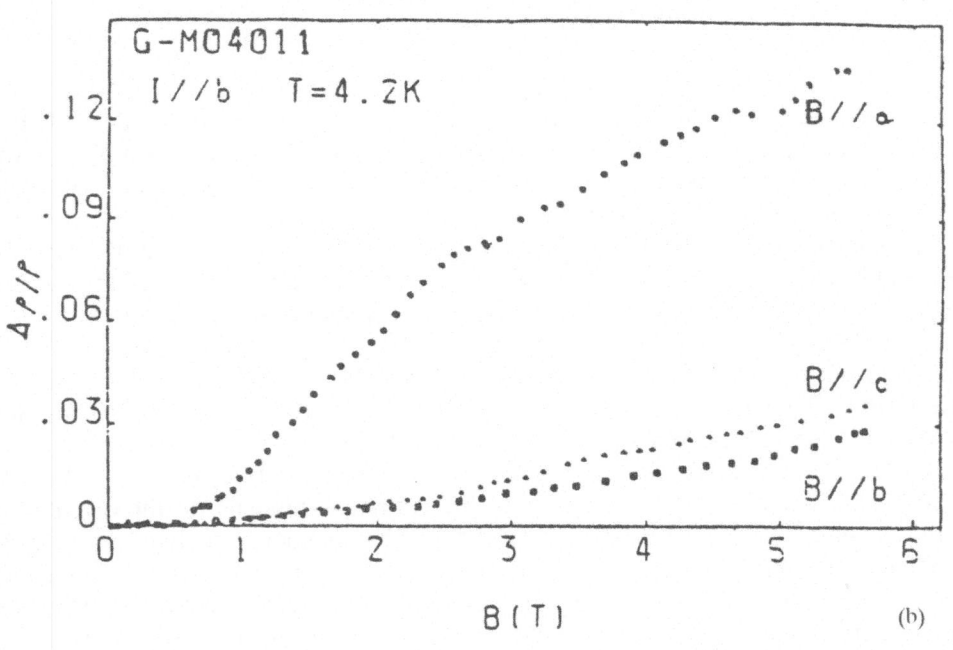

Fig. 51. (a) Magnetoresistance of η-Mo_4O_{11} as a function of magnetic field for $I\|b$ and $B\|b$, $\|c$ and $\|a^*$. (b) Magnetoresistance of γ-Mo_4O_{11} as a function of magnetic field for $I\|b$ and $B\|b$, $\|c$ and $\|a$.

TABLE III

Summary of magnetoresistance data in $K_{0.9}Mo_6O_{17}$, η-Mo_4O_{11} and γ-Mo_4O_{11} (T): transverse configuration, (L): longitudinal configuration (From [146 and 150]).

| | Experimental Configuration | | | Behaviour of $\Delta\rho$ vs B | | Values of $\Delta\rho/\rho$ at T = 4.2 K | Information on |
	Current	Magnetic field		low field (B $\leqslant$ 4 T)	high field B = 4 T	B = 4 T	the orbits
$K_{0.9}Mo_6O_{17}$	$I\|(a,b)$	$B\|c$	$B \perp I$ (T)	$\propto B$	saturation	15	Closed orbits and possible magnetic breakdown
	$I\|(a,b)$	$B\|(a,b)$	$B \perp I$ (T)	$\propto B$	no saturation	1	open orbits $\|c$
	$I\|(a,b)$	$B\|(a,b)$	$B\|I$ (L)	between $\propto B$ and $\propto B^2$	no saturation	3	open orbits $\|c$
	$I\|c$	$B\|c$	$B\|I$ (L)		no saturation	5	magnetic breakdown
	$I\|c$	$B\|(a,b)$	$B \perp I$ (T)		no saturation	2	open orbits $\|c$
γ-Mo_4O_{11}	$I\|c$	$B\|a$	$B \perp I$ (T)	$\propto B$		0.04	magnetic breakdown
	$I\|c$	$B\|c$	$B\|I$ (L)	$\propto B^2$		0.02	
	$I\|a$	$B \perp a$	$B \perp I$ (T)	$\propto B^2$		0.02	
	$I\|a$	$B\|a$	$B\|I$ (L)		saturation	0.1	
	$I\|b$	$B\|a$	$B \perp I$ (T)		saturation	0.1	
η-Mo_4O_{11}	$I\|b$	$B\|b$	$B\|I$ (L)	$\propto B$	saturation	0.5	
	$I\|b$	$B\|c$	$B \perp I$ (T)	$\propto B$	no saturation	5	
	$I\|b+c$	$B\|a^*$	$B \perp I$ (T)	$\propto B^2$	no saturation	25	open orbits in (b^*, c^*) plane

tion. Since the magnetoresistance behavior and the shape of electron orbits are strongly connected, the experimental behavior for the different configurations of the magnetic field and of the current with respect to the principal axis of the crystals, gives information on the topology of the Fermi surface in the low temperature phase, as is discussed in [153].

The results are consistent with a very anisotropic Fermi surface (quasi-cylindrical) and with open orbits along the c axis for $K_{0.9}Mo_6O_{17}$ and in the (b, c) plane for η-Mo_4O_{11} and γ-Mo_4O_{11}.

For these three quasi-two dimensional compounds, the temperature dependent behavior of the magnetoresistance is identical. As the temperature increases towards the transition temperature, the magnetoresistance decreases and becomes extremely weak and difficult to measure above 120 K [57, 146]. As for the low temperature results, the magnetoresistance of η-Mo_4O_{11} is stronger than that of γ-Mo_4O_{11}, even above 30 K, the temperature of the second transition of η-Mo_4O_{11}. This shows that, below 100 K, the carrier mobilities are higher in η-Mo_4O_{11} than in γ-Mo_4O_{11}, since the Fermi surfaces of both compounds are expected to be very similar.

In η-Mo_4O_{11}, the change in magnetoresistance below 30 K shows that the low temperature transition affects also the Fermi surface (see Section 4.5).

The linear dependence of the magnetoresistance vs field, as observed for different orientations in these compounds, is not explained by conventional theories. M. Naito and S. Tanaka [154] have examined several extrinsic mechanisms, geometrical effects, inhomogeneities in the carrier concentration, and several intrinsic explanations such as quantum limit, static skin effect, small angle phonon scattering and magnetic breakdown.

This last effect seems, as in $2H$-$NbSe_2$ and $2H$-$TaSe_2$ where CDW instabilities occur, the most likely, since CDW creates small energy gaps. The magnitude of these gaps, which depends on the CDW order parameter, is generally small. Therefore the magnetic breakdown through these gaps may occur at comparatively low fields [155]. However, one should point out that a transverse magneto-resistance large and linear in B is also found in quasi-two dimensional intercalated graphites C_8K and C_6Li, which are not known to be CDW systems [156].

(2) *Hall effect.* The Hall effect has been studied in $K_{0.9}Mo_6O_{17}$ and in both η- and γ-Mo_4O_{11}. Figure 52 shows the results obtained on the bronze between 50 K and 200 K in fields smaller than 10 kG. Above T_P, the Hall effect is n-type, as is the thermopower. It is only below ~ 70 K that it becomes positive. Figure 53a shows the Hall voltage V_H as a function of magnetic field for several temperatures $T < T_c$. For $T > 75$ K, V_H is negative and proportional to B in the field range explored. Deviation from linearity increases with decreasing temperature. V_H also changes sign in the vicinity of 70 K. The Hall constant R_H obtained at small fields is plotted as a function of temperature in Figure 53b. $|R_H|$ increases steeply at $T < T_c$, and goes through a maximum at ~ 80 K. At $T < 70$ K, R_H is clearly positive and increases dramatically with decreasing temperature. At $T = 200$ K, the Hall constant is found to be 0.7×10^{-9} m^3/C. In a free electron picture, this

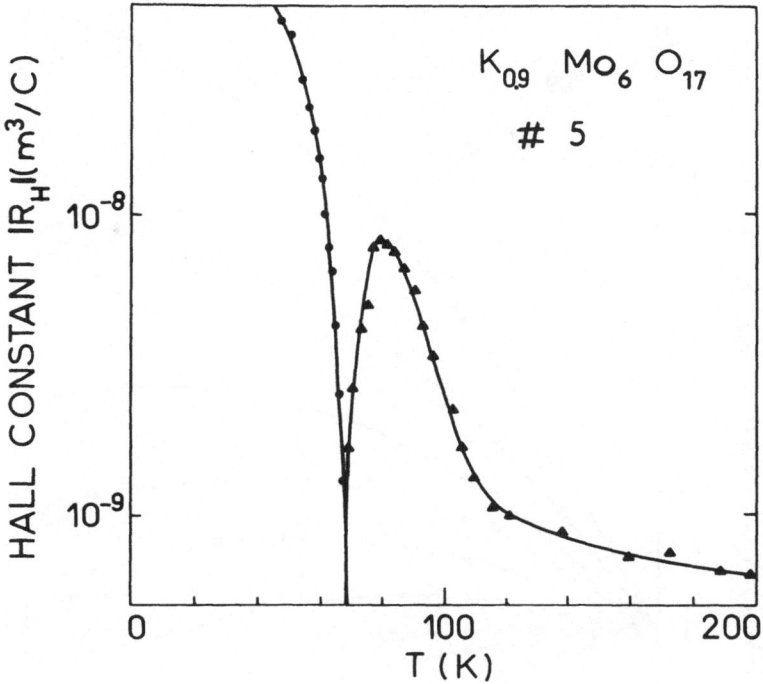

Fig. 52. Hall constant vs temperature for $K_{0.9}Mo_6O_{17}$. ▲: $-R_H$, ●: $+R_H$. (From [57].)

corresponds to an electron concentration of 8.9×10^{21} cm^{-3}, which is in good agreement with the value of 8.3×10^{21} cm^{-3} obtained from the chemical formula by assuming a complete charge transfer from the K to the conduction band. The change of sign of R_H at 70 K clearly indicates that, while the dominant carriers are electrons in the high temperature phase, the gap openings at T_c leads to the formation of both electron and hole pockets. A two band model [157] is therefore adequate to describe the Hall effect and low field magnetoresistance data. By combining both sets of data, one can evaluate the electrons (n) and holes (p) concentrations as well as their mobility, μ and v, respectively. The results are shown in Figure 54.

The steep decrease of n below T_c is consistent with gap openings on large parts of the Fermi surface at the transition. The increase of p shows that, at the same time, hole pockets are induced. These results also indicate that the gaps open down to ~ 70 K. This is consistent with the temperature dependence of the intensity of the satellites, as found by X-ray studies (see Pouget, this volume). The increase of mobility with decreasing temperature shows clearly two different regimes at $T < 70$ K and $T > 70$ K. This may be due to an increase of the electron phonon diffusion when the gap starts to close down at $T < T_c$.

Hall effect has been measured on η- and γ-Mo$_4$O$_{11}$ single crystals in low

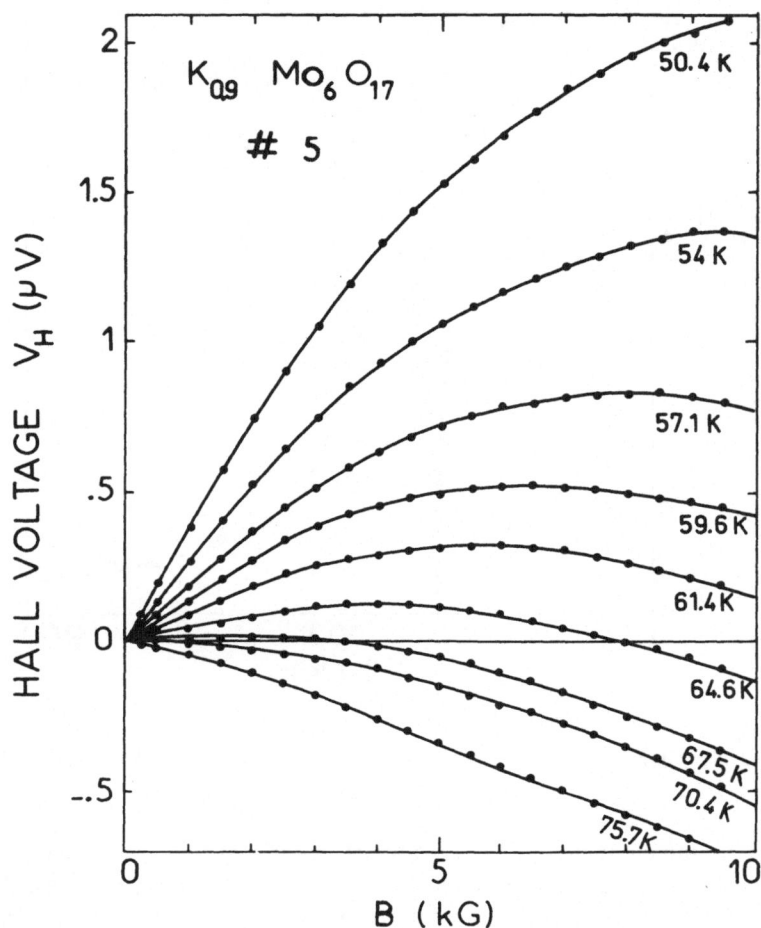

Fig. 53. Hall voltage as a function of magnetic field at different temperatures ($T < T_c$) for $K_{0.9}Mo_6O_{17}$. (From [57].)

magnetic fields ($B < 5$ T) and for B perpendicular to the layers [146]. Figure 55 shows data obtained on γ-Mo_4O_{11} for the Hall resistivity R_HB as a function of B at various temperatures above and below T_P. R_H is always negative for both compounds, as expected from thermopower data. The Hall resistivity is proportional to B up to 5.5 T. The effective concentrations of carriers have been estimated in a one-band picture, through $n_{eff} = 1/R_He$. Figure 56 shows that the CDW transitions are accompanied by decrease of carriers concentrations, as expected. In the case of η-Mo_4O_{11} (Figure 56b), a second anomaly corresponding to the low temperature transition is also apparent. Just as for $K_{0.9}Mo_6O_{17}$ the Hall constant obtained in the high temperature regime ($T > T_P$) gives a carrier concentration consistent with the value calculated from the chemical formula (see

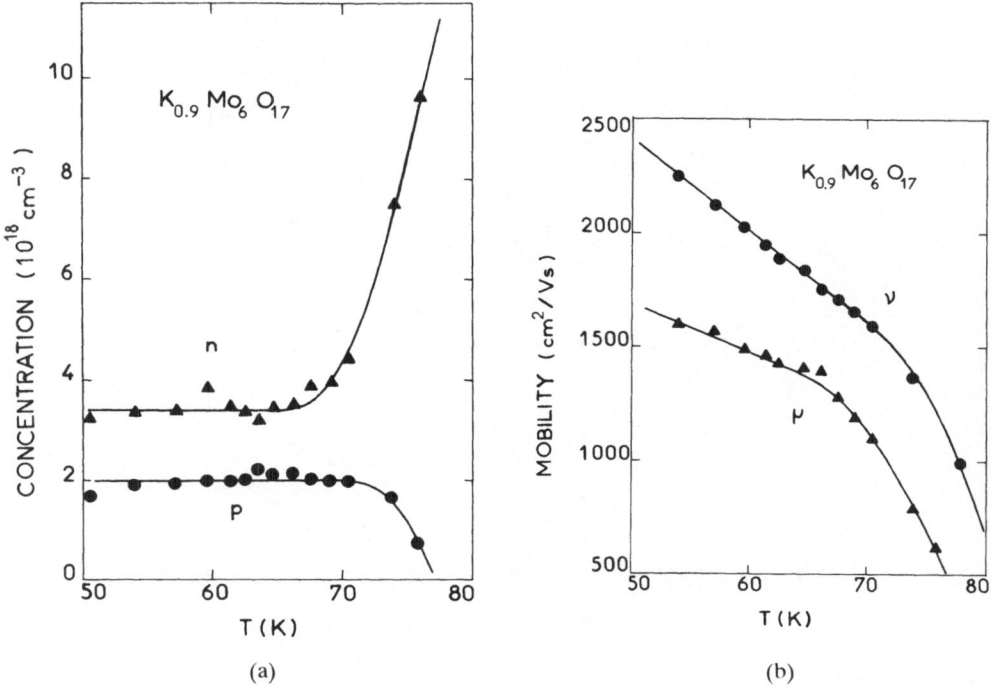

Fig. 54. Results obtained from a two-band model for $K_{0.9}Mo_6O_{17}$. (a) Electron (n) and hole (p) concentration vs temperature. (b) Electron (μ) and hole (ν) mobility vs temperature. (From [57].)

Table II). Below T_p, the electronic concentration decreases very strongly down to 4.2 K, especially for η-Mo_4O_{11}: a change of three orders of magnitude is found (9×10^{21} to 8.8×10^{18} cm^{-3} at 4.2 K). The Hall constant variation at the Peierls transition shows the decrease of the density of states at the Fermi level correlated to the opening of the Peierls gap. In the case of η-Mo_4O_{11}, below ~ 50 K, a further decrease of the carrier concentration is connected with the opening of a second gap which is responsible for the low temperature transition.

4.3. MAGNETIC SUSCEPTIBILITY

4.3.1. *Experimental Results*

The magnetic susceptibilities of large single crystals of $K_{0.9}Mo_6O_{17}$, η-Mo_4O_{11} and γ-Mo_4O_{11} have been measured against temperature for a magnetic field applied parallel or perpendicular to the MoO_6 octahedra slabs, in the directions of the crystallographic axes a, b and c. In all experiments, performed with a vibrating sample magnetometer or a SQUID, the temperature has been swept from 4.2 K up to 150 K or 300 K.

Figures 57 and 58 show, for $K_{0.9}Mo_6O_{17}$ and γ-Mo_4O_{11}, the temperature dependence of χ, obtained with a magnetic field of 1.7 T or 3 T [140, 159a]. For

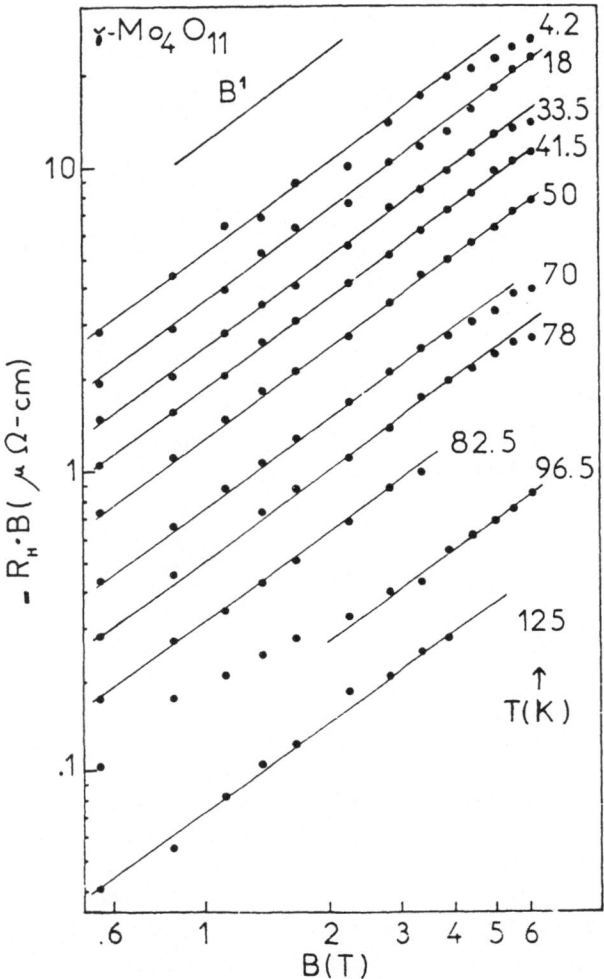

Fig. 55. Hall resistivity ($R_H B$) of γ-Mo$_4$O$_{11}$ as a function of B at various temperatures above and below T_p.

η-Mo$_4$O$_{11}$, the results are similar to those obtained with the orthorhombic phase [146, 159b]. The susceptibility of all three compounds exhibits the same behavior down to low temperature. It is weak at all temperatures ($< 2 \times 10^{-7}$ emu/g), isotropic and slightly temperature dependent in the normal metallic state. In the CDW state, χ drops and becomes anisotropic as the temperature is decreased, depending whether the field is applied parallel or perpendicular to the layers. Within the experimental accuracy, no anisotropy has been detected in the two directions of the octahedra slabs, perpendicular and parallel to the direction of nesting.

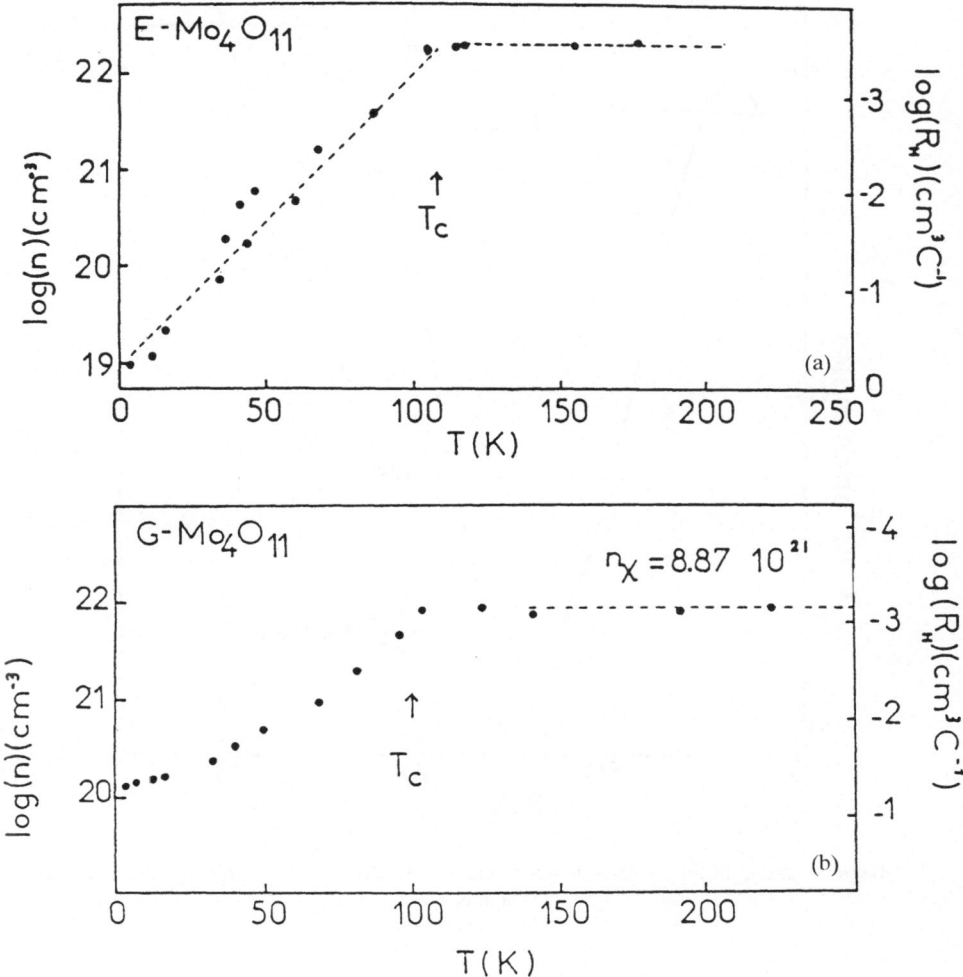

Fig. 56. Temperature dependance of the Hall constant and carrier concentration: (a) η-Mo$_4$O$_{11}$. (b) γ-Mo$_4$O$_{11}$.

At low temperature, the susceptibility of the Mo$_4$O$_{11}$ compounds increases steeply whereas that of K$_{0.9}$Mo$_6$O$_{17}$ remains finite: this difference is due to the presence of extrinsic paramagnetic centers in the η-Mo$_4$O$_{11}$ compounds. In η-Mo$_4$O$_{11}$, the Curie tail has been fitted with a Curie coefficient $C = (7.8 \pm 0.07) \times 10^{-6}$ emu K g^{-1} corresponding to a concentration of paramagnetic ($s = 5/2$, $g = 2$) impurities of $n = 2.5 \times 10^{-4}$ per Mo site. After correction from the Curie contribution, the susceptibility of η-Mo$_4$O$_{11}$ shows a plateau at low temperature and becomes similar to that of K$_{0.9}$Mo$_6$O$_{17}$. In γ-Mo$_4$O$_{11}$, the low temperature data have been analyzed by assuming that in addition to the Curie contribution,

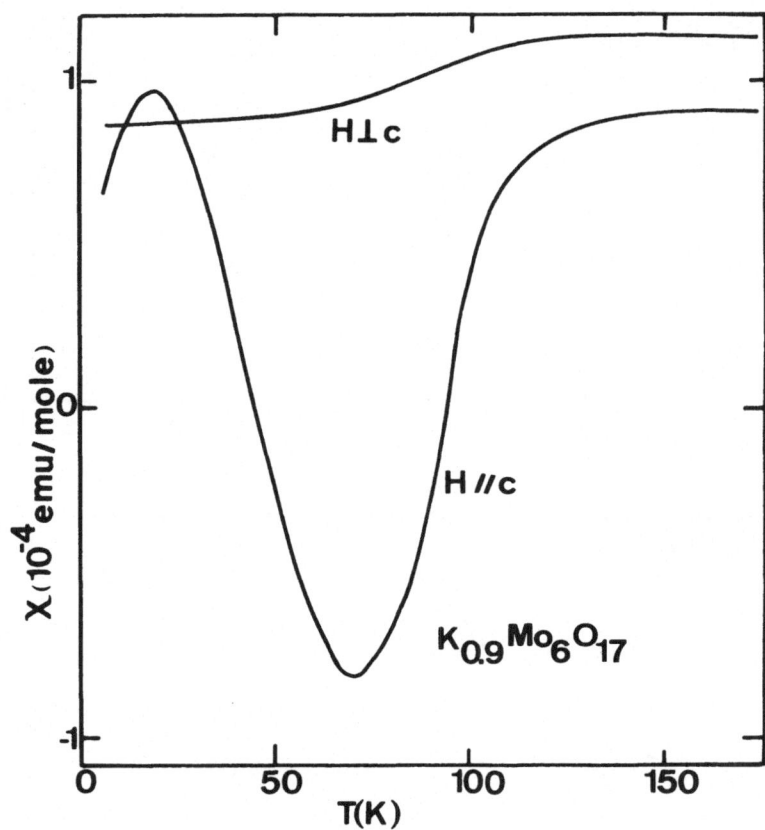

Fig. 57. Magnetic susceptibility versus temperature for $K_{0.9}Mo_6O_{17}$, for a magnetic field $B\|$ or $\perp$ c-axis.

the linear decrease of χ, observed at intermediate temperatures below the Peierls temperature, can be extrapolated down to the lowest temperatures. Under this assumption, the Curie coefficient $C = (1.10 \pm 0.03) \times 10^{-6}$ emu K g^{-1} corresponds to a concentration of paramagnetic ($s = 5/2$, $g = 2$) impurities $n = 1.1 \times 10^{-4}$ per Mo site.

The presence of paramagnetic centers in η-Mo_4O_{11} has been confirmed by measuring the variation of the magnetization with the amplitude of the magnetic field (B). Figure 59 shows this variation in the configuration $B\|a^*$, at 4.2 K. This magnetization is represented by the addition of a nonsaturating component, a saturating one, and an oscillating one. The second component has been fitted with a good accuracy by the Brillouin function $\mathscr{B}_{5/2}$. This indicates that the majority of the impurities have a spin of 5/2 and are most likely due to Fe^{3+} rather than Mo^{5+}, which has a spin of 1/2. In this context, the iron concentrations are 70 and 30 ppm in the η- and γ-Mo_4O_{11} samples, respectively. These very weak values are

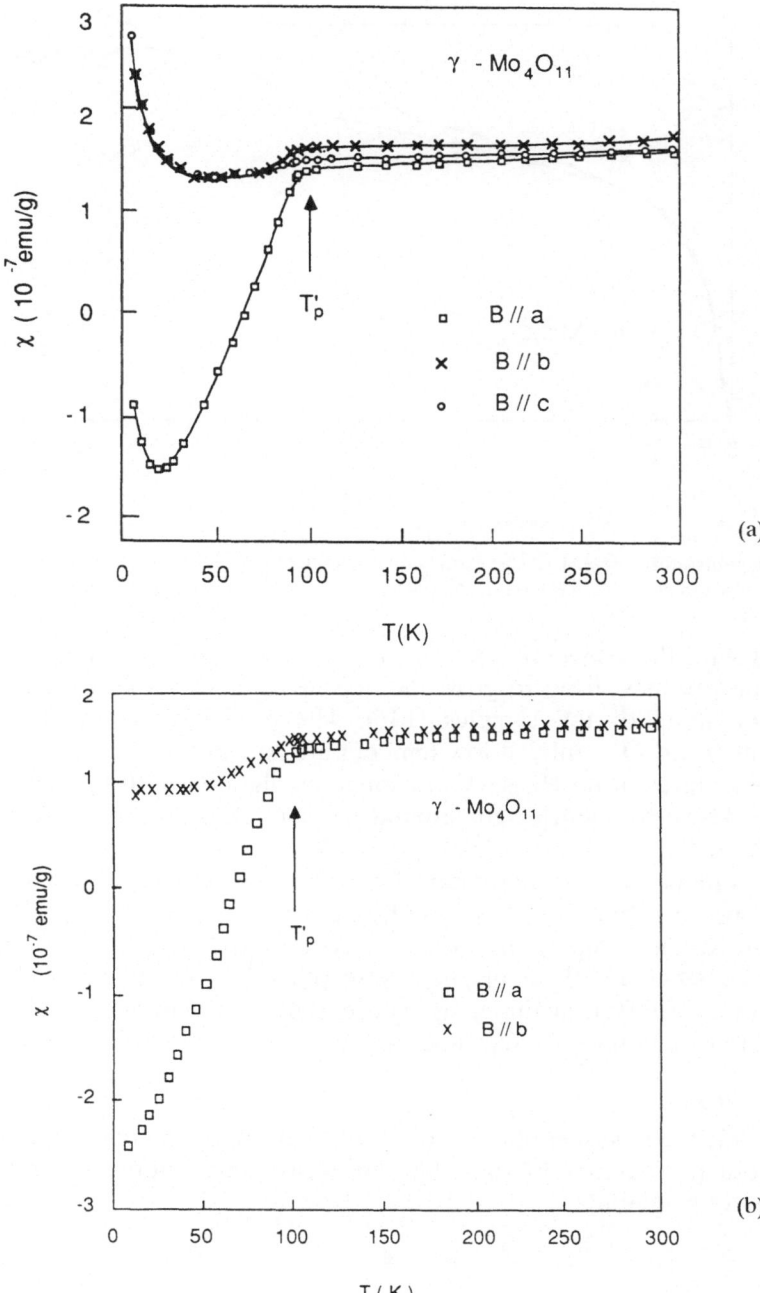

Fig. 58. (a) Magnetic susceptibility versus temperature of γ-Mo_4O_{11}, for a magnetic field: □ : ∥ a-axis; × : ∥ b-axis; ○ : ∥ c-axis. (b) Data corrected from the Curie paramagnetism.

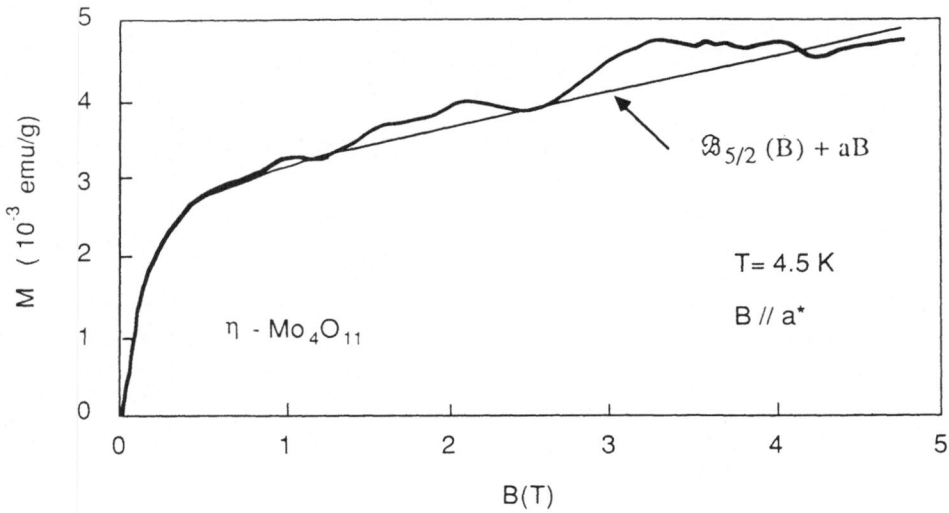

Fig. 59. Magnetization of η-Mo_4O_{11} versus a magnetic field parallel to a^*. The data are compared to the law $\mathscr{B}_{5/2}(B) + aB$, $\mathscr{B}_{5/2}(B)$ is the Brillouin function for $5/2$ spin.

consistent with the concentrations of the magnetic impurities (Fe, Cr) present in the starting products used to grow the crystals, and with the results of chemical analysis of some of the samples [146]. The oscillating component has been detected in η-Mo_4O_{11} only, at low temperature and in the configuration $B \parallel a^*$. It shows the features of de Haas—Van Alphen oscillations, with frequencies of 17 T and 50 T. These frequencies correspond to small closed orbits, with a surface of $\approx 3 \times 10^{-3}\,\text{Å}^{-2}$.

Finally, one can notice that contrarily to the resistivity or to the thermopower, the magnetic susceptibility of η-Mo_4O_{11} does not exhibit any significant change around 30 K, the temperature of the second transition. In fact, the similarity between the behaviors of the magnetic susceptibilities of both Mo_4O_{11} oxides may result from an artefact: at this temperature, the Curie paramagnetism has a large temperature variation which may hide any weak change due to the transition.

4.3.2. *Discussion*

The weak magnetic susceptibilities of these three molybdenum compounds result mainly from the balance of four different weak contributions, which are of the same order of magnitude:

$$\chi = \chi_{core} + \chi_{VV} + \chi_P + \chi_L + \chi_C$$

where χ_{core} is the core diamagnetism, χ_{VV} is the Van Vleck paramagnetism, χ_P and χ_L are the Pauli and Landau contributions of the conduction electrons and χ_C is the Curie paramagnetism of localized centers. The core diamagnetism from the inner electrons has been estimated, from the chemical formulas, with the tabulated contributions of the different ions: -3.0×10^{-7} emu/g in Mo_4O_{11} and -3.1×10^{-7} emu/g in $K_{0.9}Mo_6O_{17}$ [160a].

The important decrease of χ, which appears below the Peierls temperature in each compound, corresponds to the large change induced by the opening of the Peierls gap. One should note that this decrease takes place in the CDW state, and not above the Peierls transition, as would be expected if pretransitional fluctuations were predominant. It is directly connected to the partial opening of the Peierls gap, therefore to a decrease of the density of states at the Fermi level $g(\varepsilon_F)$ and to a remodelling of the Fermi surface (FS). The Peierls gap does not significantly affect the Van Vleck paramagnetism, which results from interband couplings. But the Pauli and Landau contributions are strongly affected by the Peierls transition.

In the quasi-free electron approximation, the Pauli diamagnetism, when it has a non-zero value, is proportional to the Pauli contribution. For a closed FS, the Landau contribution is $\chi_L = -1/3 \ (m/m_c)^2 \chi_P$, where m is the band mass and m_c the cyclotron mass of the closed orbit normal to the magnetic field. The factor m/m_c may introduce an anisotropy of χ, directly related to the anisotropy of the FS.

The presence of an anisotropy of χ in the CDW state is therefore attributed to a nonzero anisotropic contribution of the Landau diamagnetism. This is due to the existence of small closed anisotropic pockets, elongated in the direction normal to the MoO_6 octahedra slabs. These pockets, characteristic of the CDW state, very likely result from an imperfect nesting of the FS: their anisotropy is consistent with that of the high temperature FS. The absence of anisotropy of χ in Mo_4O_{11}, in the plane parallel to the octahedra slabs, indicates that the shape of the closed pockets has a predominant role, which takes over the broken symmetry induced by the CDW, as predicted by Boriack in a three-dimensional system [160b].

The change of $g(\varepsilon_F)$ between the Peierls temperature and 4.2 K is estimated to be close to 1 eV/molecule in all three compounds. Assuming that, at 4.2 K, the FS of the Mo_4O_{11} compounds consists only of one ellipsoidal pocket with a circular cross-section in the plane parallel to the octahedral slabs, the anisotropy of χ leads to an anisotropy of ≈ 90 of the pocket. This very high value, which is confirmed by the analysis of the Hall coefficient in the two Mo_4O_{11} oxides, indicates that the cross-section of the pocket is very small. This model is consistent with the observation of de Haas—Van Alphen oscillations in η-Mo_4O_{11}, in the configuration $B \| a^*$ only, associated with closed orbits of $\approx 3 \times 10^{-3} \ \text{Å}^{-2}$. This surface represents a few thousandths of the two-dimensional ($b^* \times c^*$) first Brillouin zone. Whereas no quantum effects have been detected in γ-Mo_4O_{11}, due to the low mobility of the conduction electrons even at low temperature, in $K_{0.9}Mo_6O_{17}$, the shape of the low temperature FS has been confirmed by the analysis of Shubnikov—de Haas oscillations [146].

4.4. SPECIFIC HEAT

The electric and magnetic properties of these quasi-two dimensional compounds show that at T_P changes take place in their electronic system. The specific heat measurements shown in Figures 60 and 61 for $K_{0.9}Mo_6O_{17}$ and η-Mo_4O_{11} corroborate the existence of a phase transition at T_P.

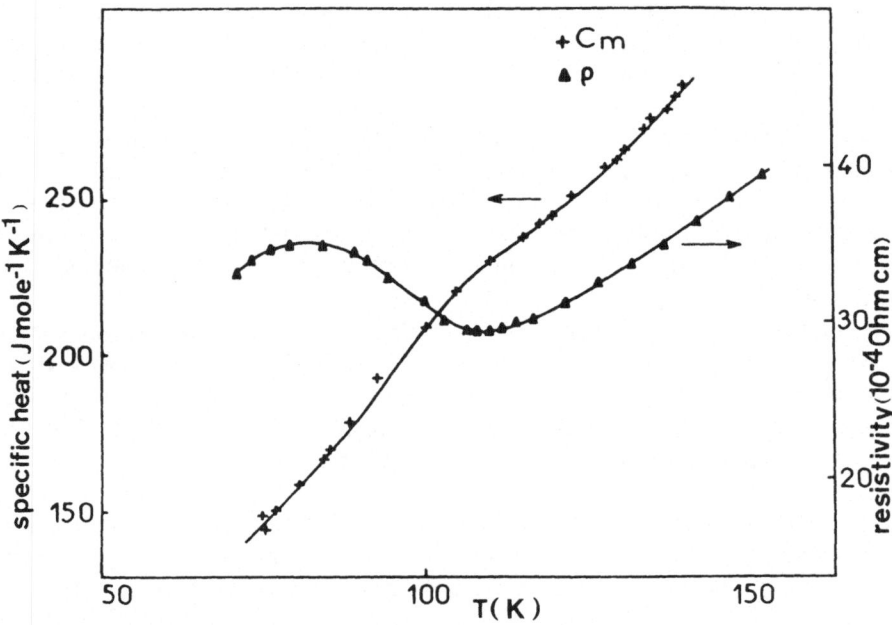

Fig. 60. Specific heat and electrical resistivity as a function of temperature for $K_{0.9}Mo_6O_{17}$ around T_p. The specific heat anomaly takes place at the same temperature as the upturn of the resistivity curve.

The results are summarized in Table IV in which the following information are listed:

— T_p, the onset temperature of the CDW transition;
— γ, the coefficient of the linear term in the low temperature heat capacity;
— θ_D, the Debye temperature obtained from the β coefficient of the T^3 term by $\beta = 12/5 \, \pi^4 Rn 1/\theta_D^3$, n is the number of atoms per molecule;
— ΔH is the integrated heat of transition, given with a large uncertainty due to the difficulties of estimating the base line;
— ΔS is the entropy of the transition, approximately equal to $\Delta H/T_p$.

The Debye temperatures, which are related to the stiffness of the interatomic force constants are similar for all these compounds and for the blue bronzes [63]. This indicates that the MoO_6 octahedra and MoO_4 tetrahedra are primarily responsible for the lattice rigidity. The role of the alkali ions does not seem to be predominant in this property. $Na_{0.9}Mo_6O_{17}$ shows the lowest θ_D and $Li_{0.9}Mo_6O_{17}$ the largest one. This may be due to an increased three dimensionality in the Li compound as compared to the Na and K ones.

The nonzero γ coefficients corroborate the metallic character of the low temperature phase of these compounds, due to the fact that the CDW does not affect the entire Fermi surface. The γ values are higher in the case of the Na and

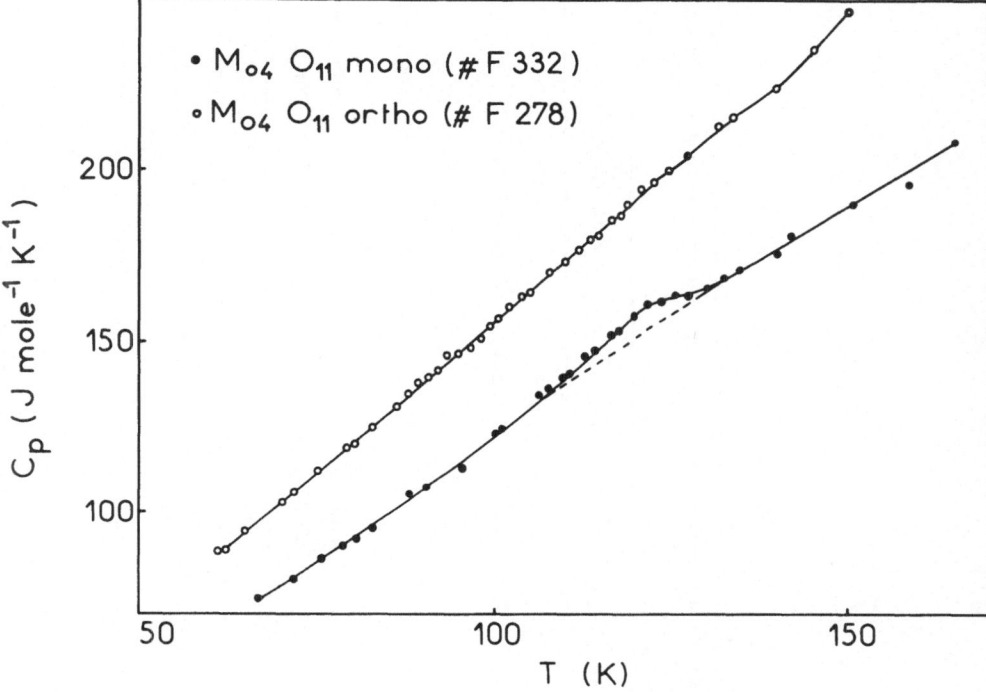

Fig. 61. Specific heat of η-Mo$_4$O$_{11}$ and γ-Mo$_4$O$_{11}$ versus temperature.

Li bronzes than for the K bronze and the Mo$_4$O$_{11}$ oxides. One can deduce, from the value of γ, the density of states at the Fermi level in the low temperature phase, $g(\varepsilon_F)_{LT}$, and compare it to the value $g(\varepsilon_F)_{HT}$ obtained from the thermopower in the high temperature one. These data are summarized in Table IV.

For the three compounds, K$_{0.9}$Mo$_6$O$_{17}$ and the two Mo$_4$O$_{11}$ oxides for which data are available, the density of states at low temperature is found to be smaller than that obtained by thermopower data at high temperature by a factor of 2 or 3. One can therefore estimate that at least 50% of the conduction electrons are condensed at the transition.

One should note the high value of γ in the case of the Li and specially Na bronzes. Also the Pauli paramagnetic susceptibility is found in the case of Li$_{0.9}$Mo$_6$O$_{17}$, one order of magnitude larger than for K$_{0.9}$Mo$_6$O$_{17}$. This might be due to the importance of electron correlations in the Li compound. This may be consistent with a larger resistivity below T_P.

All these compounds, except γ-Mo$_4$O$_{11}$, show an anomaly in the specific heat around T_P. These anomalies are weak and the transitions appear to be second order. The temperature range is much larger than what is observed on other physical properties. This indicates the existence of fluctuations more or less large above T_c, depending on the amount of short range ordering.

TABLE IV

Thermal data for the purple bronzes and the molybdenum oxides Mo_4O_{11}. γ is the coefficient of the linear term of the specific heat, θ_D the Debye temperature, T_P the Peierls temperature, ΔT is the temperature range of the specific heat anomaly, ΔS and ΔH the entropy and enthalpy changes at the transition. (From [63, 131 and 148].)

	$Li_{0.9}Mo_6O_{17}$	$Na_{0.9}Mo_6O_{17}$	$K_{0.9}Mo_6O_{17}$	$TlMo_6O_{17}$	$\gamma\text{-}Mo_4O_{11}$	$\eta\text{-}Mo_4O_{11}$
γ (mJ mole^{-1} K^{-2})	6 ± 0.5	10 ± 0.5	1.6 ± 0.2	—	2.6 ± 0.2	3 ± 0.3
θ_D (K)	365 ± 10 / 410 ± 20	290 ± 10	320 ± 10	—	410 ± 10	335 ± 10
T_P (K)	24	80	120	113	98	109 / 30
ΔT (K)	22—34	—	80—130	100—140	not measurable	100—130 / 25—30
ΔS (mJ mole^{-1} K^{-1})	240 ± 20	—	530 ± 50	314 ± 17	—	560 ± 50 / 37 ± 5
ΔH (J mole^{-1})	5.76 ± 0.5	—	55 ± 5	36 ± 2	—	70 ± 5 / 1.1 ± 0.2

By analyzing in detail the entropy change ΔS at the transition and the values of $g(\varepsilon_F)_{LT}$ and $g(\varepsilon_F)_{HT}$, one can estimate that ΔS contains both a lattice and an electronic contribution, which should be of the same order in $K_{0.9}Mo_6O_{17}$. In η-Mo_4O_{11}, the lattice contribution may be less important. In the case of γ-Mo_4O_{11}, the enthalpy of the transition is not measurable, possibly because the transition is broadened by fluctuations or by another mechanism.

4.5. BAND STRUCTURE

Band structures have been calculated for both $K_{0.9}Mo_6O_{17}$ and $Li_{0.9}Mo_6O_{17}$ by the tight binding method [161]. In the case of the potassium purple bronze, a two dimensional approximation was used [161a]. Figure 62 shows the Fermi surface and the band structure obtained in this approximation. The calculations predict a nesting wave vector in good agreement with the structural data. Experimental results obtained for the band structure by ultraviolet photoemission spectroscopy (UPS) on $K_{0.9}Mo_6O_{17}$ single crystals are consistent with a strong hybridization between the Mo $4d$ and O $2p$ orbitals [162].

In the case of $Li_{0.9}Mo_6O_{17}$, the band structure has been calculated in a pseudo 1D approximation [161b]. The crystal structure of $Li_{0.9}Mo_6O_{17}$ differs from that of $K_{0.9}Mo_6O_{17}$ by the configuration of MoO_4 tetrahedra between the MoO_6 layers. Moreover, the estimation of the Mo ions charge through a Zachariasen analysis (see the chapter by Vincent and Marezio in this volume), shows that the Mo^{5+} ions

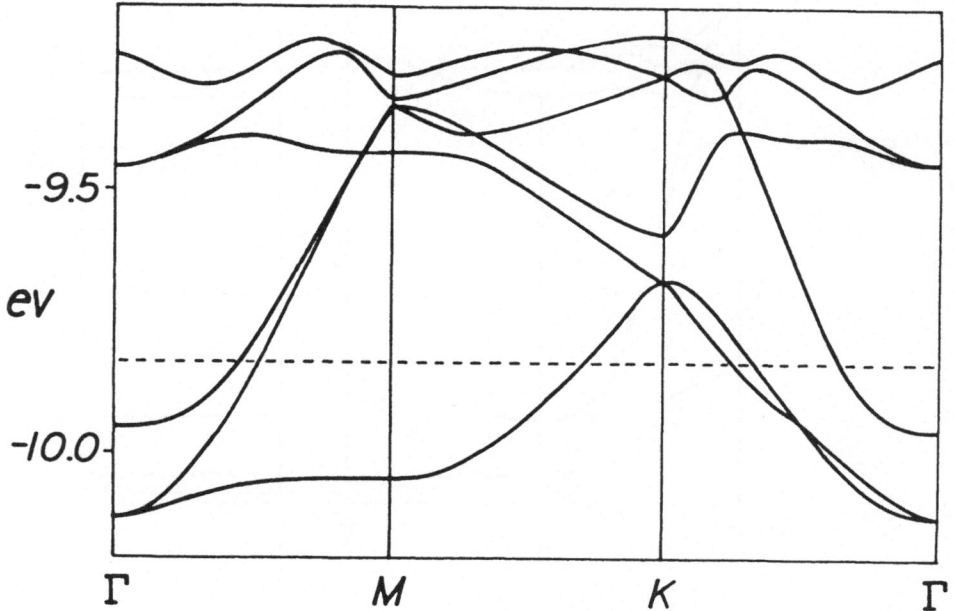

Fig. 62. Band structure of $K_{0.9}Mo_6O_{17}$. (From [161a].)

rather form infinite chains of MoO_6 octahedra in the plane of the layers. In a first approximation, the Fermi surface is then expected to be more flat than in the case of $K_{0.9}Mo_6O_{17}$. Figure 63 shows the band structure calculated along the three reciprocal directions a^*, b^* and c^* (a^* is perpendicular to the layers). The pseudo 1D character appears along ΓY and this leads to a flat Fermi surface. The calculation predicts a nesting wave vector $q = (0, 0.45b^*, 0)$. Up to now, no structural instability has been observed in $Li_{0.9}Mo_6O_{17}$ by X-ray diffraction.

Band structure calculations have also been performed on the oxides Mo_4O_{11} [163]. In the case of η-Mo_4O_{11} a direct information on the CDW gaps has been obtained by tunneling experiments [164]. Differential resistance measurements have been done on η-Mo_4O_{11}—Insulator—Lead junctions along the crystalline a^* axis (perpendicular to the layers). Two gaps are found, one opening at $T_{c_1} = 109$ K and the second one at $T_{c_2} \approx 30$ K. The nature of this instability, CDW or SDW, is not clear at present [165].

Optical reflectivity studies performed in the infrared range corroborate the anisotropy of the band structure [138]. Figure 64 shows the spectra obtained at room temperature with polarized light for various orientations of the polarization P with respect to the high conductivity b-axis. One should note a metallic behavior for $P \| b$ and a semiconductor-like one for $P \| a^*$. However, the anisotropy is non-negligible for P in the plane of the layers in the case of η-Mo_4O_{11} (Figure 64b).

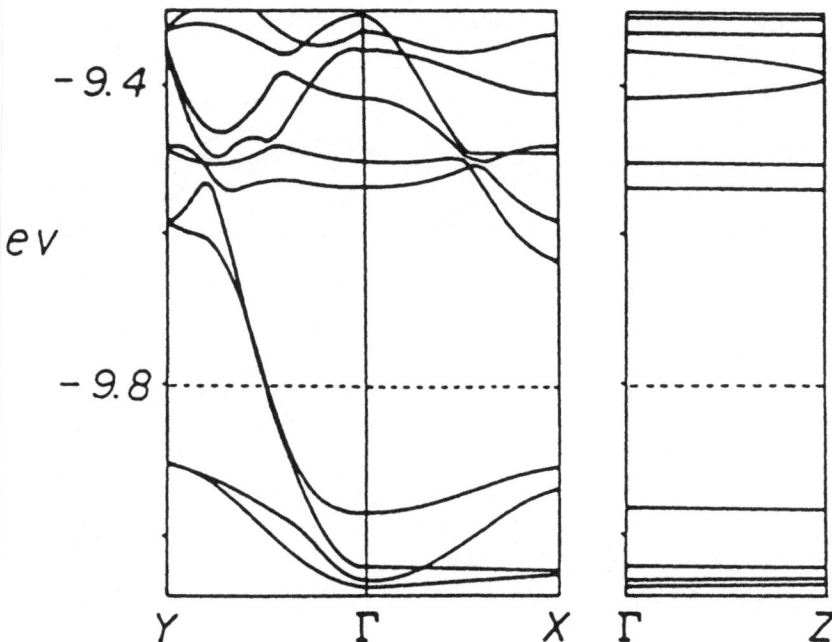

Fig. 63. Band structure calculated for $Li_{0.9}Mo_6O_{17}$ where Γ is the Brillouin zone center, $X = (c^*/2, 0, 0)$, $Y (0, b^*/2, 0)$ and $Z = (0, 0, a^*/2)$. The dashed line refers to the Fermi level. (From [161b].)

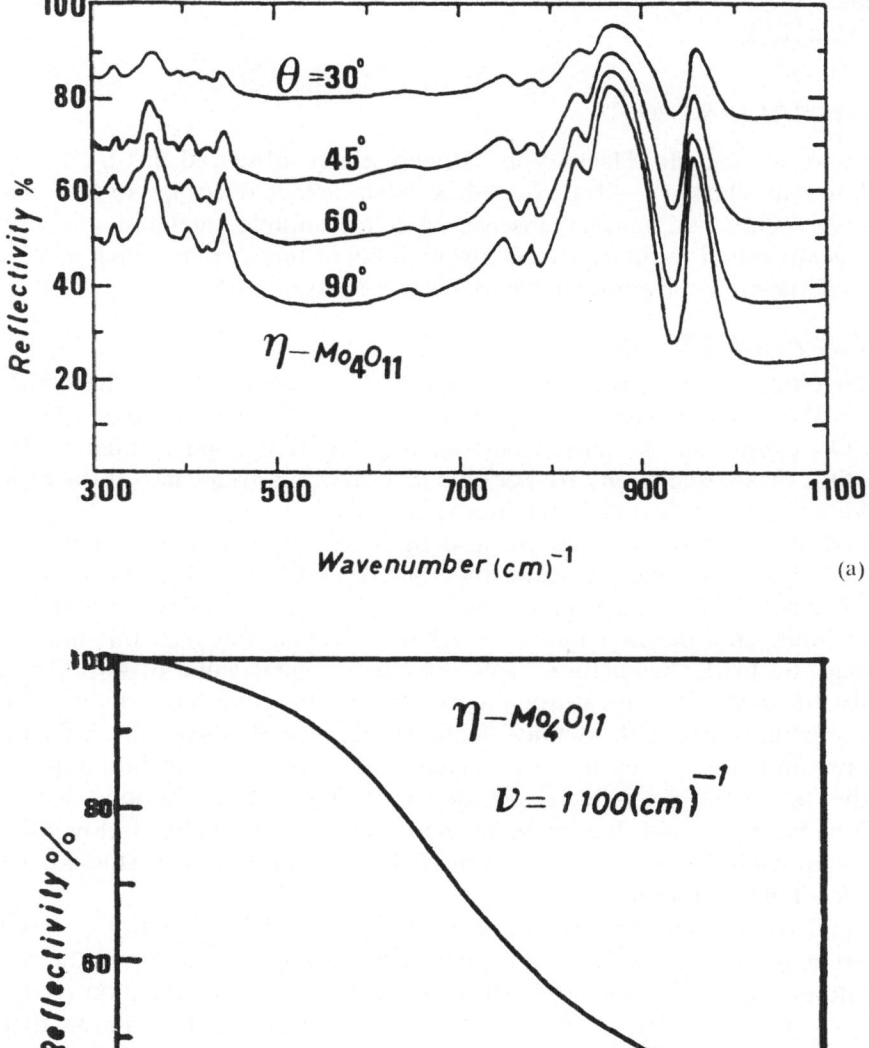

Fig. 64. (a) Optical reflectivity spectra obtained on η-Mo$_4$O$_{11}$ single crystal on the crystallographic face (001) (obtained with several crystals glued together). θ is the angle between the polarization and the high conductivity b-axis. (b) Variation of the reflectivity of η-Mo$_4$O$_{11}$ as a function of the orientation of the polarization in the (100) plane ($\theta = 0$, $P \| b$; $\theta = 90°$ $P \| c$) (wavelength 1100 cm^{-1}). (c) Same as (a) for γ-Mo$_4$O$_{11}$ for (011) crystallographic plane.

The peaks that are apparent for $P \| a^*$ have been attributed to phonons and are discussed in [146].

4.6. QUANTUM TRANSPORT

In addition to the de Haas—Van Alphen effect observed in η-Mo_4O_{11}, a Shubnikov—de Haas (S—dH) effect has been detected in η-Mo_4O_{11} and in $K_{0.9}Mo_6O_{17}$ below 4.2 K. The absence of a measurable quantum effect in γ-Mo_4O_{11} is attributed to the relatively low mobility of the carriers which is at 4.2 K an order of magnitude lower than that found in η-Mo_4O_{11} [146].

4.6.1. *Experimental Results*

In similar configurations, with the magnetic field perpendicular to the octahedra infinite slabs, the transverse magnetoresistances (TMR) of η-Mo_4O_{11} and $K_{0.9}Mo_6O_{17}$ exhibit an oscillating component at low temperature (Figure 65a). In $K_{0.9}Mo_6O_{17}$, these oscillations are periodic in $1/B$ with a frequency of the order of 10 T and their amplitudes are temperature dependent.

In η-Mo_4O_{11} the oscillations are also of S—dH type, although they are not periodic in $1/B$, very likely because they are related to small quantum numbers ($n = 3$, 2 and 1). The large divergence of the TMR above 11 T is attributed to the quantum limit, only the first Landau level $n = 0$ being filled. In this limit, for a 19 T magnetic field, the relative TMR reaches the giant value of 400. The Hall resistivity of η-Mo_4O_{11}, measured in the same configuration, exhibits also an oscillating component, with identical features. Figure 65b shows the influence of the temperature on the oscillations related to the non-zero quantum numbers: when the temperature is decreased down to 900 mK the amplitude of the oscillations increases and they move to the lower magnetic fields. Below 900 mK the quantum oscillations are frozen. The Hall resistivity has been studied only in the very low temperature range.

At liquid helium temperature, for $B = 10$ T, the TMR of η-Mo_4O_{11} has been measured as a function of θ, the angle between the magnetic field and the normal to the infinite slabs. The data are well fitted by the ellipsoidal function ($\cos^2 \theta + \varepsilon \sin^2 \theta)^{1/2}$ with an eccentricity of $\varepsilon = 0.01$ (Figure 65c). This shows that the TMR depends, in a first approximation, only on the amplitude of the component of B normal to the infinite slabs. In $K_{0.9}Mo_6O_{17}$, a simple $\cos \theta$ law describes the angular variation of the amplitudes of the oscillations when B is tilted. These results indicate that the closed pockets, responsible for the quantum effects, have a very anisotropic shape, elongated perpendicular to the infinite slabs.

4.6.2. *Discussion*

The theory of quantum effects predicts that the positions of the oscillations with a small quantum number deviate from the $1/B$-periodic law. This is due to the influence of a partial degeneracy of the first Landau levels and depends on the width of these levels and on the temperature. Although the unperturbed oscillations with higher quantum numbers are not detectable in η-Mo_4O_{11}, it was possible to calculate by a self-consistent method the pseudoperiod of the oscil-

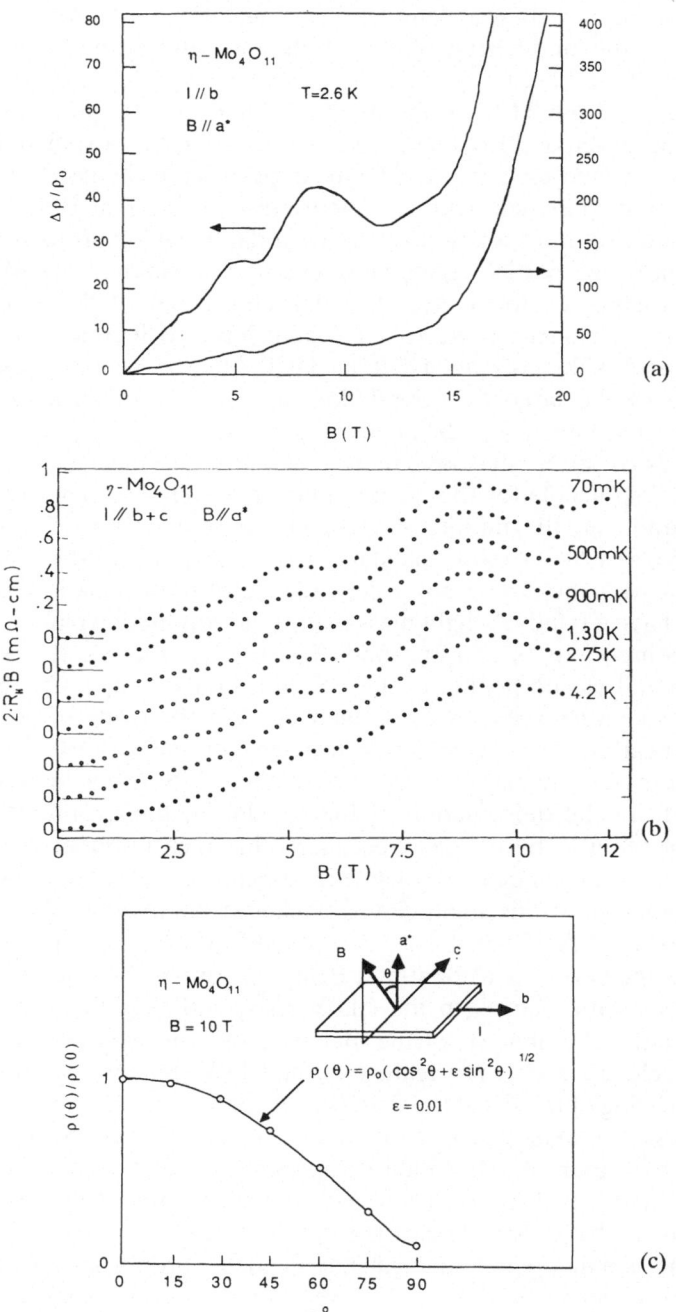

Fig. 65. (a) Transverse magnetoresistance in η-Mo$_4$O$_{11}$ vs magnetic fields up to 19 T. The upper curve corresponds to the left scale and shows the details of the oscillations. (b) Hall resistivity as a function of magnetic field for η-Mo$_4$O$_{11}$ at different temperatures. The magnetic field is normal to the infinite slabs. (c) Angular dependence of the transverse magnetoresistance in η-Mo$_4$O$_{11}$.

lating component, which was found to be equal to 9.3 ± 0.4 T. This value is consistent with the numbering of the oscillations and with the presence of the quantum limit above 11 T.

The behavior of the TMR in the quantum limit corroborates this analysis. The two minima observed at 11.6 T and 14.1 T correspond to a spin splitting over the first minimum, related to $n = 1/2$. The mean position of the non-split minimum, at 12.7 T, is in total agreement with the 'theoretical' position, at $B(1/2, 0) = 12.4 \pm 0.3$ T, calculated by interpolation of the periodic behavior up to $n = 1/2$ and by taking account of the partial degeneracy. The first maximum, related to $n = 0$ and $s = +1/2$, is expected to be near to 20 T and is out of the investigated field. Nevertheless, its presence close to 19 T contribute to increase the TMR in this region: the giant value of the TMR at 19 T ($\Delta\rho/\rho_0 = 400$) results from the superposition of the large first oscillation and of the nonoscillating component, which is expected to diverge as B^3 in this region.

The analysis of the S—dH oscillations leads to the following characterization of the electrons responsible for the quantum effect: a cyclotron mass $m_c = 0.39\ m_0$ in the plane parallel to the infinite slabs, a Dingle temperature $T_D = 2.6$ K and a Landé factor $g = 1.52 \pm 0.06$, calculated from the spin splitting [146]. These data are consistent with the temperature dependence of the oscillations and show that the electrons have a relatively light mass in the high conducting plane.

In both compounds η-Mo_4O_{11} and $K_{0.9}Mo_6O_{17}$, the S—dH oscillations are related to a small closed pocket of the FS, of a similar shape. On this pocket, the mass tensor is characterized by light eigenvalues in the plane of the infinite slabs, and an eigenvalue about 100 times heavier perpendicularly. This anisotropy calculated from the angular dependence of the TMR for a high magnetic field could result from the deformation of the FS due to the quantum state ($n = 1$). This does not seem to be the case, because other measurements performed in η-Mo_4O_{11} with a low magnetic field (magnetic susceptibility) lead to the same result.

The cross-section of this small closed pocket in the plane of the infinite slabs is of the order of 10^{-3} Å^{-2} (8.9×10^{-4} Å^{-2} in η-Mo_4O_{11}). This represents about one thousandth of the two-dimensional first Brillouin zone of that plane. This pocket is detected at low temperature, in the CDW state, and very likely results from the Peierls transition. The analysis of the magnetic susceptibility also showed that the presence of a closed pocket is a feature of the CDW state and is probably due to a non-perfect nesting of the FS at the Peierls transition.

Nevertheless, the anisotropy of this pocket is directly related to the anisotropy of the high temperature FS. Both anisotropies must therefore be of the same order of magnitude. This result is consistent with the anisotropy of the high temperature FS obtained from the electrical resistivity measurements.

Due to this small pocket, the galvanomagnetic properties of η-Mo_4O_{11} and $K_{0.9}Mo_6O_{17}$ show some similarity. However, some differences indicate that the FS of these compounds are not identical. The TMR of $K_{0.9}MoO_{17}$ does not seem to diverge at high magnetic field. This may result from the presence of two kinds of carriers, as it has been proposed to analyze the galvanomagnetic data around the

Peierls transition [150]. In this model, one should attribute de S—dH oscillations to the holes, which may have, by interpolation, a higher mobility. In the case of η-Mo_4O_{11}, the galvanomagnetic properties are interpreted with a single particle model, and the low temperature FS requires only a single closed pocket. The number of electrons remaining at low temperature, calculated from the Hall resistivity data, is compatible with a diminution of the FS towards a single closed pocket [146].

Although η-Mo_4O_{11} shows some strong anisotropic transport properties, the quantum Hall effect measured at very low temperature does not present the plateaus characteristic of the quantum Hall effect in two-dimensional systems [166a] or observed in one-dimensional compounds with spin density waves [166b]. These oscillations seem to be 'conventional'. Nevertheless, the temperature dependence of the positions of the oscillations is opposite to that expected by the theory of isotropic compounds. This may indicate that the anisotropic shape of the FS has some influence on the quantum properties or that the FS is sensitive to the magnetic field and can be slightly modified at very low temperature by high magnetic fields.

4.7. SUPERCONDUCTIVITY IN $Li_{0.9}Mo_6O_{17}$

Greenblatt et al. [127] first pointed out the existence of a superconducting transition at $T_c = 1.9$ K in a lithium molybdenum bronze single crystal grown by a temperature gradient flux technique. Our results show that the existence and the temperature of the superconducting transition are sample dependent [167]. Only one third of the samples show the superconducting transition. For those becoming superconducting, T_c ranges between 1.1 K and 1.6 K for samples grown by electrolytic reduction, when the transition exists.

Matsuda et al. [148] have measured T_c as a function of doping concentration in $(Li_{1-x}K_x)_{0.9}Mo_6O_{17}$, $(Li_{1-x}Na_x)_{0.9}Mo_6O_{17}$ and $Li_{0.9}(Mo_{1-x}W_x)_6O_{17}$ (Figure 66). If no sign of the superconducting transition was observed for all W-doped crystals down to 1.3 K, for the $(Li_{1-x}K_x)_{0.9}Mo_6O_{17}$ and $(Li_{1-x}Na_x)_{0.9}Mo_6O_{17}$, T_c seems to be insensitive to the value of x for $0 \leqslant x \leqslant 0.4$. No systematic dependence of T_c on x was found, even if the largest value of T_c has been obtained for $(K_{0.4}Li_{0.6})_{0.9}Mo_6O_{17}$. The authors underline the existence of a correlation between the increasing magnitude of the resistivity upturn and the decrease of T_c.

The ac susceptibility measurements have been performed in order to obtain the onset temperature of superconductivity and the critical field with a good accuracy [167]. Figure 67 shows the influence of a very low magnetic field on T_c. The temperature dependence of the critical fields $Hc_{2\parallel}$ and $H_{c_{2\perp}}$ (the indices $\parallel$ and $\perp$ denote the field orientation $\perp$ and $\parallel$ to the crystallographic c axis, respectively) shows a positive curvature (Figure 68). This upward curvature has been observed in the layered superconductors of intercalated $NbSe_2$ and TaS_2 [168]. In the approach developed by Klemm, Beasley and Luther [169], this can be due to a dimensional crossover effect: close to T_c, the material behaves like an anisotropic

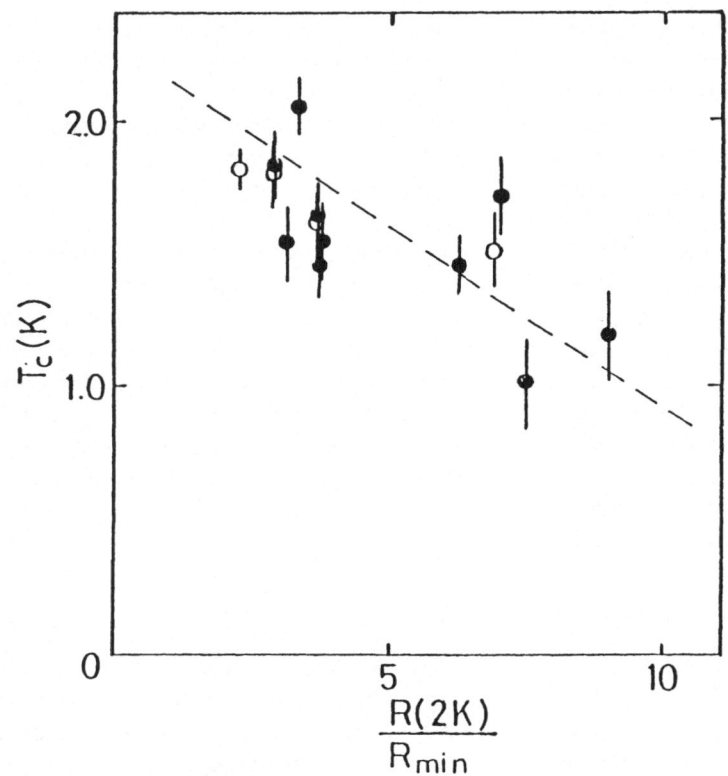

Fig. 66. Superconducting transition temperature in several crystals of $(Li_{1-x}K_x)_{0.9}Mo_6O_{17}$ and $(Li_{1-x}Na_x)_{0.9}Mo_6O_{17}$ as a function of R (2 K)$/R_{min}$. R_{min} is the resistance corresponding to the maximum of R vs T. Open and closed circles correspond to K and Na doped samples, respectively. (From [148].)

three-dimensional superconductor, while at low temperatures the behavior is determined essentially by the properties of the individual layers. However, if this were to be the case, $H_{c_{2\perp}}(T)$ should be linear and should not exhibit the same kind of curvature as $Hc_{2\parallel}$.

The anisotropy parameter $(1/\varepsilon = Hc_{2\parallel}/H_{c_{2\perp}})$ is temperature independent with a value of $1/\varepsilon = 3.9 \pm 0.2$ (insert of Figure 68b). One can then apply a Ginzburg-Landau model:

$$H_{c_2}(0) = \frac{\phi_0}{2\pi\xi_\perp(0)\xi_\parallel(0)} \quad \text{and} \quad H_{c_{2\perp}}(0) = \frac{\phi_0}{2\pi\xi_\parallel^2(0)}$$

ϕ_0 being the flux quantum, $\xi_\perp$ and $\xi_\parallel$ are the coherence lengths perpendicular and parallel to the layers. This leads to $\xi_\parallel(0) = 194$ Å and $\xi_\perp(0) = 50$ Å. As $\xi_\perp(0)$ is

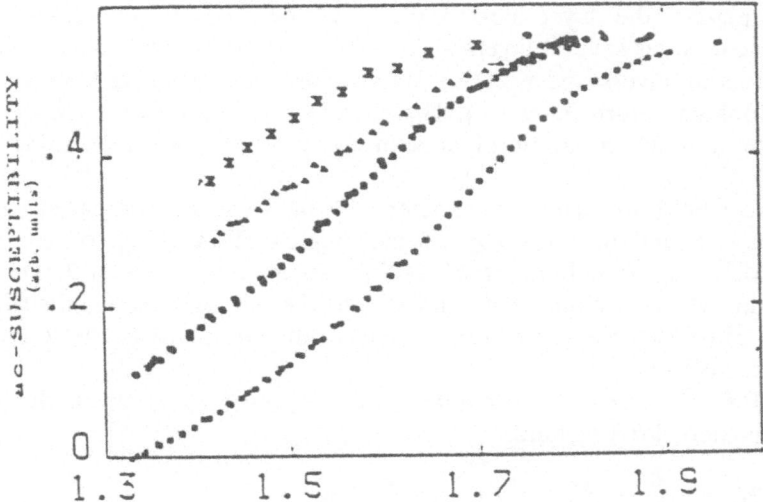

Fig. 67. ac Susceptibility as a function of temperature for a crystal of $Li_{0.9}Mo_6O_{17}$. The curves correspond to applied dc fields of 100 Oe, 30 Oe, 3 Oe and 0 Oe from top to bottom. Measuring frequency 200 Hz. (From [167b].)

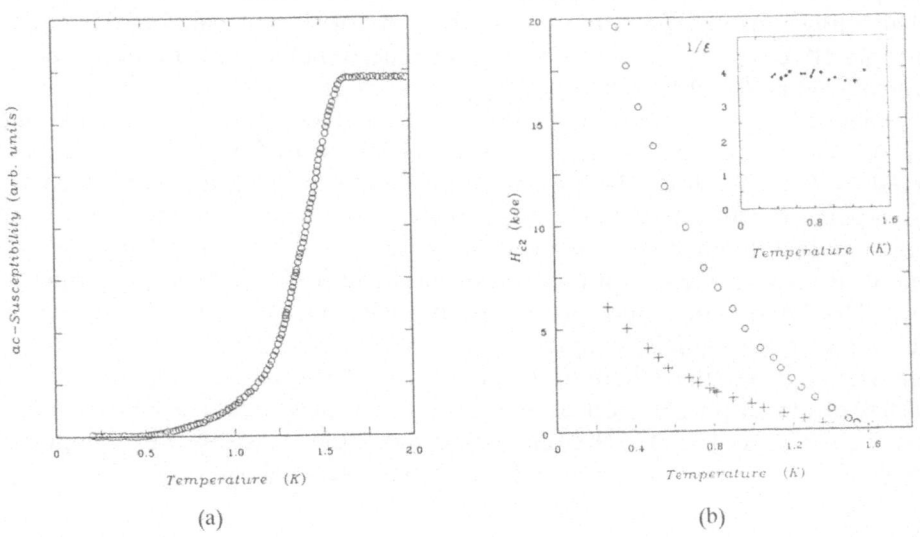

Fig. 68. (a) ac Susceptibility versus temperature in zero dc field for a single crystal of $Li_{0.9}Mo_6O_{17}$ (measuring frequency 42 kHz). (From [167a].) (b) Temperature dependence of the upper critical field H_{c_2}. 0 denotes field orientation parallel to the layers and + perpendicular. The insert shows the anisotropy parameter (see text). (From [167a].)

much larger than the layer period (12.7 Å), pairing of the superconducting electrons occurs over several layers at $T = 0$ K. This excludes a decoupling effect of the conducting layers and a 2D to 3D crossover, as speculated on the basis of the anomalous curvature in $H_{c_2}(T)$. Therefore a distribution of T_c in the sample, arising from a local variation of stoichiometry, could be responsible for this curvature.

The critical field anisotropy parameter $\varepsilon = 0.26$ can be compared to the low temperature conductivity anisotropy. According to Tilley [170], $\sigma_\parallel/\sigma_\perp = (1/\varepsilon)^2$. From the data of Greenblatt et al. [127] one obtains $\varepsilon = 0.09$. Due to the anisotropy in the basal plane [130] and also to the inhomogeneous distribution of the current caused by the anisotropic nature of the conductivity, these two values of ε compare fairly well.

At $T = 0$ K, the critical field parallel to the layers, $Hc_{2\parallel}$, exceeds the so-called Pauli paramagnetic limiting field,

$$H_p = \frac{\Delta_0}{\sqrt{2}\mu_B} = 18.6 \ T_c(kOe) = 29.4 \ kOe$$

resulting from the Pauli paramagnetism of the conduction electrons [171]. It has been shown that in the layered superconductors such as intercalated complexes of TaS$_2$ [168], the parallel critical fields may exceed the Pauli paramagnetic limit by a large amount. Very large spin orbit scattering rates are then required to account for the absence of the Pauli limit. The presence of heavy molybdenum atoms in the lithium purple bronzes provides the possibility of considerable spin orbit scattering. This allows an interpretation of H_{c_2} in terms of orbital pair breaking. Then anisotropic effective mass analysis, which was questionable with the high value of the critical field at $T = 0$ K, can be justified.

Information on the critical field could be obtained from calorimetric data. Figure 69 shows the specific heat data as a plot of C/T vs T^2. The anomaly observed by Matsuda et al. [148] or by us [167] is very small. It is very likely due to the superconducting transition. Surprisingly, it is no more pronounced in the case of a pure lithium bronze than in the case of a mixed one. No systematic specific heat measurement as a function of magnetic field has been performed up to now. The large broadening of the specific heat anomaly may indicate some distribution of T_c through the sample.

One should note that there is some analogy between the superconducting properties of the Li purple bronzes and those of the new high T_c superconducting oxides, such as YBa$_2$Cu$_3$O$_7$. The understanding of the quasi-2D Mo compounds should therefore, hopefully, help to understand these new oxides.

5. Conclusion

The considerable amount of work performed in recent years on the Mo bronzes and oxides has now established the importance of low dimensional electronic properties and charge density wave instabilities in this class of materials. Most of the properties related to the Peierls transition are now comparatively well

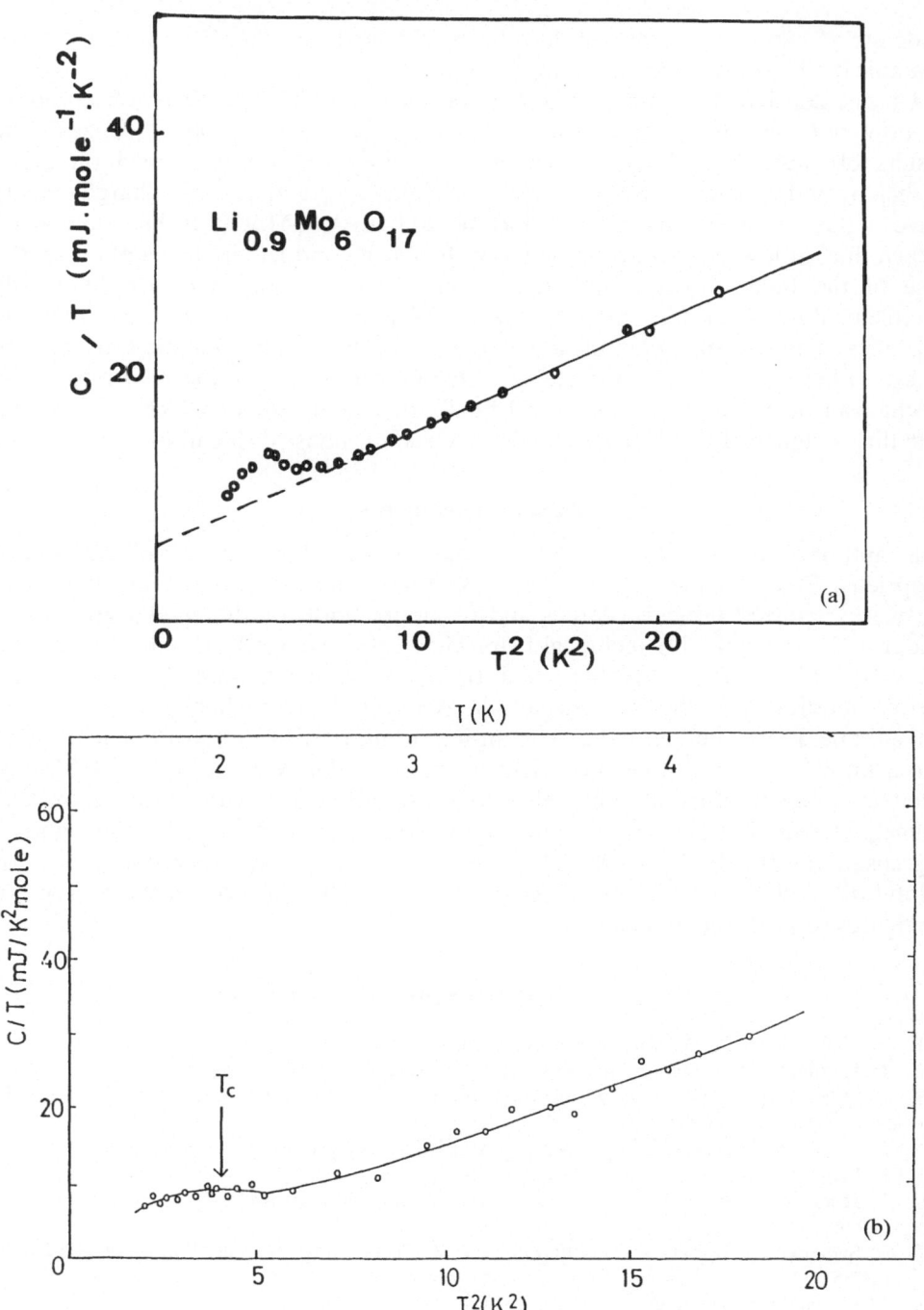

Fig. 69. Low temperature specific anomaly plotted as C/T vs T^2. (a) For $Li_{0.9}Mo_6O_{17}$. (b) For $(Li_{0.6}K_{0.4})_{0.9}Mo_6O_{17}$ (from [148].).

understood. Some information has been obtained on the Fermi surface, for example by the study of quantum properties.

Charge density wave transport due to the sliding of CDW has been observed up to now only in the quasi-one-dimensional compounds, the blue bronzes. The results obtained have largely complemented the work on the transition metal trichalcogenides. The hysteresis properties related to the metastable charge density wave states are particularly well documented in $A_{0.30}MoO_3$. It has now been shown that, at low temperature, the CDW Fröhlich conductivity takes place, in the case of the blue bronze, with an extremely low damping. Due to the nearly complete absence of free carriers, the CDW is able to slide with considerable velocities. These results should now stimulate further studies on the blue bronze and on other quasi-1D CDW materials. One may hope that in the near future, the mechanism of narrow band noise will finally be elucidated, very likely in relation with the existence of CDW structural defects, such as phase dislocations.

Acknowledgements

The authors are grateful to many colleagues at Laboratoire d'Etudes des Propriétés Electroniques des Solides, CNRS Grenoble, for their contribution to the work reported here: J. Marcus and G. Fourcaudot for the crystal growth, R. Buder and J. Devenyi for technical help. They acknowledge B. K. Chakraverty, R. Chevalier, R. Cinti, D. Feinberg and B. Horovitz, for helpful discussions and various studies. The thesis works of A. Arbaoui, P. Beauchêne, E. Bervas, K. Konate and J. Y. Veuillen have been an important part of this work. The authors also acknowledge several visiting scientists, in particular A. Janossy and G. Mihaly. They have benefited from many collaborations, particularly with M. Marezio, J. P. Pouget, M. Ghedira and H. Vincent for the structural studies, C. Berthier and P. Segransan for the NMR work. They are grateful to many other colleagues for helpful discussions at different stages of this work. They also thank the authors of the figures reproduced in this review.

References

1. J. A. Wilson and A. D. Yoffe, *Adv. Phys.* **18**, 193 (1969).
2. D. Jérome and H. J. Schulz, *Adv. Phys.* **31**, 299 (1982).
3. P. Monceau (ed.), *Electronic Properties of Inorganic Quasi-One Dimensional Compounds* (D. Reidel Publ. Co.) (1985).
4. R. Peierls, in *Quantum Theory of Solids* (Oxford University Press), (1955) p. 108.
5. H. Fröhlich, *Proc. Royal Soc. A* **223**, 296 (1954).
6. P. Monceau, N. P. Ong, A. M. Portis, A. Meerschaut, and J. Rouxel, *Phys. Rev. Lett.* **37**, 602 (1976).
7. C. Schlenker, J. Dumas, C. Escribe-Filippini, H. Guyot, and J. Marcus, *Phil. Mag. B* **52**, 643 (1985).
8. M. Greenblatt, this volume.
9. M. Marezio and H. Vincent, this volume.
10. J. P. Pouget, this volume.
11. S. Aubry, this volume.

12. J. Friedel and D. Feinberg, this volume; see also D. Feinberg and J. Friedel, *J. Physique* **49**, 485 (1988).
13. R. H. Friend and D. Jérome, *J. Phys. C — Solid State Phys.* **12**, 1441 (1979).
14. G. A. Toombs, *Physics Reports* **40**, 181 (1978).
15. M. J. Rice and S. Strassler, *Solid State Comm.* **13**, 125 (1973).
16. P. A. Lee, T. M. Rice, and P. W. Anderson, *Solid State Comm.* **13**, 1931 (1973).
17. Y. A. Firsov, V. N. Prigadin, and Ch. Seidel, *Phys. Reports* **126**, 245 (1985).
18. See for example G. Grüner, *Physica Scripta* **32**, 11 (1985); G. Grüner and A. Zettl, *Phys. Reports* **119**, 117 (1985); J. C. Gill and H. Wills, *Contemp. Phys.* **27**, 37 (1986).
19. P. Monceau in Ref. 3, p. 139; R. M. Fleming and C. C. Grimes, *Phys. Rev. Lett.* **42**, 1423 (1979).
20. H. Fukuyama and P. A. Lee, *Phys. Rev. B* **17**, 535 (1978).
21. P. A. Lee and T. M. Rice, *Phys. Rev. B* **19**, 3970 (1979).
22. D. S. Fisher, *Phys. Rev. B* **31**, 1396 (1985).
23. H. Matsukawa and H. Takayama, *Solid State Comm.* **50**, 283 (1984).
24. P. B. Littlewood, *Phys. Rev. B* **33**, 6694 (1986).
25. S. Aubry, L. de Sèze, *Festkörperprobleme*, Vol. XXV, p. 59 (Braunschweig-Vieweg) (1985).
26. S. Aubry in *Structure et Instabilités*, p. 73, C. Godrèche (Ed.) (Editions de Physique) (1986).
27. W. L. McMillan, *Phys. Rev. B* **14**, 1496 (1976).
28. L. P. Gorkov, *JETP Letters* **38**, 87 (1983); I. Batistic, A. Bjelis, and L. P. Gorkov, *J. Physique* **45**, 1049 (1984).
29. N. P. Ong and K. Maki, *Phys. Rev. B* **32**, 6582 (1985); K. Maki, in Ref. 3, p. 125.
30. J. Dumas and D. Feinberg, *Europhys. Lett.* **2**, 555 (1986).
31. J. Friedel, NATO-ASI on *Low Dimensional Conductors and Superconductors*, Aug. 1986, Magog, Quebec, Canada (Plenum) (to be published).
32. B. K. Chakraverty, *Phys. Rev. B* **34**, 5287 (1986).
33. J. Bardeen, *Phys. Rev. Lett.* **42**, 1498 (1979); **45**, 1978 (1980); Proc. Int. School of Physics, Varenne, 1983, *Highlights of Condensed Matter Theory*, LXXXIX, 349 (1985).
34. J. Bardeen, Proceedings of the Yamada Conference on the Physics and Chemistry of Quasi-One-Dimensional Solids (Lake Kawaguchi, Japan, May 1986, *Physica* **143B**, 14 (1986); in Ref. 3, p. 105.
35. S. E. Barnes and A. Zawadowsky, *Phys. Rev. Lett.* **5**, 1003 (1983).
36. (a) *Symposium on Nonlinear Transport and Related Phenomena in Inorganic Quasi-One-Dimensional Conductors*, Sapporo (Japan) (1983), Ed. Hokkaido University. (b) 'Charge density waves in solids' (Proceedings, Budapest 1984). *Lecture Notes in Physics* **217**, Ed. Gy Hutiray and J. Solyom (Springer Verlag, 1985).
37. Proceedings of the Yamada Conference on the Physics and Chemistry of Quasi-One-Dimensional Solids (Lake Kawaguchi, Japan, May 1986), *Physica B* **143** (1986).
38. Proceedings of the Intern. Conf. on Synthetic Metals (Kyoto, Japan, June 1986), *Synthetic Metals* **19** (1987).
39. (a) NATO-ASI on *Low-Dimensional Conductors and Superconductors*, Series B **155** (1987) (Plenum Publ. Corp.) Ed. D. Jerome and L. G. Caron; (b) Second European Workshop on CDW's, Aussois (France) Sept. 1987.
40. Int. Conf. on Synthetic Metals, Abano Terme, June 1984, *Molecular Crystals, Liquid Crystals* **121**, 185.
41. C. Schlenker and J. Dumas, in *Crystal Chemistry and Properties of Materials with Quasi-One-Dimensional Structures*, Ed. J. Rouxel (D. Reidel Publ. Co.) 1986, p. 135.
42. C. Schlenker, NATO-ASI on *Low-Dimensional Conductors and Superconductors*, Magog, Quebec, Canada (Aug. 1986) (Plenum Publ. Corp.) Vol. 155, p. 477 (1987).
43. P. Segransan, A. Janossy, C. Berthier, P. Butaud, and J. Marcus, *Phys. Rev. Lett.* **56**, 854 (1986).
44. C. Berthier and P. Ségransan, in Ref. 39a.
45. A. Wold, W. Kunnmann, R. J. Arnott, and A. Ferreti, *Inorg. Chem.* **3**, 545 (1964).

46. (a) J. Graham and A. D. Wadsley, *Acta Cryst.* **20**, 93 (1966); (b) M. Ghedira, J. Chenavas, M. Marezio, and J. Marcus, *J. Solid State Chem.* **57**, 300 (1985).

47. G. H. Bouchard Jr, J. H. Perlstein, and M. J. Sienko, *Inorg. Chem.* **6**, 1682 (1967).

48. D. S. Perloff, M. Vlasse, and A. Wold, *J. Phys. Chem. Solids* **30**, 1071 (1969).

49. W. Fogle and J. H. Perlstein, *Phys. Rev. B* **6**, 1402 (1972).

50. R. Brusetti, B. K. Chakraverty, J. Devenyi, J. Dumas, J. Marcus, and C. Schlenker, in *Recent Development in Condensed Matter Physics* (Eds. J. T. Deevreese, L. F. Lemmens, V. E. Van Doren, and J. Van Royen), Plenum, New York, Vol. 2, p. 181 (1981).

51. G. Travaglini, P. Wachter, J. Marcus, and C. Schlenker, *Solid State Comm.* **37**, 599 (1981).

52. J. P. Pouget, S. Kagoshima, C. Schlenker, and J. Marcus, *J. Physique Lett.* **44**, L113 (1983).

53. J. Dumas, C. Schlenker, J. Marcus, and R. Buder, *Phys. Rev. Lett.* **50**, 757 (1983).

54. R. H. Fleming, L. F. Schneemeyer, and D. E. Moncton, *Phys. Rev. B* **32**, 3568 (1985).

55. J. P. Pouget, C. Noguera, A. H. Moudden, and R. Moret, *J. Physique* **46**, 1731 (1985).

56. R. M. Fleming and R. J. Cava, this volume.

57. E. Bervas, Thèse d'Ingénieur-Docteur, Université Scientifique, Technologique et Médicale de Grenoble (1984).

58. (a) L. Forro, J. R. Cooper, A. Janossy, and K. Kamaras, *Phys. Rev. B* **34**, 9047 (1986); (b) L. Forro, J. R. Cooper, A. Janossy, and M. Maki, *Solid State Comm.* **62**, 715 (1987).

59. L. F. Schneemeyer, F. J. Di Salvo, S. E. Spengler, and J. V. Waszczak, *Phys. Rev. B* **30**, 4297 (1984).

60. K. Machida and M. Nakano, *Phys. Rev. Lett.* **55**, 1927 (1985); *Phys. Rev. B* **33**, 6718 (1986).

61. L. F. Schneemeyer, F. J. Di Salvo, R. M. Fleming, and J. V. Waszczak, *J. Solid State Chem.* **54**, 358 (1984).

62. D. C. Johnston, *Phys. Rev. Lett.* **52**, 2049 (1984).

63. K. Konaté, Thèse de 3ème Cycle, Université Scientifique, Technologique et Médicale de Grenoble (1984).

64. K. J. Dahlhauser, A. C. Anderson, and G. Mozurkewitch, *Phys. Rev. B* **34**, 4432 (1986).

65. L. C. Bourne and A. Zettl, *Solid State Comm.* **60**, 789 (1986).

66. L. F. Schneemeyer, S. E. Spengler, F. J. Di Salvo, and J. V. Waszczak, *Mol. Cryst. Liq. Cryst.* **125**, 41 (1985).

67. S. Girault, A. H. Moudden, G. Collin, J. P. Pouget, and R. Comès, *Solid State Comm.* **63**, 17 (1987).

68. H. Mutka, S. Bouffard, J. Dumas, and C. Schlenker, *J. Physique Lettres* **45**, L729 (1984).

69. H. Mutka, S. Bouffard, M. Sanquer, J. Dumas, and C. Schlenker, Proc. of the Int. Conf. on Synthetic Metals (Abano Terme, Italy, 1984), *Mol. Cryst., Liq. Cryst.* **121**, 133 (1985).

70. G. Travaglini and P. Wachter, in 'Charge Density Waves in Solids', *Lecture Notes in Physics* **217**, 115 (1985).

71. M. H. Whangbo and L. F. Schneemeyer, *Inorg. Chem.* **25**, 2424 (1986).

72. G. K. Wertheim, L. F. Schneemeyer, and D. N. E. Buchanan, *Phys. Rev. B* **32**, 3568 (1985).

73. J. Y. Veuillen, R. C. Cinti, and E. Al Khoury Nemeh, *Europhysics Letters* **3**, 355 (1987).

74. (a) M. Ganne, A. Boumaza, M. Dion, and J. Dumas, *Mat. Res. Bull.* **20**, 1297 (1985); (b) B. T. Collins, K. V. Ramanujachary, M. Greenblatt, and J. V. Waszczak, *Solid State Comm.* **56**, 1023 (1985).

75. (a) P. Beauchêne, J. Dumas, A. Janossy, J. Marcus, and C. Schlenker, Yamada Conf. on the Physics and Chemistry of quasi-one-dimensional conductors, Japan, May 1986, *Physica B* **143**, 126 (1986); C. Schlenker and J. Dumas, *Physica B* **143**, 103 (1986); (b) P. Beauchêne, Thèse, Université Scientifique, Technologique et Médicale de Grenoble (1987).

76. K. T. Tsutsumi, T. Tamegai, S. Kagoshima, and M. Sato, *J. Phys. Soc. Japan* **54**, 3004 (1985).

77. R. M. Fleming and L. F. Schneemeyer, *Phys. Rev. B* **28**, 6996 (1983).

78. (a) L. F. Schneemeyer, R. M. Fleming, and S. E. Spengler, *Solid State Commun.* **53**, 505 (1985); (b) R. J. Cava, P. B. Littlewood, R. M. Fleming, L. F. Schneemeyer, and E. A. Rietman, *Phys. Rev. B* **34**, 1184 (1986).

79. J. Y. Veuillen, R. Chevalier, J. Marcus, and C. Schlenker, *Solid State Commun.* **63**, 587 (1987).

80. J. Y. Veuillen, R. Chevalier, J. Marcus, and C. Schlenker, Yamada Conf., *Physica B* **143**, 186 (1986).
81. S. Bouffard, M. Sanquer, H. Mutka, J. Dumas, and C. Schlenker in Ref. 36b, p. 449.
82. C. H. Chen, L. F. Schneemeyer, and R. M. Fleming, *Phys. Rev. B* **29**, 3765 (1984).
83. R. M. Fleming and L. F. Schneemeyer, *Phys. Rev. B* **31**, 899 (1985).
84. R. M. Fleming, R. J. Cava, L. F. Schneemeyer, E. A. Rietman, and R. G. Dunn, *Phys. Rev. B* **33**, 5450 (1986).
85. (a) A. Maeda, T. Furuyama, and S. Tanaka, Solid State Comm. **55**, 951 (1985). (b) A. Maeda, T. Furuyama, K. Uchinokura, and S. Tanaka, *Solid State Commun.* **58**, 25 (1986); (c) A. Maeda, T. Furuyama, K. Uchiokura, and S. Tanaka, Yamada Conf., *Physica B* **143**, 108 (1986); K. Tsutsumi, *Physica B* **143**, 129 (1986); (d) G. Mihaly and P. Beauchêne, *Solid State Comm.* **63**, 911 (1987); (e) G. Mihaly, P. Beauchêne, and J. Marcus, *Solid State Comm.* **66**, 149 (1988). (f) G. Mihaly, P. Beauchêne, J. Dumas, J. Marcus, and C. Schlenker, *Phys. Rev. B* **37**, 1047 (1988). (g) T. Chen, W. P. Beyermann, L. Mihaly, D. Reagor, B. Alavi, and G. Gruner, to be published.
86. (a) L. Mihaly and G. X. Tessema, *Phys. Rev. B* **33**, 5858 (1986); (b) H. Kubota, N. Onuki, T. Masumi, A. Anzai, and M. Sato, ICSM 86, Kyoto Japan, *Synth. Metals* **19**, 944 (1987); (c) G. X. Tessema and L. Mihaly, *Phys. Rev. B* **35**, 7680 (1987); (d) L. Mihaly, M. Crommie, and G. Grüner, *Europhysics Letters* **4**, 103 (1987).
87. N. P. Ong, C. B. Kalem, and J. C. Eckert, *Phys. Rev. B* **30**, 2902 (1980).
88. G. Travaglini and P. Wachter, *Phys. Rev. B* **30**, 1971 (1984).
89. H. K. Ng, G. A. Thomas, and L. F. Schneemeyer, *Phys. Rev. B* **33**, 8755 (1986).
90. J. Dumas, C. Schlenker, P. Beauchêne, C. Filippini, R. Buder, and B. Daudin, ICSM, Kyoto, Japan, *Synth. Metals* **19**, 917 (1987).
91. P. Butaud, P. Segransan, C. Berthier, J. Dumas, and C. Schlenker, *Phys. Rev. Lett.* **55**, 253 (1985); C. Berthier, personal communication.
92. K. Aoki, T. Tobayoshi, and K. Yamamoto, *J. Phys. Colloque C7* **42**, 51 (1981).
93. R. M. Fleming, L. F. Schneemeyer, and R. J. Cava, *Phys. Rev. B* **31**, 1181 (1985).
94. R. J. Cava, R. M. Fleming, P. Littlewood, E. A. Rietman, L. F. Schneemeyer, and R. G. Dunn, *Phys. Rev. B* **30**, 3228 (1984).
95. A. Janossy, G. Kriza, S. Pekker, and K. Kamaras, *Europhysics Letters* **3**, 1027 (1987).
96. A. Janossy, G. Mihaly, S. Pekker, and S. Roth, *Solid State Comm.* **61**, 33 (1987).
97. G. X. Tessema, B. Alavi, and L. Mihaly, *Phys. Rev. B* **31**, 6878 (1985).
98. J. C. Gill in Ref. 36a.
99. B. Fisher, *Solid State Commun.* **58**, 1 (1986).
100. B. Fisher, in *Lecture Notes in Physics* **217**, 513 (1985).
101. H. Unruhe, *J. Phys. C* **16**, 3245 (1983).
102. M. P. Everson, G. Eiserman, and R. V. Coleman, *Phys. Rev. B* **32**, 541 (1985).
103. J. Dumas, A. Arbaoui, H. Guyot, J. Marcus, and C. Schlenker, *Phys. Rev. B* **30**, 2249 (1984).
104. G. Kriza and G. Mihaly, *Phys. Rev. Lett.* **56**, 2529 (1986).
105. W. Y. Wu, L. Mihaly, G. Mozurkewich, and G. Grüner, *Phys. Rev. B* **33**, 2444 (1986).
106. H. Mutka, F. Rullier-Albenque, and S. Bouffard, *J. de Physique* **48**, 425 (1987).
107. L. Mihaly, T. Chen, B. Alavi, and G. Grüner, in *Lecture Notes in Physics* **217**, 455 (1985).
108. A. Arbaoui, Thèse de Doctorat, Université Scientique, Technologique et Médicale de Grenoble (1985).
109. Z. Z. Wang and N. P. Ong, Yamada Conf, *Physica B* **143**, 100 (1986).
110. Z. Z. Wang, Thèse Université Scientifique, Technologique et Médicale de Grenoble (1985), unpublished.
111. P. Lederer, G. Montambaux, J. P. Jamet, and M. Chauvin, *J. Phys. Lett.* **45**, 4627 (1984).
112. J. Dumas, A. Arbaoui, and C. Schlenker, European Workshop on CDWs in Solids, Zagreb, Oct. 1985, unpublished.
113. J. Dumas, R. Buder, J. Marcus, C. Schlenker, and A. Janossy, Yamada Conf. 1986, *Physica B* **143**, 183 (1986).
114. R. M. Fleming and L. F. Schneemeyer, *Phys. Rev. B* **33**, 2930 (1986).

115. M. Ido, Y. Okajima, and M. Oda, *J. Phys. Soc. Japan* **55**, 2106 (1986).
116. R. J. Cava, R. M. Fleming, E. A. Rietman, R. G. Dunn, and L. F. Schneemeyer, *Phys. Rev. Lett.* **53**, 1677 (1984).
117. T. Tamegai, K. Tsutsumi, S. Kagoshima, Y. Kanai, M. Tani, H. Tomozawa, M. Sato, K. Tsuji, J. Harada, M. Sakata, and T. Nakajima, *Solid State Comm.* **51**, 585 (1984).
118. (a) R. M. Fleming, R. G. Dunn, and L. F. Schneemeyer, *Phys. Rev. B* **31**, 4099 (1985); (b) L. Mihaly, K. B. Lee, and P. W. Stephens, *Phys. Rev. B* **36**, 1793 (1987).
119. (a) T. Haga, Y. Abe, and Y. Okwamoto, *Phys. Rev. Lett.* **51**, 678 (1983); (b) M. Nunez-Regueiro, B. Daudin, M. Dubus, and C. Ayache, *Solid State Commun.* **54**, 457 (1985).
120. Y. Abe, T. Haga, T. Kimura, Y. Tajima, and K. Imai, in Ref. 36a, p. 223.
121. (a) B. Daudin, M. Dubus, J. Dumas, and J. Marcus, *J. de Physique* **48**, 1779 (1987); (b) B. Daudin, M. Dubus, J. Dumas, and J. Marcus, Second European Workshop on CDW, Aussois, France (1987).
122. (a) J. Y. Veuillen, Thèse de Doctorat de l'Université Scientifique, Technologique et Médicale de Grenoble (1986); (b) J. Y. Veuillen, R. Chevalier, J. Marcus, and C. Schenker, *Solid State Commun.* **63**, 587 (1987).
123. G. Bang and G. Sperlich, *Z. Phys. B* **22**, 1 (1975).
124. (a) J. Dumas, C. Schlenker, J. Y. Veuillen, R. Chevalier, J. Marcus, R. Cinti, and E. Al Khoury Nemeh, *Synthetic Metals* **19**, 937 (1987); (b) J. Dumas, B. Laayadi, and J. Marcus, Second European Workshop on CDW, Aussois, France (1987).
125. (a) K. Nomura, K. Kume, and M. Sato, *J. Phys. C* **19**, L289 (1986); (b) P. Butaud, P. Ségransan, C. Berthier, J. Dumas, and C. Schlenker, *Phys. Rev. Lett.* **55**, 253 (1985); (c) D. C. Douglass, L. F. Schneemeyer, and S. E. Spengler, *Phys. Rev. B* **36**, 1831 (1987).
126. A. Janossy, C. Berthier, P. Ségransan, and P. Butand, *Phys. Rev. Lett.* **59**, 2348 (1987); P. Ségransan, A. Janossy, C. Berthier, J. Marcus, and P. Butaud, *Phys. Rev. Lett.* **56**, 1854 (1986).
127. M. Greenblatt, W. H. McCaroll, R. Niefeld, M. Croft, and J. V. Waszczak, *Solid State Commun.* **51**, 671 (1984).
128. N. C. Stephenson, *Acta Cryst.* **20**, 59 (1966).
129. H. Vincent, M. Ghedira, J. Marcus, J. Mercier, and C. Schlenker, *J. Solid State Chem.* **47**, 112 (1983).
130. M. Onoda, K. Toriumi, Y. Matsuda, and M. Sato, *J. Solid State Chem.* **69**, 67 (1987).
131. M. Ganne, M. Dion, A. Boumaza, and M. Tournoux, *Solid State Commun.* **59**, 137 (1986).
132. L. Kihlborg, *Arkiv. for Kemi* **21**, 365 (1963).
133. M. Ghedira, H. Vincent, M. Marezio, J. Marcus, and G. Fourcaudot, *J. Solid State Chem.* **56**, 66 (1985).
134. A. Magneli, *Acta Chem. Scand.* **2**, 501 (1948); **2**, 861 (1948).
135. H. Fujishita, M. Sato, S. Sato, and S. Hoshino, *J. Solid State Chem.* **66**, 40 (1987).
136. C. Escribe-Filippini, K. Konaté, J. Marcus, C. Schlenker, R. Almairac, R. Ayroles, and C. Roucau, *Phil. Mag. B* **50**, 321 (1984).
137. H. Guyot, C. Schlenker, J. P. Pouget, R. Ayroles, and C. Roucau, *J. Phys. C — Solid State Phys.* **18**, 4427 (1985).
138. M. Greenblatt, W. H. McCarroll, R. Neifelt, M. Croft, and S. V. Waszczak, *Solid State Commun.* **51**, 671 (1984).
139. H. Guyot, E. Al Khoury, S. Jandl, J. Marcus, and C. Schlenker (unpublished).
140. R. Buder, J. Devenyi, J. Dumas, J. Marcus, J. Mercier, C. Schlenker, and H. Vincent, *J. de Physique Lettres* **43**, L59 (1982).
141. K. V. Ramanujachary, B. T. Collins, and M. Greenblatt, *Solid State Commun.* **59**, 647 (1986).
142. H. Guyot, C. Schlenker, G. Fourcaudot, and K. Konaté, *Solid State Commun.* **54**, 909 (1985).
143. M. Sato, K. Nakao, and S. Hoshino, *J. Phys. C — Solid State Phys.* **17**, L817 (1984).
144. H. Guyot, C. Escribe-Filippini, G. Fourcaudot, K. Konaté, and C. Schlenker, *J. Phys. C — Solid State Phys.* **16**, L1227 (1983).
145. H. Gruber, E. Krautz, and H. P. Fritzer, *Phys. Stat. Sol.* (a) **65**, 589 (1981).

146. H. Guyot, Thèse de Doctorat d'Etat, Université Scientifique, Technologique et Médicale de Grenoble (1986).
147. H. Gruber, E. Krautz, H. P. Fritzer, K. Gatterer, and A. Pofitseh, *Phys. Stat. Sol.* (a) **86**, 749 (1984).
148. Y. Matsuda, M. Sato, M. Onoda, and K. Nakao, *J. Phys. C — Solid State Phys.* **19**, 6039 (1986).
149. M. Sato, H. Fujishita, S. Sato, and S. Hoshino, *J. Phys. C — Solid State Phys.* **19**, 3059 (1986).
150. E. Bervas, R. W. Cochrane, J. Dumas, C. Escribe-Filippini, J. Marcus, and C. Schlenker, in Ref. 36b, p. 144.
151. K. Hirunua, G. Kido, and N. Mima, *J. Phys. Soc. Japan* **51**, 3278 (1982).
152. H. Gruber and E. Krautz, *Phys. Stat. Sol.* (a) **62**, 615 (1980).
153. E. Fawcett, *Adv. Phys.* **13**, 139 (1964).
154. M. Naito and S. Tanaka, *J. Phys. Soc. Japan* **51**, 288 (1982).
155. A. W. Overhauser, *Phys. Rev. B* **3**, 3173 (1971).
156. S. Tanuma, *Lecture Notes in Phys.* **177**, 177 (1983).
157. See for example J. M. Ziman, *Principles of the Theory of Solids*, Cambridge Univ. Press (1972).
158. J. Dumas, C. Escribe-Filippini, J. Marcus, and C. Schlenker, Proc. of the NATO Davy Advanced Study Institute, *Physics and Chemistry of Electrons and Ions in Condensed Matter*, Plenum (1983) **130**, 571 (1983).
159. (a) C. Schlenker, S. S. P. Parkin, and H. Guyot, *J. Magn. Magn. Mat.* **54**, 1313 (1986); (b) H. Guyot, C. Schlenker, and G. Fourcaudot, in Ref. 36b, p. 133.
160. (a) P. W. Selwood, *Magnetochemistry*, Interscience Publ., New York, 78 (1964); (b) M. L. Boriack, *Phys. Rev. Lett.* **44**, 208 (1980).
161. (a) M. H. Whangbo, E. Canadell, and C. Schlenker, *J. Am. Chem. Soc.* **109**, 6308 (1987); (b) M. H. Whangbo and E. Canadell, *J. Am. Chem. Soc.* **110**, 358 (1988).
162. K. Ohtake, H. Matsuoka, R. Yamamoto, M. Doyama, H. Sakamoto, T. Mori, K. Soda, and S. Suga, *J. Phys. C — Solid State Phys.* **19**, 7207 (1986).
163. E. Canadell, M. H. Whangbo C. Schlenker, and C. Escribe-Filippini, *Inorg. Chem.* (1989) (to be published).
164. A. Fournel, J. P. Sorbier, M. Konczykowski, and P. Monceau, *Phys. Rev. Lett.* **57**, 2199 (1986).
165. A. Fournel, J. P. Sorbier, H. Guyot, G. Fourcaudot, and C. Schlenker (to be published).
166. See for example M. Ribault, J. Cooper, D. Jérome, D. Mailly, A. Moradpour, and K. Bechgaard, *J. de Physique Lettres* **45**, L 935 (1985).
167. (a) C. Schlenker, H. Schwenk, C. Escribe-Filippini, and J. Marcus, *Physica* **135B**, 511 (1985); (b) C. Escribe-Filippini, H. Schwenk, J. L. Tholence, J. Marcus, and C. Schlenker, Proceedings RCP-CNRS Magnétisme, Liquide de Fermi, Supraconductivité (1985).
168. D. E. Prober, R. E. Schwall, and M. R. Beasley, *Phys. Rev. B* **21**, 2717 (1980).
169. R. A. Klemm, M. R. Beasley, and A. Luther, *J. Low Temp. Phys.* **16**, 607 (1974).
170. D. R. Tilley, *Proc. Phys. Soc.* **86**, 289 (1965).
171. A. M. Clogston, *Phys. Rev. Lett.* **3**, 266 (1962).
172. B. Horovitz and J. A. Krumhansl, *Phys. Rev. B* **29**, 2109 (1984).

FREQUENCY-DEPENDENT CONDUCTIVITY IN $K_{0.30}MoO_3$

ROBERT M. FLEMING and ROBERT J. CAVA

AT&T Bell Laboratories, Murray Hill, New Jersey 07974, U.S.A.

1. Introduction

1.1. CONCEPTUAL MODELS

Although conductivity from moving charge density waves (CDWs) was first envisioned in the 1950s [1, 2], it has only been in the last ten years that materials with a conductivity mechanism similar to the sliding 'Fröhlich-mode' have been investigated. Thus far, about ten compounds show nonlinear conductivity from a moving CDW, and four compounds, $K_{0.30}MoO_3$, $NbSe_3$, TaS_3, and $(TaSe_4)_2I$ have been extensively studied [3—6]. The physical properties of materials with moving charge-density waves (CDWs), while similar to predictions based on the original Peierls—Fröhlich ideas, are in reality considerably more rich than would be predicted from the original model. In many anisotropic metals the ground state is periodic distortion of the host structure consisting of two parts, a periodic modulation of the electron density and a periodic lattice distortion [7]. The CDW, which can be thought of as a charged lattice, can couple to an applied electric field and, in the absence of pinning, it can move and carry current. In the original Peierls—Fröhlich model, the lattice is neglected and unattenuated conductivity arises because of the uniform translation of a rigid, one-dimensional charge-density wave that has translational invariance. Non-attenuated conductivity, however, no longer occurs if one includes the pinning effects of a commensurate lattice or impurities. Lee, Rice and Anderson [8] showed that such pinning effects shift the oscillator strength of the CDW sliding mode from zero to finite frequency. This finite frequency mode is referred to as the pinned phase mode, and in models which do not allow for deformations, it is the only mode expected at frequencies less than the single particle gap. Thus, in a model which includes pinning to the lattice, dc conductivity is only achieved by applying a sufficiently large dc field to overcome the pinning potential. In highly one dimensional systems dominated by strong pinning, the threshold field, E_T, required to induce dc CDW motion is expected to be large [9].

Recent experimental work on materials with moving CDWs has shown that the real response is considerably more complex than the models involving the translation of a rigid CDW would predict. First, the threshold fields needed to induce CDW motion can be surprisingly small (< 10 mV/cm) [10]. Second, both the ac and the dc response are more complex than would be expected from the dynamical predictions based on a model of a rigid CDW. The ac response, which is the primary focus of this chapter, shows a frequency-dependent response at frequencies far from the pinning energy. The dc response shows hysteresis as one

C. Schlenker (Ed.), Low-Dimensional Electronic Properties of Molybdenum Bronzes and Oxides,
259—294.
© 1989 *Kluwer Academic Publishers.*

changes both the temperature and the field. The hysteresis indicates that the conformational energy of the pinned CDW has many metastable minima corresponding to local deformations of the CDW. Therefore, at frequencies where metastability plays a significant role (low frequencies) the CDW cannot be treated as a rigid entity. Another unusual feature which is often observed in small samples is a periodic response to a dc bias, [10] (sometimes called 'narrow-band noise'). This effect requires that the CDW have long-range coherence extended over the dimensions of the sample. The oscillatory voltage results as the periodic CDW moves past either pinning centers or the contacts.

This chapter will be an experimental review of ac conductivity the 'blue bronze', $K_{0.30}MoO_3$, which we take to be a prototypical material for a discussion of CDW motion. The bulk of the review will concentrate on low-frequency measurements made in our laboratory; however, in some cases measurements of the ac conductivity in other materials will be cited. Three frequency regimes will be discussed. The first regime is termed the dielectric relaxation regime (dc—10 MHz) where the dynamics can be described as the decay of an induced polarization of the CDW resulting from from local deformations of the CDW. In this frequency regime one can neglect the effects of inertia and take the CDW mass to be zero. The dominant energy loss mechanism results because of screening of the CDW deformations by the normal electrons. The second frequency regime (10—100 GHz) is dominated by the pinned phase mode of the CDW. This is the pinned response predicted by Lee, Rice and Anderson [8] and in this regime one must include inertia in describing the CDW dynamics. The effective mass of the CDW causes an inertial roll-off in the ac response with a frequency dependence which is approximately described by the response of a rigid CDW [11—15]. At frequencies near the pinning energy, electrostatic screening is less important as a damping mechanism and deformations of the CDW do not occur. Damping mechanisms in this frequency realm are temperature independent and are less well-understood. The third frequency regime is the far-infrared frequency region between the pinned phase mode and the single electron energy gap. This regime has been studied less intensely than the two low frequency regimes, however large resonances in infrared (FIR) measurements of $(TaSe_4)_2I$ and $K_{0.30}MoO_3$ suggesting further modes may be present in CDW materials [16, 17].

This chapter is not intended to be a comprehensive review of CDW dynamics. We will concentrate on the ac conductivity and will not address issues such as the periodic response to a dc bias and combined ac/dc experiments (mode locking). The issue of metastability is central to understanding the low frequency dynamics, however we will not review the literature of the metastable and hysteretic CDW response in detail. Similarly, this is intended to be an experimental review and the various microscopic models and the controversy between them will not be addressed in detail.

1.2. UNIFORM PINNING MODEL

Attempts to theoretically describe the microscopic details of CDW motion have been difficult, primarily due to the complexity of dealing with the random pinning and CDW deformations. There has been considerable controversy over this issue, particularly in the frequency regime dealing with dynamics of the pinned phase mode. The basic discussion is whether or not the dynamics can be described by a classical model involving a deformable CDW [9, 18], a tunneling model involving the coherent tunneling of the entire condensate [19], or models invoking strong pinning and a single degree of freedom [20]. This review will not address the issue of the validity of the various models, however all of the data presented in this chapter can be explained and quantitatively reproduced by using the classical, deformable model discussed in the next section. Before we go to that discussion it is useful to first outline the description of a rigid CDW, a CDW that moves with no deformations [21—23]. Although the rigid CDW model has severe limitations, it remains a useful pedagogical tool and is also useful in its ability to define parameters in terms of experimentally measurable quantities. This model assumes that the CDW is rigid with the only degree of freedom being the phase. In the literature, this is sometimes referred to as the 'single-particle model', however a more descriptive term is the 'uniform pinning model'. For small amplitude displacements the equation of motion can be written as [24]:

$$m^* \ddot{u} + \lambda_0 \dot{u} + V_0 \sin{(\mathbf{Q} \cdot \mathbf{u})} = \rho_c E. \tag{1}$$

Here m^* is the collective mass density of the CDW, λ_0 is the effective CDW viscosity, V_0 is the pinning force per unit volume, r_c is the collective CDW charge density and E is the electric field. Using the relation $\sigma(\omega) = i\omega\varepsilon(\omega)/4\pi$ the frequency dependent conductivity can be derived from Equation 1 to be:

$$\sigma(\omega) = \left[\frac{\rho_c^2}{m^*} \right] \frac{i\omega}{[\omega_0^2 - \omega^2 + i\omega/\tau_0]} \tag{2}$$

where the pinning frequency is given by $\omega_0^2 = |Q| V_0/m^*$ and the inertial damping frequency by $1/\tau_0 = \lambda_0/m^*$. In the overdamped limit where there is no inertia ($m^* \approx 0$), it useful to rewrite Equation 2 as the dielectric function:

$$\varepsilon(\omega) = \left[\frac{4\pi\rho_c^2}{|Q| V_0} \right] \frac{1}{1 + i\omega\tau_R} \tag{3}$$

where $\tau_R = \lambda_0/|Q| V_0$. For comparison with experiment, the pinning potential V_0 can be approximated as $V_0 \approx \rho_c E_T$. Equation 3 is the form of a Debye relaxation originally used to describe the dielectric relaxation of polar molecules [25]. The relaxation frequency $1/\tau_R$ marks the point where Re ε relaxes from its zero-frequency value $\approx 4\pi\rho_c/|Q| E_T$ to zero. There is a corresponding peak in Im ε at $1/\tau_R$. For the case of the conductivity described by Equation 2 with small but finite effective mass, the conductivity first rises at the relaxation frequency $1/\tau_R$ and saturates with a maximum value of $\sigma_{max} = \rho_c^2/\lambda_0$. The conductivity remains saturated until the inertia forces a roll-off of the conductivity at a frequency of $1/\tau_0$. One expects a peak in the imaginary part of σ at the relaxation frequency and a negative peak in Im σ at the roll-off frequency. Note that the relaxation frequency can be expressed in terms of the roll-off frequency and pinning frequency, $1/\tau_R = \omega_0^2\tau_0$. In the literature $1/\tau_R$ is sometimes referred to as the 'crossover frequency'.

We emphasize that the uniform pinning model has severe limitations in describing the actual CDW response. For example, it predicts the wrong shape of the dc $I-V$ response at threshold. It cannot account for the metastable behavior or the dielectric relaxation at low frequencies.

1.3. RANDOM PINNING MODEL

A microscopic model that addresses the issue of random pinning of the CDW has been developed by Fukuyama and Lee [18] and by Lee and Rice [9] (FLR). FLR consider pinning by random impurities and allow for phase fluctuations of the charge-density wave while ignoring amplitude fluctuations. This approach is different from the uniform pinning model since it describes the motion of a deformable object. The FLR model inherently allows for many degrees of freedom and thus it can be used to describe a hysteretic response. FLR write a Hamiltonian as

$$H = \int \left[\frac{1}{2} \kappa |\nabla\phi|^2 + \sum_i V_i(\mathbf{r} - \mathbf{R}_i) \rho(r) - \rho_c E\phi(\mathbf{r})/Q \right] d^3r. \qquad (4)$$

Here κ is the CDW elasticity, V_i is the pinning of an impurity at R_i and $\rho(r) = \rho_0 \cos[\mathbf{Q} \cdot \mathbf{r} + \phi(\mathbf{r})]$ is the CDW modulation. Because of the nonlinear nature of the response, Equation 4 does not easily lead to a general analytic solution for $\sigma(\omega)$. Efforts toward an analytical solution have focused on small systems [26], one-dimensional systems [27, 28] as well as three-dimensions [29]. Several workers [30−32] have investigated the dynamics using the overdamped equation of motion

$$\lambda_0 \, d\phi/dt = -\partial H/\partial\phi \qquad (5)$$

in the limit of large electric fields $\gg E_T$. A mean-field theory valid for many degrees of freedom has also been proposed [33]. In addition, numerical models based on the response of a CDW moving in a one-dimensional FLR Hamiltonian using a variety of pinning potentials have also been proposed to describe the CDW response at intermediate fields [34−39]. For a review of the application of the FLR model to problems in systems with moving CDWs see Coppersmith [40].

The FLR Hamiltonian ignores amplitude fluctuations of the CDW on the grounds that these are not energetically favorable unless one is very close to the CDW onset temperature. In certain situations however, amplitude fluctuations of the CDW probably play an important role. Ong and Maki [41] and Gorkov [42] have used amplitude fluctuations to describe the CDW a mechanism for current oscillations near the current contacts, however experimental verification of this model remains controversial. Another example is that of materials where the CDW depins via a 'switching' mechanism [43]. The term 'switching' refers to the abrupt transition from the pinned state to the moving state observed in some samples. Switching has been shown to result from CDW phase slips localized near strong pinning centers [44] and the dynamics of motion in this situation can be modeled using amplitude fluctuations of the CDW [45]. The experiments outlined in this chapter, for the most part, only describe the response of non-switching samples. One should keep in mind, however, that amplitude fluctuations may also be important in $K_{0.30}MoO_3$ and other CDW materials where it has been shown that only a fraction of the sample volume participates in the CDW transport [46]. The interface between moving and non-moving parts of the CDW must be a point where the CDW amplitude collapses. We also note that strong pinning has been used in describing the dynamics of CDW motion in both the tunneling [20] and classical models [47].

2. Dielectric Relaxation Regime

From the analysis of low frequency and dc conductivity data it is clear that uniform pinning does not give a complete description of the low frequency dynamics. For example, dc experiments demonstrate that the normal conductivity and the transient response are a function of the history of the sample. Such hysteresis implies that the CDW does not relax to the ground state following a displacement, rather it is able to distort and relax to a metastable state with an energy higher than the ground state. Experiments indicate that there are a large number (perhaps an infinite number) of metastable states closely spaced in energy. Such behavior is not possible if the only degree of freedom is the CDW phase.

A numerical study of CDW motion based on the FLR Hamiltonian shows that a metastable state can be visualized as a local displacement fo the CDW [30]. Local displacements can occur at fields less than the threshold for non-linear conductivity because of the random pinning. Since a metastable state involves a charge displacement, it is accompanied by a polarization of the CDW that may or may not be externally observed depending on the number of normal carriers available for screening. In this model, if the sample is biased in the non-linear regime and the current is switched off, the CDW stops in the highest energy (the most distorted) metastable state. It can decay toward the ground state, but the ground state is never reached because each barrier is higher than the last. This leads to a 'glass-like' behavior of the CDW which is particularly pronounced at low frequencies and low temperatures [48].

The physics of the low-frequency response of the CDW is contained in the description of the relaxation of the CDW following a small displacement. We assume that the amplitude of the applied field is sufficiently small that the system is in a linear response regime, that is to say, it relaxes within a local metastable state and does not move to a neighboring state. It has been suggested on theoretical grounds that because of the large number of metastable states spaced close in energy no applied field may be 'sufficiently small', however as we will discuss below, experiments suggest that the response is linear for electric field less than about 10% E_T. This problem has other physical analogs such as the response of glasses, however a proper theoretical treatment of the dynamical response usually leads to analytical solutions only in certain limiting cases. For this reason, the characterization of experimental data is made easier if one uses semi-empirical formulas derived, for example, for conducting polymers. While such a characterization is not intended to be a microscopic, physical description of the dynamics, it allows one to compare different materials and characterize the nature of the dielectric relaxation process.

In this section we first preset a discussion of phenomenological formulas used to describe the low frequency response. The experimental sections can be divided into two categories, ac measurements at zero dc bias that probe the linear response of the pinned CDW (these include the chemical doping experiments), and measurements during or shortly after the application of a dc bias that probe the relation between metastable states and the low frequency dynamical response.

2.1. PHENOMENOLOGICAL DESCRIPTION

The following phenomenological description of dielectric relaxation was initially developed to analyze conductivity data from conducting polymers where the dominant loss mechanism is the movement or rotation of a polar molecule [49]. The analysis begins by assuming that one has a dielectric material polarized by the presence of an electric field. If, when the field is turned off, the rate of decay of the polarization depends only on the instantaneous magnitude of the polarization, the polarization decays in an exponential manner that can be expressed as

$$P(t) = P(0) \exp(-\tau/\tau_R) \tag{6}$$

where τ_R is a characteristic time that we will call the dielectric relaxation time. Note that this process is, by definition, over-damped. There is no inertia, and consequently no oscillatory response to switching off the field. The complex dielectric function associated with the polarization described in Equation 6 was first calculated by Debye and can be written as a complex dielectric function $(\varepsilon(\omega) = \varepsilon'(\omega) - i\varepsilon''(\omega))$:

$$\varepsilon(\omega) = \varepsilon_{HF} + \frac{(\varepsilon_0 - \varepsilon_{HF})}{1 + i\omega\tau_R}, \tag{7}$$

where ε_0 is the dielectric constant for $t \gg \tau_R$ and ε_{HF} is the dielectric constant for $t \ll \tau_R$. One can see that Equation 7 is the dielectric function predicted by the uniform pinning model in the over-damped limit given in Equation 3. As discussed above, Equation 7 has only one degree of freedom and cannot describe a system with metastable states and a hysteretic response. In the language of dielectric relaxation, this is equivalent to saying that the decay of the polarization does not follow a single characteristic time but rather a distribution of times. Recall that the uniform pinning model gave $\tau_R = \lambda_0/|Q| V_0$, illustrating that a concept of a distribution of 'relaxation times' is the same as a 'distribution of pinning energies'. Throughout the following discussion, one can think of a distribution of relaxation times as a distribution of pinning energies resulting from the random pinning of the CDW by impurities. Calling the distribution of relaxation times $g(\tau)$, Equation 6 becomes:

$$P(t) = P(0) \int g(\tau)\, e^{-t/\tau_R}\, d\tau \tag{8}$$

and the complex dielectric function becomes:

$$\varepsilon(\omega) = \varepsilon_{HF} + (\varepsilon_0 - \varepsilon_{HF}) \int \frac{g(\tau)\, d\tau}{1 + i\omega\tau} \tag{9}$$

with the requirement that $\int_0^\infty g(\tau)\, d\tau = 1$. A considerable body of literature exists concerning the characterization of $g(\tau)$ and $\varepsilon(\tau)$ for systems with relaxation processes displaying a distribution of relaxation times [49].

In general, description of the dielectric response for systems with a distribution of relaxation times has followed two different strategies. In the first strategy one devises a functional form for $g(\tau)$ based on elementary principles and calculates an $\varepsilon(\omega)$ to compare with experiment. In the second approach, one fits the data with an empirical form of $\varepsilon(\omega)$ and calculates the functional form of $g(\tau)$. In previous studies of polar relaxation, the second case has been used with more success than the first. In the case of charge-density waves, we have found one of the empirical descriptions of $\varepsilon(\omega)$ based on a modification of the Debye formalism to be an excellent description of the dielectric data over wide ranges of frequency and temperature.

The key to the selection of the correct formalism is found in the shape of the dielectric loss function, $\varepsilon''(\omega)$, for $\omega \ll \omega_R$ and $\omega \gg \omega_R$ ($\omega_R = 1/\tau_R$). For the single relaxation case (rigid CDW), the loss function derived from Equation 7 is given by:

$$\varepsilon''(\omega) = \frac{(\varepsilon_0 - \varepsilon_{HF})\, \omega\tau_R}{1 + \omega^2 \tau_R^2}. \tag{10}$$

The low and high frequency limiting behavior of $\varepsilon''(\omega)$ is given by $\varepsilon''(\omega) \propto \omega^n$ and $\varepsilon''(\omega) \propto \omega^{-n}$, ($n = 1$) for $\omega \ll \omega_R$ and $\omega \gg \omega_R$ respectively. The fact that the low and high frequency dependence of $\varepsilon''(\omega)$ both have $n = 1$ is a result of the single degree of freedom in the model. In a distributed process (a nonrigid CDW response), the exponent n willll be less than one for one or both exponents. Thus, the presence of additional relaxational processes is immediately apparent if the loss function shows limiting behavior with the exponent $n < 1$. In addition, the loss function can be used to infer information about the shape of relaxation time distribution, $g(\tau)$. For any distribution function which is logarithmically symmetric about the mean relaxation time, $\varepsilon''(\omega)$ will also be symmetric, i.e. the magnitude of the exponents, n, will be equal. (Logarithmically symmetric means that $g(\tau)$ has equal values for times $\tau = x\tau_R$ and τ_R/x.) For CDW systems, the data show that $\varepsilon''(\omega)$ is *not* logarithmically symmetric about τ_R, i.e. $\varepsilon''(\omega) \propto \omega^n$ for $\omega \gg \omega_R$ and $\varepsilon''(\omega) \propto \omega^{-m}$ for $\omega \ll \omega_R$ with $\eta \neq m$ and $m, n \leq 1$. This means that any

$g(\tau)$, e.g. Gaussian or a block function, that will result in a symmetric form of $\varepsilon''(\omega)$ will not describe the data. For similar reasons, the empirical descriptions known in the literature as Cole–Cole [50], Cole–Davidson [51], Fuoss–Kirkwood [52] and Williams–Watts [53] equations are not sufficiently general to describe the data. Jonscher [54] has proposed an asymmetric loss function, however the limiting behavior of that function does not appear consistent with the results on CDW materials. The temperature dependence of the dielectric relaxation in $K_{0.30}MoO_3$ indicates a gradual evolution from a Debye-like single particle response at high temperatures to a broadened response at low temperatures that can be analyzed in terms of a modification of the Debye formalism.

Data for the CDW relaxation in $K_{0.30}MoO_3$ and related compounds are described to an excellent approximation (within a few percent) be a generalization of Equation 7 introduced by Havriliak and Negami [55] to describe dielectric relaxation in polymers:

$$\varepsilon(\omega) = \varepsilon_{HF} + \frac{(\varepsilon_0 - \varepsilon_{HF})}{[1 + (i\omega\tau_R)^{1-\alpha}]^\beta} , \tag{11}$$

where ε_{HF} is the value of $\varepsilon'(\omega)$ for $\omega \gg \omega_R$, ε_0 is the value of $\varepsilon'(\omega)$ for $\omega \ll \omega_R$ and α and β are exponents describing the shape of $\varepsilon''(\omega)$ and $\varepsilon'(\omega)$. For frequencies much less than ω_R, the loss function is given by $\varepsilon''(\omega) \propto [\omega/\omega_R]^{1-\alpha}$, whereas for frequencies much greater than ω_R, $\varepsilon''(\omega) \propto [\omega/\omega_R]^{\beta(1-\alpha)}$. From the limiting values of the loss function one can see that the parameter α determines the breadth of the distribution of relaxation time while β determines the asymmetry. (Here, and in much of the discussion that follows, τ_R will be used to refer to the mean relaxation time since $\alpha > 0$ or $\beta < 1$ implies a distribution of relaxation times.) Equation 11 reduces to the symmetric Cole–Cole [50] form for $\beta = 1$, the asymmetric Cole–Davidson [51] form for $\alpha = 0$, and the Debye form for $\alpha = 0$ and $\beta = 1$, and thus Equation 11 is a general form of the dielectric function allowing the description of a wide range of deviations from the single Debye relaxation. The strength of this empirical description of the $\varepsilon(\omega)$ data lies in its use of a few well-defined parameters. Since it is not derived from a microscopic model, its purpose is not to verify the correctness of a physical description of CDW dynamics. Rather it forms a basis that any microscopic model should duplicate to be correct.

One can derive the implied distribution of relaxation times from Equation 11. Taking $y = \tau/\tau_R$, one has:

$$g(y) = \left[\frac{1}{\pi}\right] \frac{y^{\beta(1-\alpha)} \sin \beta\theta}{D} , \tag{12}$$

where

$$D = [y^{2(1-\alpha)} + 2y^{(1-\alpha)} \cos \pi(1-\alpha) + 1]^{\beta/2} \tag{13}$$

and

$$\theta = \tan^{-1}\left[\frac{\sin \pi(1-\alpha)}{[y+\cos \pi(1-\alpha)]}\right]. \tag{14}$$

We have employed this distribution function to elucidate interesting aspects of the temperature dependence of the CDW response in K$_{0.30}$MoO$_3$. Determining the distribution of relaxation times $g(\tau)$ is equivalent to determining the frequency dependent density of metastable states $n(\nu)$ for the pinned CDW [56]. Considering the response of the CDW as being the sum over the density of states, then the dielectric constant is given by:

$$n(\nu) = \frac{\sin \beta\theta}{\pi D}, \tag{15}$$

using $y = \nu\tau_R$ in the definition of D (Equation 13). A normalized expression for $n(\nu)$ such that $\int_0^\infty n(\nu)\,d\nu = 1$, for $\varepsilon_{HR} \ll \varepsilon_0$, is given by

$$n'(\nu) = \left[\frac{\varepsilon_0}{4\pi\sigma_\infty}\right] n(\nu), \tag{16}$$

where σ_∞ is the limiting high frequency (field) conductivity. We note finally that the relationships between $\varepsilon(\omega)$, $g(\tau)$, and $n'(\nu)$ (Equations 11, 12 and 16) are exact. Therefore, errors in $g(\tau)$ or $n'(\nu)$ derive only from how well the $\varepsilon(\omega)$ data are described by Equation 11. For the CDW response in K$_{0.30}$MoO$_3$ and related materials at temperatures where τ_R is finite, those errors are smaller than a few percent.

Before we present experimental results on CDW materials, we would like to emphasize the importance of analyzing the low-frequency data as the dielectric function rather than the complex conductivity. If the relaxation is Debye-like (Equation 7), the distinction is not so important because there is a peak in $\sigma''(\omega)$. If however, the relaxation has an asymmetric distribution of relaxation times (Equation 11), and in addition there is oscillator strength at high frequencies (the phase mode) giving a finite value of ε_{HF}, neither $\sigma'(\omega)$ nor $\sigma''(\omega)$ can be intuitively recognized as being characteristic of a dielectric relaxation. Analyzing the data as the complex conductivity obscures the close relation of the distributed relaxation times and the Debye relaxation.

2.2. DIELECTRIC RELAXATION: ZERO DC BIAS

The real and imaginary parts of $\sigma(\omega)$, normalized to σ_{dc} at 300 K, for fifteen temperatures between 40 and 101 K for a sample of K$_{0.30}$MoO$_3$ are presented in Figure 1 [57]. Similar data have been obtained for NbSe$_3$ [57], TaS$_3$ [58—60], (TaSe$_4$)$_2$I [61, 62] and (NbSe$_4$)$_{10/3}$I [62]. Inspection of the data indicates significant variation of both $\sigma'(\omega)$ and $\sigma''(\omega)$ within the observed frequency range. Useful

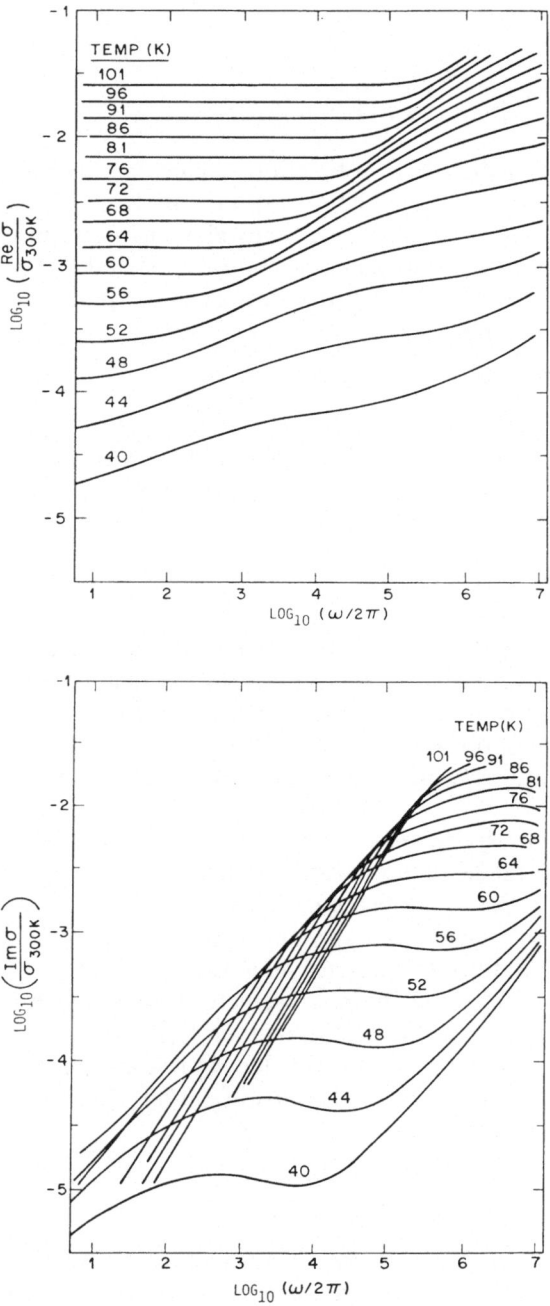

Fig. 1. The conductivity in $K_{0.30}MoO_3$ as a function of frequency for temperatures between 40 and 101 K showing (a) the real part and (b) the imaginary part.

preliminary analysis of the data can be made by consideration of the sample response in the complex impedance plane. For electric fields less than threshold, the dc conductivity of the sample is due to electrons activated above the single particle Peierls gap. The total sample admittance results from the dc conduction process, in parallel with the dielectric relaxation of the pinned CDW. In the case of uniform pinning (rigid CDW), the conductivity can be written as $\sigma(\omega) = \sigma_{dc} + \sigma_{dielec}(\omega)$ where

$$\sigma_{dielec}(\omega) = \frac{\omega^2 \tau_R(\varepsilon_0 - \varepsilon_{HF})}{1 + \omega^2 \tau_R^2} + i\omega\varepsilon_{HF} + \frac{i\omega(\varepsilon_0 - \varepsilon_{HF})}{1 + \omega^2 \tau_R^2}. \tag{17}$$

This single-relaxation response can be modeled with linear circuit elements as shown in the insert of Figure 2. The equivalent linear circuit consists of parallel

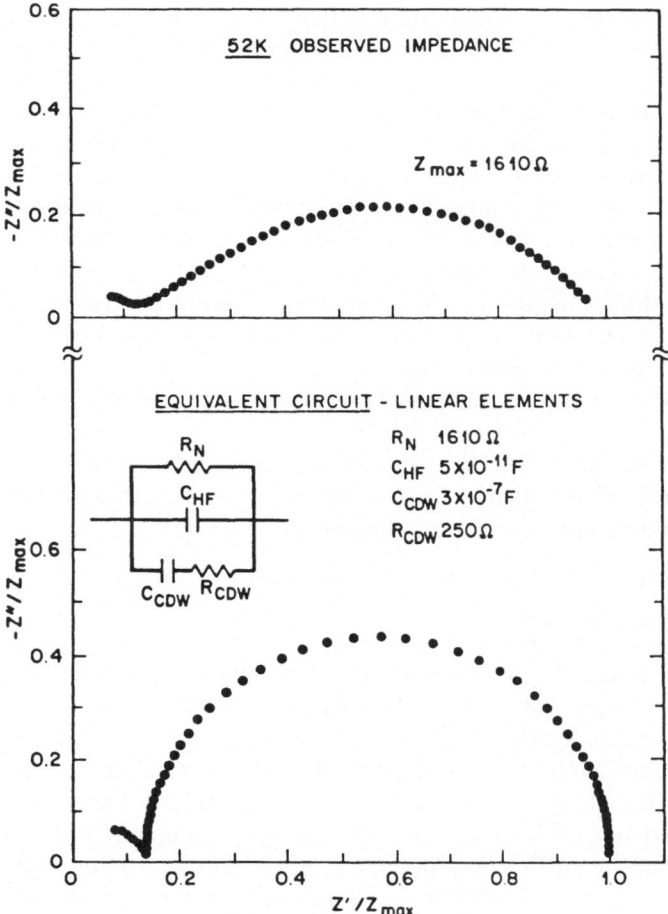

Fig. 2. Response of $K_{0.30}MoO_3$ at 52 K at frequencies between 5 Hz and 12 MHz in the complex impedance plane (upper figure) and comparison to the response for the equivalent linear circuit modeling dielectric relaxation in parallel with a conductive process (lower figure).

circuit elements composed of: (1) a single capacitance representing high-frequency processes C_{HF}, (2) a series combination of a capacitor and resistor representing the moving CDW, $C_{CDW}-R_{CDW}$, and (3) a resistor R_N, representing the contribution of the thermally excited normal electrons. A distributed relaxation rate would require multiple parallel circuits. A model of the actual CDW response would have nonlinear elements in place of R_{CDW} and C_{CDW}.

The response of $K_{0.30}MoO_3$ at 52 K is compared in the complex impedance plane ($Z''(\omega)$ versus ($Z'(\omega)$)) to the response for the equivalent circuit composed of linear elements in Figure 2. Low frequency points are found at large Z' and high frequency points at low Z'. The low frequency extrapolation of the low frequency arc to the real axis yields the dc resistance of the sample. In contrast to the equivalent circuit, the actual sample response shows an asymmetric and depressed arc at low frequencies indicating that the CDW response is not that of an ideal Debye relaxation, but is rather that characteristic of a relaxation process with a distribution of relaxation times. The impedance plane plots allow one to eliminate the possibility that ionic conductivity contributes significantly to the sample impedance. The indium contacts are expected to be blocking to the motion of potassium ions, and ionic motion would result in a straight line in the impedance plane at low frequencies. Analysis of sample response in the complex impedance plane is of also great advantage in determining the quality of the contacts. Poor contacts have a significant resistive component as well as a capacitive component. As a result, additional low frequency arcs appear in the impedance-plane plots, one for each additional process (e.g. capacitive coupling between the sample and the contacts). In one sample we were able to eliminate the low frequency arc by reapplying the contacts. Subsequently, we used the impedance plane plots to screen the quality of the contacts and all samples that showed low frequency arcs were rejected.

For reasons discussed in the preceding section, the detailed analysis of the pinned CDW response was made in terms of the complex dielectric function which was determined from measured quantities by the relations:

$$\varepsilon'(\omega) = 4\pi\sigma''/\omega \tag{18}$$

$$\varepsilon''(\omega) = 4\pi(\sigma'(\omega) - \sigma_{dc})/\omega \tag{20}$$

The dc conductivity, σ_{dc} was obtained from measured values of $\sigma'(\omega)$ at frequencies where $\sigma(\omega)$ was independent of ω. Analysis of the complex dielectric constants for pure $K_{0.30}MoO_3$ was limited to the temperature range of 60—101 K since at 60 K, $\sigma(5\ Hz) > \sigma_{dc}$ and above 101 K the dielectric response occurs at frequencies higher than those measured here. Fits to the dielectric function data were performed by minimization of the average agreement index, as defined in Table I, by variation of the parameters τ_R, ε_0, α and β. For undoped $K_{0.30}MoO_3$, $\varepsilon''(\omega)$ was found to be $< 0.005\ \varepsilon_0$ at all temperatures measured. It was therefore set to zero and not varied. The values determined for τ_R, ε_0, α and β are presented in Table I for each temperature, along with agreement index. Estimates of the errors are ± 0.01 for α, between ± 0.01 and ± 0.02 for β, ± 0.02 for ε_0 and

TABLE I

Parameters describing the electrical response of the charge density wave in $K_{0.30}MoO_3$ where $\varepsilon(\omega) = \varepsilon_0/[1 + (i\omega\tau_R)^{1-\alpha}]^{\beta}$.

Temperature (K)	ε_0	α	β	$\beta(1-\alpha)$	τ_R (sec)	$\omega_R/2\pi$ (Hz)	Agreement[a] $(R, \%)$
101	3.89×10^6	0.07	0.76	0.71	2.30×10^{-7}	6.9×10^5	0.9
96	4.45×10^6	0.08	0.77	0.71	3.20×10^{-7}	5.0×10^5	0.7
91	5.18×10^6	0.11	0.80	0.71	4.50×10^{-7}	3.5×10^5	1.6
86	6.10×10^6	0.13	0.81	0.70	7.10×10^{-7}	2.2×10^5	1.4
81	7.75×10^6	0.17	0.85	0.71	1.20×10^{-6}	1.3×10^5	1.7
76	1.06×10^7	0.18	0.85	0.70	2.65×10^{-6}	6.0×10^4	2.7
72	1.34×10^7	0.20	0.87	0.70	4.80×10^{-6}	3.3×10^4	1.7
68	1.84×10^7	0.22	0.90	0.70	9.55×10^{-6}	1.7×10^4	1.7
64	2.42×10^7	0.24	0.93	0.71	2.00×10^{-6}	8.0×10^3	2.5
60	3.46×10^7	0.25	0.95	0.71	4.90×10^{-6}	3.2×10^3	2.0

[a] $R = \Sigma_\omega \left[\left| \varepsilon'_{OBS}(\omega) - \varepsilon'_{CALC}(w) \right| + \left| \varepsilon''_{OBS}(\omega) - \varepsilon''_{CALC}(\omega) \right| \right] / \Sigma_\omega \left| \varepsilon'_{OBS}(\omega) + \varepsilon''_{OBS}(w) \right|$

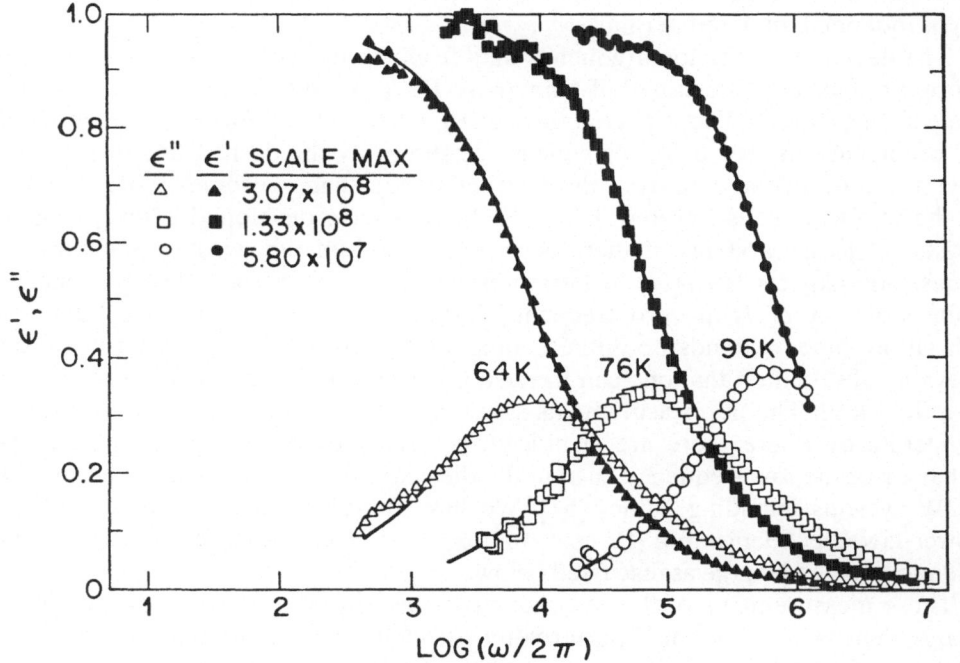

Fig. 3. The real (solid) and the imaginary (open) parts of the dielectric function as a function of the log of the frequency in $K_{0.30}MoO_3$ at three temperatures. Solid lines are from fits of Equation 11 to the data.

± 0.03 for τ_R. The agreement values, which are between 1 and 3%, indicate that Equation 11 describes the data to an excellent approximation. Figure 3 presents the dielectric function as a function of frequency for three temperatures and illustrates the increase in the relaxation frequency as the temperature is increased. The solid lines are from Equation 11 using the parameters presented in Table I. Inspection of the loss function, $\varepsilon''(\omega)$, in Figure 3 reveals the characteristic shape, discussed in the previous section, resulting from a distribution of relaxation times. (Note that $\varepsilon''_{max} < \varepsilon_0/2$.) Furthermore, the asymmetry of the loss peak indicates an asymmetry in the distribution of relaxation times about ω_R. The parameters that describe the temperature dependence of the CDW relaxation show some interesting trends. The characteristic frequency of the relaxation, τ_R, decreases from 6.9×10^5 Hz to 3.2×10^2 Hz between 101 and 60 K, frequencies several orders of magnitude lower than the 'crossover' frequencies extracted from higher frequency measurements of the pinned phase mode in $NbSe_3$. The dielectric constant, ε_0, increases with decreasing temperature. Both τ_R and ε_0 are thermally activated with activation energies of 829 K (0.071 eV) and 334 K (0.029 eV) respectively. In the same temperature range the CDW bandgap is approximately 1020 K (0.088 eV). The distribution of relaxation times for $\omega < \omega_R$ (as indicated by the temperature dependence of α) becomes broader with decreasing temperature. In contrast, the distribution of relaxation times for $\omega > \omega_R$ (as indicated by $\beta(1 - \alpha)$) is independent of temperature.

The deviation of the low-frequency data from the predictions of Equation 3 are more dramatically illustrated if one plots Re ε versus frequency on a linear scale as in Figure 4. The expected form of Re ε from the uniform pinning model is a Lorentzian centered at zero frequency as shown by the dashed line in Figure 4. The deviation of the data from the Lorentzian line-shape becomes more dramatic as the temperature is lowered. By 60 K the data have developed a 'cusp' characteristic of glassy systems. if there is an energy gap in the density of metastable states, one expects Re $\varepsilon(\omega)$ to be a constant at low frequency. The presence of finite slope of $\varepsilon(\omega)$ at zero frequency indicates that the broad distribution of relaxation times extends to infinite time [30]. This is in agreement with our previous observation that one can freeze a polarization into the material by cooling in a field [63]. The ac measurements show the existence of a polarization even at temperatures where there are significant screening electrons. The same general behavior of the low frequency cusp in the dielectric function is found in all moving CDW systems including $NbSe_3$ [51]. We have found no evidence of a reported power-law dependence [64] (Re $\varepsilon \sim \omega^a$, $a > 1$) except when the ac amplitude is intentionally made large as discussed below.

These measurements of the ac conductivity measure the response of the pinned charge-density wave to small perturbations from a local, metastable minimum. As the applied signal level is increased, and the CDW is moved farther from a local minimum (and perhaps to a new metastable state), a nonlinear response of the CDW can occur. The nonlinearity is expected to increase with increasing ac signal level. To investigate the limits of the small signal linearity regime, we measured $\varepsilon(\omega)$ in $K_{0.30}MoO_3$ at 77 K for ac signal amplitudes between 2% and 18% of E_T

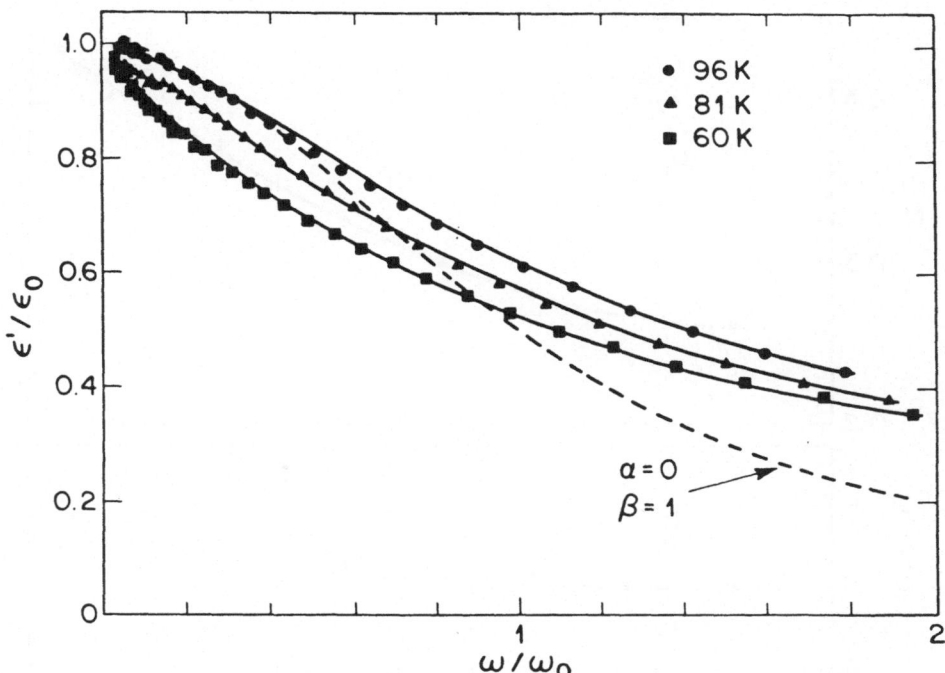

Fig. 4. A linear plot of the real part of the dielectric function of $K_{0.30}MoO_3$ at three temperatures. The dashed line is the Lorentzian line shape predicted by the uniform pinning model (Equation 3) and the solid lines are fits of Equation 11 to the data.

[65]. Similar results have been obtained on TaS_3 [58, 59]. We found that the nonlinear CDW response becomes significant for ac signal amplitudes on the order of 10% of E_T. The nonlinear response is reflected in an increasing proportion of low frequency (long time) relaxations that effectively increase the mean relaxation time. The results of the study are summarized in Figure 5 where the mean relaxation time τ_R, the breadth of the distribution, α, and the static dielectric constant, ε_0 are plotted. The flattening of the curves at low signal levels allows us to empirically define a small signal linear response region for amplitudes less than 5—10% of E_T.

2.3. CHEMICAL DOPING: ZERO BIAS

An interesting aspect of studying low frequency CDW dynamics in the 'blue bronzes' is the varied chemical sites available for doping. In the blue bronzes the conduction electrons come from the Mo d-band and the primary role of the alkali atom is to be an electron donor to the conduction band. $K_{0.30}MoO_3$ and $Rb_{0.30}MoO_3$ are isostructural and show a CDW response which is practically identical. Therefore, samples of $K_{0.30}MoO_3$ with rubidium partially substituted for potassium at the compositions $K_{0.25}Rb_{0.05}MoO_3$ and $K_{0.15}Rb_{0.15}MoO_3$ were taken

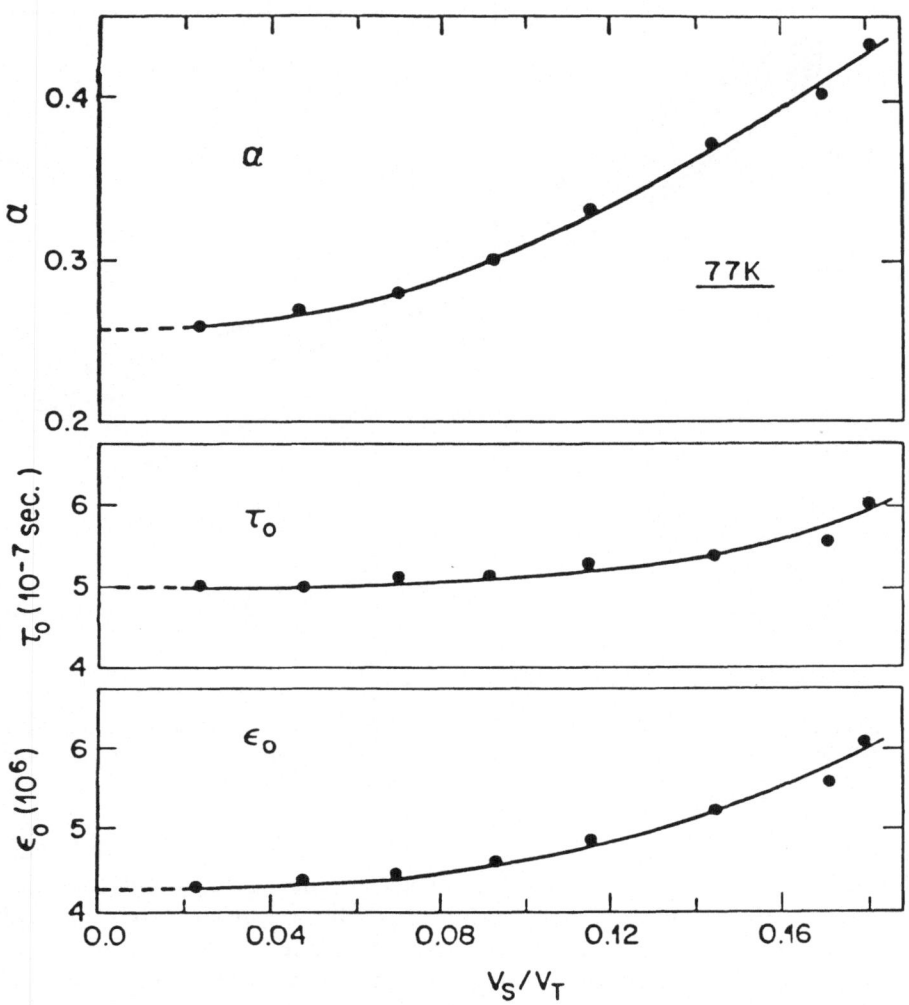

Fig. 5. Variation of the shape of the relaxation distribution function (α), the mean relaxation time (τ_0) and the static dielectric constant (ε_0) with ac signal amplitude.

as examples of weak pinning impurities. Support for this view comes from measurements of the threshold field that varies as the square of the the Rb concentration [66] in agreement with the Lee and Rice [9], prediction for weak pinning. On the other hand, samples with tungsten partially substituted for molybdenum at the compositions $K_{0.30}Mo_{0.996}W_{0.004}O_3$ and $K_{0.30}Mo_{0.99}W_{0.01}O_3$ were taken as examples of samples with strong pinning impurities since the threshold field was observed to vary linearly with the W concentration [66]. These chemical substitutions allows one to study the CDW dynamics as the strength and the type of pinning is changed [67].

As in pure material, measurements of the ac conductivity in doped samples of $K_{0.30}MoO_3$ are fit very well by the dielectric function in Equation 11 with average agreement between 0.7 and 3.3%. Again, the mean relaxation time τ_R is strongly temperature dependent in all materials, varying between 4.1×10^{-4} sec ($\omega_R/2\pi = 388$ Hz) and 6.15×10^{-9} sec ($\omega_R/2\pi = 29.9$ MHz) in the temperature range 25—90 K. The results and comparison to data for undoped $K_{0.30}MoO_3$ are summarized in Figure 6 and 7 where the mean relaxation time and the static dielectric constant are plotted as a function of $1000/T$ for all materials. For a particular temperature, the impurities dramatically decrease the mean relaxation time of the CDW, with the tungsten substitutions having the largest effect. At 65 K, for instance, the mean relaxation time of the CDW is approximately 1.6×10^{-5} sec for the undoped bronze, 9.3×10^{-7} sec for 17% Rb substitution, 1.5×10^{-7} sec for 50% Rb substitution and 1.3×10^{-8} sec for 0.4% tungsten substituion

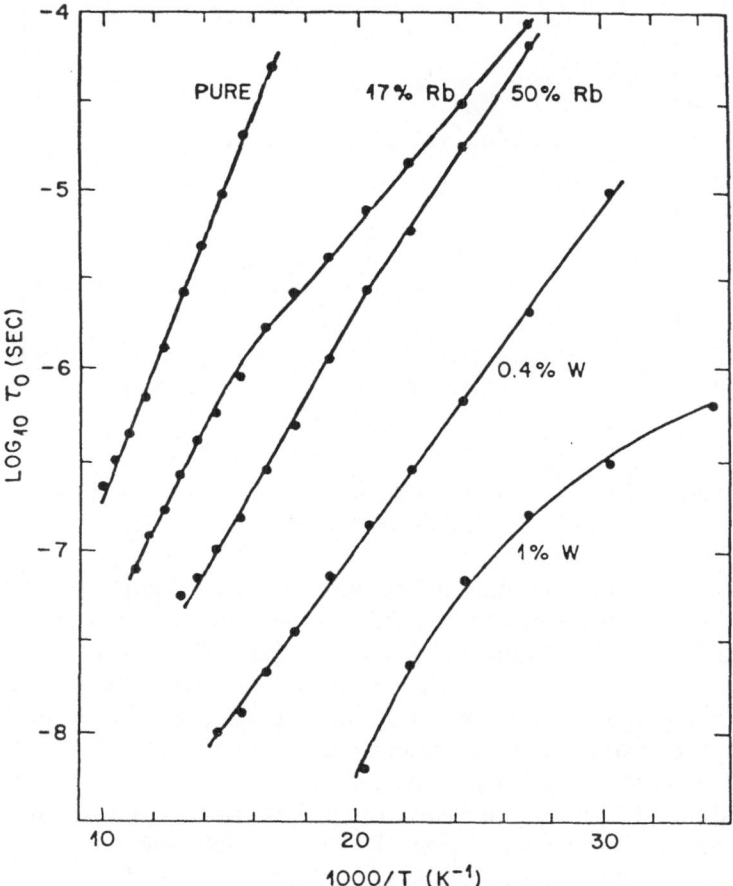

Fig. 6. Temperature dependence of the mean relaxation time, τ_0 in pure and impurity doped blue bronzes.

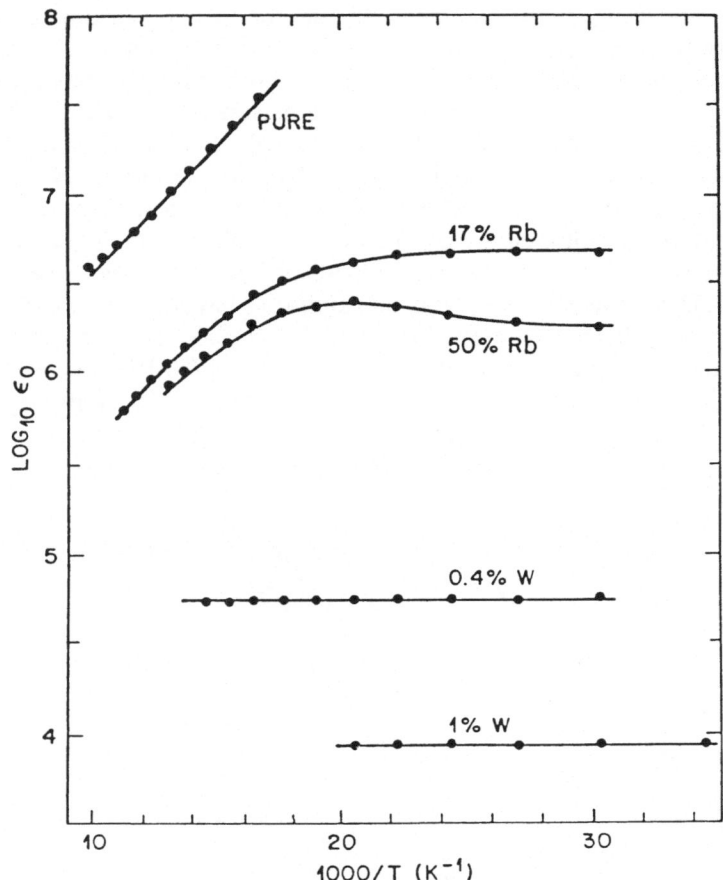

Fig. 7. Temperature dependence of the static dielectric constant, ε_0 in pure and impurity doped blue bronzes.

with the relaxation time too short to measure by our technique for 1% tungsten substitution. Thus a difference of three orders of magnitude is observed in τ_R while the threshold field for nonlinear conductivity has changed by one order of magnitude or less. The data for the doped samples were taken over a wider temperature range than those of the study on pure $K_{0.30}MoO_3$ and generally show a more complex temperature dependence of τ_R than that seen in $K_{0.30}MoO_3$. However, subsequent measurements indicate that the change of slope seen in the Rb substituted samples also seems to occur in pure $K_{0.30}MoO_3$ at low temperature.

As with the mean relaxation time, The static dielectric constant ε_0 is also dramatically affected by impurities. The trends in the observed dielectric constants are presented in Table II and summarized in Figure 7. Selecting a temperature of 65 K for comparison purposes, we find static dielectric constants of 2.2×10^7 for the pure material, 2.1×10^6 for the 17% Rb substitution, 1.5×10^6 for the 50%

TABLE II

Characteristics of the CDW Dielectric Response in Various Blue Bronzes at 77 K.

Sample	ε_0	$\omega_R/2\pi$ (Hz)	E_T (mV/cm)	α
$K_{0.30}MoO_3$	3.48×10^6	6.34×10^4	61	0.20
$K_{0.30}MoO_3$[a]	8.47×10^6	7.58×10^4	113	0.18
$K_{0.30}MoO_3$[b]	4.31×10^6	3.17×10^5	130	0.26
$K_{0.25}Rb_{0.05}MoO_3$	1.12×10^6	5.98×10^5	680	0.08
$K_{0.15}Rb_{0.15}MoO_3$	8.80×10^6	2.79×10^6	1640	0.25

[a] Used in analysis of the temperature dependence of the dielectric function.
[b] Used in studies of thermally stimulated depolarization, the effect of the ac signal level on the dielectric response and the inductive ringing in response to a dc pulse.

Rb substitution, 5.6×10^4 for the 0.4% W substitution and 8.6×10^3 for the 1% W substitution. As in the case of τ_R, the static dielectric constant varys nearly three orders of magnitude over the range of impurity concentrations studied. The temperature dependence of ε_0, however, is different from τ_R. For both tungsten substitutions, ε_0 is independent of temperature, whereas for the 17% Rb substitution, ε_0 at low temperatures has also been seen.

The impurities also have a significant effect of the distribution of relaxation times. The distribution function for the 17% Rb substituted material displays the same general characteristics as the undoped material, α is small (narrow distribution) at high temperatures and increases as the temperature is lowered. The exponent $\beta(1 - \alpha)$ describing the distribution of relaxations for $\omega \gg \omega_R$ is essentially independent of temperature at high temperatures, as in the undoped material, but changes significantly below about 50 K. For the high impurity substitutions, the width of the distribution is less temperature dependent, but the value of α increases as the impurity level is raised.

2.4. DIELECTRIC RELAXATION: FINITE DC BIAS

Bias dependent measurements of the dielectric response of the CDW at fields both below and above E_T yield further information on the dynamics of CDW motion [57]. For threshold fields less than E_T, the dielectric response is still that of the pinned CDW but there are significant changes in the characteristics of the relaxation process. The dielectric response of the CDW for one sample of $K_{0.30}MoO_3$ at 77 K is presented in Figure 8. The data are presented in the Cole–Cole representation which plots $\varepsilon''(\omega)$ versus $\varepsilon'(\omega)$ for each ω. Low frequency data are at high values of $\varepsilon'(\omega)$. For a single relaxation process, a Cole–Cole plot is a expected to be semicircular but for a distributed relaxation time process the plot is expected to be a depressed arc that intercepts the $\varepsilon'(\omega)$ axis at angles of $\frac{1}{2}\pi(1 - \alpha)$ and $\frac{1}{2}\beta(1 - \alpha)$ at the low and high frequency ends respectively. From the intercept of the arcs with the ε' axis at low frequencies, the increase of ε_0 as the

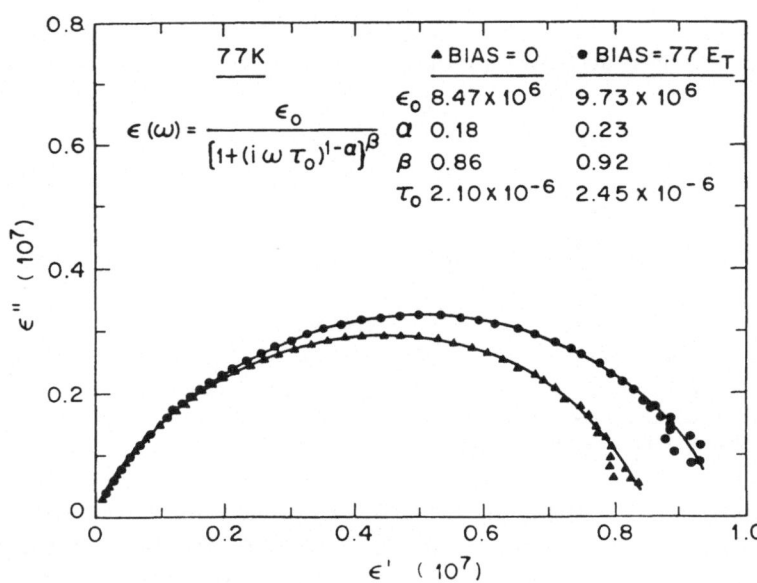

Fig. 8. Cole—Cole representation of the complex dielectric function of $K_{0.30}MoO_3$ for two biases below threshold. The solid lines are fits of Equation 11 to the data.

bias approaches E_T is easily seen. Further, the bias broadens the distribution of relaxation times, with long relaxation time states increasingly occupied, whereas the high frequency response is not greatly affected. The mean relaxation time τ_R also increases with increasing bias, reflecting an increased population of long-lived metastable states. This behavior, as well as the increase in ε_0 in the presence of the bias suggests critical behavior in the CDW dynamics as the bias approaches E_T from below. The magnitude of the 'critical behavior' differs from sample to sample, with some showing only a very minor effect. The peak in E_T is also sensitive to temperature and bias history, as will be described below.

The form of $\varepsilon(\omega)$ for bias fields above threshold is different from that of the pinned CDW. In contrast to the results at zero dc field, the data indicate the presence of a lossy resonance which occurs at a frequency that increases with increasing field. A resonance absorption process above E_T is expected to occur as the driving frequency of the ac signal matches the internal 'washboard frequency' of the moving CDW, and such an effect is seen in all CDW materials. Figure 9 presents the results of experiments on $K_{0.30}MoO_3$ at 77 K where the real part of the dielectric function measured a fixed frequencies is plotted as a function of bias [68]. Note the large negative dielectric values of $\varepsilon'(\omega)$ at the prominent minima. The inductive dips in ε' occur at biases where the washboard frequency and the driving frequency are equal. One can also discern subharmonic structure at biases which correspond to the washboard frequency of $\omega/2$, in both the data shown in Figure 9 and in the transient voltage oscillations observed in this sample [68]. Also observable just below E_T is the divergent dielectric constant as $E \rightarrow E_T$ as

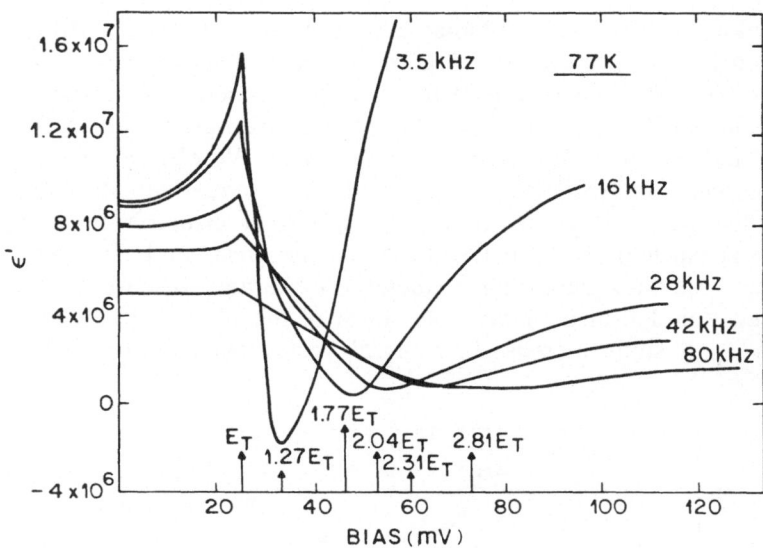

Fig. 9. The bias dependence of the real part of the dielectric function at fixed frequencies. Minima occur when the bias-dependent 'washboard frequency' of the specimen and the driving frequency are equal.

described above. In $NbSe_3$ crystals displaying current oscillations, sharp minima also occur in ε' at bias values where the fixed frequency is equal to the fundamental noise frequency [69]. In our $K_{0.30}MoO_3$ samples we do not generally observe current oscillations due to the moving CDW for dc biases above threshold. This is consistent with the notion of a longer CDW phase coherence in $NbSe_3$. The observation of a resonance-absorption process at the washboard frequency above E_T in the presence of an ac driving field does not require intrinsic phase coherence of the sliding CDW, whereas the observation of current oscillations in the presence of a dc bias alone does.

2.5. DIELECTRIC RELAXATION TEMPERATURE DEPENDENCE: FINITE DC BIAS

A metastable state is a property of the pinned CDW characterized by a local distortion of the CDW. Normally one would assume that the configuration of the pinned CDW would have little to do with CDW dynamics, at least at long time scales. We have seen however that τ_R, the time scale governing the relaxation into a metastable state diverges as the temperature is lowered [56]. There exists therefore an intermediate temperature regime, between about 80 K and 30 K in the blue bronze, where τ_R is sufficiently long that the dielectric response is both time and history dependent. Below 80 K two effects occur in bipolar scans which are characteristic of metastable pinned CDW states: (1) an increase in the low frequency dielectric constant as $E \rightarrow E_T$ from below, and (2) a zero field dielectric constant whose magnitude is dependent on whether zero field is approached from above or below. The results of bipolar bias scans at 60 K for $K_{0.30}MoO_3$ at three fixed frequencies are presented in Figure 10. Each measurement was made two seconds after setting the bias, a time that was determined to be sufficient to reach equilibrium at 60 K. The 1 kHz ε' and σ' data represent the static dielectric constant ε_0 and the dc conductivity σ_0 respectively. We observe hysteresis in the dielectric response for $-E_T < E < E_T$ with the conductivity showing a smaller effect. The divergence of the low frequency dielectric constant as E_T is approached from below is particularly dramatic. Both the hysteresis and divergent dielectric response decrease with increasing frequency. As we have discussed, this is due to the increased population of metastable states of the CDW with relaxation times longer than τ_R. A small increase in ε' as E_T is approached from below has also

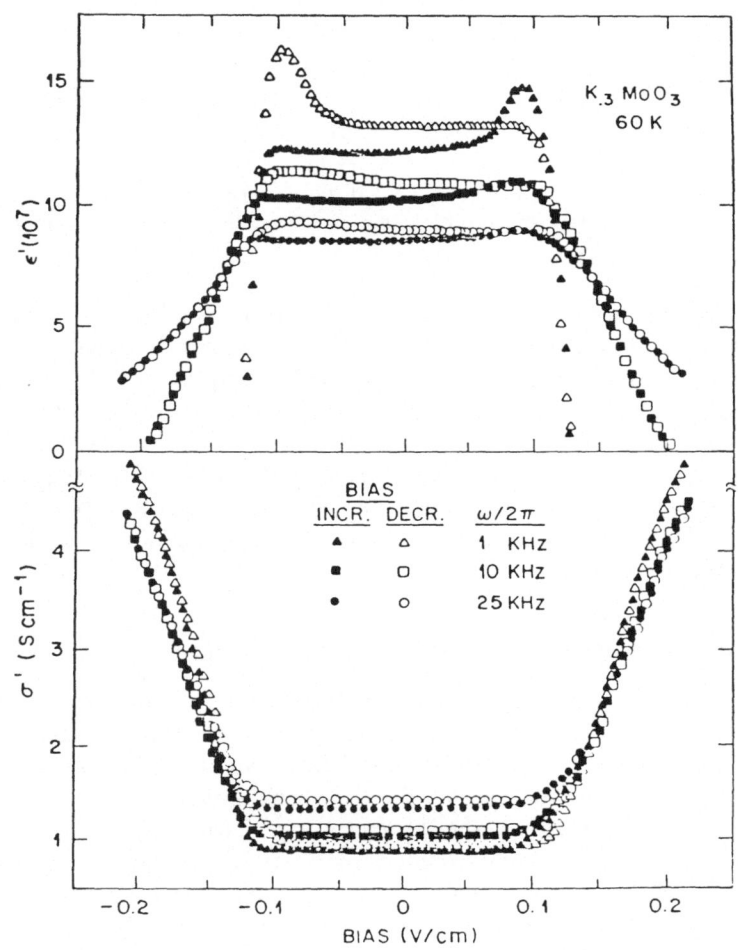

Fig. 10. Bipolar bias scans in $K_{0.30}MoO_3$ at 60 K, 2 sec/point, at three fixed frequencies. Solid symbols bias increasing, open symbols bias decreasing.

been observed in bias dependent dielectric measurements on TaS_3. In TaS_3 hysteresis is not observed for $|E| > |E_T|$ at this temperature.

The bias dependent dielectric response of the CDW for temperatures below 45 K is dominated by long lived metastable states resulting in a response that is time as well as history dependent. The results of measurement of the dielectric response and conductivity in bipolar bias scans similar to those performed at 60 K are presented in Figure 11 for a temperature of 32 K and a frequency of 50 Hz. At 32 K the dc $I-V$ characteristics are dependent on the bias sweep rate. The data in Figure 12 were taken with 2 sec and with 50 sec between setting bias and the dielectric response measurement. The response at the 2 sec rate is significantly different from that for the longer equilibrium times. As can be seen in Figure 11, there is no indication of the presence of a sharp threshold field in the low frequency dielectric constant and conductivity at 32 K for bipolar bias sweeps as

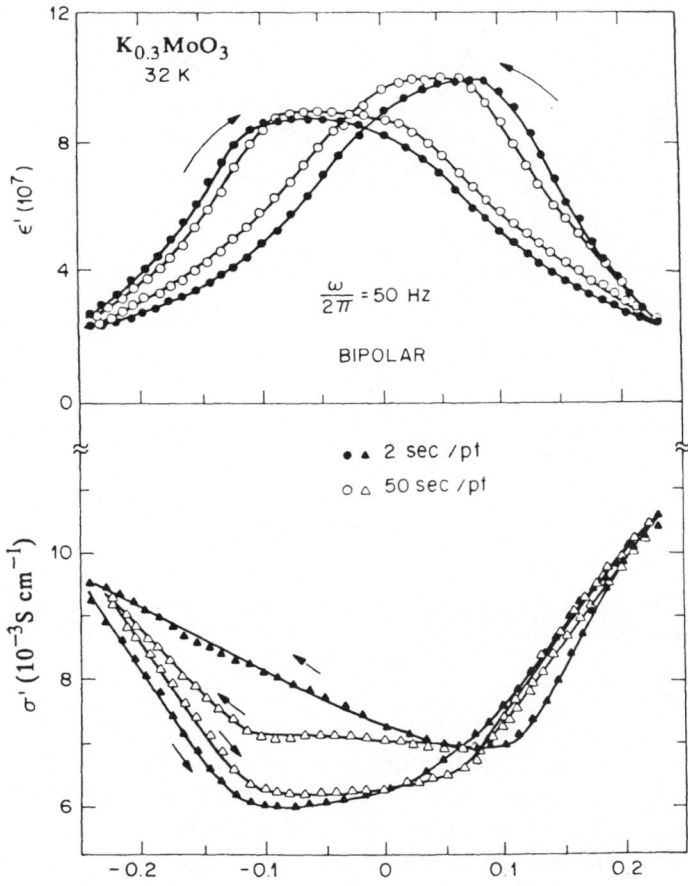

Fig. 11. Bipolar bias scans in $K_{0.30}MoO_3$ at 32 K at a fixed frequency of 50 Hz. Solid symbols 2 sec/point, open symbols 50 sec/point.

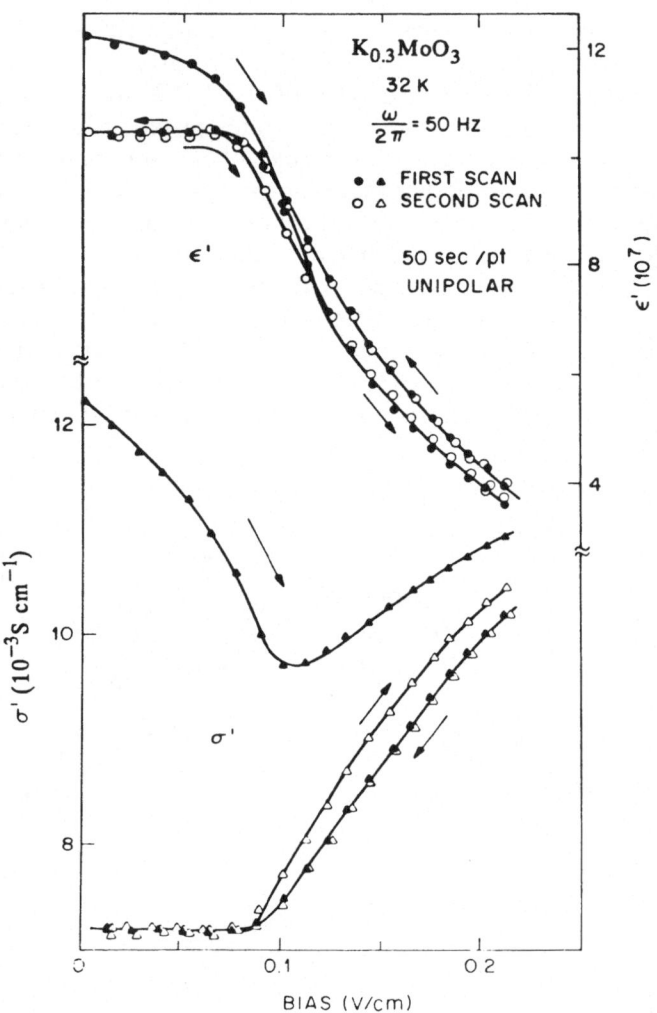

Fig. 12. Unipolar bias scans of the real part of the dielectric function and the conductivity at 32 K, 50 Hz, 50 sec/point. Solid symbols are the initial scan after cooling from 110 K, open symbols are the subsequent scan.

slow as 50 sec/point. A sharp freature at threshold at 32 K at the 50 sec/point rate depends on whether the CDW has been biased beyond negative E_T. Also significant is the offset between positive and negative bias sweeps. The offset increases further from zero bias with increasing sweep rate, indicating relaxation processes involving the CDW with decay times on the order of seconds. The offset is a measure of the hysteresis loop of the polarization in dc bias.

To further investigate history-dependent phenomena we performed unipolar bias scans on the sample cooled in zero field from 110 K to below 90 K. Initially we expect the sample to be unpolarized, but a polarization will be induced following the first positive sweep. If the polarization is sufficiently long lived, it should be observable as a difference in measured ε and σ. Indeed, we find dramatic differences in the response of the system between the first sweep and second and subsequent sweeps. Figure 12 shows the results of a unipolar scan bias at 32 K taken with 50 seconds between data points. The initial scan is taken after the sample has been cooled from 110 K to 32 K and has been at 32 K for about 5 minutes. On increasing the field, both ε' and σ' decrease smoothly until approximately 0.1 V/cm where σ' begins a smooth rise. On decreasing the field, a well-defined knee in both curves is seen at a bias of approximately 0.08 V/cm. In subsequent unipolar sweeps between 0 and 0.2 V/cm ε' and σ' behave differently from the initial sweep, instead showing more conventional behavior, including a sharp feature at threshold on increasing and decreasing bias. Both Figures 11 and 12 indicate that at 32 K one has very slow relaxation to dynamical equilibrium at fields above threshold E_T indicating than on a time scale of 50 s, the CDW dynamics is a function of the pinned state.

The results of a series of experiments where the sample was cooled in zero field from 100 K to temperatures between 90 and 30 K are shown in Figure 13. After a ten minute equilibration period we measured ε_0 and σ_0 in the 'as-cooled' state at zero field (labeled ε_0 and σ_0 in Figure 13). We then raised the bias above E_T and remeasured ε_0 and σ_0 after returning the bias to zero (labeled ε_R and σ_R in Figure 13). The difference between the temperature quenched and field polarized metastable states was strongly temperature dependent. The temperature quenched conductivity is larger than that obtained after polarization of the CDW by biasing above E_T and becomes relatively larger as the temperature is decreased. By 30 K the temperature quenched conductivity is 100% larger than that obtained after biasing above E_T. The dielectric constant of the temperature quenched state, on the other hand, is less than that found after biasing above E_T. The difference in static dielectric constants between the two states increases until about 50 K where the measurements of ε' at our lowest measurement frequency are no longer representative of pure dc response. Therefore the dielectric constant was obtained by fits to the frequency dependent response extrapolated to zero frequency at 60, 40 and 30 K. These data are shown by solid squares in Figure 13 (ε_{0C}, ε_{0R}). The data indicate a change in sign of the ratio of the temperature quenched and field relaxed dielectric constants near 30 K. This surprising effect is likely to be real, but because we do not have data for $\omega \ll \omega_R$ at 30 K we believe that further measurements at low frequencies would be necessary to confirm it. The temperature quenchable metastable states give rise to history dependent conductivity and dielectric states which become very much different from those obtained after pulsing above E_T at low temperatures. These long-lived temperature dependent metastable states apparently disappear at temperatures near 100 K as confirmed by X-ray measurements of the CDW coherence perpendicular to the direction of current flow [70]. The same temperature, 100 K, has also been associated with peaks in the threshold electric field for non-linear dc conductivity, at least in some samples [71]. In addition, 100 K has been identified as the point where the CDW wave vector becomes equal to 3/4 in the b^* direction [72].

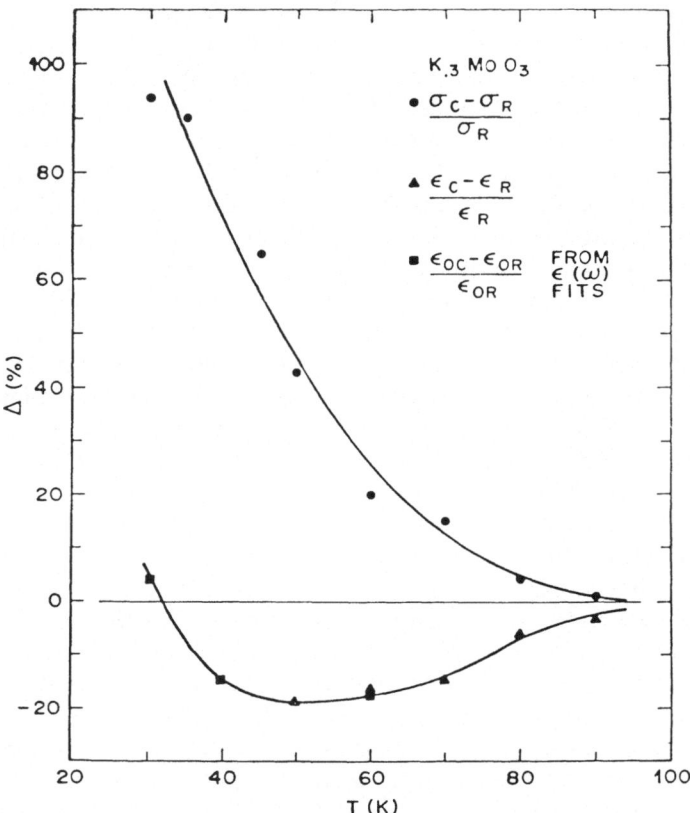

Fig. 13. The difference in the conductivity and the low frequency dielectric constant before and after a bias above the threshold field plotted as function of temperature. Each measurement was made after cooling from 110 K.

Measurements of the frequency dependent dielectric function were also made on the 'as-cooled' and the 'field-polarized' CDW states obtained as described above [56]. The parameters used to fit Equation 11 to the data were then used to obtain a density of pinned CDW states as described in Equation 16. The results at 60 K are plotted in Figure 14. Both the temperature quenched and the field polarized CDW states show a broad density of states, but on polarizing the spectral weight of the CDW density of states shifts dramatically toward lower frequencies with a much greater population of low frequency modes than in the original state. The enhanced response at low frequencies after pulsing with an electric field is also seen in numerical simulations of a one-dimensional model of random pinning. This behavior is characteristic of the CDW being driven into a high energy metastable state by the application of a bias field exceeding threshold.

Like the case of dc measurements, measurements of the dynamical response of the CDW at temperatures where τ_R is becoming large show hysteretic and history-

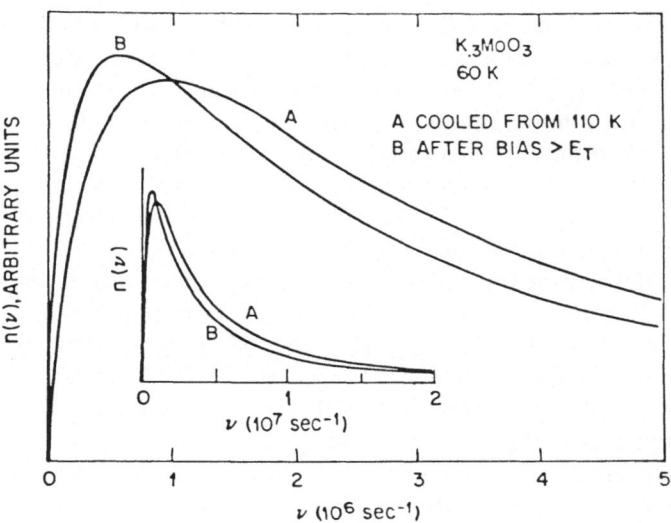

Fig. 14. Low frequency density of states as a function of frequency at 60 K after cooling from 110 K (curve A) and after biasing above threshold (curve B). The insert is a larger frequency range. The curves are calculated from fits of Equation 11 to the data.

dependent phenomena that are consistent with the existence of metastable states of the pinned CDW. As the temperature is lowered, the predominant effect is a slowing down of the characteristic time scales that can be attributed to an increase in the viscous drag forces acting to oppose the CDW motion [72]. Empirically, the increase in the viscosity is directly proportional to the normal-carrier resistivity and clearly represents a strong dynamic coupling between the low-frequency motion of the CDW and the normal carriers. This divergence in the CDW viscosity as the temperature is lowered can lead to a frozen-in polarization of the CDW in field-cooled samples [63, 73]. It is important to distinguish, however, between the existence of frozen metastable states of the CDW and their manifestation as a frozen electrical polarization. Frozen metastable states exist throughout the temperature range below 80 K for electric fields below threshold. A frozen-in polarization can only appear at low temperatures when the relaxation time for normal carrier motion through the sample is slow enough that the internal electric fields cannot be effectively screened on measurable times scales. The screening of the CDW involves macroscopic charge flow, thus the relaxation process is a dissipative one involving resistive loss. Such a dissipative mechanism is the likely origin of the apparent increase in the CDW viscosity at low temperatures as discussed by Sneddon [74].

3. Phase Mode Regime

We use the term 'phase mode regime' to refer to the response at frequencies near the CDW pinning frequency, ω_0. This is the only frequency regime where one expects an ac response if the CDW is assumed to be rigid (uniform pinning), and the presence of such a mode is clear evidence of collective, vice single electron motion of the CDW. (The pinning energy, $\hbar\omega_0$ is orders of magnitude smaller than kT). In the uniform pinning model, there is no screening and all parts of the CDW see the same applied field. Recall that Equation 2 predicts a resonance in $\sigma'(\omega)$ at ω_0, a 'crossover' from low to high frequency behavior at $\omega = 1/\omega_0^2 \tau_0$ and a roll-off due to the inertia at $\omega = 1/\tau_0 = \lambda_0/m^*$ where λ_0 is the damping from sources other than screening and m^* is the CDW effective mass.

The ac conductivity in the phase mode varies significantly among the various CDW materials, both in the magnitude of the damping and the pinning energy [11–14]. For example, in $(TaSe_4)_2I$ [11] the damping $(1/2\pi\tau_0 = 21$ GHz$)$ is significantly smaller than in TaS_3 [14] $(1/2\pi\tau_0 = 125$ GHz$)$ or $NbSe_3$ [14] $(1/2\pi\tau_0 = 70$ GHz$)$. $K_{0.30}MoO_3$ is unusual in that the pinning frequency is higher than the other CDW materials [11, 14, 15]. Because of the high pinning energy in $K_{0.30}MoO_3$, extensive ac conductivity data over a wide range of frequencies spanning the pinning frequency are not available in that material. As a comparison, however, it is interesting to look at the response of TaS_3 where such measurements have been made. Figure 15 shows the real and the imaginary part of the conductivity in TaS_3 plotted as a function of $\log(\omega)$ [14]. (The response of $NbSe_3$ is qualitatively the same [13].) The solid line is a fit of the uniform pinning model (Equation 2) to the data. The dashed line is a fit assuming a rigid CDW with a distribution of pinning energies. The distribution of pinning energies within the uniform pinning model broadens the crossover frequency, which depends on ω_0, but does not broaden the inertial roll-off which depends only on the damping. Notice that the high frequency data are well-fit by the uniform pinning model indicating that the high frequency response can be described with a single degree of freedom (the CDW phase). The low frequency data on the other hand differ markedly from the uniform pinning model as discussed in Section 2. Bleher has shown that the frequency dependence of the conductivity can be modeled using an analytic solution to the FLR model and a explicitly added damping term [28]. In this model the effect of the random impurities is to introduce an effective damping that is frequency dependent. Note that this differs from the approach used in [13 and 14] in that the CDW is assumed to be deformable with spatially random pinning rather than rigid with a distribution of pinning energies.

The inertial roll-off of the pinned phase mode at high frequencies allows one to estimate an effective mass of the CDW of 940 m_e for TaS_3 [14], 270 m_e for $NbSe_3$ [14], and 10^4 m_e for $(TaSe_4)_2I$ [11]. High frequency data on $K_{0.30}MoO_3$ are less complete, however, at low temperatures far infrared measurements [15] show a underdamped resonance near 2.8 cm^{-1} with a damping frequency of 175 GHz and an effective mass of about 10^4.

In the previous section we emphasized that the low frequency damping results from normal electron screening. This mechanism is not available at high frequencies and the origin of the damping at frequencies near the pinning energy is not presently understood. The differences in the low and high frequency damping is illustrated in Figure 16 where we show the real part of the conductivity of TaS_3

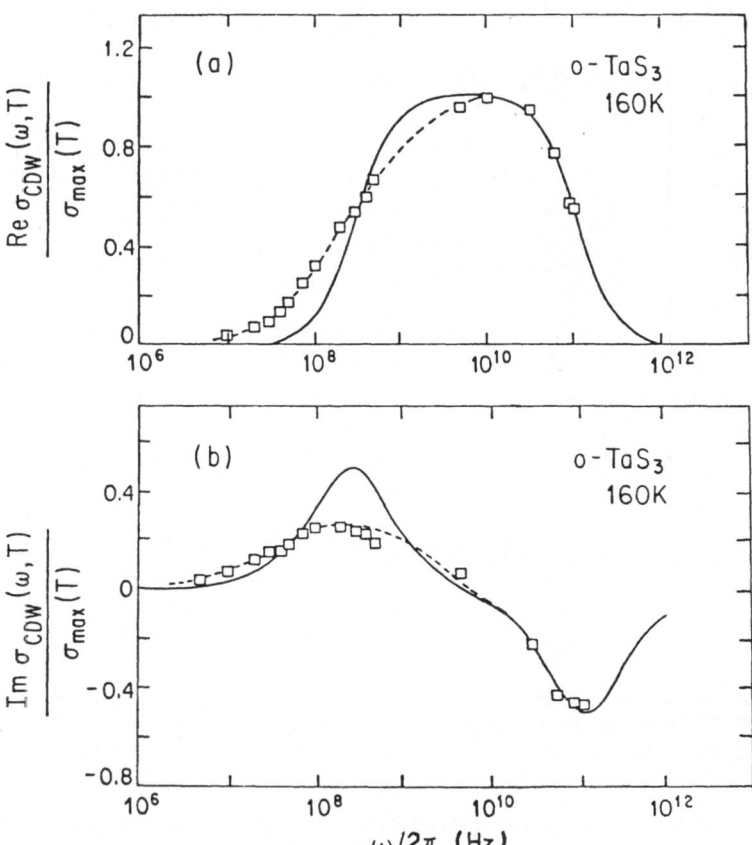

Fig. 15. Frequency dependence of the real and imaginary parts of the conductivity of TaS_3 at 160 K. (From [14]). The solid line is a fit of the uniform pinning model (Equation 2) to the data. (From [14]).

plotted as a function of the inverse temperature for several frequencies between dc and 95 GHz. Note that at 4.5 GHz the data scale with the dc response over a range of temperature. The conductivity for frequencies above 32 GHz, on the other hand, is nearly temperature independent indicating little temperature dependence of the damping. The scaling of low frequency conductivity with the dc conductivity and the temperature independence of the high frequency conductivity has been explicitly shown in TaS_3 [14] and TaSe [11]. For $K_{0.30}MoO_3$ [75, 76], the conductivity at about 2 GHz has been shown to scale with σ_{dc} for temperatures above about 45 K. The temperature independent damping at high frequencies is unusual because models of CDW damping that are based on scattering by thermally activated quasi-particles would be expected to predict a damping that would decrease as the temperature is decreased [77].

The differences in the frequency dependence of the conductivity in the dielectric

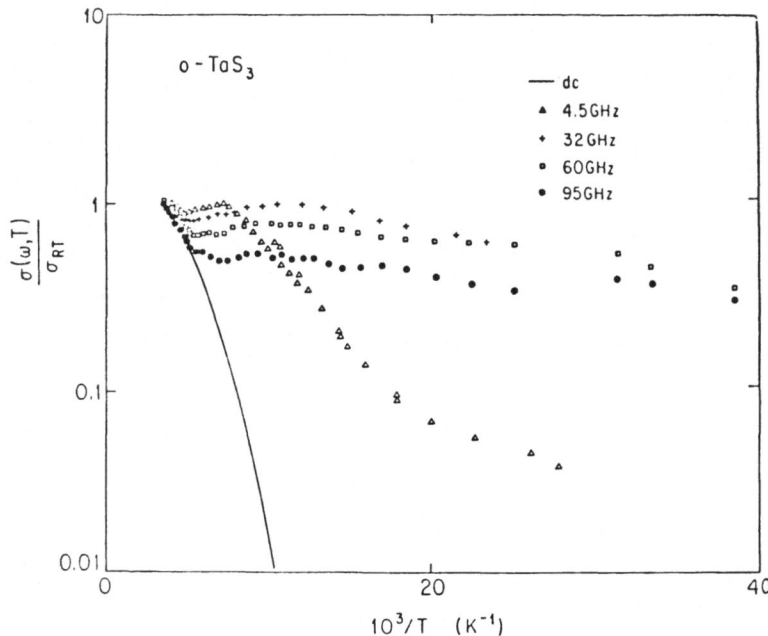

Fig. 16. The temperature dependence of the conductivity of TaS$_3$ measured at fixed frequencies. For frequencies of 32 GHz and above there is little temperature dependence of the damping. For frequencies of 4.5 GHz and below, the ac conductivity scales with the dc conductivity. (From [14]).

regime and frequencies near the pinning energy can be compared with differences in the expected dc response at high and low electric fields. At low velocities the moving CDW is accompanied by time dependent fluctuations and back-flow currents of normal electrons. At high velocities the normal electrons cannot screen the deformations and a transition to a state characterized by the motion of a rigid CDW is predicted [78]. This state is similar to the original 'sliding CDW' model proposed by Fröhich, however at temperatures where non-linear transport via the deformable CDW occurs (typically > 40 K in K$_{0.30}$MoO$_3$), the high-velocity region is not experimentally accessible. At low temperatures, on the other hand, the low conductivity allows one to apply large electric fields and a striking non-linearity is observed in K$_{0.30}$MoO$_3$ with a threshold about an order of magnitude higher than the one seen at higher temperatures [80, 81]. A similar feature can be observed in NbSe$_3$ at high magnetic fields where the large magnetoresistance allows one to apply electric fields [82]. The transition is from an insulating state to a highly conducting state, a change of eight to ten orders of magnitude. The transition into the highly conducting state is characterized by the differential conductivity increasing proportionally with current at low current levels. At higher currents a region of negative differential conductivity (bistability) is observed [83]. Current oscillations are measured with a linear frequency to current dependence over five orders of magnitude. The results can be qualitatively accounted by the onset of Fröhlich-mode conductivity, a model describing the translation of a rigid charge-density-wave (CDW).

4. Far Infrared Regime

One of the many surprises found in CDW materials was the occurrence of a dielectric relaxation at low frequencies below the pinned phase mode. It now appears that the pinned phase mode and the dielectric relaxation may not account for all of the oscillator strength of the collective response of the CDW. Recent experiments in the frequency regime between the pinned phase mode and the Peierls gap have revealed the presence of additional modes. The most clearly observed is a giant resonance in $(TaSe_4)^2I$ [16] at 38 cm^{-1}. A similar mode (originally identified as the pinned phase mode) is also seen [17] in $K_{0.30}MoO_3$ at 15 cm^{-1}. From standard theory [8] one would expect that the phase mode the CDW is IR-active and the amplitude mode is Raman active. This is not consistent with the observation of two IR active modes in $(TaSe_4)_2I$ and $K_{0.30}MoO_3$ (the phase mode and the new, high frequency mode). The resolution of this paradox is not clear; however, three possibilities have been suggested [16]. The first possibility has been identified the mode as an 'optical phason' or 'first harmonic phason'. [84]. This model has an analogy with ionic crystals such as NaCl where IR active potical modes can be observed. The second possibility is that the mode is an asymmetric phase mode associated with an optical phonon that has been folded into the zone center by the CDW distortion. The third possibility is that the observed resonance is the amplitude mode of the CDW. Normally the amplitude mode would not be IR active, but in this case one assumes that the amplitude mode is asymmetric. Finally, we note that the presence of a large FIR mode in other CDW materials such as $NbSe_3$ and TaS_3 is not established, so the extent to which the FIR mode is a generic feature of all CDW materials remains unclear.

5. Normal-Electron Screening

A key feature of the dynamics of the CDW at low frequencies is the deformability of the CDW caused by the random pinning. In the pinned state these deformations give rise to metastability and hysteresis in the static properties of the CDW. Similarly, the 'sliding state' produced by a dc field above E_T is characterized by time-dependent fluctuations of the CDW phase. Both of these situations produce gradients of the phase of the CDW that, when viewed locally, are indistinguishable from a shift of the CDW wave vector or a change in the local CDW energy gap. This means that the introduction of a distortion in the CDW phase will be accompanied by a back flow of normal electrons into or out of the local area. This flow of normal electrons will produce dissipation and experimentally, this has been shown to be the dominant energy loss mechanism at low frequencies. This loss mechanism was omitted from the uniform pinning model (Equation 1) where we assumed that the unscreened field is seen by the CDW.

The first study of CDW screening was a time-dependent Landau theory by McMillan [85], a theory completed before the observation of CDW charge transport. McMillan calculated a damping parameter which scales as ρ_c^2/σ, however he considered coupling to long wavelength phason modes and did not include coupling to collective modes of the CDW electrons. Collective motion of the CDW was explicitly treated by Sneddon who independently developed a hydrodynamic model (Q, $\omega = 0$) based on the FLR Hamiltonian and dissipation via the normal carriers [72]. Sneddon reached a similar conclusion to McMillan, the damping due to screening is temperature-dependent and scales as ρ_c^2/σ. Screening has also been phenomenologically addressed within the tunneling model of CDW transport by Tucker, et al. [47].

Measurements of the dc conductivity indicate that the CDW viscosity is diverging at low temperatures at a rate proportional to $1/\sigma$ [72]. High frequency measurements, on the other hand, present a different picture [11—15]. The high frequency damping ($1/\tau_0 = \lambda_0/m^*$ from Equation 2) is nearly temperature independent at frequencies above about 4 GHz in all CDW materials. In $(TaSe_4)I_2$ [11] and $K_{0.30}MoO_3$ [15], low temperature conductivity measurements show that the response is dominated by a high-frequency weakly damped mode. This suggests that the damping mechanism is frequency dependent with the temperature dependence of the damping scaling as $1/\sigma$ at low frequencies and a constant at high frequencies. The overall CDW response therefore consists of two parts, a weakly damped high frequency mode at the pinning frequency and a heavily damped low frequency mode that breaks off from the pinned phase mode and moves to lower frequencies as the temperature is lowered. The oscillator strength in the low frequency mode as well as the frequency $1/\tau_R$ scales as $1/\sigma$ due to screening.

The connection between the heavily damped low-frequency CDW motion and the weakly damped high-frequency motion has been addressed by Littlewood [86] who distinguishes between longitudinal and transverse modes of the CDW. In a normal metal the dynamics of the longitudinal modes, which couple to an electrostatic potential, are expected to be influenced by Coloumb correlations and screening. An example of a longitudinal mode in a metal is a plasmon mode which is typically probed by an electron energy loss experiment. Transverse modes on the other hand, which couple to the transverse field provided by electromagnetic radiation, are not expected to be affected by screening. A confusing aspect of the experimental configurations in CDW experiments is that both the low frequency measurements made with leads on the sample and high frequency measurements performed in a microwave cavity have the electric field parallel to the nonlinear current flow and would be expected to excite only longitudinal modes of the CDW. Littlewood points out that the nonuniform pinning (and to some extent the anisotropy of the conductivity) acts to mix the longitudinal and the transverse response. The result is that the dielectric relaxation at low frequencies is produced by coupling to longitudinal CDW modes with damping provided by normal electron screening while the phase mode seen at the pinning frequency results from coupling to transverse CDW modes that are unscreened. A key feature of the

Littlewood model is a plateau in $\varepsilon'(\omega)$ and $\sigma'(\omega)$ as a function of frequency at frequencies intermediate between the dielectric relaxation frequency and the pinning frequency. Unfortunately, little data exists in this frequency interval, however indirect measurements of the non-linear dc conductivity are consistent with this prediction [72]. In addition, evidence of the low frequency mode moving away from the phase mode in (TaSe$_4$)$_2$I may have been observed [87].

6. Summary

The ac conductivity of K$_{0.30}$MoO$_3$, which may be taken as a prototypical material with a moving CDW shows a rich variety of phenomena resulting from the collective motion of the electrons comprising the CDW. K$_{0.30}$MoO$_3$ undergoes a metal-semiconducting transition (Peierls distortion) at 180 K and a single electron energy gap appears in the band structure. As originally predicted by Lee, Rice and Anderson [8], a resonance in the conductivity is observed at a pinning frequency that is much smaller than the single electron gap with a damping that is nearly temperature independent. In this frequency regime, the response of the CDW can be modeled by the motion of rigid CDW and the pinning can be viewed as uniform.

The ac response of CDW materials at low frequencies is dramatically different from the simple resonance predicted for a rigid CDW. The low frequency response can be characterized as a dielectric relaxation with a frequency $1/\tau_R$ that scales as $1/\sigma$. The dielectric relaxation results because of local deformations of the CDW brought about by non-uniform pinning of the CDW. In contrast to the mode at the pinning frequency, the damping at low frequencies mode is dominated by the dissipation from screening currents of the normal electrons which move to screen the CDW distortions.

A direct consequence of the deformability of the CDW phase is the occurrence of metastable and history dependent phenomena. A metastable CDW state occurs because of local changes in the phase of the pinned CDW due to the spatially varying random pinning. A metastable state results in a macroscopic polarization of the CDW and the decay of this polarization may be infinitely long at low temperatures. In addition to changing static properties such as the resistance, metastable states can also affect the ac response when measured on time scales short compared to the dielectric relaxation time. Depending on the temperature, the dipole moment associated with the polarization produced by a metastable state may be screened by the normal carriers.

Finally we note that the additional higher frequency modes have been observed within the single-electron energy gap. In particular, a large resonance has been observed in (TaSe$_4$)$_2$I [16] at 38 cm^{-1} and in K$_{0.30}$MoO$_3$ [17] at 15 cm^{-1}. The origin of these additional modes is not clear and it is not certain whether they occur in all CDW materials. The presence of these additional modes emphasizes the richness of the physical phenomena associated with the dynamics of the collective CDW mode in this class of materials.

Acknowledgements

We would like to thank our collaborators for their valuable contributions to the research reviewed here: R. G. Dunn, P. B. Littlewood, S. Martin, E. A. Rietman and L. F. Schneemeyer. We would like to thank S. B. Coppersmith and P. B. Littlewood for numerous discussions on moving charge-density waves.

References

1. R. E. Peierls, *Quantum Theory of Solids* (Oxford University Press), 1955.
2. H. Fröhlich, *Proc. Roy. Soc. London* **A233**, 296 (1954).
3. Gy. Hutiray and J. Solyóm (Eds.), *Charge Density Waves in Solids*, (Springer-Verlag, Berlin), 1985.
4. Proceedings of the Yamada Conference XY on the Physics and Chemistry of Quasi One-Dimensional Conductors, Lake Kawaguchi, Japan, May 1986, *Physica* **143B** (1986).
5. P. Monceau in *Electronic Properties of Quasi-One-Dimensional Materials*, Vol. II, P. Monceau (Ed.) (Reidel, Dordrecht, The Netherlands), 1985.
6. C. Schlenker and J. Dumas in *Crystal Chemistry and Properties of Materials with Quasi-One-Dimensional Structures*, J. Rouxel (Ed.) (Reidel, Dordrecht, The Netherlands), 1986.
7. For a general review of CDWs in transition metal compounds see F. J. DiSalvo in *Electron—Phonon Interactions and Phase Transitions*, Tormed Riste (Ed.) (Plenum, New York), 1977, p. 107.
8. P. A. Lee, T. M. Rice, and P. W. Anderson, *Solid State Commun.* **14**, 703 (1974).
9. P. A. Lee and T. M. Rice, *Phys. Rev. B* **19**, 3970 (1979).
10. R. M. Fleming and C. C. Grimes, *Phys. Rev. Lett.* **42**, 1423 (1979).
11. D. Reagor, S. Sridhar, M. Maki, and G. Gruner, *Phys. Rev. B* **32**, 8445 (1985).
12. S. Sridhar, D. Reagor and G. Gruner, *Phys. Rev. Lett.* **55**, 1196 (1985).
13. D. Reagor, S. Sridhar, and G. Gruner, *Phys. Rev. B* **34**, 2212 (1986).
14. S. Sridhar, D. Reagor, and G. Gruner, *Phys. Rev. B* **34**, 2223 (1986).
15. H. K. Ng and G. A. Thomas, and L. F. Schneemeyer, *Phys. Rev. B* **30**, 8755 (1986).
16. M. S. Sherwin, A. Zettl, and P. L. Richards, *Phys. Rev. B* **36**, 6708 (1987).
17. G. Travaglini and P. Wachter, *Phys. Rev. B* **30**, 1971 (1984). 6708 (1987).
18. H. Fukuyama and P. A. Lee, *Phys. Rev. B* **19**, 3970 (1979).
19. John Bardeen, *Phys. Rev. Lett.* **42**, 1498 (1979), **45**, 1978 (1980), **55**, 1010 (1985); R. E. Thorne, J. H. Miller, Jr., W. G. Lyons, J. W. Lyding, and J. R. Tucker, *Phys. Rev. Lett.* **55**, 1006 (1985); John Bardeen and J. R. Tucker, in Ref. 3, p. 155; John Bardeen, *Physica* **143B**, 14 (1986); John Tucker, *Physica* **143B**, 19 (1986); J. Bardeen, J. W. Lyding. W. G. Lyons, J. H. Miller, R. E. Thorne, Jr., and J. R. Tucker, *Synth. Met.* **19**, 1 (1987).
20. R. E. Thorne, J. R. Tucker, and John Bardeen, *Phys. Rev. Lett.* **58**, 828 (1987) and R. E. Thorne, W. G. Lyons, J. W. Lyding, J. R. Tucker and John Bardeen, *Phys. Rev. B, Phys. Rev. B* **35**, 6360 and 6348 (1987).
21. G. Grüner, A. Zawadowski, and P. Chaikin, *Phys. Rev. Lett.* **46**, 511 (1981).
22. P. Monceau, J. Richard and M. Renard, *J. Phys. C* **15**, 931 (1982).
23. W. Wonneberger and F. Gleisberg, *Solid State Commun.* **23**, 665 (1977).
24. One should be aware of two assumptions in this model which are incorrect, uniform pinning and the lack of screening. The model assumes that the field, E, producing the CDW displacement is the same as the applied field.
25. P. Debye, *Polar Molecules* (Chemical Catalog Co., New York) 1929, Chapter V.
26. R. A. Klemm and J. R. Schrieffer, *Phys. Rev. Lett.* **51**, 47 (1983); M. O. Robbins and R. A. Klemm, *Phys. Rev. B* **34**, 8496 (1986).
27. M. V. Feigelman and V. M. Vinokur, *Phys. Rev. Lett.* **87A**, 53 (1981); *Solid State Commun.* **45**, 603 (1983).
28. M. Bleher, *Solid State Commun.* **63**, 1071 (1987).

29. S. Abe, *J. Phys. Soc. Jpn.* **54**, 3494 (1985).
30. L. Sneddon, M. C. Cross, and D. S. Fisher, *Phys. Rev. Lett.* **49**, 292 (1982).
31. L. Sneddon, *Phys. Rev. B* **29**, 719 (1984), **29**, 725 (1984).
32. S. N. Coppersmith and P. B. Littlewood, *Phys. Rev. B* **31**, 4049 (1985).
33. D. S. Fisher, *Phys. Rev. Lett.* **50**, 1486 (1983) and *Phys. Rev. B* **31**, 1398 (1985).
34. P. B. Littlewood, in Ref. 3, p. 369, also *Physica* **23D**, 45 (1986); also *Phys. Rev. B* **33**, 6694 (1986).
35. S. N. Coppersmith and D. S. Fisher, *Phys. Rev. B* **28**, 2566 (1983).
36. S. N. Coppersmith, *Phys. Rev. B* **30**, 410 (1984).
37. S. N. Coppersmith and P. B. Littlewood, *Phys. Rev. Lett.* **57**, (1986).
38. S. Abe, *J. Phys. Soc. Japan* **55**, 1987 (1986).
39. H. Matsukawa and H. Takayama, *Physica* **143B**, 30 (1986), and **143B**, 80 (1986).
40. S. N. Coppersmith in *Low Dimensional Conductors and Superconductors*, D. Jeróme and L. G. Caron (Eds.), Nato ASI Series Vol. 155, (Plenum, New York, 1987), p. 425.
41. N. P. Ong, C. Verma, and K. Maki, *Phys. Rev. Lett.* **52**, 663 (1984); N. P. Ong and Kazumi Maki, *Phys. Rev. B* **32**, 6582 (1985).
42. L. P. Gorkov, *JETP Lett.* **38**; I. Batistic, A. Bjelis and L. P. Gorkov, *J. de Phys. (Paris)* **45**, 1049 (1984).
43. A. Zettl and G. Gruner, *Phys. Rev. B* **26**, 2298 (1982).
44. R. P. Hall, M. F. Hundley, and A. Zettl, *Phys. Rev. Lett.* **56**, 2399 (1986); R. P. Hall, M. F. Hundley and A. Zettl, *Phys. Rev. B*, to be published; R. P. Hall, and A. Zettl, *Phys. Rev. B*, to be published.
45. M. Inui, R. P. Hall, S. Donicach, and A. Zettl, *Phys. Rev. B*, to be published.
46. G. Mihály and P. Beauchene, *Solid State Commun.* **63**, 911 (1987).
47. J. R. Tucker, W. G. Lyons, and G. Gammie, *Phys. Rev. B* **38**, 1148 (1988).
48. P. B. Littlewood and R. Rammal, *Phys. Rev. B* **38**, 2675 (1988).
49. For a review of dielectric relaxation see for example C. J. F. Böttcher and P. Bordewijk, *Theory of Electric Polarization*, Vol. II (Elsevier, Amsterdam, 1978).
50. K. S. Cole and R. H. Cole, *J. Chem. Phys.* **9**, 341 (1941).
51. D. W. Davidson and R. H. Cole, *J. Chem. Phys.* **18**, 1417 (1950) and **19**, 1484 (1951).
52. R. M. Fuoss and J. G. Kirkwood, *J. Am. Chem. Soc.* **FB63**, 385 (1941).
53. G. Williams and D. C. Watts, *Trans. Faraday Soc.* **66**, 80 (1970).
54. A. K. Johnscher, *Colliod Pol. Sci.* **253**, 231 (1975).
55. S. Havriliak and S. Negami, *J. Polymer Sci.* **C14**, 99 (1966).
56. R. J. Cava, P. B. Littlewood, R. M. Fleming, L. F. Schneemeyer, and E. A. Rietman, *Phys. Rev. B* **34**, 1184 (1986).
57. R. J. Cava, R. M. Fleming, P. Littlewood, L. F. Schneemeyer, and R. G. Dunn, *Phys. Rev. B* **30**, 3228 (1984).
58. R. J. Cava, R. M. Fleming, R. G. Dunn, and E. A. Rietman, *Phys. Rev. B* **31**, 8325 (1985).
59. J. P. Stokes, Mark O. Robbins, and S. Bhattacharya, *Phys. Rev. B* **32**, 6939 (1985).
60. N. P. Ong, D. D. Duggan, C. B. Kalem, T. W. Jing, and P. A. Lee, Ref. 3, p. 387; C. B. Kalem, N. P. Ong and J. C. Eckert, unpublished.
61. R. J. Cava, P. Littlewood, R. M. Fleming, R. G. Dunn, and E. A. Rietman, *Phys. Rev. B* **33**, 2439 (1986).
62. Tomoyuki Sekine, Tomonobu Tsuchiya, and Etsutuki Matsuura, *Physica* **143B**, 158 (1986).
63. R. J. Cava, R. M. Fleming, E. A. Rietman, R. G. Dunn, and L. F. Schneemeyer, *Phys. Rev. Lett.* **53**, 1677 (1984).
64. Wei-yu Wu, L. Mihály, George Mozurkewich, and G. Gruner, *Phys. Rev. Lett.* **52**, 2382 (1984).
65. R. J. Cava, R. M. Fleming, R. G. Dunn, E. A. Rietman, and L. F. Shcneemeyer, *Phys. Rev. B* **30**, 7290 (1984).
66. L. F. Schneemeyer, R. M. Fleming, and S. E. Spengler, *Solid State Comm.* **53**, 505 (1985); L. F. Schneemeyer, S. E. Spengler, F. J. DiSalvo and J. V. Waszczak, *Mol. Cryst. Liq. Cryst.* **125**, 41 (1985).
67. R. J. Cava, L. F. Schneemeyer, R. M. Fleming, P. B. Littlewood, and E. A. Rietman, *Phys. Rev. B* **32**, 4088 (1985).

68. R. M. Fleming, L. F. Schneemeyer, and R. J. Cava, *Phys. Rev. B* **31**, 1181 (1985).
69. A. Zettl and G. Gruner, *Phys. Rev. B* **29**, 755 (1984).
70. R. M. Fleming, R. G. Dunn, and L. F. Schneemeyer, *Phys. Rev. B* **31**, 4099 (1985).
71. See for example the review of C. Schlenker, in *Low Dimensional Conductors and Supercon-ductors*, D. Jérome and L. G. Caron (Eds.), Nato ASI Series Vol. 155, (Plenum, New York, 1987), p. 477.
72. R. M. Fleming, R. J. Cava, L. F. Schneemeyer, E. A. Rietman, and R. G. Dunn, *Phys. Rev. B* **33**, 5450 (1986).
73. G. Kriza and G. Mihály, *Phys. Rev. Lett.* **56**, 2529 (1986); P. B. Littlewood and R. Rammal, *Phys. Rev. Lett.* **58**, 524(C) (1987).
74. Leigh Sneddon, *Phys. Rev. B* **29**, 719 (1984).
75. R. P. Hall, M. S. Sherwin, and A. Zettl, *Solid State Commun.* **54**, 683 (1985).
76. G. Mihaly, J. Dumas, and A. Jannosy, *Solid State Commun.* **60**, 785 (1986).
77. S. Takada, K. Y. M. Wong, and T. Holstein, Ref. 3, p. 227.
78. P. B. Littlewood, *Solid State Commun.* **65**, 1347 (1988).
79. W. Fogle and J. H. Perlstein, *Phys. Rev. B* **6**, 1402 (1972).
80. A. Maeda, T. Furuyama, and S. Tanaka, *Solid State Comm.* **55**, 951 (1985) and A. Maeda, T. Furuyama, K. Uchinokura and S. Tanaka, *Solid State Comm.* **58**, 25 (1986).
81. L. Mihály and G. X. Tessema, *Phys. Rev. B* **33**, 5858 (1986).
82. M. P. Everson, G. Eiserman, A. Johnson, and R. V. Coleman, *Phys. Rev. B* **32**, 541 (1985).
83. S. Martin, R. M. Fleming, and L. F. Schneemeyer, *Phys. Rev. B* **38**, 5733 (1988).
84. M. B. Walker, *Can J. Phys.* **56**, 127 (1978).
85. W. L. McMillan, *Phys. Rev. B* **12**, 1197 (1976).
86. P. B. Littlewood, *Phys. Rev. B* **36**, 3108 (1987).
87. S. Sridhar, unpublished.

BREAKING OF ANALYTICITY IN
CHARGE DENSITY WAVE SYSTEMS:
PHYSICAL INTERPRETATION AND CONSEQUENCES

SERGE AUBRY* and PASCAL QUEMERAIS**

* Main address:
Laboratoire Léon Brillouin,
(Laboratoire commun CEA-CNRS),
C.E.N. Saclay,
F-91191-Gif-sur-Yvette Cedex, France

and

Institut Laue-Langevin   *Center for Non-Linear Studies*   *Institute for Theoretical Physics*
Avenue des Martyrs, 156X,   *Los Alamos National Laboratory*   *University of California*
F-38042-Grenoble Cedex, France   *Los Alamos, N.M 87545, U.S.A.*   *Santa Barbara, CA 93106, U.S.A.*

** *Laboratoire de Physique Cristalline*
Université de Nantes
2 rue de la Houssinière
F-44072-Nantes, France

1. Introduction: The Peierls Instability in 1D Conductors

The aim of this paper is to review results of the authors and collaborators concerning the theory of the transition by breaking of analyticity in incommensurate systems with a special emphasis on charge density wave (CDW) systems. New developments concerning the application of this approach to the phenomena observed in many CDW systems and particularly those considered in this book are suggested at the end of the paper (for other reviews about CDW systems see [75, 87, 88, 90]). On some points, more detail can be found in the Ph.D. dissertation of P. Quemerais (1987) [1]. These applications are currently being developed.

Although much further works are needed, the important conclusion suggested by the present works is that any real CDW system can be considered as a pinned *bipolaron superlattice* (see [89] for information about bipolarons in real systems). Unlike most early bipolaron theories, these bipolarons do not form a band of extended states but remains *localized in the real space* (because of the 'breaking of analyticity' of the CDW). The interacting potential between the bipolarons is much larger than their kinetic energy and prevents their quantum tunnelling through the PN potential. They can be considered as classical particles submitted to a strong lattice potential (Peierls—Nabarro potential or PN potential) [2]. At zero degree K (0 K), these bipolarons form an ordered structure in real space (the CDW) which is *strongly pinned* to the lattice sites and makes the CDW structure insulating. At finite temperature, the Ohmic part of the electric transport is due to the hopping of bipolarons. An extra nonlinear contribution is possible if the phase domain walls move under the effect of the electric field. The CDW critical

295

C. Schlenker (Ed.), Low-Dimensional Electronic Properties of Molybdenum Bronzes and Oxides,
295—405.
© 1989 *Kluwer Academic Publishers.*

temperature corresponds to the melting of the ordered bipolaron structure. This interpretation provides a new basis for finding more appropriate physical models (incommensurate lattice gas models) which could globally interpret the whole set of puzzling properties of CDW compounds.

In this paper, we review the concept of 'breaking of analyticity' on several simple models with an incommensurate structure in order to explain these new physical ideas before to apply them to quasi one-dimensional conductors (the Peierls—Fröhlich conductors) [3, 4] far from the standard small *electron—phonon coupling* limit. When the electron—phonon coupling exceeds a critical value which is reasonably large, the physical behavior of a CDW becomes sharply different from what is expected from the standard theories. We try to clarify this physical concept by many intuitive and empirical arguments illustrated by some numerical experiments. For the purely mathematical aspects concerning the exact results of the FK model, the interested reader is referred to other publications.

The molybdenum bronzes studied in this book (see also [90]) are examples of quasi one-dimensional CDW compounds which, below some critical temperature, spontaneously develop a periodic modulation of their electronic density associated with a periodic lattice distortion (PLD). Generally, the period of the CDW—PLD is not rationally related to the periods of the undistorted lattice so that the resulting structure is called 'incommensurate'.

Let us start by a brief description of the well-known arguments for the Peierls—Fröhlich [3, 4] instability which produces a CDW. The arguments of this instability apply at 0 K in systems where the electrons mostly propagate on one dimensional chains (with a small interchain exchange). Then the Fermi surface is almost flat and perpendicular to the chains. In addition, the electrons do not interact with each other so that in the ground state they always occupy the electronic eigenstates by pairs with opposite spins. The electrons only interact with the atomic lattice distortion (phonons), which is supposed to be classical (adiabatic approximation). Considering a periodic lattice distortion (PLD) the wave vector of which is twice the Fermi wave vector, it is found that because of a gap opening at the Fermi surface, the electrons gain an energy per site proportional to $-\Delta^2 \ln \Delta^2$ (the half width of the gap is Δ), while the cost in elastic energy due to the lattice distortion is proportion to Δ^2 (cf. Figure 1). Since for small Δ, the electronic energy gain is always larger than the elastic energy loss and the system at 0 K is always unstable with respect to such a PLD. This PLD induces a periodic distortion of the electronic density with the same period (CDW). When $2k_F$ is incommensurate with the lattice wave vectors, the crystal structural loses its translational invariance and the crystal is an incommensurate structure.

A simple exact calculation on a discrete model for a Peierls chain in the dimerized case (which we shall study intensively in the incommensurate case in this paper) illustrates and confirms the existence of the Peierls instability. Let us assume that the propagation of the electrons along the chains are described by the tight binding eigenequation in one dimension:

$$-T\psi_{n+1} - T\psi_{n-1} + \Delta_n \psi_n = E\psi_n \tag{1a}$$

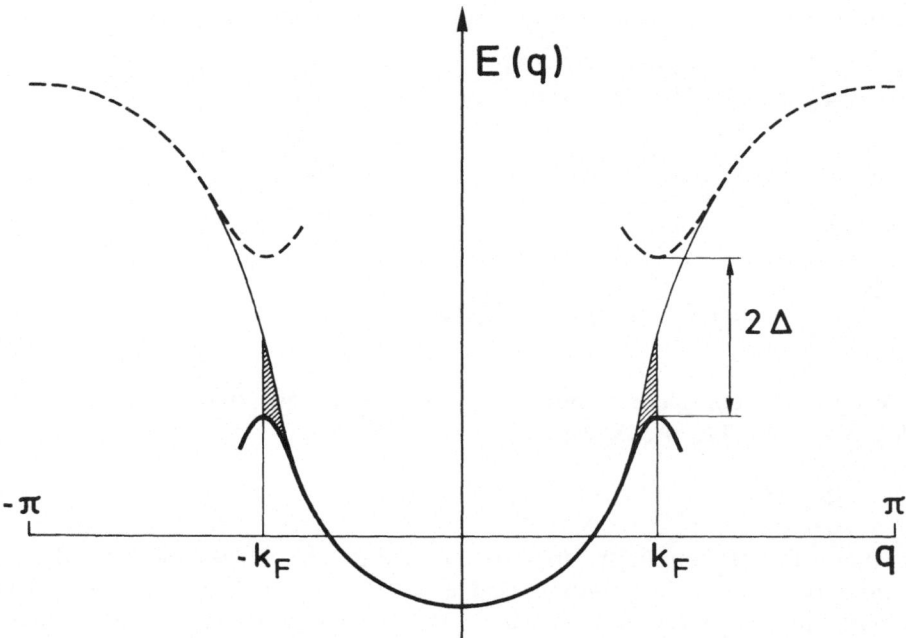

Fig. 1. Electronic energy as a function of the wave vector in the absence of CDW (thin line) and with a CDW (thick line). The opening of a gap at the Fermi energy produces an electronic energy gain (hatched area) which is proportional to $-\Delta^2 \ln \Delta^2$ while the elastic energy loss is proportional to Δ^2.

where T is the exchange constant between neighbouring sites, E the eigenenergy and Δ_n is the variation of the on-site energy at site n due to the lattice distortion. Suppose that the band is half-filled so that the $2k_F$–PLD just corresponds to a dimerized chain (commensurability $1/2$):

$$\Delta_n = \Delta^* + (-1)^n \Delta. \tag{1b}$$

Δ^* corresponds to a (trivial) constant distortion of the lattice while Δ characterizes the on-site potential modulation associated with the dimerization. Due to the Bloch theorem, the amplitude ψ_n of the eigenwave functions at site n has the form

$$\psi_n = \exp(iqn)\chi_n \tag{2a}$$

where χ_n has period 2. The eigenenergies of the electrons are simply given by

$$E(q) = \Delta^* \pm (4T^2 \cos^2 q + \Delta^2)^{1/2}. \tag{2b}$$

As expected, there exist two energy bands which correspond to a minus sign and to a plus sign in (2b), respectively. The energy gap between these two bands is equal to 2Δ. The filled band corresponds to a minus sign in (2b). Assuming a double occupany of each filled electronic state, the electronic energy is

$$\Phi_{el}(\Delta) = \frac{2}{\pi} \int_0^{\pi/2} [\Delta^* - (4T^2 \cos^2 q + \Delta^2)^{1/2}] \, dq$$

$$= \frac{2}{\pi} \int_0^{\pi/2} [\Delta^* - (4T^2 + \Delta^2 - 4T^2 \sin^2 q)^{1/2}] \, dq$$

$$= \Delta^* - \frac{2}{\pi} (4T^2 + \Delta^2)^{1/2} \mathbf{E} \left(\frac{2T}{(4T^2 + \Delta^2)^{1/2}} \right) \qquad (3a)$$

where $\mathbf{E}(x)$ is the complete elliptic integral of the second kind (Gradshteyn and Ryshik [5]). For small Δ, $\Phi_{el}(\Delta)$ expands as

$$\Phi_{el}(\Delta^*, \Delta) \approx \Delta^* - 4T + \frac{1}{4\pi T} \Delta^2 \ln \Delta^2 \qquad (3b)$$

For the sake of simplicity, the on-site energy Δ_n is assumed to depend proportionally on a single phonon variable u_n. Its Fourier transform being $u(0)$ and $u(\pi)$ at the wave vector zero and at the wave vector corresponding to the dimerization, respectively, we have

$$\Delta^* = \lambda^* u(0) \quad \text{and} \quad \Delta = \lambda u(\pi) \qquad (4a)$$

where λ^* and λ are electron—phonon coupling constants.

The elastic energy of the lattice is then equal to

$$\Phi_{elastic}(\Delta^*, \Delta) \approx \frac{1}{2} \frac{K^*}{\lambda^{*2}} \Delta^{*2} + \frac{1}{2} \frac{K}{\lambda^2} \cdot \Delta^2 \qquad (4b)$$

where K^* and K are the elastic constants of the lattice for the deformations with wave vector 0 and π respectively. By minimizing the total energy $\Phi_{el}(\Delta^*, \Delta) + \Phi_{elastic}(\Delta^*, \Delta)$ it is clear that Δ is *nonzero* which confirms that this one dimensional mode is always unstable with respect to a dimerization of the lattice. A complete calculation of the dimerized model for all Δ is given in Feinberg and Ranninger [6].

Although this instability argument is not sufficient for proving that the ground state of this model is necessarily a dimerized chain, many numerical investigations (for *usual simple models*) including our studies described in this review (Section 4), confirms that the ground state remains modulated with the wave vector $2k_F$ at large electron—phonon couplings in agreement with the Peierls predictions.

Let us also note that there exist some discrete incommensurate Peierls models solved by Brazovskii *et al.* [7] in which the electron—phonon coupling and the elastic energy have been specially chosen as particular nonlinear functions of the phonon variables in order that the ground state of the model be explicitly calculable. These models also confirm the Peierls predictions. However, the *integrability* of these models appears to be an *exceptional situation*, as will appear in this paper. Similar situations are usual and well understood in the theory of dynamical systems (see e.g. [8]). For this reason, these integrable models do not exhibit any TBA although most Peierls models do and therefore are *non-generic*.

At finite temperature, it is essential to know the low-lying energy excitations of the Peierls chain in order to get a correct prediction of its thermodynamical behaviour. In usual theory, the considered excitations only consist in electrons which jump from the filled band into the empty band. When the thermal energy $k_B T$ becomes comparable to the electronic gap at 0 K, there are too many electronic excitations so that electronic energy gain due to the distortion tends to vanish. The PLD—CDW is then usually obtained by minimizing the sum of the electronic free energy and of the lattice energy with respect to the amplitude Δ of the gap. It is found that beyond some critical temperature T_{MF}, the Peierls instability disappears (see e.g. [6]). The relation between Δ (at 0 K) and T_{MF}, which is identical to the relation between the critical temperature of a BCS super-conductor and its gap at 0 K (Bardeen *et al.* [9]), is given by

$$2\Delta \cong 3.52\, k_B T_{MF}. \tag{5}$$

In fact, the measured coefficient in (5) is much larger than 3.52 in most real CDW compounds. It generally ranges around 10 or 12 but it is often much larger, as has been pointed out by Monceau [10]. This theory has neglected essential physical features.

The first argument is that a strictly one dimensional system does not exhibit any real transition at T_{MF} but rather a smooth crossover. However, a transition should exist at a temperature smaller than but close to T_{MF} because any weak elastic interchain coupling makes the system tridimensional. In fact, only the electrons are supposed to be one-dimensional, the phonons in CDW compounds are three-dimensional and create a significant elastic interchain coupling. The second argument is that at finite temperature the atomic distortion is supposed to remains periodic without any thermal (and also quantum) fluctuations. Our claim is that this assumption is not valid in most CDW compounds because the electron—phonon coupling is too large.

When the electron—phonon coupling ranges beyond some critical value, the CDW is *nonanalytic*. Then, the physical description of the CDW must be changed drastically because the *configurational* defects (instead of the electronic excita-tions) become the essential low-lying energy excitations. Our interpretation is physically different of the theory developed in the strong electron—phonon coupling regime by Varma and Simons [11] which is based on the *softening* of the phase fluctuations with a gapless phason mode. By contrast, we predict a finite phason gap for the CDW but also a possible (but incomplete) softening of the pinned phase mode at T_c. Within the approach which we suggest, configurational defects also produce phase fluctuations but these have a much longer relaxation time compared to the relaxation time of the phase fluctuations due to phonons (essentially the phason branch). Nevertheless, the theory of Varma *et al.* confirms that the Peierls transition temperature could become much smaller than the temperature given by (5) at large electron—phonon coupling.

The new results of our theory compared to standard theories concern *the nature and the properties of the low-lying excitations* of an incommensurate CDW. The phase excitations of an incommensurate CDW are generally considered as phonons (i.e. phasons). It is generally considered to be obvious that the gap of these excitations is zero because the energy of the system is invariant by any global phase shift of the CDW. (As we will see below, this argument implicitly assumes the analyticity of the physical functions involved in the CDW, which is not necessarily fulfilled.)

This standard assumption, has the consequence that under any small applied electric field, this incommensurate CDW should carry an electric current corresponding to the uniform translation of its phase. This is the model which was initially proposed by Fröhlich [4]. It was rejected for the interpretation of superconductivity after that the BCS theory [9] was discovered but this theory is generally considered as valid for CDW systems. Later, the Fröhlich theory has been improved by Lee, Rice and Anderson [12] by involving the effects of the impurities or defects of the system which create a weak pinning of the CDW. This theory is often considered as the basic model for the interpretation of the observed electric field threshold and related phenomena in the CDW compounds like those which are extensively described in this book.

In contradiction with this standard theory, in the adiabatic limit, our theory *proves* that *the gap of the phason excitations of an incommensurate CDW systems is finite* for strong electron—phonon coupling. In that case, the existence of several incommensurate periods (those of the lattice and of the CDW) suppress this zero frequency sliding mode and pins the CDW to the lattice. More precisely when the electron—phonon coupling increases, a transition of the system occurs from a regime with a zero gap phason to a regime where this gap is finite. We called this transition: transition by breaking of analyticity (TBA).

As for usual phase transitions, which are described by the breaking of some symmetry group, the TBA corresponds to the breaking of the gauge group, the elements of which transforms the incommensurate structure into an equivalent one by rotation of its phase. Beyond the TBA, the gauge group which transforms the incommesurate CDW into an equivalent one, is reduced to a subgroup of the initial one, and loses its compactness [13]. Because of this universal character, it is not surprising that the TBA is now found in a large number of models for incommensurate structures in one and *several dimensions*. However, the determination of universality classes for the TBA in the usual sense (which, for example, fixes the critical exponents) not only depends on the model dimensionality but also depends of parameters like the *incommensurability ratios*. In spite of some preliminary results, the general problem is still an open question.

The new nonanalytical regime predicted beyond the TBA is characterized by the existence of low-lying energy *configurational* excitations ('discommensurations' of the *incommensurate* structure) which play *the essential physical role* for the thermodynamical behavior at low temperature. *A priori*, the analytic regime for a

CDW could physically exist if the electron—phonon coupling is small enough. In fact, we argue in the final section that an analytic CDW shoud be unstable with respect to quantum lattice fluctuations. Then, the classical localized bipolarons are expected to become extended by quantum lattice effects (Cooper's pairs). The CDW state should disappear and be replaced at 0 K by a superconducting state. For this reason, we claim that *all incommensurate CDW in real compounds have nonanalytic incommensurate ground states*. Particularly, the compounds for wich nonlinear conductivity is observed should belong to the class of nonanalytic CDW systems.

When such a situation occurs, the lattice pinning of the CDW is found to be of several orders of magnitude larger than the observed electric field threshold, which prevents the CDW from moving as a whole under any physically reasonable electric field. However, the system *is not a usual insulator*. A different mechanism for the electric conductivity by *domain wall propagations* then becomes possible. (Note that this new type of conductivity was not mentioned when we wrote our most recent paper concerning this subject [14]). We assumed there that, in the absence of Fröhlich mode, the system should be an insulator at 0 K and that this property should be maintained at low temperature.) On the contrary, in this paper, we claim that although the dc conductivity of the CDW is zero at 0 K, the thermal fluctuations play an essential role at finite temperature. Under an applied electric field, the phase of the CDW may increase by the nucleation and the propagation of phase domain wall. Their propagation produces an electric current by a mechanism which is analogous to the polarization current in a ferroelectric. But unlike a ferroelectric, where this current is transitory and ends upon the total polarization of the sample, in a nonanalytic CDW compounds, this polarization current reaches a stationary regime because there is no limitation to the increase of the phase of the CDW. This mechanism requires an electric field threshold which is physically accessible in certain compounds only. It is briefly described here (see further details in [1]).

2. The Transition by Breaking of Analyticity (TBA) in the Discrete Frenkel—Kontorova (FK) Model

In this model, in which the TBA was first discoverred, a simple intuitive understanding of the transition is easier than in more complex models for CDW. A detailed description of this model is useful because the physical behavior of Peierls chain will appears to be *qualitatively the same*. This model can be derived from phenomenological models for incommensurate structures where only phase fluctuations of the order parameter are taken into account and where the amplitude fluctuations are neglected. It has a simple representation as a discrete elastic chain of atoms i located at the abcissa u_i and submitted to a periodic potential with energy

$$\Phi(\{u_i\}) = \sum_i [W(u_{i+1} - u_i) + k.\, V(u_i) - \mu.\,(u_{i+1} - u_i)]. \tag{6a}$$

The elastic constant of the harmonic interaction potential

$$W(x) = \tfrac{1}{2}x^2 \tag{6b}$$

is unity by an appropriate choice of the energy unit. The amplitude of the sine potential

$$V(x) = (1 - \cos x) \tag{6c}$$

is k. μ is a constant which in the absence of periodic potential ($k = 0$) fixes an atomic mean distance

$$l = \lim_{N - N' \to \infty} \frac{u_N - u_{N'}}{N - N'} \tag{7a}$$

which can be either commensurate or incommensurate with 2π, or in other words that the incommensurability ratio

$$\zeta = \frac{l}{2\pi} \tag{7b}$$

is either a rational or an irrational number.

The 2π-periodic potential in (6) represents the physical force from which the incommensurate modulation originates. In particular, within a naive (non-self-consistent) approach of models for Peierls chains, it can be considered as the potential resulting from the Fermi electron gas which produces the lattice distortion. The amplitude k of this periodic potential corresponds to the electron—phonon coupling. Then, the parameter ζ corresponds to the ratio of the period l of the lattice to the half-Fermi wavelength $2\pi = l/c$. In other words, it is the concentration c of electronic pairs per site.

The rigorous aspects of the theory of the ground states and metastable states of the FK model are rather sophisticated and take into account all possible cases including pathological and exceptional situations [15]. They are connected to the theory of nonintegrable Hamiltonian dynamical systems which may exhibit both Kolmogorov—Arnol'd—Moser (KAM) trajectories and chaotic trajectories. We do not discuss this theory here, since it is explained in detail elsewhere [8]. We essentially focus on the description and the physical interpretation of the results which are obtained in the most general case.

It is proved that the classical ground state of model (6) (obtained by the absolute minimization of the energy functional with the condition (7)), is always either 'commensurate' or 'incommensurate' (and never chaotic). In the absence of periodic potential ($k = 0$), the model is trivial and the atoms in the ground state are equidistant. The distance between subsequent atoms being fixed by the boundary condition (7a) to the arbitrary value l, the atom i is located at $il + \alpha$ where α is an arbitrary phase. This case corresponds to the unmodulated crystal structure.

When this potential is switched on, the atom i of the unperturbed chain, located at $il + \alpha$ is displaced at $il + \alpha + \delta_i$. It is then proved, for a given k and l, that the modulating function $g(x)$ defined by

$$\delta_i = g(il + \alpha) \tag{8a}$$

is well defined. This function which is called the *hull function of the modulation* with commensurability into ζ, determines the whole set of ground state configurations $\{u_i\}$ which are obtained *for all phases* α. In this FK model we also consider for convenience the *hull function of the atomic positions* defined by

$$f(x) = x + g(x). \tag{8b}$$

The *analyticity* property of the hull function $f(x)$ (or equivalently $g(x)$) means that the coordinates of the configuration vary analytically with respect to *phase variations*. This mathematical property has important consequences on the physical properties of the ground state. In the incommensurate case, this function may be a usual smooth analytical function or a discontinuous funtion with infinitely many discontinuities (Do not confuse this curve with a devil's staircase, which is a continuous function) or possibly at the critical point between these two regimes a 'singular continuous function'. In the commensurate case, this function is a 'harmless' staircase that is a piecewise constant function with a discrete set of discontinuities.

In order to understand the TBA which occurs in the incommensurate case which is the most interesting one, let us first examine the simpler commensurate case.

2.1. COMMENSURATE GROUND STATES

When ζ is fixed rational equal to $\zeta = r/s$ where r and s are two irreducible integers, it is proven that the ground-state configuration $\{u_i\}$ fulfills for all i

$$u_{i+s} = u_i + 2r\pi \tag{9}$$

which means that the chain consists of a super unit cell of s atoms and size $2r\pi$, repeated periodically as for a usual commensurate crystal.

The *effective phase* ϕ takes into account the average phase shift due to the periodic potential and is defined as

$$\phi(\alpha) = \alpha + \frac{1}{s} \sum_{i=1}^{s} \delta_i. \tag{10}$$

This definition corresponds to the usual phase for a modulation considered in the litterature. In the commensurate case, it is *different* than the phase α defined in the absence of periodic potential ($k = 0$). The energy per unit cell of s atoms, of the commensurate configuration which minimizes the energy (6) with the constraints (9) and (10) is a function $V(\phi)$ of this effective phase ϕ. This *pinning potential* $V(\phi)$, due to the discreteness of the lattice, has no reason to be a constant, and so it is usually found not to be a constant.

The transformations g_0 and g_1 defined by $g_0(\{u_i\}) = \{u_i + 2\pi\}$ and $g_1(\{u_i\}) = \{u_{i+1}\}$ map the configuration space onto itself and let the energy (6) be invariant. It is found that these transformations change the phase ϕ of the commensurate configurations into $\phi + 2\pi$ and $\phi + 2\pi r/s$, respectively. Consequently, $V(\phi)$ is periodic with both periods 2π and $2\pi r/s$ so that $V(\phi)$ has the period $2\pi/s$ because r and s are irreducible integers. The commensurate ground state of model (6) is obtained by minimizing $V(\phi)$ with respect to the phase ϕ which then is fixed modulo $2\pi/s$ to a series of equidistant discrete values. Because, in general, $V(\phi)$ is not constant, the commensurate ground states are pinned by the lattice.

The commensurate ground states have s possible phase determinations modulo 2π of ϕ which are physically equivalent. They are obtained from the s possible labellings of the first atom of the super unit cell of s atoms (cf. Figure 2). Because the initial phase α is a continuous variable, the effective phase ϕ of the ground state (and also the coordinates $u_i(\alpha) = f(il + \alpha) = il + \alpha + g(il + \alpha)$ of the corresponding configuration $\{u_i(\alpha)\}$) varies as a function of α as a piecewise constant staircase with equal step widths $2\pi/s$. In this simple commensurate case, the hull function $f(x)$ (cf. Figure 3), which is a harmless staircase, is *not analytic*.

This property is related to the physical property of *phase defectibility*. It can be proved that there exists configurations which are identical to a commensurate ground state for $i \rightarrow -\infty$ and for $i \rightarrow +\infty$ but with different phases ϕ^- and ϕ^+ respectively. In some finite region of the chain, these configurations are different of a commensurate ground state and exhibit a localized 'defect' (cf. Figure 4). They are called phase defects or *discommensurations* [83]. These defects are topologically stable only when the phase shift $\delta = \phi^+ - \phi^-$ is equal to $+2\pi/s$ (elementary *advanced* discommensuration) or $-2\pi/s$ (elementary *delayed* discommensuration). This topological stability means that it is impossible to suppress such a defect (or to decay the energy of the system) by any finite displacement of a finite number of atoms. For larger phase shifts $\delta = \phi^+ - \phi^- = m2\pi/s$ where m is an integer and $|m| > 1$, the system always gains energy in splitting the phase defect far apart into m elementary discommensurations, which are advanced or delayed according to the sign of m.

These discommensurations are often described as sine-Gordon solitons [83] but it is the result of an approximation where the discreteness of the lattice is neglected. This approximation is only valid in the limit where the solitons are thick compared to the size s of the unit cell of the commensurate ground state. In fact, in a discrete lattice, an advanced (or delayed) discommensuration appears as the jump of a series of atoms from a well of the periodic potential to the next one (cf. Figure 4). When the height of the potential barriers becomes large enough, it is clear, that the discommensurations become pinned by the periodic potential. The minimum of the energy which must be provided for moving the discommensuration is the Peierls—Nabarro (PN) energy barrier E_{PN} of this considered discommensuration. E_{PN} decays exponentially as a function of the inverse of the discommensuration thickness when this one diverges [2].

The finite PN energy for the discommensurations has the consequence that there exists *chaotic metastable configurations* which are obtained from the ground

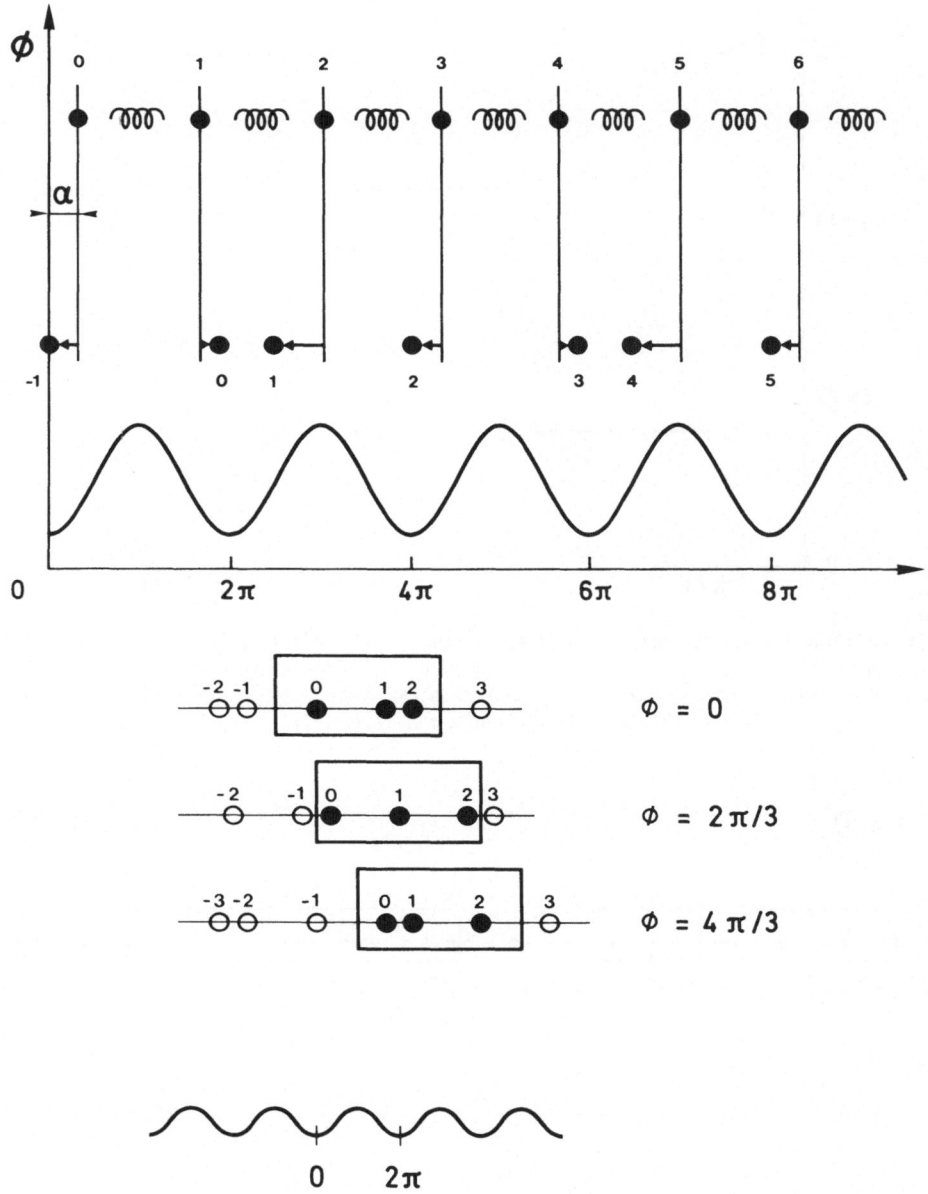

Fig. 2. Ground state configurations of the FK chain in a commensurate case with $\zeta = 2/3$. The unperturbated chain is shown above with equidistant atoms. The phase α is the coordinate of the atom 0. The same chain is shown below after relaxation in the sine potential. Each atom has been displaced toward the bottom of the well in which it was initially. Note that the effective phase ϕ which takes into account the average displacement of the atoms, is different of α. The inset shows the three possible unit cells modulo 2π of this commensurate ground state which are obtained for different values of α. They correspond to different well occupancy the three possible values modulo 2π of the effective phase ϕ.

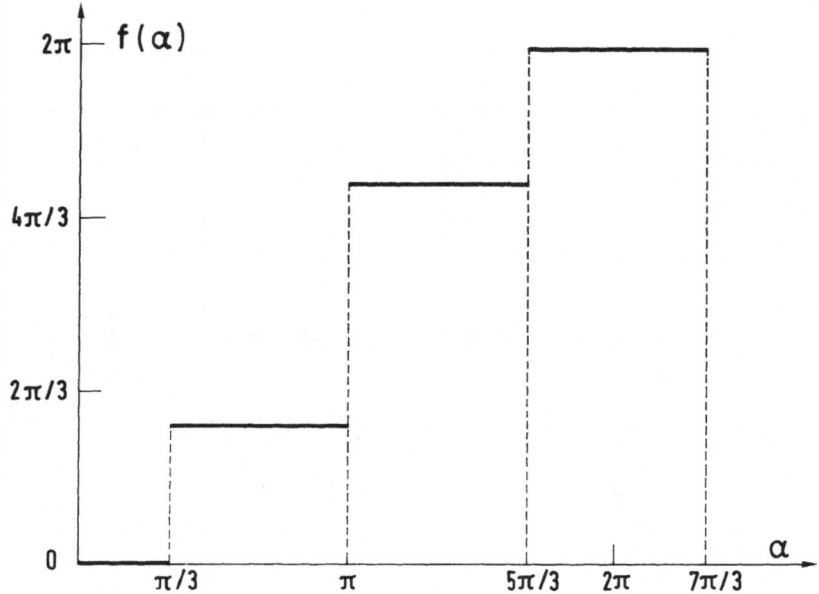

Fig. 3. Hull function $f(\alpha)$ of the commensurate ground state of Figure 2. It is given by the position $u_0(\alpha) = f(\alpha)$ of the atom 0 as a function of the initial phase α. This curve $u_0(\alpha)$ which is constant in each of the intervals $-\pi/3 < \alpha < \pi/3$, $\pi/3 < \alpha < \pi$, $\pi < \alpha < 5\pi/3$, etc. is a 'harmless' staircase.

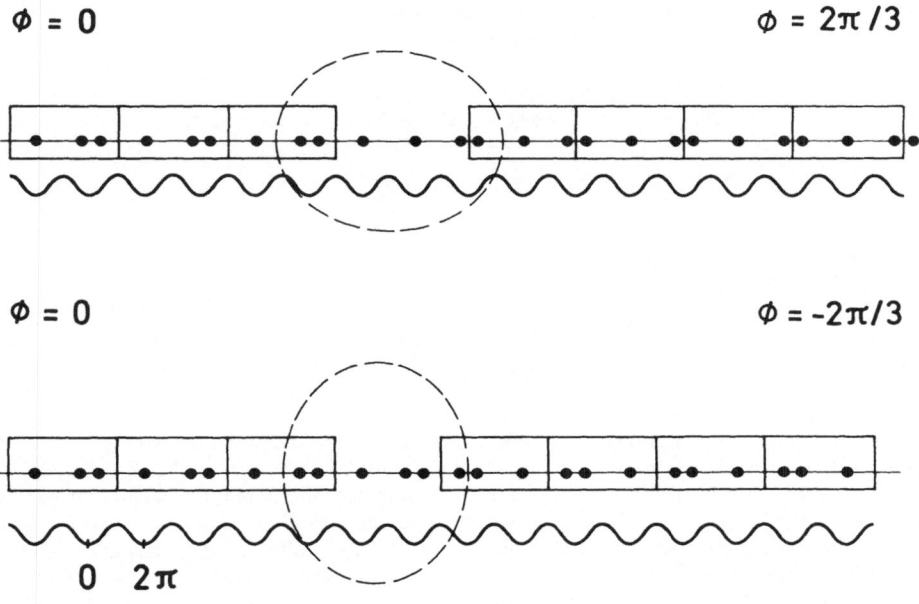

Fig. 4. Configurations of an advanced elementary discommensuration (a) and of a delayed elementary discommensuration (b) of the commensurate ground state shown on Figure 2.

state by a random distribution along the chain of advanced and delayed discommensurations with equal concentrations. Since these discommensurtaions are pinned by the periodic potential, if there are distributed sufficiently far apart, this configuration may remain locally stable (see Figure 5). Infinitely many such chaotic metastable configurations exist and fulfill the same boundary condition (7a).

Another property related to the nonanalyticity of the hull function $f(x)$ is the existence of a gap for the phason excitations. The dynamical motion equation of the FK model is

$$v_i''(t) = -\frac{\partial \Phi(\{v_i\})}{\partial v_i}$$

$$= v_{i+1}(t) + v_{i-1}(t) - 2v_i(t) - k \sin v_i(t) \tag{11a}$$

where a mass of unity has been assumed for the atoms. In the neighborhood of a commensurate ground state $\{u_i\}$, the small motion equation which yields the phason spectrum, is obtained by setting the dynamical coordinate $\varepsilon_i(t) = v_i(t) - u_i$ and by linearizing (11a) with respect to $\varepsilon_i(t)$

$$\varepsilon_i''(t) = \varepsilon_{i+1}(t) + \varepsilon_{i-1}(t) - 2\varepsilon_i(t) - k \cos(u_i)\varepsilon_i(t). \tag{11b}$$

The time Fourier transform of (6b) yields the eigenfrequency equation

$$\omega^2 \varepsilon_i = -\varepsilon_{i+1} - \varepsilon_{i-1} + 2\varepsilon_i + k \cos(u_i)\varepsilon_i \tag{11c}$$

where for convenience ε_i also represents the time Fourier transform of $\varepsilon_i(t)$. The eigenmode with the smallest frequency ω_G^2 determines the gap of the phason. For a commensurate structure, this gap (which is necessarily positive or zero because the ground state is stable) is finite and *nonzero*.

Another consequence of a nonanalytic hull function $f(x)$, is that a commensurate ground state exhibits a finite coherence length. Suppose that the position u_n of some arbitrary atom n is fixed at a position $u_n + \delta_n$ which is different from its equilibrium position in the ground state (for example due to an impurity or a defect), then the FK chain relaxes in a new minimum energy configuration, the positions of the other atoms will be displaced at $u_i + \delta_i$. This displacement δ_i behaves roughly as $\delta_n \exp(-\gamma |i - n|)$ where $\xi = 1/\gamma$ is the coherence length over which the local perturbation extends.

Since the commensurate ground state is pinned to the lattice, there exists *a finite minimum E_{PN}* for the extra energy which must be provided to the ground state for

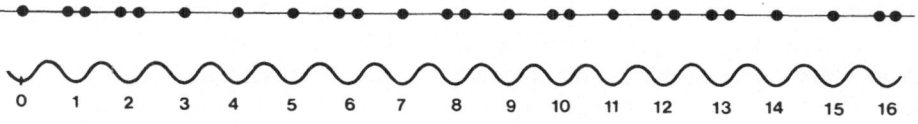

Fig. 5. A chaotic metastable configuration. The atoms are randomly distributed in the wells with a mean number $3/2 = 1/\zeta$ of atoms per well. It fulfills the same boundary conditions as the commensurate ground state of Figure 2 but it has more energy. It can be also viewed as a random distribution of advanced and delayed discommensurations shown in Figure 4.

moving its phase by $2\pi/s$ in order to overcome the energy barrier due to the lattice. This energy barrier is larger than the Peierls—Nabarro barrier of a single discommensuration. The path which requires the smallest extra energy for moving the phase of a commensurate structure by $2\pi/s$ consists in creating locally a pair of advanced and delayed discommensurations and in moving these discommensurations far apart. It is the nucleation process for the phase translation of the commensurate structure which requires the smallest extra energy. The Peierls—Nabarro energy barrier of the commensurate ground state is approximately the energy of a pair of advanced and delayed discommensurations plus their Peierls—Nabarro energy barrier.

A closely related physical quantity is the depinning field. Suppose that a constant field F is adiabatically applied to each atom of the FK chain in its ground state. For sake of simplicity, we assume that the atomic motion is overdamped with the equation:

$$\Gamma v_i'(t) = - \frac{\partial \Phi(\{v_j\})}{\partial v_i} + F \tag{12}$$

(Γ is the damping constant). There exists a critical field F_c such that for $F < F_c$, the atomic positions converge to a metastable configuration (under a finite field) and such that for $F > F_c$, there exist no limit configuration. Then, the FK chain reaches a dynamical stationary regime with a well defined average velocity. In the underdamped limit ($\Gamma = 0$) with finite atomic mass, one would also find a critical force which is roughly a half the previous one.

In summary, we have exhibited in the simple commensurate case, important physical properties which are connected with the non analyticity of the hull function. These are: 1 — the phase defectability, 2 — a finite phason gap; 3 — a finite coherence length; 4 — a finite Peierls—Nabarro barrier; and 5 — a finite depinning field. The same properties studied in the incommensurate case will exhibit a critical behavior at the transition by breaking of analyticity.

2.2. INCOMMENSURATE GROUND STATES

In the simpler theories, it is usually believed that the physical quantities mentioned above always vanish for incommensurate structures. This result is found within approximations which either neglect the discreteness of the lattice or only take incomplete account of it.

In contradiction with these usual theories, it is rigorously provided for the FK model (and extensions) (see e.g. [8 and 16]) and numerically found in other models ([17—21]), that there exists a critical value $k_c(\zeta)$ for the amplitude k of the periodic potential responsible for the incommensurate modulation. For $k > k_c(\zeta)$, *despite the fact that the ground state is incommensurate, many properties of commensurate structures are preserved.* The physical quantities above discussed for the commensurate ground states, are nonzero and their magnitude is comparable to those obtained for the lowest order commensurability.

The transition by breaking of analyticity occurs at $k = k_c(\zeta)$ where these physical quantities are critical. These remain constant and zero below the transition for $k < k_c(\zeta)$. In this regime only, we find again the qualitative phenomenology of incommensurate structures predicted by the standard theories.

The signature of the TBA is given by the analytical properties of the hull functions defined by (8) which describe the whole configuration. At the incommensurate limit ($s \to \infty$), the average the atomic displacement $\langle \delta_i \rangle$ becomes zero. This result comes out from global symmetry considerations. Then, the hull function $f_\zeta(x)$ defined in (8b) is the limit of the staircase hull functions $f_{\zeta_r}(x)$ (as shown in Figure 3) obtained for $\zeta_r = r/s \to \zeta$. This limit is proved to exist and to remain a monotonous increasing function.

This limit $f_\zeta(x)$ may be a smooth continuous (and analytical) function. In that case it is unique. The mathematical proof of this result is a consequence of the Kolmogorov—Arnol'd—Moser theorem [8]. Then we observe that the amplitudes of the discontinuities of $f_{\zeta_r}(x)$ vanish when $\zeta_r = r/s \to \zeta$. But $f_\zeta(x)$ may also be a discontinuous function. In that case, there exist two possible determinations of this function: it can be either left continuous or right continuous. In numerical experiments, we obesrve that the amplitudes of the largest discontinuities of the staircases $f_{\zeta_r}(x)$ do not converge to zero. Since the number s of discontinuities of $f_\zeta(x)$ in an interval, the width of which is 2π, diverges when $\zeta_r = r/s \to \zeta$, $f_\zeta(x)$ has infinitely many discontinuity points (however, this set is countable).

Figure 6, which has been calculated in [22], shows two examples of hull functions $f(x)$ corresponding to the first case and to the second case respectively. The basic principle of the numerical method consists in performing numerically the integration of the equation:

$$\frac{\partial u_i(\tau)}{\partial \tau} = - \frac{\partial \Phi(\{u_j(\tau)\})}{\partial u_i} \tag{13a}$$

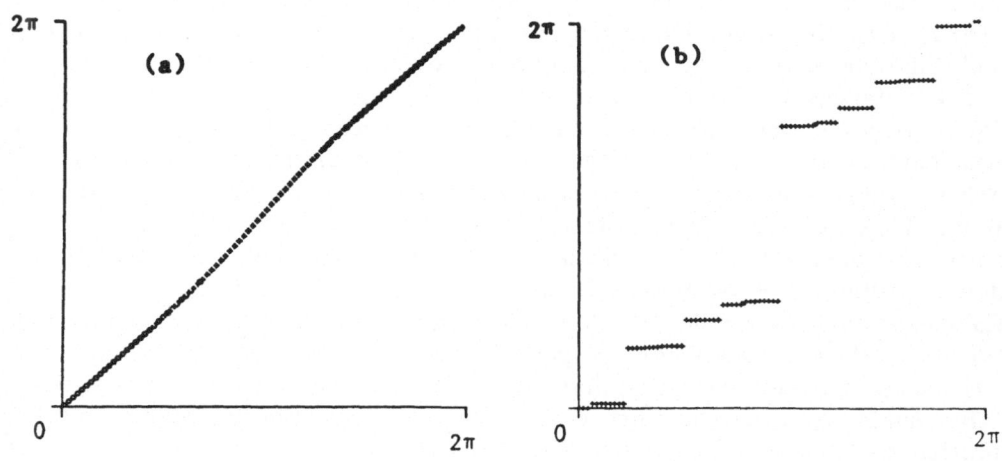

Fig. 6. Two hull functions $f(x)$ as for Figure 3 for $k = 0.4935$ (a) and $k = 1.9739$ (b) but for the rational commensurability ratio $\zeta_r = 34/89$. This rational which is a close approximation of the irrational $\zeta = (3 - \sqrt{5})/2$ practically yields the hull function in this incommensurate limit. Note that the first hull function looks smooth while the second hull function exhibits many large discontinuities (from [22] Figure 1).

τ is a fictional time. Taking the unperturbed chain configuration as the initial configuration $\{u_i(0) = il + \alpha\}$, the limit configuration $u_i(\infty)\} = \{u_i\}$ which is obtained for large enough τ fulfills the equation of stationarity

$$0 = -\frac{\partial\Phi(\{u_j\})}{\partial u_i}$$

$$= u_{i+1} + u_{i-1} - 2u_i - k\sin u_i. \tag{13b}$$

As a consequence of the exact results obtained in [15], the configuration which is obtained by this method is the ground state and not a metastable state. It corresponds to the unperturbed chain which has been relaxed in the periodic potential as shown in Figure 2. In practice, this numerical technique has the advantage of being safe but it converges rather slowly in the neighbourhood of the limit configuration, particularly in the critical regions. Therefore, when the convergence rate slows down, we terminate the convergence process to the required precision by using a Newton method [23, 24] for finding the zero points of Equation (13b). However, this method, which converges very rapidly, can only be used close to the solution which we wish to reach. If it is used at the beginning of the convergence process, this algorithm could also yield an unstable configuration which corresponds to a solution of (13b).

For obvious practical reasons, because we can only study finite-sized systems, these calculations were done in commensurate cases with $\zeta = r/s$ with a large enough commensurability ratio s (for the example $\zeta = 34/89$ of Figure 6). Although a TBA cannot exist in principle in these cases, in practice a sharp crossover is numerically observed around a fairly well defined value of k: $k_{co}(\zeta_r)$. For sufficiently high order commensurability (typically $s \geqslant 10$) and for k below $k_{co}(\zeta_r)$, the physical quantities such as the phason gap, the coherence length etc. almost vanish and within the numerical accuracy are not distinguishable from zero while for k above $k_{co}(\zeta_r)$ they have finite values which can be accurately measured.

The commensurability ratio s turns out to play the role of the finite size of a system for usual phase transition. Indeed, finite size scaling theories based on this idea, have been developed for the transition to stochasticity in the dynamical system (standard map) which is associated to this FK model and is closely related to the TBA [25, 26]. The existence of the TBA can be understood just by considering the position of a single atom of the FK chain. When ζ is irrational, the atomic positions $il + \alpha$ modulo 2π are uniformly dense on the interval $[0, 2\pi]$. There exist atoms which lie very close to a maximum of the periodic potential. Let us consider the atom which for the uperturbed chain is exactly at a maximum of the periodic potential (for example with $\alpha = \pi$, $u_0(0) = \pi$). Two situations may occur.

For small k, the elastic force coming from the neighbouring atoms may be sufficient for maintaining this atom 0 after relaxation at this maximum $u_0 = \pi$ (cf. Figure 7a). In this situation, it is observed that the atomic positions u_i modulo 2π remain dense everywhere on the interval $[0, 2\pi]$ and that the hull function of the incommensurate structure is necessarily continuous.

By contrast, for large k, the atom 0 cannot stay at this maximum. Because of

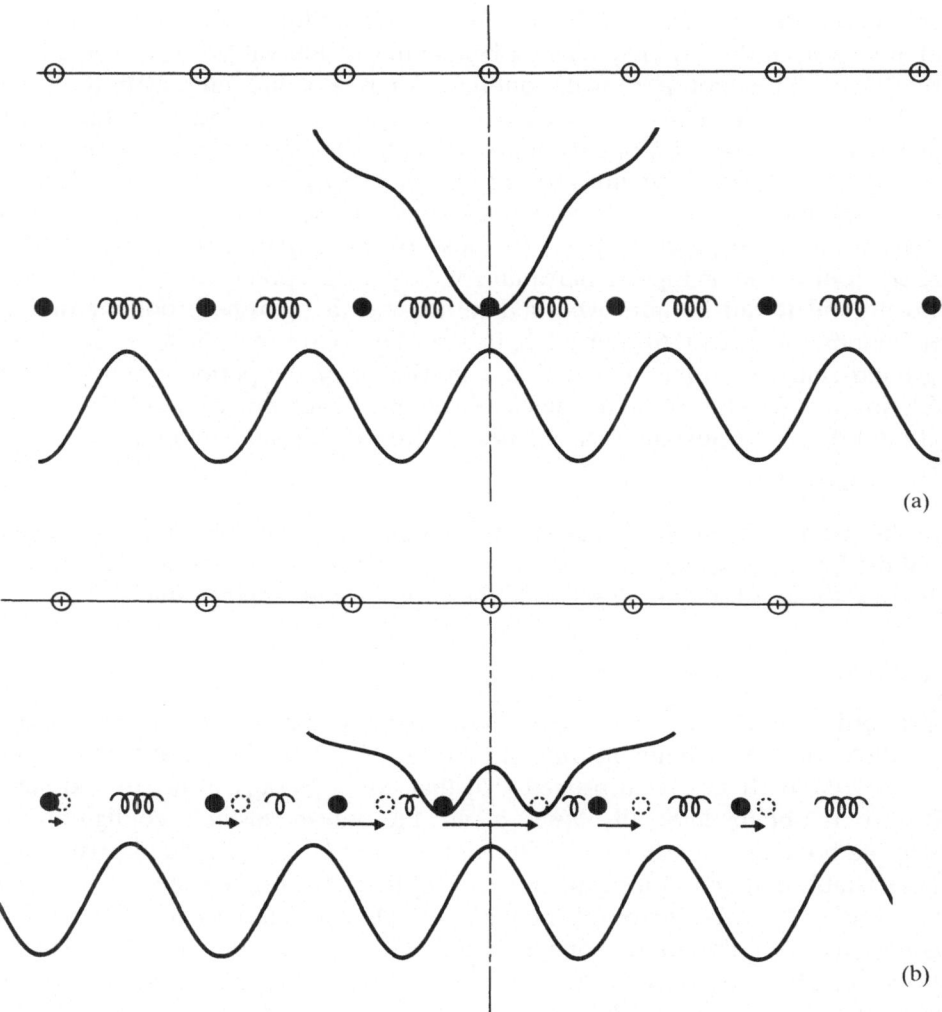

Fig. 7. Scheme showing the breaking of analyticity. On Figure 7a the atom which is at the top of the periodic potential for the unperturbed chain, remains at the top after relaxation. The resulting potential which is the sum of the elastic potential (with fixed neighbours) and of the periodic potential is single well when k is small enough. On Figure 7b, the atom which is at the top of the periodic potential for the unperturbed chain may relax either onto the left side (black dots) or onto the right side (open dots). If the atomic position of this atom is fixed in between these two positions, the energy of the ground state is increased and reaches its maximum when this atom is fixed at the top. The height of the potential barrier shown on the figure is the Peierls—Nabarro energy barrier of the incommensurate ground state.

symmetry considerations, it can relax either to the right at $\pi + \Delta$ or to the left at $\pi - \Delta$ (cf. Figure 7b). Consequently, there exist *two ground states with the same phase* α which either correspond to the right continuous determination or to the left continuous determination of the hull function. In addition, the position modulo 2π of any atom of the FK chain cannot belong to the interval $]\pi - \Delta, \pi + \Delta[$. As a result, the hull function $f_\zeta(x)$ is discontinous for $x = \pi$ and jump from the value $\pi - \Delta$ to $\pi + \Delta$. This parameter $\Delta(k)$ has been considered as the order parameter of the TBA by Coppersmith *et al.* [23]. This discontinuity implies that $f\zeta(x)$ is also strictly discontinuous for all phases $x = \alpha_{m,n} = \pi + ml + n2\pi$ and therefore has infinitely many discontinuities on a dense set of points (cf. Figure 7b). These phases $\alpha_{m,n}$ correspond to those of the unperturbed configurations where an atom just sit on top of a maximum of the periodic potential.

This argument can be improved and turned into a rigorous proof for finding upper bounds for $k_c(\zeta)$ [16]. For all ζ, it is readily found that for $k > 2$ and for any ground state, no atom can stand at a maximum of the periodic potential. By considering the stability of larger blocks of atoms, better exact bounds have also been found. The accurate numerical studies by Greene [27] yield for all ζ:

$$\text{Sup}_\zeta \, k_c(\zeta) \approx 0.9716. \tag{14}$$

For the FK model, the inverse golden mean number $\zeta_g = (\sqrt{5} - 1)/2$ is supposed to yield the largest possible value $k_c(\zeta_g) \approx 0.9716$. (But note, because of the model symmetries, for $\zeta = 1/\zeta_g$ and $\zeta = 1 - \zeta_g$, the critical values are the same.)

2.3. CRITICAL BEHAVIOR AT THE TBA

The properties which have been considered for the commensurate structures are also studied for the incommensurate structures. The most important one is the *phase defectibility*. It has been proved [16] that, for $k < k_c(\zeta)$, the ground state configurations obtained for all phases α, are the only metastable configurations fulfilling the boundary condition (7). On contrary, for $k > k_c(\zeta)$, there exist many other metastable configurations with more energy than the ground state.

This property is easy to check numerically. One just has to find metastable configurations with the same boundary conditions (7) as the incommensurate ground state. The procedure is the same as for finding the ground state, but the initial configuration which is chosen in Equation (13a) is arbitrarily random. The limit configuration which is obtained may be either the ground state or another configuration which is a local minimum of the functional energy (6), which is a metastable configuration.

For $k < k_c(\zeta)$, we found that, whatever the initial configuration, the limit configuration which is obtained is *always an incommensurate ground state*, the phase α of which only depends on the initial configuration. By contrast, for $k > k_c(\zeta)$, we found that *there exists many other metastable configurations* which have more energy than the incommensurate ground state. These configurations have been described by Peyrard *et al.* [22] as the incommensurate ground state with a random distribution of phase defects or nonelementary discommensurations (cf. Figure 8).

Since for commensurate configurations of order s the phase shift associated with the phase defects is at most $1/s$. (In that case, they are elementary discommensurations), the minimum phase shift associated with these phase defects

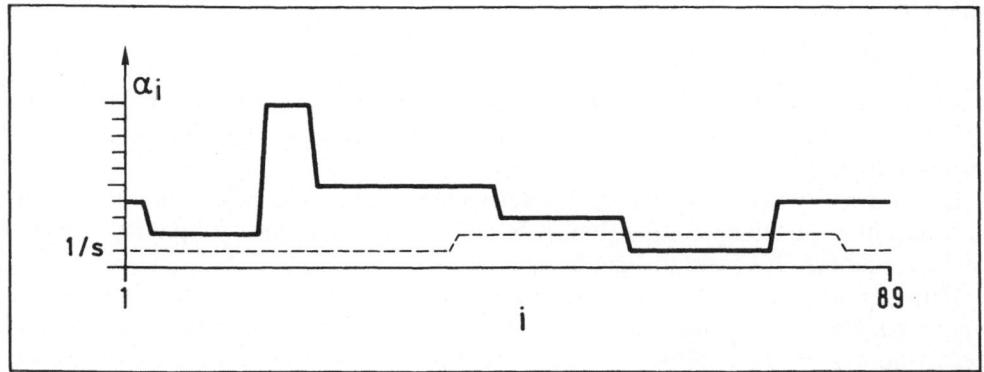

Fig. 8. Variation of the phase α_i for two metastable configurations $\{v_i\}$ obtained for $k > k_c(\zeta)$. The incommensurability ratio is $\zeta = 34/89 \approx (3 - \sqrt{5})/2$, $k = 1.9739$. The phase α_i is defined at site i by $v_i = f(il + \alpha_i)$. One of the two configurations (broken line) exhibits two discommensurations with phase shift $1/s$. The other configuration exhibits 6 discommensurations which corresponds to larger phase shifts of various amplitudes.

becomes arbitrarily small ($s \rightarrow \infty$) in the incommensurate limit. For a zero phase shift, the defect energy becomes strictly zero. Indeed, we have seen above that for non-analytic ground states and for a countable and dense set of phase α, there exists two ground states. Each of them can be considered as the zero energy elementary discommensuration of the other.

Otherwise, the phase shift δ cannot be too large in order that the corresponding phase defect remains metastable. If it is too large, it could spontaneously splits into several discommensurations with smaller phase shifts, the total phase shift being the same. Although, they clearly are not elementary discommensurations, the density of state of these discommensurations as a function of their energy has been studied in details in [28, 29] for a precise analysis of the thermodynamic properties of the nonanalytic one-dimensional FK model. The support of this density of state close to the energy zero is a zero-measure Cantor set with infinitely many gaps which are determined by the coefficients of the continued fraction expansion of ζ. In addition, these discommensurations have been proved to be the relevant low-lying energy excitations of the nonanalytic FK chain which determine its low temperature behavior. This physical behavior sharply contrasts with the analytic regime where the low-lying energy excitations are the phonons associated with the phase fluctuations (phasons).

Let us now examine the phonons excitations of the incommensurate ground state. The hull function $f(x)$ necessarily fulfills the functional equation obtained from (13b) by substitution of (8)

$$f(x + l) + f(x - l) - 2f(x) - k \sin f(x) = 0 \qquad (15a)$$

When $k < k_c(\zeta)$, this equation can be differentiated with respect to x and yields

$$f'(x + l) + f'(x - l) - 2f'(x) - k \cos f(x) f'(x) = 0 \qquad (15b)$$

which shows that

$$\varepsilon_i = f'(il + \alpha) \tag{15c}$$

is a time independent solution of the small motion Equation (11b) or (11c) with $\omega = 0$. This static mode is the phason mode. Consequently, the phason gap ω_G is zero (see Figure 9a).

By contrast, when $k > k_c(\zeta)$, the derivative of $f(x)$ which has infinitely many discontinuities, *is always zero* when it is defined [16]. No zero frequency solution can be found for (11c) and the gap ω_G is finite.

This gap as a function of k, which has been numerically calculated, is shown in Figure 10. For $\zeta = \zeta_g$, ω_G behaves as $(k - k_c(\zeta))\chi$ in the upper neighborhood of $k_c(\zeta)$ where $\chi \cong 1.02$. However, this behavior is *not universal* in the usual sense. This critical exponent is known to be well defined only when the integer coefficients n_i of the continued fraction expansion of the incommensurability ratio ζ becomes periodic at large i which is equivalent to the property that ζ is the root of an algebraic equation of degree 2 with integer coefficients (quadratic numbers). More precisely, there exist p and i_0 such that for $i > i_0$, $n_{i+p} = n_i$ in the expansion:

$$\zeta = n_0 + \cfrac{1}{n_1 + \cfrac{1}{n_2 + \cfrac{}{\ddots \quad n_{i-1} + \cfrac{1}{n_i + \zeta_i}}}} \tag{16}$$

The critical exponent depends on the sequence of integers $\{n_i, n_{i+1}, \ldots, n_{i+p}\}$ of this period ([25, 26, 30, 31, 32]). It has been numerically found that this critical exponent fluctuates very little (a few %) when ζ varies. When, ζ is not a quadratic number, it might be possible that a unique critical exponent could be defined for most the irrational numbers but this is not proven.

Because the phason gap vanishes for $k < k_c(\zeta)$, the inverse coherence length $\gamma(k)$ which is defined as for the commensurate case, also vanishes in the same region, as shown Figure 11. In the upper neighborhood of $k_c(\zeta)$, $\gamma(k)$ behaves as $(k - k_c(\zeta))^\nu$ with a critical exponent ν ($\nu \cong 0.9874$ for $\zeta = \zeta_g$) which is only well defined when χ also is well defined. The Peierls–Nabarro energy barrier E_{PN} and the depinning field F_c defined as for the commensurate case also vanish for $k < k_c(\zeta)$ and are finite above $k_c(\zeta)$.

This PN barrier is found to be the (finite) difference between the energy of the ground state and the energy of the configuration which minimizes the energy of the chain with an atom fixed on top of the periodic potential at $u_0 = \pi$ (the same boundary conditions are fulfilled). The depinning field F_c is also found when the atom which sits the closest to a top of this periodic potential is moved at this top by the external field. $E_{PN}(k)$ and $F_c(k)$ behaves as $(k - k_c(\zeta))^\psi$ and $(k - k_c(\zeta))^\psi$ where ψ and $\boldsymbol{\psi}$ are well defined critical exponents ($\psi \cong 3.0117$ and $2.85 < \boldsymbol{\psi} < 3.05$ for $\zeta = \zeta_g$) again only when ζ has a periodic continued fraction expansion (cf. Figure 12). Then we found the scaling relation:

$$2\chi + \nu = \psi = \boldsymbol{\psi}. \tag{17a}$$

ENERGY

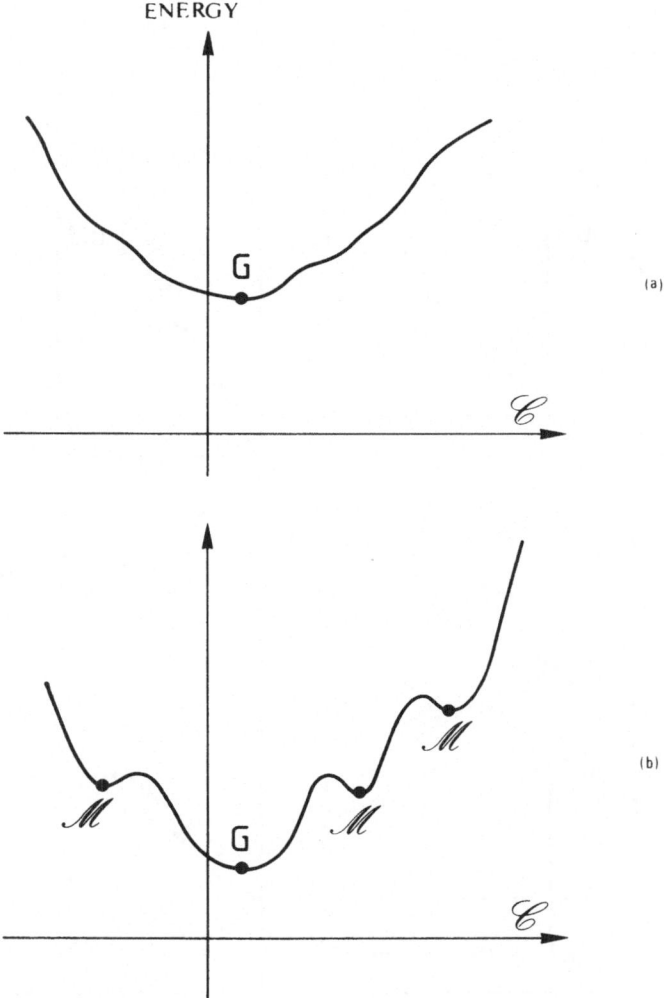

(a)

(b)

Fig. 9. Schemes of the functional energy (6) versus the 'configuration'. This infinitely many dimensional 'variable' is symbolically represented as a one-dimensional variable. When the hull function of the ground state is analytic ($k < k_c(\zeta)$) this functional has a single minimum (Figure 9a). When it is nonanalytic ($k > k_c(\zeta)$), there exists infinitely many other local minima which correspond to excited configurations with discommensurations (Figure 9b).

Far above $k_c(\zeta)$, the atoms lie close to the periodic potential minima and then it is easy to find the behavior $\omega_G(k) \cong \sqrt{k}$, $\gamma(k) \cong \ln k$, $E_{PN}(k) \cong 2k$ and $F_c(k) \cong k$ which shows that these pinning quantities tend to be independent of the incommensurability ratio ζ *whether it is rational or not.*

There also exists physical quantities which are finite below the critical value $k_c(\zeta)$ and vanish above. The elastic constant $C(k)$ is defined as the second derivative of the mean energy per atom $\Psi(\zeta)$ as a function of the incommen-

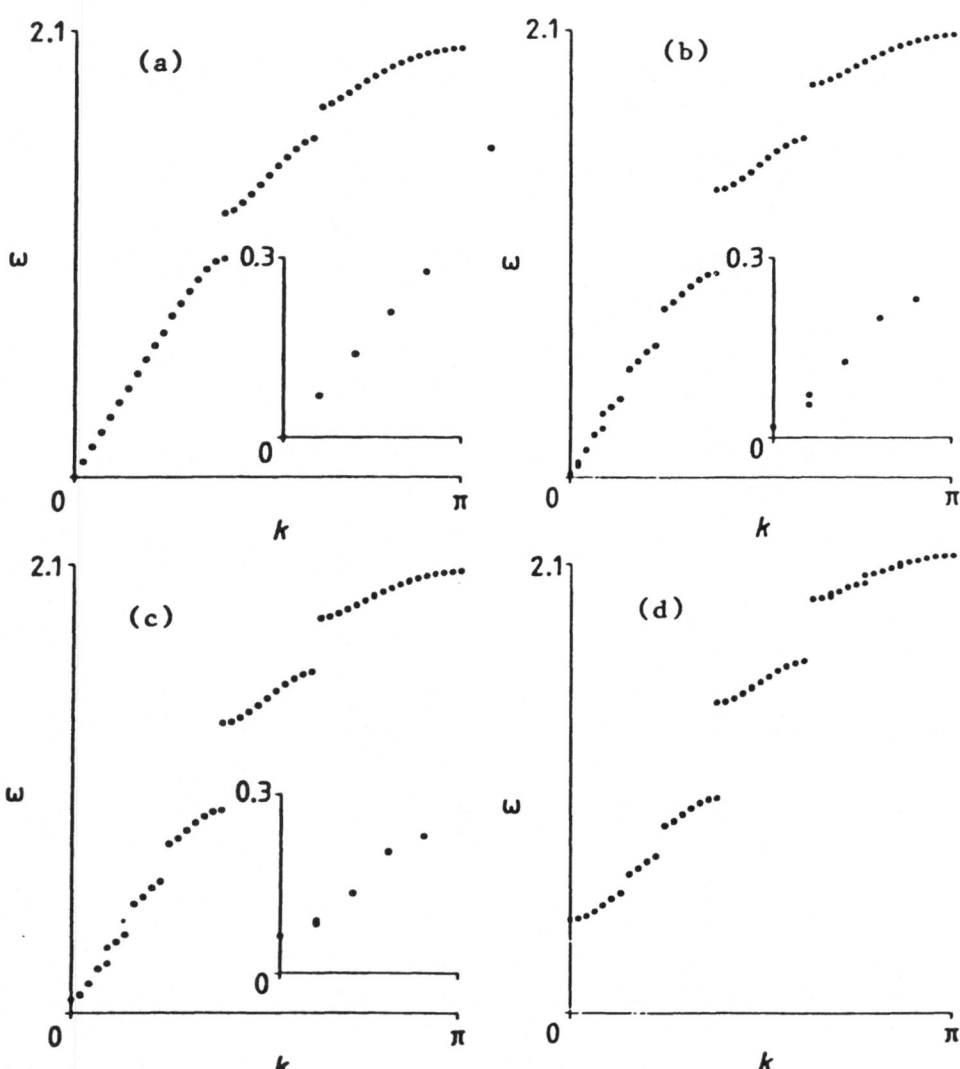

Fig. 10. Phonon eigenfrequencies of the FK chain versus the density of nodes of the corresponding eigenmode for $\zeta = 34/89 \approx (3 - \sqrt{5})/2$ and for four values $k = 0.4935, 0.9870, 1.0363$ and 1.4804 shown on Figures a, b, c, d respectively. (In 1D nearest neighbour systems, the density of nodes is equal to the wave vector of the eigenmode when it is defined). The insets show enlargements of the curves at the origin. For $k = 0.4935$, the gap ω_G remains practically zero and as soon as $k > k_c(\zeta) \cong 0.9716$, the gap clearly becomes finite.

surability ratio ζ (Figure 13). This derivative is proved to be only defined for ζ irrational (it is infinite for ζ rational because the response of a commensurate configuration to an external force is not linear but consists in the creation of elementary discommensurations or 'solitons'). For $k = 0$, $C(k)$ is equal to 1 which

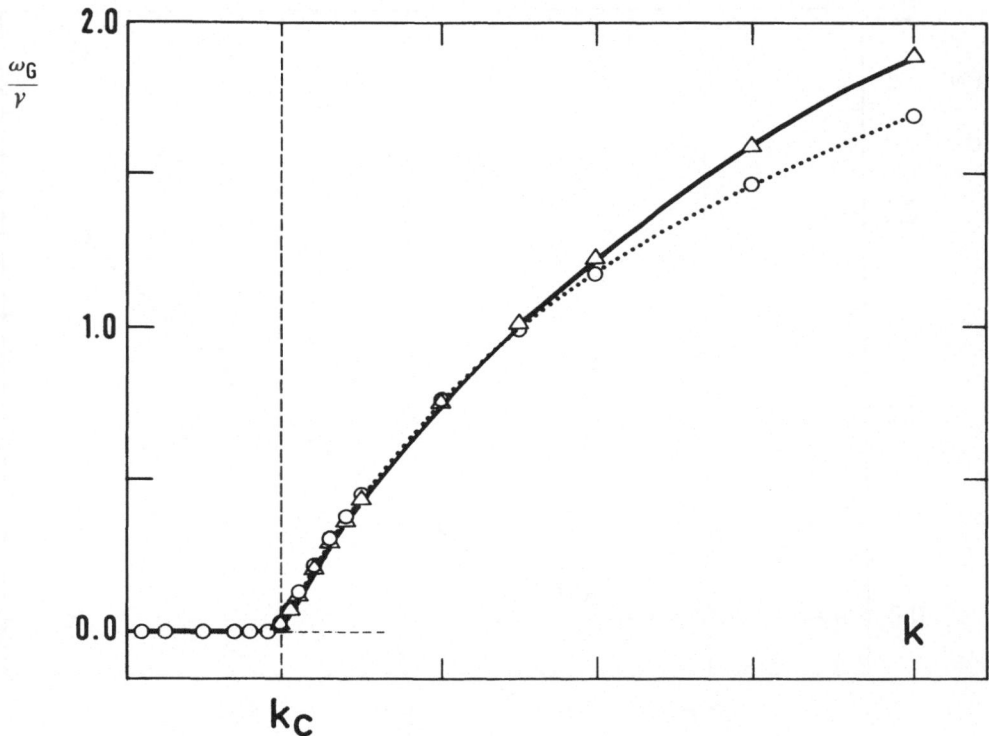

Fig. 11. Variation of the gap (dotted line) and of the inverse coherence length (full line) versus k for $\zeta = 34/89 \cong (3 - \sqrt{5})/2$. These quantities vanish for $k < k_c(\zeta)$ and are critical for $k \to k_c(\zeta)$ with $k > k_c(\zeta)$ ([22]).

is the elastic constant of the unperturbed chain (see (6)). Figure 12 from [24] shows the variation of $C(k)$ as a function of k. It is found that $C(k)$ decays as a function of k for $k < k_c(\zeta)$. It has a sharp but continuous variation for k close to $k_c(\zeta)$ and vanishes at $k = k_c(\zeta)$. Let us note that this elastic constant was studied earlier by Pokrovsky [33]. He did indeed find $C(k)$ as a decaying function of k but he also found that $C(k)$ should become negative beyond a certain critical value of k. Then he concluded that the incommensurate structure should become unstable for large values of k, which is a wrong result, as we have proved.

For $k > k_c(\zeta)$, $C(k)$ remains strictly zero but the FK chain is not unstable because $\Psi(\zeta)$ always remains a convex function. This result only means that $\Psi(\zeta)$ cannot be expanded analytically in the neighborhood of any point which is a property related to the existence of a complete devil's staircase as shown in [24]. In fact, $\Psi(\zeta')$ can be approximate in the vicinity of an irrational number ζ by a smooth function with an essential singularity at $\zeta' = \zeta$ [29]. (Let us note that for the calculation of this elastic constant, we assume that, after a small change of length, the chain does not remains pinned in some metastable state, but relaxes to its minimum energy configuration.)

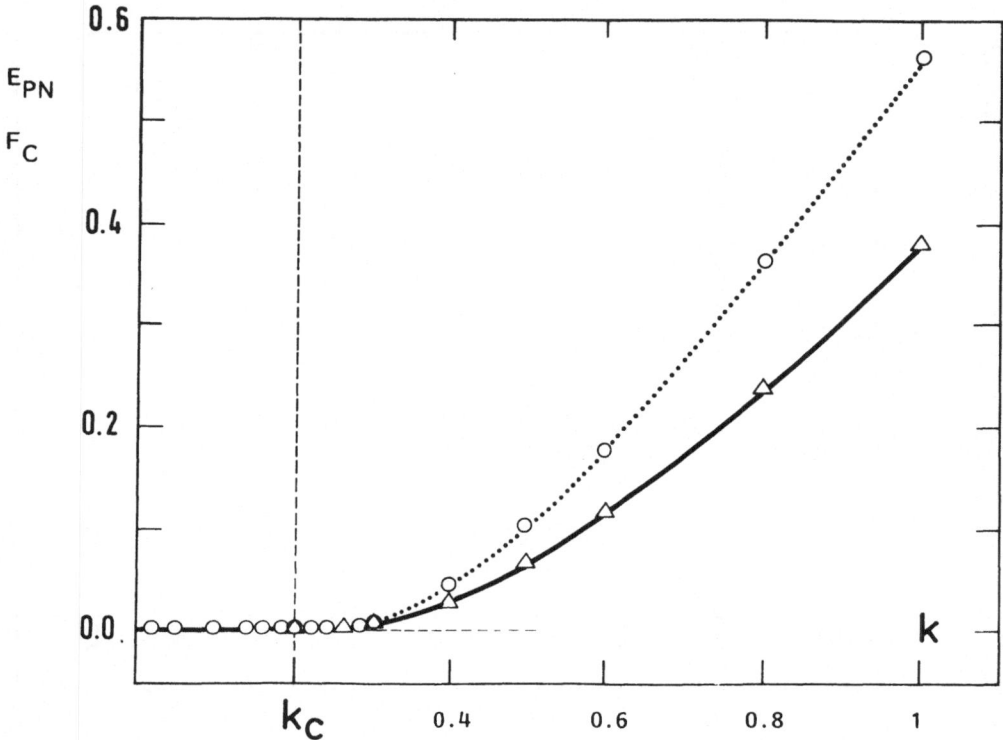

Fig. 12. Variation of the Peierls—Nabarro energy barrier $E_{PN}(k)$ (dotted line) and of the depinning field $F_c(k)$ (full line) versus k for $\zeta = 34/89$. These quantities vanish for $k < k_c(\zeta)$ and are critical for $k \to k_c(\zeta)$ with $k > k_c(\zeta)$ ([22]).

As for the other critical exponents, the critical behavior of $C(k) \approx (k_c(\zeta) - k)^\zeta$ determines a critical exponent ζ which is only well defined for an incommensurability ratio with a periodic continued expansion and depends on the sequence of its periodic coefficients. For the golden mean $\zeta_g = (\sqrt{5} + 1)/2$, we foudn $\zeta \cong 0.04927$ and for $\zeta = \sqrt{3} - 1$ and $\zeta = (\sqrt{3} - 1)/2$, we found $\zeta \cong 0.0985$, which is approximately double. A scaling law

$$\zeta = 2(\chi - \nu) \tag{17b}$$

relates this critical exponent below $k_c(\zeta)$ to other critical exponents above $k_c(\zeta)$ whatever the incommensurability ratio.

Another quantity which has some physical interest for the dynamical behavior of the incommensurate structure in the unpinned regime is the effective viscosity (unpublished results). For $k < k_c(\zeta)$ the FK chain can be translated by any finite and uniform external field F. The dynamical equation of the FK chain is (11a) where one assumes in addition that each atom i is submitted to a damping force proportional to its velocity $-\Gamma_0 \, du_i/dt$ (Γ_0 is the damping constant) and to the constant force F. For a slow transition at velocity v, the time evolution of the

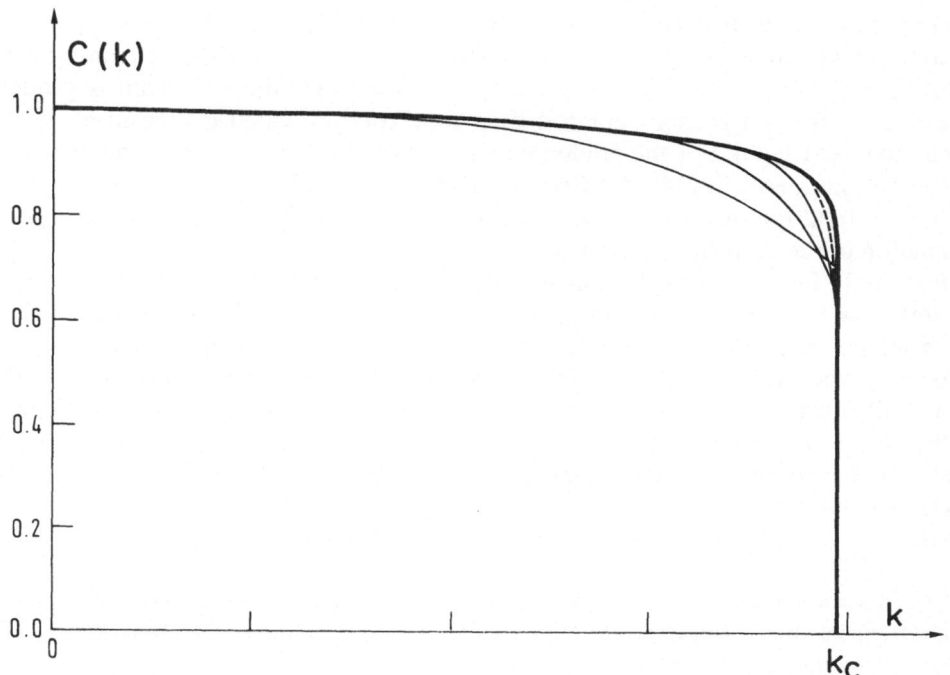

Fig. 13. Variation of the elastic constant $C(k)$ versus k for $\zeta = 34/89 \cong (3 - \sqrt{5})/2$. Unlike the critical quantities of Figures 10 and 11, this quantity vanishes for $k > k_c(\zeta)$ and is critical for $k \rightarrow k_c(\zeta)$ with $k < k_c(\zeta)$.

configuration $\{u_i(t)\}$ of the FK chain can be assumed to be stationary. Then, the chain motion can be described by a hull function $f_v(x)$ as

$$u_i(t) = f_v(il + vt + \alpha) \tag{18a}$$

where $f_v(x) - x$ is 2π-periodic and fulfills the functional equation

$$v^2 f_v''(x) = f_v(x + l) + f_v(x - l) - 2f_v(x) - \\ - k \sin f_v(x) - \Gamma_0 v f_v'(x) + F. \tag{18b}$$

The dissipated energy $U(v)$ per atom and per unit of time is obtained by multiplying both members of (18b) by $vf'(x)$ and by performing their integration for $0 < x < 2\pi$. Since most of the integrals vanish except two, one obtains

$$U(v) = Fv = \Gamma_0 \langle f_v'^2 \rangle v^2 \tag{18c}$$

where $\langle f_v'^2 \rangle$ is the average of $f_v'^2(x)$ over one period 2π. For $v \rightarrow 0$, the hull function $f_v(x)$ goes to the static analytic hull function $f(x)$ defined in (8). Then for small v, the effective damping constant of the chain is

$$\Gamma(k) = \Gamma_0 \langle f'(x)^2 \rangle. \tag{19a}$$

It increases as a function of k because of the growing of the harmonics of

$g(x) = f(x) - x$. It diverges at $k = k_c(\zeta)$ when the derivatives of $f(x)$ which is still continuous becomes zero for most values of x and infinite for the remaining zero measure set. $\Gamma(k)$ is not defined for $k > k_c(\zeta)$ because the FK chain is pinned. The viscous friction is then replaced by a solid friction which is measured by the depinning field $F_c(k)$. For the golden mean, we found by inspection of the available numerical data used in [24] that $\Gamma(k)$ calculated by (19a) behaves as $(k_c(\zeta) - k)^{-\eta}$ with $\eta \cong 0.029$. However, further specific calculations should be done for a good estimation of the accuracy of this result.

For analysing the critical behavior of the phase fluctuations (phasons), let us assume now that there is no damping and no external force ($\Gamma = F = 0$) in (18b). Then we get back to (11a). When for $k < k_c(\zeta)$, the FK chain is moving at low velocity v, the atomic motion $v_i(t)$ can be described by (18a) where $f_v(x)$ is the static hull function $f(x)$. The kinetic energy of the whole FK chain is then the sum of the atomic kinetic energies $1/2 \, m_0 v^2 \sum_i f'(il + vt + \alpha)^2$ (the atomic mass m_0 is unity). The average kinetic energy per atom becomes $1/2 \, m^*(k)v^2$ where the effective mass of the atoms

$$m^*(k) = m_0 \langle f'(x)^2 \rangle \tag{19b}$$

has been renormalised in the same way as the effective damping constant (19a). This effective mass diverges at $k_c(\zeta)$ (with the same critical exponent η as $\Gamma(k)$ when ζ is a quadratic number).

For $k < k_c(\zeta)$, in the vicinity of the zero frequency mode $\omega_G = 0$, a wave-vector q can be ascribed to each eigen solutions of (11c) with eigen frequency $\omega(q)$ despite the system is not translationally invariant. For small q, theses modes which essentially correspond to long wave length phase fluctuations, have a frequency $\omega(q)$ which is a linear function of the wave vector q. The slope $c_p(\zeta)$ defined by $\omega(q) \cong c_p(\zeta)q$ is the phason velocity. The effective atomic mass $m^*(k)$ and the effective elastic constant $C(k)$ above defined can be used for calculating $c_p(\zeta)$ because the corresponding phase variations are slow both in space and time. The atomic motions $v_i(t) = f(il + \alpha(i, t))$ define the local phase fluctuation $\alpha(i, t)$ but since the length scale of these phase fluctuations is supposed to be much larger than the lattice spacing, $\alpha(i, t)$ can be considered as a continuous function $\alpha(x, t)$ of space x and time t. Therefore, the energy of small amplitudes phase fluctuations is the sum of the space integral of the elastic energy $1/2 \, C(k) \, (\partial\alpha/\partial x)^2$ and of the space integral of their kinetic energy $1/2 \, m^*(k) \, (\partial\alpha/\partial t)^2$. These fluctuations are described by the continuous Hamiltonian $\int 1/2 \, m^*(k)(\partial\alpha/\partial t)^2 + 1/2 \, C(k) \, (\partial\alpha/\partial x)^2) \, dx$ which yields eigenmodes similar to acoustic phonons, which are here the phasons. The phason velocity is then

$$c_p(\zeta) = \sqrt{\frac{C(k)}{m^*(k)}} \tag{19c}$$

Since for $k \to k_c(\zeta)$, $C(k) \to 0$ and $m^*(k) \to \infty$, $c_p(\zeta)$ has to go to zero. When ζ is a quadratic number, its critical behavior should exhibit the exponent $(\zeta + \eta)/2$. When $\zeta = (3 - \sqrt{5})/2$, this exponent should be approximately 0.039 but this result has not been numerically checked up to now (note that the accuracy of the calculations shown in Figure 10 were by far not sufficient to check the existence of such a small exponent).

For $k > k_c(\zeta)$, the effective mass as well as the phason velocity cannot be defined because the FK chain is pinned. In addition, we also believe that the phonon-like phase excitations do not have a well defined wave vector (see the next section). (Do not confuse phonon-like phase excitations with the configurational phase excitations which are phase defects). Then, although the adiabatic elastic constant is zero, it does not correspond dynamically to any zero phason velocity.

There exists other critical quantities. For example the gap Δ of the hull function $f(x)$ which vanishes below $k_c(\zeta)$. For the golden mean ζ_g, it behaves above $k_c(\zeta)$ as $(k - k_c(\zeta))^{\sigma}$ (with $\sigma \cong 0.721$ for $\zeta = \zeta_g$) [23]. There still exists no renormalization theory which analytically predicts all these exponents and their scaling relations.

2.4. INCOMMENSURATE STRUCTURE AS AN ARRAY OF EQUIDISTANT DISCOMMENSURATIONS

An incommensurate structure can be viewed as an array of equidistant discommensurations. For example, these discommensuration can be referred to the registered ground state obtained for $\zeta_0 = 0$. Since an advanced discommensuration corresponds to a lengthening of the chain of 2π, the density per site of discommensurations which is required for having the incommensurability ratio $\zeta = l/2\pi$ is ζ. Since the lattice is discrete, each of these discommensurations is submitted to a PN potential. In addition, there exists an interacting potential between these discommensurations which essentially depends on their overlapping. The TBA can be interpreted as a consequence of the competition between these two competing potentials.

These discommensurations behave as classical particles with interacting potentials which approximately determine a new FK model. This new model can be considered as the first step of a renormalization process of the initial model [34]. Although other more accurate methods of renormalization have been developed for the transition to stochasticity which is associated in the standard map or similar dynamical models [30, 31, 26] this qualitative description appears to be more transparent in our physical context.

The finite PN barrier which is found for a single discommensuration corresponding to $\zeta_0 = 0$ is roughly proportional to $\exp(-\kappa/\sqrt{k})$ (where κ is some constant) see Nabarro [2], and [35, 36, 37] which is an exponential of their width $1/\sqrt{k}$. It is the amplitude $k' \approx \exp(-\kappa/\sqrt{k})$ of the 1-periodic PN potential to which the discommensurations are submitted. The interaction energy between two consecutive discommensurations decay exponentially as a function of their distance $d = 1/\zeta$ as $\exp(-d/a) = \exp(-\sqrt{k}/\zeta)$. The new elastic constant C' for small variation of the distance between consecutive discommensurations is roughly proportional to $\exp(-\sqrt{k}/\zeta)$.

The physical arguments developed in Section 2.2 clearly shows that an upper value of $k_c(\zeta)$ for all ζ, is obtained when the energy $2k$ of the barrier of the periodic potential become comparable to the elastic energy $1/2 C(0)\pi^2$ for relative displacements of consecutive atoms by half a period of the periodic potential. Therefore, when the two parameters of the FK model which describes the discommensuration array becomes roughly equal that is $\exp(-\kappa/\sqrt{k}) \approx \exp(-\sqrt{k}/\zeta)$, the incommensurate phase is nonanalytic. Such a condition yields an upper bound for $k_c(\zeta)$ which is proportional to ζ. Since the numerical value of $k_c(\zeta)$ is known for $\zeta = (3 - \sqrt{5})/2$, we expect

$$k_c(\zeta) \approx \ < 2.54|\zeta|. \tag{20a}$$

The same arguments are applicable when ζ_0 is any integer which yields

$$k_c(\zeta) \approx \ < \mathrm{Env}(k_c(\zeta)) = 2.54 \, \mathrm{Inf}_{n\,\mathrm{integer}}|\zeta - n|. \tag{20b}$$

A more accurate estimation of the details of the function $k_c(\zeta)$ could be obtained by considering the incommensurate structure with incommensurability ratio ζ as a higher order commensurate structure where $\zeta_r = r/s$ is different than an integer with a superimposed array of equidistant elementary discommensurations. However, it is simpler to iterate the renormalization procedure. the first step of which has been described above. This work has been reported in [34].

It appears that $k_c(\zeta)$ as a function of ζ is a highly pathological function because t vanishes for ζ rational and is finite for most ζ irrational. However, it is important to note that for the commensurate cases the TBA is replaced by a sharp crossover as we noticed above and therefore $k_c(\zeta)$ can be physically considered as a rather smooth and continuous function of ζ. The values below which the pinning energies of these commensurate structure become physically neglegible, determines the order s at which the details due to the high order commensurability must be neglected.

Figure 14 shows a scheme of function $k_c(\zeta)$ versus ζ generated as the lower envelop of the inequalities similar to (20) generated by the rationals up to order 5.

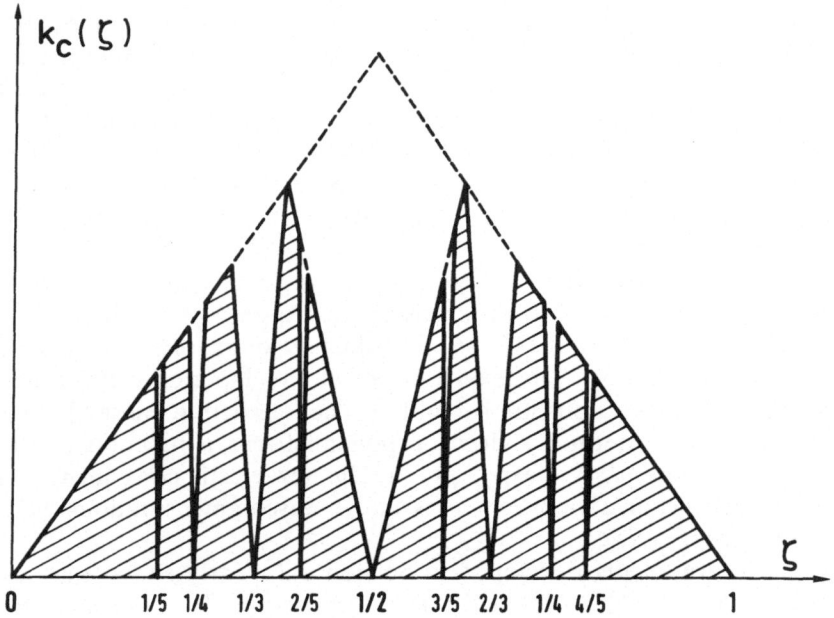

Fig. 14. Scheme showing the variation of $k_c(\zeta)$ as a function of ζ. It is the lower envelope of infinitely many pairs of curves starting from the ζ axis at each rational point. On this scheme, the commensurability effects have been taken into account up to order 5.

2.5. ISING REPRESENTATION OF A NONANALYTIC INCOMMENSURATE STRUCTURE

The TBA occurs in incommensurate structures of the FK model because the potentials $V(x)$ (6c) and $W(x)$ (6b) compete with each other. $W(x)$ favors the 'analyticity' and $V(x)$ the 'nonanalyticity'. The *dominant potential* in the energy functional is determined both by the incommensurability ratio ζ and by the amplitude k. Let us be more precise.

When the incommensurate ground-state is analytic, the *elastic contribution* to the energy functional (6a) is the dominant term because its configuration can be obtained by considering the periodic potential as a perturbation of the chain of equidistant atoms. The hull function $g(x)$ is expanded as a series of k:

$$f(x) = x + g(x) = x + \sum_{n=1}^{\infty} g_n(x)k^n. \tag{21}$$

As function $g(x)$, functions $g_n(x)$ are 2π-periodic. By substitution of (21) into Equation (15a), the formal calculation of the Fourier series of $g_n(x)$ can be done recursively. (The reader can perform this calculation explicitly to the two first orders by using Bessel functions). For $\zeta = r/s$ rational, the reader can check that the vanishing of denominators, hinder a proper definition of $g_n(x)$ for $n \geqslant 2$. This observation is confirmed by the KAM theorem which asserts that for most irrational ζ and for $k < k_c(\zeta)$ only, the series (21) is defined and *converges* for all orders n. Then, the incommensurate configuration is determined unambiguously.

By contrast, for $k > k_c(\zeta)$ or for ζ rational (and also for the irrational Liouville numbers which have a zero measure), the formal series (21) *diverges* and cannot be used for finding the ground state. Then, the periodic potential becomes the dominant term in the Hamiltonian (6a). The incommensurate ground state and also the other metastable configurations can be found by considering that the elastic coupling constant of the FK chain is small compared to k and by performing the expansion of the configuration coordinates with respect to $1/k$.

The ground state configuration $\{u_i^\infty\}$ of the FK chain with no elastic coupling is degenerate. Each atom i lies arbitrarily at any minima of the periodic potential $u_i^\infty = 2m_i\pi$ (m_i is an integer which corresponds to the minima number). By switching the elastic constant to unity, this configuration becomes $\{u_i\} = \{u_i^\infty + \delta_i\}$ and fulfills

$$\delta_{i+1} + \delta_{i-1} - 2\delta_i - k \sin \delta_i = -2\pi[m_{i+1} + m_{i-1} - 2m_i]. \tag{22a}$$

If $|\delta_i|$ is small, it can be obtained from the linearized Equation (23a) as

$$\delta_i \cong \pi/\sqrt{k + k^2/4} \sum_{n=1}^{\infty} \eta^n(m_{i+n} + m_{i-n} - 2m_i) \tag{22b}$$

where

$$\eta = 1 + k/2 - \sqrt{(k + k^2/4}. \tag{22c}$$

The series expansion of $\{\delta_i\}$ with respect to $1/k$ is, at the lowest order,

$$\delta_i \cong \frac{\pi}{8}\, k^2(m_{i+1} + m_{i-1} - 2m_i). \tag{23}$$

For a given set $\{m_i\}$, the analytic continuation of the solution $\{u_i^\infty\}$ which is obtained for $1/k \to 0$ ends at some critical value $1/k_c(\{m_i\})$ of k. For $k < k_c(\{m_i\})$, there exists atoms which cannot stay in equilibrium in their ascribed potential well and collapse discontinuously in neighboring wells. The relaxed configuration does not fulfills anymore the condition that each atom i belongs to the ith well. For k large enough, the proof of the existence of chaotic states has been recently obtained [91].

The physical behavior of the incommensurate structure for large k is well described by the variation of the FK model with a piecewise parabolic periodic potential,

$$V(x) = \tfrac{1}{2}(x - 2\pi m(x))^2 \tag{24a}$$

with

$$m(x) = \mathrm{Int}(x + \pi) \tag{24b}$$

($\mathrm{Int}(x)$ is the integer part of x) instead of (6b) in functional (6a). Then formula (23) becomes exact.

This model makes it clear that any metastable configuration can be *characterized* by a sequence of integers $\{m_i\}$. This sequence physically determines the occupied wells of the periodic potential. On the other hand, for a given sequence $\{m_i\}$, the corresponding configuration $\{u_i\}$ does not necessarily exist because, for consistency, one must have for all i, $|\delta_i| < \pi$. However, there still exist infinitely many choice for $\{m_i\}$ with a chaotic behavior which fulfills this condition (e.g. [8, 38, 39]).

In the nonanalytic regime, the energies of the metastable configurations are functions of $\{m_i\}$. Model (6a) with potential (24) yields a lattice gas model which has the energy

$$\Phi(\{m_i\}) = \frac{1}{2} \sum_{i,j} K(i - j)\,(m_i - m_j)^2 - \sum_i 2\pi\mu.(m_{i+1} - m_i) \tag{25}$$

with

$$K(n) = K(0)\eta^{|n|} \tag{26a}$$

where

$$\eta = 1 + \frac{k}{2} - \frac{1}{2}\sqrt{4k + k^2} \tag{26b}$$

and

$$K(0) = \frac{2\pi^2 k^2}{(4k + k^2)^{1/2}}. \tag{26c}$$

The atomic positions $\{u_i\}$ are linear functions of $\{m_j\}$ and are given by

$$u_i = \frac{2\pi}{K(0)} \frac{1-\eta}{1+\eta} \sum_{j=-\infty}^{+\infty} K(n)m_{i+n}. \tag{26d}$$

This model can be transformed into a spin model by setting $m_{i+1} - m_i = \sigma_i$ and by restricting the chaotic configuration space to those where $\sigma_i = 0$ or 1 [38, 39]. It clearly appears that model (6a) with potential (24) does not exhibit any TBA. In fact, the analyticity of the hull function of the ground state is always broken simply because the analyticity of the energy functional is already broken in the definition (24). The critical quantities which characterize the pinning of the incommensurate ground state are exactly calculated in [38] and are found to remains finite for any k.

Because of the pathology of the model, these quantities do not have the usual behavior at small k. For example, the PN barrier of single discommensurations is much larger than for an analytic periodic potential (6b). This PN barrier behaves as *the cube of the inverse width* of the discommensuration instead of an exponential. The empirical arguments developed in the above subsection would predict a TBA which in fact does not exist.

In FK models which exhibits a TBA, it has been rigorously proved that the hull function $f(x)$ of a nonanalytic incommensurate ground state defined by (8b) is a discrete function with a dense set of discontinuities [40, 38, 41]. Discarding possible exceptional cases, numerical calculations have shown that the location of all the discontinuities of $f(x)$ are obtained from the knowledge of the location x_0 of only one discontinuity by the formula $x_{m,n} = x_0 + 2m\pi + nl$ where m and n are arbitrary integers. Then it is straightforward to show that all the incommensurate ground states with incommensurability ratio $l/2\pi$ can be *exactly* written with the form

$$u_i = u + \sum_{j=-\infty}^{+\infty} f_n m_{i+n} \tag{26e}$$

where f_n is the strictly positive amplitude of the discontinuity of $f(x)$ at $x = x_{m,n}$ (which because of the periodicity of $g(x) = f(x) - x$ is independent of m), u is some constant and $m_n = m(il + \alpha)$ where $m(x)$ is given by (24b) and α is an arbitrary phase. (26e) can also be written as

$$u_i = u + \sum_{j=-\infty}^{+\infty} b_n \sigma_{i+n} \tag{26f}$$

with the pseudospin definition $m_{i+1} - m_i = \sigma_i = 0$ or 1 and $b_{n-1} - b_n = f_n$. This representation clearly proves the surprising result that the nonanalytic incommensurate ground states can be exactly represented as *linear combination* of discommensurations located at the sites i determined by $\sigma_i = 1$. Their 'shape' is determined by $\{b_n\}$. b_n converge exponentially either to zero or to 2π for $n \rightarrow$

$-\infty$ or $n \rightarrow +\infty$ respectively with a characteristic length which is exactly, the coherence length ξ. This discommensuration does not correspond to a isolated single discommensuration, but to an 'effective discommensuration' in the presence of the others. A rigorous proof of these results has been obtained for potentials $V(x)$ in (6a) with a single minima per period [91].

Therefore, if we already know that the incommensurate ground state is nonanalytic (for example for $k > 0.97$ in the FK model), the minimization of the energy of the FK model can be done without any approximation by using for $\{u_i\}$ the form (26f). Knowing the sequence $\{\sigma_i\}$, the problem of finding the coefficients b_n or f_n can be considered as the 'dual' problem of finding the harmonics of $g(x)$ in the analytic regime. If the coefficients b_n are known, an Ising Hamiltonian (with multispin interactions) is found in principle which has the same ground state as the FK model. For large k, this Ising model is well approximate by the Ising model obtained from the piecewise parabolic model (24) above described and allows also a good description of the metastable states which are not ground states. But as k approaches $k_c(\zeta)$ from above, the extension ξ of $\{b_n\}$ diverges and the coefficients of multispin interaction terms becomes very important in the Hamiltonian. At $k = k_c(\zeta)$ and beyond the TBA, in the analytic regime for $k \leqslant k_c(\zeta)$, this definition for this pseudospin Hamiltonian becomes impossible.

Then, the idea that emerates is that a nonanalytic incommensurate ground state of the FK model can be exactly represented as the ground state of a certain Ising spin Hamiltonian. When the incommensurate ground state returns to the analytic regime, the existence of this Hamiltonian simultaneously breaks down.

Here certainly is the basic physical idea concerning the concept breaking of analyticity which we will use again in the following of this paper (Section 4 and 5) for models where a rigorous proof, as in the FK model, is still lacking. However, some rigorous results are now available which confirms early conjectures [91].

2.6. EXTENDED FK MODELS AND THERMAL FLUCTUATIONS

The TBA has been found with the same qualitative physical features in extended FK models with long range atomic interaction, with several sublattices and with several dimensions. The concept of analyticity with respect to the phase of the incommensurate ground state, appears to be useful for a large class of models. The important difference between the analytic and nonanalytic incommensurate structures concerns their *low-lying energy excitations* which govern the low temperature behavior of the incommensurate structures which as we emphasized, is related to the existence (or not) of an associated Ising Hamiltonian.

Although we have not analysed the consequence of the breaking of analyticity at finite temperature, let us speculate on reasonable consequences which should be found.

The *analytic* incommensurate structures are essentially well described by zero gap phase excitations which are phonon-like. *Continuous* models where the effects of the discreteness of the lattice is neglected seems to be appropriate for describing the loss of long range order of the analytic incommensurate phases. As for standard displacive structural phase, the transition should exhibit *a soft mode* in the disordered phase the gap of which vanish at the transition temperature. This soft mode gives rise to both the phason mode and the ampliton mode. This picture corresponds to the standard description of the incommensurate structures.

By contrast, when the thermal temperature $k_B T$ does not exceed the Peierls—Nabarro energy barrier $E_{PN} = k_B T_{PN}$, the *nonanalytic* incommensurate structures are better described by lattice gas or, equivalently, by pseudospin models. (Let us note, however, that if the temperature T becomes larger than T_{PN}, the effect of the lattice pinning of the phase, disappears and one should return to a displacive model). When T_{PN} is large enough, the order temperature T_i of the incommensurate structure is smaller than T_{PN}. The phase transition is described by a pseudospin model. Like standard order—disorder structural phase transitions, one should expect a transition *without a soft mode* but with a diffusive central peak. The other important consequence of the breaking of analyticity but in CDW compounds only, concerns the transport of the electric current carried by the phase translation.

Recent numerical simulations of structural models for incommensurate structures at finite temperature by Parlinski [42] indeed confirms the existence of these two 'displacive' and 'order-disorder' regimes.

3. Another Transition by Breaking of Analyticity: The Localization Transition of Electrons in an Incommensurate Potential

By contrast with the previous section which was devoted to a *classical model*, this section now consider *a quantum problem*: an electron propagation in an incommensurate CDW structure. This propagation is described by a Schroedinger equation with a quasi-periodic potential (i.e. incommensurate) which is determined self-consistently by the electron phonon interaction. In this section, the form of this potential is supposed to be given. A review paper of the existing results for such one-dimensional Schroedinger equations with a quasi-periodic potential, has been recently given by Sokoloff [43]. However in this section, we shall rather discuss this problem in terms of 'breaking of analyticity'.

The more complex self-consistent problem which concerns the CDW will be studied numerically in the next section on two models of one-dimensional Peierls chains at 0 K (in the adiabatic limit) and will exhibit again a transition by breaking of analyticity.

The standard theory of Bloch waves implies that the eigenstate of a quantum particle moving in a *periodic potential* has two essential properties: 1 — it is not square summable, which physically means that the particle probability density extends all over the sample; 2 — it carries a *finite momentum* (except the eigenstates at the energy band edges). This second property is physically the most important because it enables electric current transportation by the quantum particle. As we will see in this section, these two properties are not equivalent.

Otherwise, it is well known that the propagation of a quantum particle in a random potential is radically different from the propagation in a periodic potential. Three dimensional eigenstates in a random potential with a small enough amplitude are extended if their eigenenergies lie in the interval between the two mobility edges. In contrast to the case of a periodic potential, these extended states have no well defined wave vector. When their eigenenergies lie outside the mobility interval, they are exponentially localized (square summable). For large random potentials, the mobility interval shrinks to zero and all eigenstates become exponentially localized. For one-dimensional random potentials (and probably for two-dimensional random potentials) all the eigenstates are localized with probability 1.

In some sense, incommensurate potentials are intermediate between periodic and random potentials. But despite the fact that there also exists a localization transition for a quantum particle in an (analytic) incommensurate potential as a function of its amplitude, this transition is *qualitatively very different* from the corresponding transition in a random potential. In the incommensurate case, the eigenstates generally have two possible behaviors. The eigenstates can be extended with a well defined wave vector and can be represented *by 'quasi-Bloch' wave functions* (which are plane waves modulated by the quasi-periodic potential). This form just extends the form of the Bloch eigenstates in periodic systems and then their coordinates depend *analytically* on each of the *phase* parameters of the quasiperiodic potential. For this reason, these eigenstates are called *analytic*. In addition, these analytic eigenstates carry a *finite momentum*. Consequently, when the Fermi energy corresponds to such an eigenstate, the resistivity of this incommensurate system vanishes at zero degrees, unlike random systems, where it remains finite. This property is due to the fact that the characteristic length of the phase coherence (mean free ptah) of this extended eigenstates is infinite.

The second possible behavior for the eigenstates in a quasiperiodic potential corresponds to an exponential localization. Then, the coordinates of the eigenstates *do not depend analytically* on the phase parameters of the quasi-periodic potential, but as functions *with infinitely many discontinuities* which form a dense set in the phase parameter space. These eigenstates are called 'nonanalytic'. The system is insulating at 0 K.

A marginal behavior for the eigenstates in a quasi-periodic potential is also possible, for example at the localization transition in incommensurate models which exhibit a localization transition or in some particular incommensurate models for all values of their parameters. These eigenstates are extended although they are not quasi-Bloch waves. They do not depend analytically on the phase parameters of the quasiperiodic potential but discontinuously and they do not carry neither any global finite momentum nor a local momentum. The associated resistivity is infinite at absolute zero as for localized eigenstates.

Unlike the extended states in quasiperiodic potentials, the extended states in random potentials do not carry any global momentum but exhibits locally finite but random momenta. For this reason, the extended states in a random potential can carry a current (at absolute zero) but with a nonzero and *finite resistivity*.

3.1. DESCRIPTION OF THE BREAKING OF ANALYTICITY OF THE EIGENWAVES OF A QUASI-PERIODIC SCHROEDINGER EQUATION

The discrete FK model and models for a quantum particle propagating in an incommensurate potential both exhibits a transition by breaking of analyticity. This similarity can be exhibited by considering for example the Schroedinger equation of the wave function $\Psi(x)$ of an electron

$$-\frac{d^2\Psi(x)}{dx^2} + V(x)\Psi(x) = E\Psi(x) \tag{27a}$$

in a one-dimensional incommensurate potential

$$V(x) = A\cos x + k\cos(Qx + \alpha) \tag{27b}$$

which is the sum of two periodic potentials, the wave vector of which are 1 and Q, respectively. The ratio $|Q|$ of the length of these two wave vectors is assumed to be an irrational number in order that $V(x)$ be quasi-periodic and not periodic with some period multiple of 2π.

To be rigorous, we should also assume that Q is a 'good irrational number' which is a property fulfilled by most irrational numbers with probability one. Since we avoid mathematical considerations which are too complex in the framework of this paper, we do not make the meaning of this term explicit here, which would not be useful for the present physical applications. The interested reader is referred for example to [44].

In Equation (27), let us vary k from the zero value. According to the Bloch theorem, when k is zero, the eigenwaves of this equation are plane waves with the form

$$\Psi(x) = \exp(iqx)\chi(x) \tag{28}$$

where $\chi(x)$ has the period 2π of $V(x)$. As a consequence of the Dinaburg and Sinai theorem [45] applied to this example, for most wave vectors q (except for a zero measure set), there exists a finite bound $k_c(q)$ such that for

$$|k| < k_c(q) \tag{29a}$$

there exists an eigensolution of (27) which has the form

$$\Psi(x) = \exp(iqx)\chi(x, Qx + \alpha). \tag{29b}$$

Function $\chi(x, y)$ in (29b) is analytic and 2π-periodic with respect to each of its variables x and y. It is a quasiperiodic function with the same periods and the same phases as the quasiperiodic potential (27b). Eigenwave (29b) which appears as the extension of the Bloch form of the eigenwaves obtained for periodic potential (27), is called a quasi-Bloch wave.

The bound $k_c(q)$ generally depends on q and vanishes for all wave vectors

$$2q_{m,n} = mQ + 2\pi n \tag{30}$$

where m and n are arbitrary integers. These wave vectors are 'resonant' with the wave vectors of the incommensurate potential. Such a result is reminiscent of the behavior of function $k_c(\zeta)$ described in the previous section (cf. Figure 14). When k is larger than $k_c(q)$, then the exponential localization of the eigenstates may occur. As for the FK model, this transition is a TBA *with respect to the phase α of the perturbative potential.* (But recall that the eigenfunctions $\Psi(x)$ nevertheless remain analytic with respect to the space variable x for all analytic potentials $V(x)$ although they do not necessarily remain analytic with respect to the phase α).

The degeneracy between the plane waves, the wave vectors of which are $+q_{m,n}$ and $-q_{m,n}$ is raised by the perturbative potential $k \cos(Qx + \alpha)$ which opens a gap at the corresponding energy in the unperturbed band spectrum. Since there exists infinitely many resonant vectors $q_{m,n}$ which form a dense set on the real axis, the resulting spectrum does not contain any band with a finite width and has infinitely many gaps. Its support is a Cantor set (which generally has a *finite measure*). Such a spectrum has been calculated by Hofstadter [46] for the Harper equation (but at the critical point where it has zero measure). For this last equation, the localization transition can be analytically exhibited because of a duality property. We present this analysis.

3.2. EXACT RESULTS FOR A SELF-DUAL MODEL

The Harper equation in a particular case was initially proposed as a model describing an electron in a two dimensional periodic potential and a magnetic field. It is a one-dimensional tight binding model where the exchange parameter between nearest neighbor sites is constant. We take it as the energy unit (the unperturbed band width is 4). The other exchange parameters are zero. The on-site energy only is modulated as a sine periodic function of the site with the wave vector Q, the amplitude $2k$ and the phase α

$$\Psi_{n+1} + \Psi_{n-1} + 2k \cos(Qn + \alpha)\Psi_n = E\Psi_n. \tag{31}$$

This is likely the simplest equation with a quasi-periodic potential but note that for the original Harper equation which was proposed for describing the motion of an electron in a magnetic field $k = 1$. The amplitude of the eigenwave at site n is Ψ_n and Q is an irrational number.

This equation has the *exceptional* property that at the same critical value $k_c = 1$ of k, all the eigensolutions of this equation exhibit *simultaneously* a transition by breaking of analyticity with respect to the phase variation [40] and become exponentially localized for $k > 1$.

The Dinaburg and Sinai theorem suggests the possible existence of eigenwaves for (31) which have the form of generalized Bloch waves

$$\Psi_n(q) = \exp(iqn)\chi_q(Qn + \alpha)$$

$$= \exp(iqn) \sum_m \chi_m^{(q)} \exp(im(Qn + \alpha)) \tag{32}$$

where $\chi_q(x)$ is a 2π-periodic function which can be expanded as a Fourier series, the coefficients of which are $\chi_m^{(q)}$. Let us note that the wave vector of this solution is not uniquely defined. q can be changed into $q + rQ + 2s\pi$ where r and s are arbitrary integers which implies that $\chi_q(Qn + \alpha)$ is changed into $\exp(ir\alpha)$ $\exp(-ir(Qn + \alpha))\chi_q(Qn + \alpha)$.

The formal equation which is fulfilled by the coefficients $\chi_m^{(q)}$ is obtained by substitution of the form (32a) into (31) and yields

$$\chi_{m+1}^{(q)} + \chi_{m-1}^{(q)} + 2k^* \cos(Qm + \alpha^*)\chi_m^{(q)} = E^*\chi_m^{(q)} \tag{33a}$$

with

$$k^* = 1/k \tag{33b}$$

$$E^* = E/k \tag{33c}$$

and

$$\alpha^* = q. \tag{33d}$$

Equation (33a) is called the *dual eigenequation* of (31) because two successive applications of this duality transformation gives back the initial equation. It also corresponds to an eigenequation with an incommensurate potential. This definition can be easily generalized to any eigenequation with a quasiperiodic on-site potential at any dimension. This particular Equation (31) has the exceptional property that the dual Equation (33) has the same form as the initial one by changing the potential amplitude k into $1/k$, the eigenenergy E into E/k and the phase α into the wave vector q. Consequently, this tight-binding model (31) is called *self-dual*.

If a quasi-Bloch eigenwave (32) exists for Equation (31) which depends *analytically* on the phase α, the series $\sum_m |\chi_n^{q)}|^2$ is necessarily exponentially convergent. Then Equation (33) has an *exponentially localized* eigenwave $\{\chi_n^{(q)}\}$. Conversely, if Equation 33 has an exponentially localized eigensolution, Equation 31 has a quasi-Bloch eigensolution defined by (32).

The Thouless formula [47], which only applies to nearest neighbor one-dimensional models, relates the inverse localization length $\gamma(E) = 1/\xi(E)$ of the eigenstate at energy E (if any) to the density of state $dN(x)$. It was initially proved for random potentials, but it remains valid for any potential. This coefficient $\gamma(E)$ is analogous to the Lyapounov coefficient which corresponds to the inverse coherence length in the FK model. In the case of Equation 31 where the exchange constant is unity, this formula yields

$$\gamma(E) = \frac{1}{\xi(E)} = \int_{-\infty}^{+\infty} \ln|E - x|\, dN(x). \tag{34}$$

For a large system of n sites, the integrated density of states $N(x)$ is the number of eigenstates of Equation (31) the energy of which is smaller than x divided by the system size n. For n going to infinity, $N(x)$ is proved to be uniquely defined.

Since $\gamma(E)$ is the inverse of a physical coherence length, formula (34) necessarily yields a positive or zero value. If the eigenstate at the energy E is extended (not quare summable) or has a square summable wave-function which decays slower than any exponential at infinity, $\gamma(E)$ is zero. On the other hand, $\gamma(E)$ is strictly positive for all exponentially localized eigenstates.

We have to mention that the Thouless formula is not a rigorous formula. From the mathematical point of view, it can be only considered as 'generally exact'.

There exists exceptional values of $Q/2\pi$ which are Liouville numbers where the conclusions which are derived from this formula are wrong. Then formula (34) only yields an upper bound for the inverse localization length. Nevertheless, these situations have zero probability to occur. In addition, the possible failure of this formula cannot be *numerically* exhibited on any example by a direct calculation. The argument is that we necessarily study large but finite system where the relation between the density of state and $\gamma(E)$ is still mathematically exact. The possible failure of this formula only occurs at the limit of an infinite system. This point will be discussed again with physical arguments at the end of this subsection. Our claim is that for physical applications, *the Thouless formula can be considered as exact.*

In the incommensurate case, the energy density of states becomes independent of the phase α and the duality relation between Equations (31) and (33) allows one to establish the *exact relation*

$$N_k(x) = N_{1/k}(x/k) \tag{35}$$

between the integrated density of states $N_k(x)$ of Equation (31) and the integrated density of states $N_{1/k}(x)$ of its dual Equation (33). This equality is proved by considering first the case where $Q/2\pi$ is a rational number r/s (r and s are irreducible integers). Then the Bloch theorem applies to Equation (31) and yields that its eigenwaves are described by the form (32) where the series is truncated to s terms only, because the potential $V_n = 2k\cos(Qn + \alpha)$ has period s. The coefficients $\{\chi_n^{(q)}\}$ ($n = 0$ to $s - 1$) fulfill the dual Equation (33) but with the periodic boundary condition $\chi_s^{(q)} = \chi_0^{(q)}$ and $\chi_{s-1}^{(q)} = \chi_{-1}^{(q)}$. Each eigenstate of the initial equation corresponds to an eigenstate of the dual closed loop and *vice versa*. Then it is easy to show in the commensurate case that $|N_k(x) - N_{1/k}^{(s)}(x/k)| < 1/s$ where $N_{1/k}^{(s)}(x/k)$ is the integrated density of state of the dual finite loop. When the commensurability order s diverges, the length s of the closed loop of the dual model also diverges and $N_{1/k}^{(s)}(x/k)$ goes to the integrated density of state of $N_{1/k}(x/k)$ of the incommensurate dual model which proves (35).

This Equation (35) used in the Thouless formula yields

$$\gamma_k(E) = \gamma_{1/k}(E/k) + \ln(k) \tag{36}$$

which relates the inverse localization length of the dual models (31) and (33). Since the localization length is necessarily positive or zero, we have

$$\gamma_k(E) \geq \ln(k) > 0 \quad \text{when} \quad k > k_c = 1. \tag{37}$$

As a result, all the eigenstates $\{\Psi_n^\mu\}$ of Equation (31) are exponentially localized for $k > k_c = 1$ and for all phases α. They form a countable base in the Hilbert space l_2 of square summable $\{\Psi_n\}$.

By contrast, for $k < k_c = 1$, Equation 31 does not possess any localized eigenstate because, if one were to exist, it should be possible by using Equation 32, to construct an extended quasi-Bloch eigenwave for Equation 33 where $k^* > k_c$ which is impossible since $\gamma_{k^*}(E) > 0$ for all E.

For $k < 1$, the countable set of localized eigenstates $\{\Psi_n^\mu\}$ with eigenenergy E^μ of the eigenequation (33a) where $k^* > k_c$ determines by Equation 32, a set of eigenstates of the same Equation (31) for $k < 1$. These states are quasi-Bloch waves with wave vector q. Then, considering all possible values of the phase $\alpha^* = q$, the whole set of quasi-Bloch waves of Equation (31) is obtained. Unlike the case $k > 1$, this set of eigenstates is uncountable for $k < 1$ because the set of values of q is continuous.

Each localized eigenstate $\{\Psi_n^\mu\}$ of Equation 31 where $k > 1$, corresponds by (32) to an extended quasi-Bloch wave which obviously, is not square summable. Consequently, $\gamma_{1/k}(E^\mu/k)$ necessarily vanishes in formula (36). Therefore, this formula yields the *exact localization length* $\xi(k)$ of $\{\Psi_n^\mu\}$

$$\xi(k) = 1/\ln(k). \tag{38}$$

Curiously, this length is independent of the energy E^μ of this localized state, but this property will appear to be a peculiarity of this self-dual model.

Therefore, when $k < 1$ each quasi-Bloch eigenwave $\{\Psi_n(q)\}$ of (31) has a *hull function* $\chi_q(x)$ with Fourier coefficients $\chi_n^{(q)}$ which go to zero as $\exp(-|n| \ln k)$ for $n \to \pm\infty$. A consequence of this property is that *this hull function* $\chi_q(z)$ *is analytic* with respect to the complex variable z (the phase) which belongs to the domain of the complex plane defined by $|\text{Im } z| < -\ln k$. When k approaches $k_c = 1$ by lower values, the domain of analyticity for $\chi_q(z)$ shrinks to zero. At $k = k_c$, singularities of $\chi_q(z)$ accumulates on the real axis.

The breaking of each quasi-Bloch wave at the localization transition then corresponds to the breaking of the analyticity of its periodic hull function $\chi_q(z)$ *due* to the raising of the high order harmonics χ_m of its Fourier series. This mechanism is qualitatively similar to the breaking of analyticity of the hull function $f(x)$ at $k = k_c(\zeta)$ discussed for the incommensurate ground states of the FK model. Unlike the TBA in the FK model beyond which the hull function $f(x)$ survives as a discontinuous real function, this hull function $\chi_q(z)$ of the eigenstates becomes undefined beyond its TBA.

When $k > 1$, it is interesting to note that any localized eigenstate $\{\Psi_n\mu\}$ of (31) cannot be a continuous function $\{\Psi_n^\mu(\alpha)\}$ of the phase α of the quasiperiodic potential in (31). This result can be simply proved by assuming that there exists such a continuous function. Then the quasi-Bloch wave defined by $\Psi_n(q) = \exp(iqn) \sum_m \Psi_m^\mu(q) \exp(im(Qn + \alpha))$ which is an eigensolution of (31) for $k^* = 1/k < 1$, is a continuous function of q. Consequently, the corresponding eigenenergy $E(q)$ is a continuous function of q. But $E(q)$, which cannot be a constant, cannot also be a continuous function because the set of eigenvalues of (31) has infinitely many gaps and is a Cantor set (see e.g. Hofstader, 1976).

We expect a similar property concerning the electron density for $k > k_c = 1$. For any electron filling concentration strictly between 0 and 1, the hull function $\rho(x)$ of the resulting charge density defined by $\rho_n = \sum_{\mu \text{ occ}} |\Psi_n^\mu(\alpha)|^2 = \rho(Qn + \alpha)$ has infinitely many *discontinuities* (although the global energy density of states is unchanged). In other words, any continuous change of the phase α, produces discontinuous changes of the electron density. A similar hull function is calculated in the next section for the Holstein model (see Figure 22). We did not prove this result, but this phenomenon is easily understood in the limit of large k, because each localized eigenstate becomes essentially localized on a single site. In this limit, the on-site energies being

$$E_n = 2k \cos(Qn + \alpha),$$

the electron density at site n is

$$\rho_n = \rho(Qn + \alpha) = Y(E_F - E_n)$$

where E_F is the Fermi energy and $Y(x)$ is the step function which is zero for negative x and 1 for positive x. Clearly, $\rho(x)$ is a discontinuous function of $x = Qn + \alpha$ providing that E_F belongs to the interval $]-2k, +2k[$ in order that the band be not completely filled or not completely empty. The phase translation of the quasiperiodic potential is associated with infinitely many discontinuous jumps of electrons from a site to another site which is far away, but not to a continuous transformation of their eigenwaves. By contrast, when $k < k_c$, since the quasi-Bloch waves are analytic functions of the phase α of the quasi-periodic potential, the charge density wave $\rho_n = \rho(Qn + \alpha)$ is a continuous function of the phase α.

Then the hull function of the charge density wave, which is defined for k as well as below and above $k_c = 1$, exhibits a TBA associated with the localization transition which is qualitatively identical to that of the hull function of an incommensurate FK chain at the TBA.

It has been proved that when $Q/2\pi$ belongs to a certain class of Liouville numbers [44] that none of the eigenstates of Equation 31 can be localized and even can converge to zero at infinity. Let us recall that Liouville numbers form a subset of the irrational numbers which has zero measure but is dense on the real axis. This result apparently sharply contradicts our prediction for the existence of a localisation transition at $k = k_c = 1$. However, by most infinitely small perturbations of $Q/2\pi$, this number becomes a non-Liouville irrational number. Then if $k < 1$, all the eigenstates suddenly becomes quasi-Bloch waves and if $k > 1$ all the eigenstates suddenly becomes exponentially localized state with a *finite* localization length given by (38). Such a sensitivity suggests that most small perturbations of the system sharply destroy the singular continuous spectrum and restore the localization transition which we predicted. This example shows that, in some cases, rigorous mathematical results becomes meaningless for physical purposes while numerical simulations, although they always are approximations, could provide more reliable informations on the real physics.

This model (31) has also been studied by renormalization methods [48, 49] which confirm the existence of a transition localization at $k = k_c = 1$ and also yields information on the self-similarity of its Cantor spectrum. There have been many further numerical studies on the structure of this Cantor spectrum and on the structure of the associated eigenstates, and recent rigorous results have been obtained (ref. 92, 93, 94).

3.3. SOME NUMERICAL INVESTIGATIONS OF THE SELF-DUAL MODEL AND OTHER NON-SELF-DUAL MODELS IN ONE DIMENSION

We present here the few numerical investigations which we performed [40] in

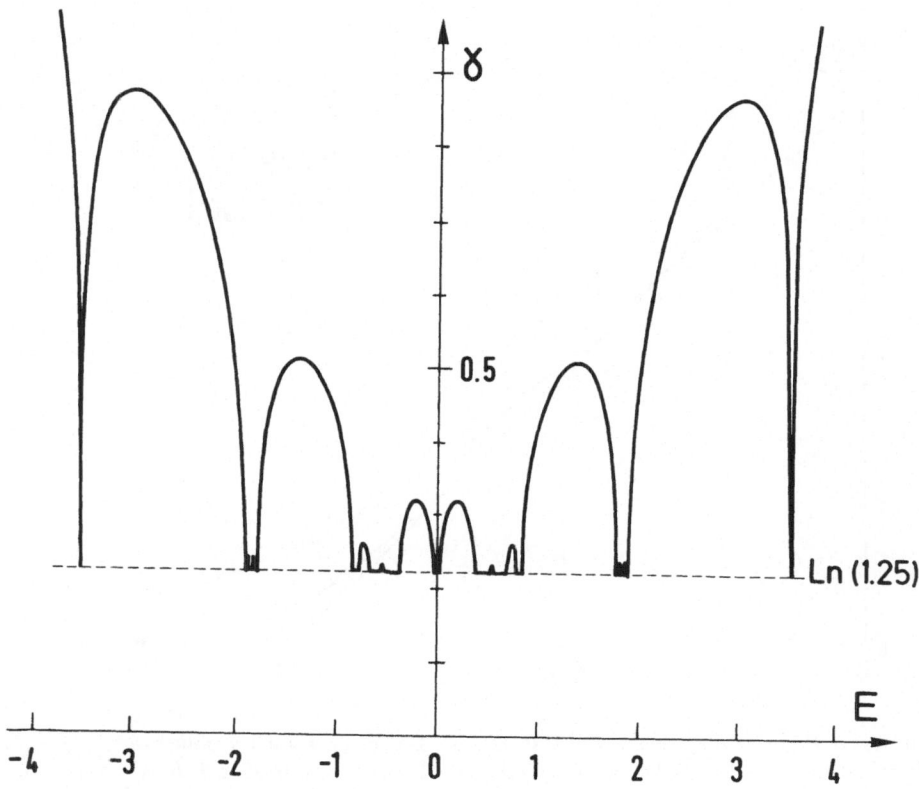

Fig. 15. $\gamma_k(E)$ for the self-dual model Equation (31) for $k = 1.25$ and a high order commensurate modulation wave vector $Q = (158/1003) \times 2\pi$ which physically is indistinguishable from a true incommensurate one. The spectrum of Equation (31) corresponds to the abscissa of the cusps of this curve. At these energies E, the inverse localization length $\xi_k(E) = 1/\gamma_k(E)$ is found to be a constant and is equal to $0.223 \approx \ln 1.25 = \ln k$ in lattice spacing units in agreement with the prediction of formula (38).

order to understand the peculiarities of the self-dual model compared to more general models.

These studies were based on the analysis of the localization length given by the Thouless formula (34). We note that $\gamma_k(E)$ is singular at each point E which belongs to the support of the density of state $dN_k(E)$ because $\ln |x - E|$ is divergent for $x \to E$. These points can be numerically found by finding the cusps of $\gamma_k(E)$. The closed set of the singular points of function $\gamma_k(E)$ is the support of the density of state $dN_k(E)$. Figure 15 shows $\gamma_k(E)$ calculated for $k = 1.25$. It appears that all these points have the constant y-coordinate $\ln(k) = 1/\xi(k)$ predicted by (38). The closure of these singular points suggests a Cantor-like structure because many new smaller gaps appear after magnification of the curve.

Incidentally, we studied the measure $\mu(Q/2\pi, k)$ of this Cantor set, which is generally non-zero. The numerical calculation has been first performed for rational

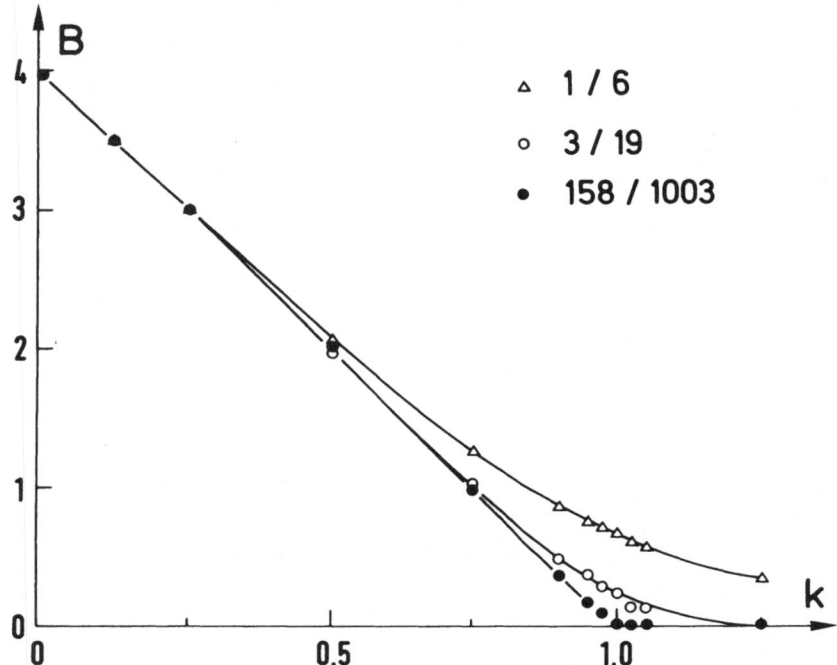

Fig. 16. Total $B_\zeta(k)$ of the s band widths of model (31) versus k for a commensurate $Q = 2\pi\zeta_c$ with $\zeta_c = r/s$ for $r/s = 1/6$, $3/19$ and $158/1003$. For $\zeta_c = r/s \rightarrow \zeta$ irrational, $B_\zeta(k)$ always goes to the same curve which is piecewise linear and fulfills (39). The vanishing of $B_\zeta(k)$ coincides with the localization of the eigenstates of (31).

values r/s of $Q/2\pi$ because in that case the support of the density of state consists of s bands. It is then easy to calculate the measure of the set of eigenvalues of the commensurate Hamiltonian which is closed. When r/s goes to ζ irrational for $s \rightarrow \infty$, it is observed that $\mu(\zeta, k)$ becomes independent of ζ and that within the numerical errors (cf. Figure 16) we have

$$\mu(\zeta, k) = 4|1 - k| \quad \text{for} \quad k \leqslant k_c = 1. \tag{39}$$

When $k > k_c = 1$, this figure suggests $\mu(\zeta, k) = 0$ which is a consequence of the localization transition. Then, the measure of the set of eigenvalues which is countable, is obviously zero. In fact, this formula (39) can also be considered as valid for $k > k_c = 1$ because of the above duality argument but it must be understood this measure $\mu(\zeta, k)$ is then the measure of the *closure* of the set of eigenvalues (which would be zero otherwise because this set is countable). The remarkable result is that the localization transition just occurs when the measure of the Cantor set which is the support of the density of state vanishes. We have no proof of this statement.

The Lyapounov number $\gamma_k(E)$ for other non-self-dual models has been studied in order to point out the difference with this self-dual model. For example, we

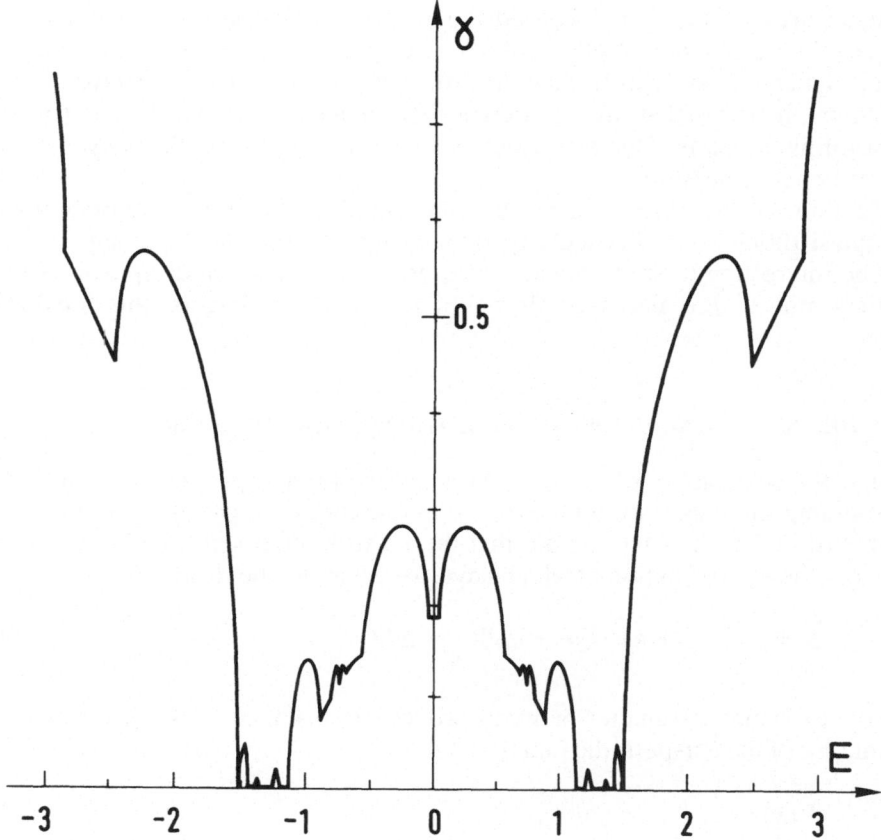

Fig. 17. $\gamma_k(E)$ for model (40) for $k = 0.55$ and $Q = (158/1003)2\pi$. Inspection of $\gamma_k(E)$ at the cusps of this curve shows that a part of the eigenstates are extended while the other part corresponds to localized eigenstates with an inverse localization length which depends on the eigenenergy.

calculated $\gamma_k(E)$ for the tight-binding Schroedinger equation

$$\Psi_{n+1} + \Psi_{n-1} + 2k[\cos(Qn + \alpha) + \cos 3(Qn + \alpha)]\Psi_n = E\Psi_n \qquad (40)$$

(see Figure 17 where $Q/2\pi = 158/1003$ and $k = 0.55$). Unlike the self-dual model (31), it is found that there exist eigenenergies E for which the corresponding eigenstates are extended (and are quasi-Bloch waves with the form (32)) and other eigenenergies for which the corresponding eigenstates are exponentially localized.

Several mobility edges are found (a mobility edge is an eigenvalue which separates a region with localized eigenstates from a region with extended eigenstates). The Dinaburg and Sinai theorem [45] predicts the existence of quasi-Bloch states for eigenenergies which belongs to a Cantor set. Although it does not give any proof, it might be expected that for eigenenergies in the gaps of this Cantor set, (which are associated with the resonant wave vectors $q_{m,n}$ given by (30), the eigenstates are exponentially localized. Thus, this localization would occur in a way similar to the occurrence of chaos which takes place of the broken resonant KAM tori in a dynamical system. In fact, according to T. Spencer or F. Delyon (private discussions), it seems that there is only a *finite* number of mobility edge (or none) but not infinitely many. This resonance phenomena is more subtle as expected and has to be better understood.

In Section 3.6, we started some numerical studies concerning the breaking of a single quasi-Bloch wave. Particularly, by analogy with the standard map studies, it would be interesting to study the last quasi-Bloch waves which disappears. Further numerical studies are necessary for a better understanding of this localization process.

3.4. OTHER SELF-DUAL MODELS IN ONE AND SEVERAL DIMENSIONS

Although the self-duality of model (31) is an exceptional property it is useful for understanding the localization mechanism in quasiperiodic potentials in more than one dimension. But first let us note that there exists other self-dual models in one dimension, the eigenequation of which have, for example, the form

$$\sum_m V_m \Psi_{n+m} + kV(Qn + \alpha)\Psi_n = E\Psi_n \tag{41a}$$

where the exchange parameter between two sites at distance m is equal to the mth harmonic V_m of the 2π-periodic function $V(x)$

$$V(x) = \sum_m V_m \exp(imx). \tag{41b}$$

This form (41a) can be extended to d dimensional self-dual models by considering that the index m is a d component index vector $\mathbf{m} = (m_1, \ldots, m_i, \ldots, m_d)$, that the phase α is a phase vector $\boldsymbol{\alpha} = (\alpha_1, \ldots, \alpha_i, \ldots, \alpha_d)$ with d components and that the modulation vector Q is a *symmetric $d \times d$ square matrix* $\mathbf{Q} = \{Q_{ij}\}$. This matrix is represented as $(Q_1, \ldots, Q_i, \ldots, Q_d)$ where $\mathbf{Q}_i = (Q_{i,1}, \ldots, Q_{i,j}, \ldots, Q_{i,d})$ is the d component vector corresponding to the ith line of $\mathbf{Q}$ (then $Q_{i,j} = Q_{j,i}$ for all i, j). In addition, for having a quasiperiodic potential with $2d$ independent periods, we assume that there exists no rational relation between these d vectors which means that there exists no system of relations

$$\sum_i m_i Q_{i,j} = 2\pi n_j \tag{42}$$

where m_i and n_j are integers. The potential $V(x)$ is replaced by a function $V(\mathbf{x})$ of the d component vector $\mathbf{x} = \{x_i\}$ with $x_i = \mathbf{Q}_i \mathbf{n} + \alpha_i$ which is 2π-periodic with respect to each of its variables x_i. The Fourier expansion of $V(\mathbf{x})$ determines the coefficients $V_\mathbf{m}$ of $\exp[i\mathbf{m} \cdot \mathbf{x}]$ which are equal to the coupling constants of the tight binding model.

For these general self-dual models, there exists no useful generalized form of the Thouless formula which is applicable and thus, the localization proof used above does not work. However, numerical tests on several such self-dual models confirm that for $k < 1$, all the eigenstates are quasi-Bloch waves and for $k > 1$ all the eigenstates are localized. The localization transition also occurs at the fixed point of the duality transformation. The fact that the localization transitions occurs simultaneously for all the eigenstates at a single value for the parameter k, confirms the peculiarity of self-dual models. As for the simple Equation (31), this localization transition is a TBA with a similar qualitative behavior as in 1 dimension (see the next subsection). These arguments show that the localization transition in an incommensurate potential is not a purely one-dimensional phenomena but exists for any dimension.

3.5. QUESTIONS AND REMARKS CONCERNING DISCONTINUOUS QUASI-PERIODIC POTENTIALS

When the quasiperiodic potential $V(x)$ is not an analytic function, the eigenstates are always nonanalytic functions of the phase variation. However, the physically important fact is not the breaking of analyticity but the breaking of continuity. We could have used this term instead of breaking of analyticity, but in reasonable models where the initial potentials are analytic, this transition appears to be also a breaking of analyticity.

Therefore, with discontinuous quasiperiodic potentials $V(x)$, the process of localization of the eigenstates is expected to be drastically different than the cases where $V(x)$ is a periodic smooth function (which is at most uniformly continuous). We did few numerical studies on small systems concerning tight binding Schroedinger equations with a discontinuous 2π-periodic square function

$$V(x) = k \operatorname{sign}(\cos x)$$

which apparently showed the localization of eigenstates at least for large k. It has been also checked [40] that the expansion at first order with respect to the amplitude k of the potential of a quasi-Bloch wave $\Psi_n = \exp(iqn)\chi_n = \exp(iqn)$ $(1 + k\chi_n^{(1)} + \ldots)$ yields a diverging form for $\chi_n^{(1)}$. The standard perturbation calculation of $\chi_n^{(1)}$ yields a series where the terms have small denominators (eigenenergy differences) proportional to $(\cos q - \cos(q + Qn))$ which 'compete' with numerators which are the nth Fourier coefficients V_n of $V(x)$. Because these numerators which decays as $1/n$ at order n, there exists a subsequence of terms in the series not converging to zero. Consequently the sum $\chi_n^{(1)}$ cannot be defined. We physically interpret this situation by the fact that any plane wave eigenstate at $k = 0$ is broken as soon $k \neq 0$.

Again a strong analogy appears between this problem and the problem of calculating to first order the configuration of modified FK models where the periodic potential has a discontinuous derivative (see the previous section). In this second problem, the divergence of this calculation is related to the fact that the TBA just occurs at zero amplitude potential. On the basis of this analogy, we speculated that the absence of quasi-Bloch waves should imply the localization of all the eigenstates for all nonzero values of the amplitude k of the discontinuous potential.

This conjecture has been indeed proved for an example of discontinuous quasipotential with hull function $V(x) = k \tan(x)$ (which also diverges at the discontinuities). Prange *et al.* have proved that all the eigenstates are exponentially localized for any value of k [50]. Nevertheless, our argument based on perturbation theory may seem to be wrong in some cases.

In the case where $V(x) = k \operatorname{sign} (\cos x)$, our conjecture is not mathematically exact. Indeed, it has been proved by Delyon and Petritis [51] for the discontinuous quasiperiodic potentials $V(x)$ which take a finite number of values that, for any value of k, there cannot exist any eigenstate $\{\Psi_n\}$ of (38) which goes to zero for $|n| \to \infty$. Then, in the particular case $V(x) = k \operatorname{sign} (\cos x)$ the existence of localized states is impossible. We expect that the essential consequence of the discontinuity of the periodic potential is that the eigenstates are *nondegenerate* although not square summable. (This assertion implies that the eigenstates cannot be doubly degenerate quasi-Bloch waves because the complex conjugate $\{\Psi_n^*\}$ of a quasi-Bloch wave $\{\Psi_n\}$ would also be an eigenstate). They correspond to what is called in mathematics a singular continuous spectrum. We can consider that we have a 'marginal localization' because these eigenstates exhibit an *intermitent behavior* at infinity. In other words, their amplitude $A_n^2 = |\Psi_{n+1}|^2 + |\Psi_n|^2$ does not converge to zero but has a lower limit: $\lim \inf_{n \to \infty} A_n^2$ which is zero. They are in some sense intermediate between extended states and localized states although they do not have a smooth power law behavior at infinity. Studies by Kohmoto *et al.* [52] and Otslund *et al.* [53] with different but similar square potentials support this picture.

The application of the Kubo—Greenwood formula [54] described in the next subsection implies that whatever the spectrum at the Fermi energy of the quasi-periodic potential be discrete (localized eigenstates) or singular continuous ('marginal localization'), the infinite system is an insulator (zero conductivity) at absolute zero. Consequently, the breaking of the quasi-Bloch waves can be considered as a TBA even if a strict localization is not obtained beyond the critical point. This TBA essentially corresponds (at 0 K) to a transition from a conducting state to an insulating state.

3.6. COMPARISON BETWEEN EXTENDED STATES IN QUASI-PERIODIC AND RANDOM POTENTIALS

We end this section concerning the localization transition in quasiperiodic potentials by describing some numerical experiments, in order to point out the important physical difference which exists between the extended states in quasi-periodic potentials and those in random potentials. We analysed [56] some tight binding Hamiltonian H with first neighbor coupling exchange on one, two, or three dimensional lattices with the form (in one dimension):

$$H|\Psi\rangle = \Psi_{n+1} + \Psi_{n-1} + kV_n\Psi_n \tag{43}$$

where V_n is either a quasiperiodic potential as in (31) or a random potential. k is the amplitude parameter which allows one to pass the localization transition, if any. In more than one dimension, the same Hamiltonian is extended on two and three dimensional square lattices with on-site potentials V_n which are quasiperiodic or random (**n** being then a 2 or 3 dimensional integer vector) and with first neighbor exchange coupling, which is unity in all space directions.

The motivation of this study was initiated after a useful discussion with J. Friedel for a better understanding of the conductivity of these systems. Since this study is connected with electron transport properties, let us recall the Kubo—Greenwood formula.

3.6.1. *The Kubo—Greenwood Formula*

The conductivity $\sigma_E(\omega)$ of a system at frequency ω and temperature T can be expressed from the knowledge of the electron eigenfunctions by the standard Kubo—Greenwood formula (see for example Mott and Davis [54]). (It is obtained as an application of the theory of linear response to an applied electric field.). At zero frequency (dc electric field) and at zero Kelvin, this formula becomes

$$\sigma_E(0) = \frac{2\pi e^2 \hbar \Omega}{m^2} N(E)^2 |P(E)|_{av}^2 \tag{44a}$$

where e is the charge of the electron, Ω is the volume of the sample, $\hbar$ is the Planck constant, m the mass of the electron. $N(E)$ is the density of state at the Fermi energy E per unit volume and $|P(E)|_{av}^2$ is the average over all possible pairs of normalized electronic eigenstates $\Psi_\alpha(\mathbf{r})$ and $\Psi_{\alpha'}(\mathbf{r})$ at the Fermi energy of

$$P(E) = \int \Psi_{\alpha'}^*(\mathbf{r}) p_x \Psi_\alpha(\mathbf{r}) \, d\mathbf{r} \tag{44b}$$

where $p_x = i\hbar \, \partial/\partial x$ is the variable conjugate to x which is the space component of **r** in the direction of the applied electric field. As a consequence of this formula, it is shown [54] that when the eigenstates are localized at the Fermi surface, $P(E)$ and consequently the dc conductivity is zero at 0 K. This property is due to the fact that the localized eigenstates are nondegenerate and consequently real. Then we necessarily have $\alpha = \alpha'$ and $\Psi_\alpha^*(\mathbf{r}) = \Psi_\alpha(\mathbf{r})$. By performing the integration in (44b), it comes out that $P(E) \equiv 0$ and $\sigma_E(0) = 0$. By contrast, when the eigenstates are plane waves with a mean free path L, it is shown that the conductivity is finite and proportional to L. (This mean free path is physically interpreted as the characteristic length beyond which the phase coherence of the eigenstates at the Fermi energy disappears.) In the case of a perfectly periodic system with an infinite volume Ω, L is infinite so that the conductivity $\sigma(E)$ becomes also infinite.

As a consequence, the Kubo—Greenwood formula shows that the ability of a system to carry a finite current at small electric field and at 0 K, is essentially due to *the existence of degenerate eigenstates* at the Fermi energy.

3.6.2. *The Numerical Technique*

In one dimensional systems with first neighbor exchange coupling, this degeneracy is at most two so that the average in formula (44a) is trivial. Therefore, it appears interesting to calculate numerically a quasi-Bloch wave with a given wave vector q and to follow the variation of its momentum P as a function of the amplitude k of the quasiperiodic potential. This momentum is proportional to the electric current carried by this eigenstate.

The Hamiltonian (43) is written under the form $H = H_0 + kV$ where $H_0|\Psi_n\rangle \geq |\Psi_{n+1}\rangle + |\Psi_{n-1}\rangle$ is the unmodulated part of the Hamiltonian, the eigenstates of which are planewaves $|\Psi_n\rangle = |\exp(iqn)\rangle$ and $V|\Psi_n\rangle = |2\cos(Qn + \alpha)\Psi_n\rangle$ is the operator associated to the quasiperiodic potential which breaks the translational invariance. By analogy with the KAM theory, H_0 is considered as an integrable Hamiltonian and V is considered as a perturbation. Then, the planewaves play the role of the invariant tori of an integrable dynamical system while the perturbation V will tend to break up this solution.

For studying the transformation of a planewave under the action of the perturbation kV, we did not diagnolize the whole Hamiltonian for each value of k, but we use a safe 'gradient method' similar to the one which was used for the FK model when calculating its ground state and other metastable configurations. The eigenstates of H correspond to the zero minima of the square energy fluctuation

$$F(\{\Psi\}) = \langle\Psi|(H - \langle\Psi|H|\Psi\rangle)^2|\Psi\rangle)$$
$$= \langle\Psi|H^2|\Psi\rangle - (\langle\Psi|H|\Psi\rangle)^2 \tag{45a}$$

in the space of all normalized $|\Psi\rangle$ (with $\langle\Psi|\Psi\rangle = 1$). The numerical procedure consists in performing the integration from a given initial normalized state $\{\Psi_n(0)\} = |\Psi(0)\rangle$ of the first order equation

$$\left|\frac{d\Psi(\tau)}{d\tau}\right\rangle = -|\nabla F(\{\Psi\})\rangle + |\Psi\rangle\langle\Psi|\nabla F(\{\Psi\})\rangle \tag{45b}$$

where $\nabla F(\{\Psi\}) = \{\partial F(\{\Psi\})/\partial\Psi_n\}$. The second term of the second member of (45b), which corresponds to a Lagrange multiplier, is such that

$$\frac{d\langle\Psi|\Psi\rangle}{d\tau} = \left\langle\Psi\left|\frac{d\Psi(\tau)}{d\tau}\right\rangle + \left\langle\frac{d\Psi(\tau)}{d\tau}\right|\Psi\right\rangle = 0 \tag{45c}$$

in order that $\langle\Psi(\tau)|\Psi(\tau)\rangle = 1$ remains a constant. The limit $\{\Psi_n(\infty)\}$, is a minimum of $F(\{\Psi\})$ in the space of normalized state. If this minimum is zero, it is a normalized eigenstate of H.

For following continuously an eigenstate of H as a function of k, the parameter k is increased from zero by small steps δ. At each step, we check that the minima of $F(\{\Psi\})$ remains zero which insures that the solution obtained for $\{\Psi\}$ is really an eigenstate of H for the corresponding value of k. Let us emphasize that the eigenenergy of this eigenstate is not fixed in this procedure, and indeed this *eigenenergy does vary* as a function of k. Generally, we have chosen $\delta = 0.05$ (but this sep has been chosen to be smaller in 'critical regions' where the convergence of Equation 45 is more difficult). For each value of k, one calculates the new eigenstate with an initial state which is the eigenstate obtained at the previous step for $k - \delta$ and so on. For big computers, this method is much less efficient than many other methods but its advantage is that it can easily be performed on minicomputers because it only requires the storage of few vectors the size N of which is the number of sites of the system instead of matrices $N \times N$.

Since the models which we study are tight binding models, the real eigenwaves are

$$\Psi(\mathbf{r}) = \sum_n \Psi_n \phi(\mathbf{r} - n\mathbf{a}) \tag{46a}$$

where $\phi(\mathbf{r} - n\mathbf{a})$ is the on-site real eigenwave at site n which is localized at $n\mathbf{a}$ ($\mathbf{a}$ is the unit spacing along the chain). The momentum of $\Psi(\mathbf{r})$ is

$$P = i\hbar \int \Psi^*(\mathbf{r}) \frac{\partial \Psi(\mathbf{r})}{\partial x} \, d\mathbf{r} = i\hbar \sum_{m,n} \Psi_n^* \Psi_m C_{n-m} \tag{46b}$$

where

$$C_{n-m} = -C_{m-n} = \int \phi(\mathbf{r} - n\mathbf{a}) \frac{\partial \phi(\mathbf{r} - m\mathbf{a})}{\partial x} \, d\mathbf{r} \tag{46c}$$

Since the tight binding model which we consider assumes that the exchange coupling between localized wave functions at sites n and m are zero for $|m - n| > 1$, for consistency we also assume that $C_{m-n} = 0$ for $|m - n| > 1$. Then the momentum of $\Psi(\mathbf{r})$ is

$$P = 2\hbar C_1 \sum_n \mathrm{Im}(\Psi_{n+1}^* \Psi_n) \tag{47a}$$

where C_1 is some constant. Because of the form (41), in the one-dimensional Hamiltonian H, $\mathrm{Im}(\Psi_{n+1}^* \Psi_n))$ is independent of n for any eigenstate $\{\Psi_n\}$ of H. For convenience, we set

$$P = 2\hbar C_1 p \tag{47b}$$

where p is given by

$$p = \sum_n \mathrm{Im}(\Psi_{n+1}^* \Psi_n) \tag{47c}$$

and is proportional to the current which is carried by this eigensolution.

For a Hamiltonian in three dimensions, the current on the bound in the direction x (resp. y, z) with unit vector $\mathbf{a}$ (resp. $\mathbf{b}$, $\mathbf{c}$) at the site $\mathbf{n}$ is proportional to

$$p_{\mathbf{n}, x} = \text{Im}(\Psi_{\mathbf{n}+\mathbf{a}}^* \Psi_{\mathbf{n}})). \tag{47d}$$

It depends on the direction x, y or z and on the site $\mathbf{n}$. However, it fulfills the current conservation equation at each site $\mathbf{n}$

$$p_{\mathbf{n}, x} + p_{\mathbf{n}, y} + p_{\mathbf{n}, z} - p_{\mathbf{n}-\mathbf{a}, x} - p_{\mathbf{n}-\mathbf{b}, y} - p_{\mathbf{n}-\mathbf{c}, z} = 0. \tag{48}$$

The global current carried by this eigenstate is then a vector $\mathbf{p}$

$$\mathbf{p} = p_x \mathbf{a} + p_y \mathbf{b} + p_z \mathbf{c} \tag{49a}$$

where p_x (resp. p_y, p_z) are the sum over all sites $\mathbf{n}$

$$p_x = \sum_{\mathbf{n}} p_{\mathbf{n}, x}. \tag{49b}$$

We calculated the current modulus $|\mathbf{p}|$ as a function of k for the complex eigenstate which is a continuous function of k and which is a given plane wave for $k = 0$. In practice, we studied a periodic system where this eigenstate necessarily remains a Bloch wave. Then this current $\mathbf{p}$ is finite (providing that the wave vector of the corresponding wave function is not a multiple of the wave vector associated with this periodicity or that in other words the corresponding eigenenergy does not belong to the edge of a gap).

When the period of the system diverges in order that the limit system be quasi-periodic or random, it is found that either $\mathbf{p}$ remains finite or goes to zero. When $\mathbf{p}$ remains finite, there still exists a complex eigenstate for the infinite system and the conductivity for a Fermi energy equal to the corresponding eigenenergy, should be infinite according to the Kubo—Greenwood formula.

By contrast, when $|\mathbf{p}|$ goes to zero, we found that the conductivity may be either zero (localized eigenstates) or finite (extended eigenstates in random potentials). In this second case, it is found that although the average (49) becomes zero, the root mean square of $p_{\mathbf{n}, i}$ is not zero. This eigenstate can be interpreted as a glass of current.

3.6.3. *Results for a Quasi-Periodic Potential in One and Two Dimensions*

We first studied the Harper Equation (31) for $Q = 2\pi r/s$ where r/s are successive rational best approximations of the inverse golden mean. The initial eigensolution is the plane wave

$$\Psi_n = \frac{\exp(iqn)}{\sqrt{s}} \quad \text{for} \quad k = 0 \tag{50}$$

where q has been chosen in order to not fulfill Equation 30. The global momentum $\mathbf{p}(k)$ is a scalar $p(k)$. Then in the absence of quasiperiodic potential $(k = 0)$, $p(k) = \sin q$ is independent of the system periodicity s. When s increases, the curve $p_s(k)$ converges to a limit curve $p(k)$. This curve clearly vanishes at the critical value $k = k_c = 1$ (see Figures 18a and 18b) and remains zero beyond this critical point. For this 1-d model where the current flow is preserved, a zero momentum implies for all n

$$\text{Im}(\Psi_{n+1}^* \Psi_n) = 0 \quad \text{for} \quad k > k_c \quad (51)$$

which means that the real part and the imaginary part of $|\Psi\rangle$ are proportional and thus are no more independent eigenstates. This feature is associated with the fact that the localized eigenstates are necessarily nondegenerate in one dimension.

Clearly, the critical behavior of $p(k)$ for $k < k_c$ depends on the choice of q (compare Figures 18a and 18b). We recently found that it is determined by the sequence of integers m_i associated with the sequence of 'best approximations' Δ_i of $2q$ by $mQ \mod 2\pi$ [57]. Considering the set of numbers $\Delta_{m,n} = mQ - 2n\pi$, a best approximation Δ_i to $2q$ is defined as a number $\Delta_{m,n}$ which fulfills the condition

$$|2q - \Delta_{m,n}| < |2q - \Delta_{m',n'}| \quad (52)$$

for all n and for all m' such that $|m'| < |m|$. It is the best approximation up to order m.

Note that when $q = 0$, this definition yields the well-known sequence of best approximations n_i/m_i of $Q/2\pi$ which is derived from the continued fraction expansion of $Q/2\pi$. For example, when $Q/2\pi = \zeta_g = (\sqrt{5} - 1)/2$ and $q = 0$, this sequence is determined by $m_i = F_i$ and $n_i = F_{i-1}$ where F_i is the well-known Fibonacci sequences. We found a generalized algorithm for finding this sequence. (The same problem was also encountered in the study of the Fourier spectrum of certain quasiperiodic bound modulated structures [57, 58].) Similarly to the breaking of KAM tori in the standard map, well-defined scaling properties at the localization transition should only exist for special choices of the couple of numbers $Q/2\pi$ and $2q$ which must belong to a periodic cycle of a certain transformation. In particular, this condition implies that $Q/2\pi$ has a periodic continued fraction expansion. Our prediction yields that the simplest choice for observing a well defined critical exponent for $p(k)$ at $k = k_c$ on Figures 18, should be $Q/2\pi = (\sqrt{5} - 1)/2$ (the inverse golden mean) and $2q/2\pi = 1/2$. (The case $q = 0$ or π looks simpler but yields a zero momentum.) This case has not yet been tested numerically. Preliminary studies [95] seems to confirm this conjecture.

The same numerical study can also easily be done in two or more dimensions. We studied the extended self-dual Harper equation (see Section 3.4)

$$\Psi_{n+1,m} + \Psi_{n-1,m} + \Psi_{n,m+1} + \Psi_{n,m-1} + kV_{n,m}\Psi_{n,m} = E\Psi_{n,m} \quad (53a)$$

with

$$V_{n,m} = \cos(Q_{11}n + Q_{12}m + \alpha_1) + \cos(Q_{21}n + Q_{22}m + \alpha_2) \quad (53b)$$

where $Q_{12} = Q_{21}$ and we found features similar to those obtained in one dimension. The nonresonant eigenstates in two-dimensions have a 2-component wave vector $\mathbf{q} = (q_1, q_2)$ which is nonresonant with the wave vectors $\mathbf{Q}_1 = (Q_{11}, Q_{12})$ and $\mathbf{Q}_2 = (Q_{21}, Q_{22})$ of the two dimensional quasiperiodic potential. This

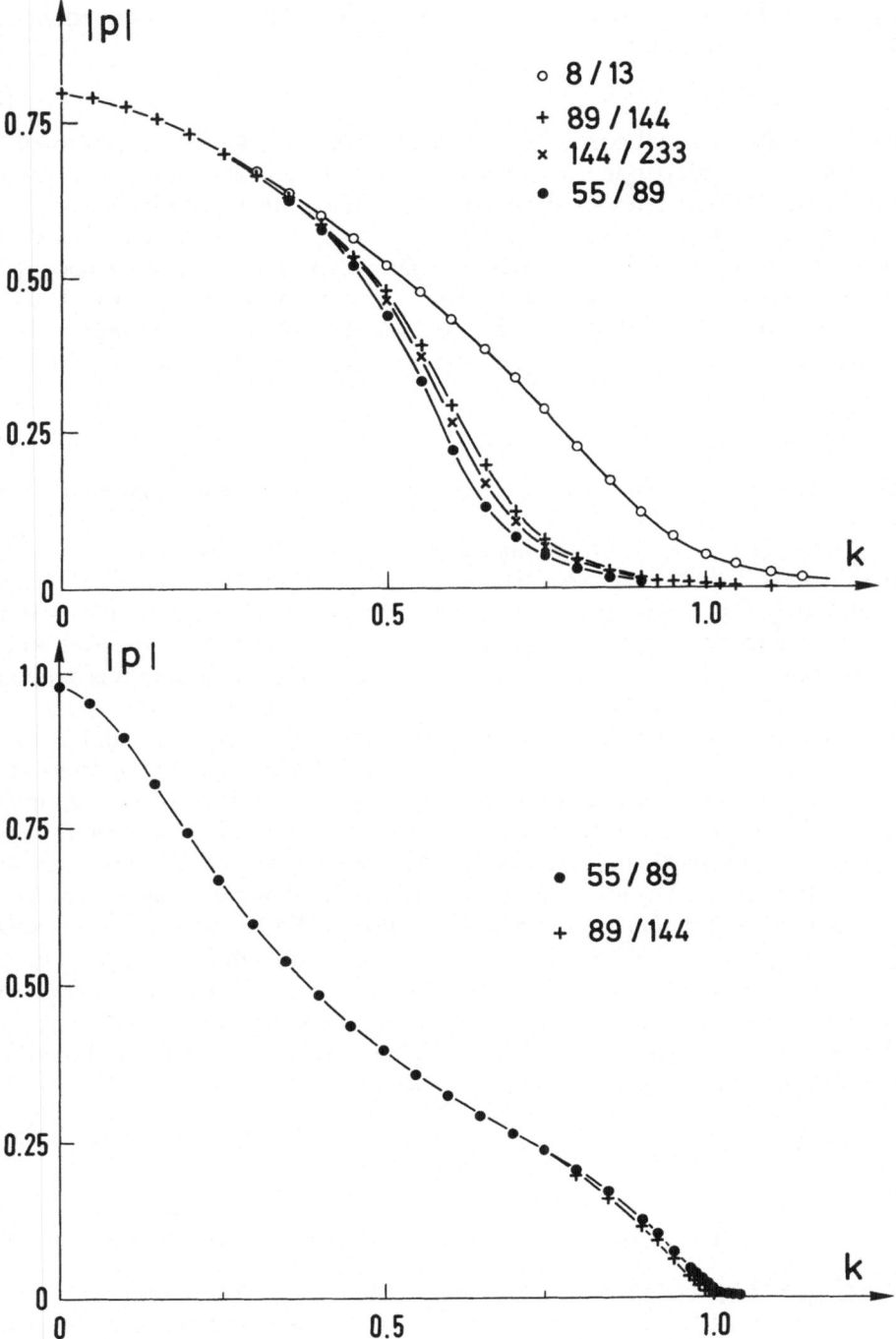

Fig. 18. Momentum $P(k)$ of the eigenstate with wave vector $q = \pi/\sqrt{2}$ (Figure 18a) and $q = \pi/\sqrt{3}$ (Figure 18b) versus k for $Q/2\pi = 8/13$, 55/89, 89/144 and 144/233 (best approximations of the inverse golden mean).

means that the inequality

$$2\mathbf{q} \neq m_1\mathbf{Q}_1 + m_2\mathbf{Q}_2 - 2n\pi \tag{54}$$

is fulfilled for any integers m_1, m_2 and n. As for the 1D case, we study periodic systems with a square unit cell $s \times s$ such that for large s, the on-site potential V_n of the Hamiltonian converges to a quasiperiodic one.

The momentum $\mathbf{p}(k)$ is equal to (sin q_1, sin q_2) for $k = 0$. Figure 19 shows an example of a variation of $|\mathbf{p}(k)|$ as a function of k. Although no rigorous proof was given that a localization transition should occur at the fixed point of the dual transformation which corresponds to $k = k_c = 1$, we found numerically that localization does indeed occur for this k value (see Figure 19). For $k > k_c$ it has been found for the largest system (19×19) that the real and imaginary part of the wave function becomes proportional and that they apparently localize. Again the critical behaviour of $|\mathbf{p}(k)|$ is expected to depend on the wave vector $\mathbf{q}$. We have not yet found the rule for the choice of $\mathbf{Q}$ and $\mathbf{q}$ which yields well defined scaling properties. Further, more precise numerical studies would be useful.

No experiments were done in three dimensions on corresponding models, but it is reasonable to expect the same qualitative results as in one and two dimensions.

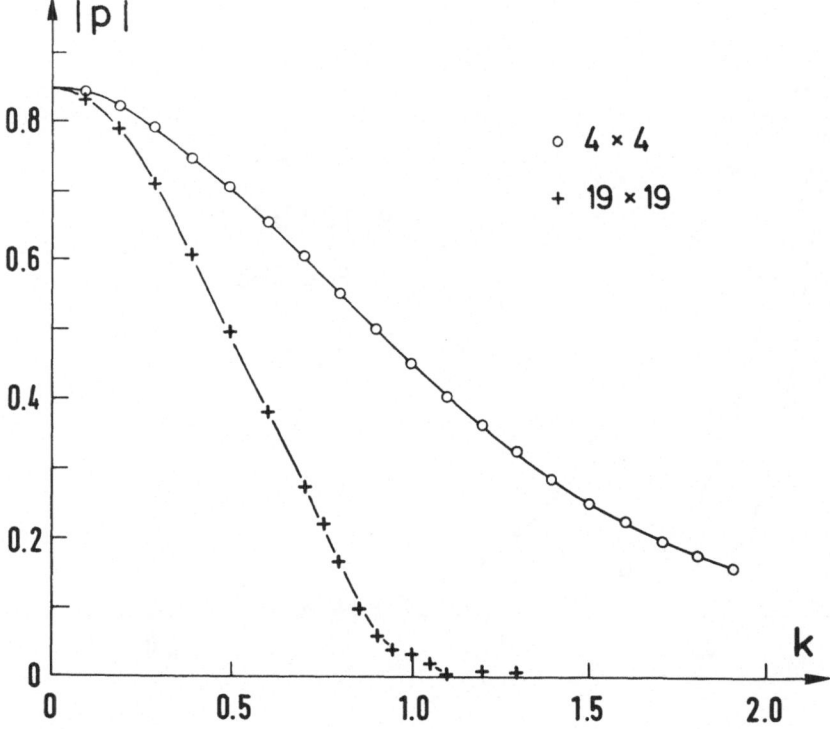

Fig. 19. Modulus of the momentum versus k for the eigenwave with wave vector $q_1 = 0.90319\pi$ and $q_2 = 0.70711\pi$ for a 2 dimensional self-dual Equation (42) in two dimensions for $Q_{11} = (3/19) \times 2\pi$, $Q_{12} = Q_{21} = (7/19) \times 2\pi$ and $Q_{22} = (11/19) \times 2\pi$.

3.6.4. *Results for a Random Potential in One and Three Dimensions*

The same numerical technique used for a quasiperiodic potential can be applied to the tight binding Hamiltonian with the form (43) in one dimension or its generalized form in two and three dimensions with a random on-site potential V_n. We have chosen the Anderson potential [59] where V_n has a uniform distribution on the interval $[-1, 1]$. We study in fact a periodic system with a finite unit cell with period s in all space direction, where the eigenstates are necessarily Bloch waves. We choose as the initial eigenstate at zero random potential ($k = 0$), an arbitrary Bloch wave $\Psi_n \propto \exp(i\mathbf{q}\mathbf{n})$ and study the global momentum.

In one dimension, the program did converge efficiently (by checking that the numbers of nodes of the real and imaginary part of the eigenstate remain constant and equal for all values of k, we can be sure that we follow the same eigenstate as a function of k). We found that $p_s(k)$ tends to vanish for all finite values of k when s diverges (see Figure 20). As for the quasiperiodic case, when $p(k)$ vanishes, the real and imaginary parts become proportional and then the eigenstate is localized.

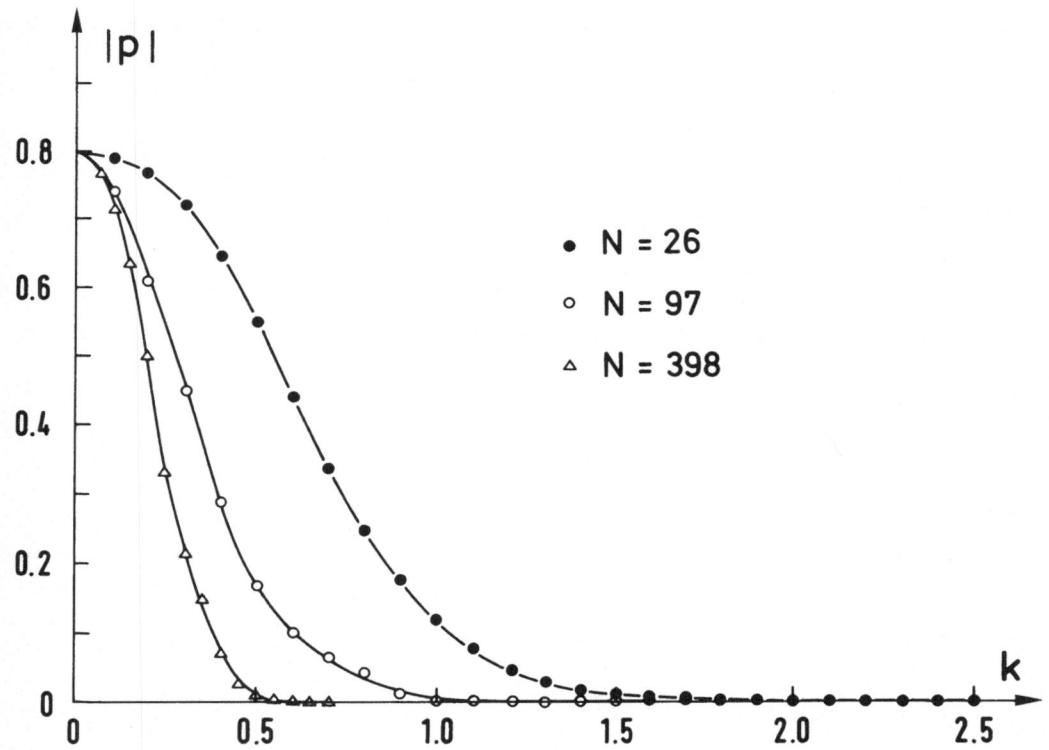

Fig. 20. Momentum $P(k)$ of the eigenstate with wave vector $q = \pi/\sqrt{2}$ for a periodic one dimensional system of period s, with a random set of s on-site energies. $s = 26, 97$ and 398.

For finite s, the crossover at which $p(k)$ sharply decreases depends on the size of the system s and can be estimated within perturbation theory with respect to the random potential V_n. Since the system is periodic with period s, one expects that a gap in the density of states of the unperturbed system ($k = 0$) is opened at each q value equal to $q_m = m\pi/s$. The amplitude of this gap δ_m is given by

$$\delta_m = 2|\langle\Psi|kV|\Psi\rangle| = \frac{2k}{s}\left|\sum_{n=1}^{s}\exp\left(2inm\,\frac{\pi}{s}\right)V_n\right| \tag{55a}$$

where Ψ is the unperturbed plane wave with wave vector q_m. Since V_m are random coefficients between -1 and $+1$, the typical gap width is

$$\langle\delta_m^2\rangle^{1/2} = 2k(1/s^2)(s\langle V_m^2\rangle)^{1/2} = (2k/\sqrt{s})(\langle V_m^2\rangle)^{1/2} = 2k/\sqrt{3s}. \tag{55b}$$

When s goes to infinity, the width of each gap goes to zero as $1/\sqrt{s}$ but their number diverges proportionally to s. Let us consider an energy interval $\Delta E = 2\sin q\,\Delta q$ corresponding to the wave vector interval $[q, q + \Delta q]$. This interval is supposed to be small compared to π but large compared to π/N. For the consistency of this perturbation theory, the total width of the opened gap in this interval which is estimated to be

$$\Delta G \sim \frac{s\,\Delta q}{\pi}\,\frac{2k}{\sqrt{3s}} \tag{56a}$$

must be much smaller than the interval width ΔE. This condition yields that, for a given k, the period of the considered system must not be too large

$$s \ll 3\pi^2\,\frac{\sin^2 q}{k^2}. \tag{56b}$$

When s becomes comparable to the second member of (56b), the perturbation theory becomes inaccurate because the Bloch eigenstate of the s-periodic system starts to differ significantly from a planar sine wave. This situation occurs when s becomes of the same order of or larger than the localization length ξ of the eigenstate of the infinite system at the same energy. Then we obtain from (56b) a rough estimation of this localization length in lattice units

$$\xi \approx 3\pi^2\,\frac{\sin^2 q}{k^2} \tag{57}$$

valid for small k or equivalently for $\xi \gg 1$. Conversely, for a given system size s, we estimate the crossover value $k_{co}(s) \approx \pi\sqrt{3/s}\sin q$ at which the momentum $p(k)$ of the Bloch wave with wave vector q starts to vanish. The curves of Figure 20 roughly agree with this estimation.

Similar numerical studies as in 1D could be done in 3D. However, for an infinite system, the limit for $k \to 0$ of an extended eigenstate in a random potential cannot be a single sine wave but is a combination of sine waves with the same eigenenergy. (By contrast, for quasiperiodic potentials, this limit is a single sine wave for the extended eigenstates which are quasi-Bloch waves). Consequently, for a random potential in more than 1 dimensions, we should take as an initial state instead of a single sine wave, an eigenstate of the random potential operator restricted to an eigenspace (at $k = 0$) corresponding to a given eigenenergy.

In practice for the numerical tests which we did, we have chosen as initial eigenstate a single sine wave. In principle, for a *periodic* system for which all the eigenstates are necessarily Bloch waves, we should be able to follow a given Bloch wave from an initial sine wave continuously as a function of k. The program which we used for relatively small sizes succeeded in following an eigenstate of the periodic 3D Hamiltonian as a function of k but we cannot guarantee that no discontinuous jump from one eigenwave to another one occured.

Despite the fact that our numerical results obtained on a small computer could be improved, we nevertheless always obtained a complex eigenstate of the 3D Hamiltonian. Figure 21a shows the variation of the global momentum $|\mathbf{p}(k)|$ of such an eigenstate as a function of k. Clearly, as the size s increases, the global current carried by this eigenstate vanishes for all nonzero values of k. This result differs sharply from the result obtained for a quasiperiodic potential in one or two dimensions.

In fact, we can examine the local density of current carried by this extended eigenstate, which is defined as

$$p_{lc} = s^3 [\langle p_{m,n}^2 \rangle_{\langle m,n \rangle}]^{1/2} \tag{58}$$

where $\langle m, n \rangle$ represents an oriented bond between neighboring sites and $p_{m,n} = \mathrm{Im}(\Psi_n^* \Psi_m)$ is the local momentum of the complex eigenstate on this bound. $p_{m,n}^2$ is averaged over all bounds. We plotted p_{lc} as a function of the random potential amplitude k (cf. Figure 21b), in order to have a physical quantity independent of s in the limit of large s. In order to reduce our numerical error, we averaged the variation curves obtained for six different random samples with size $7 \times 7 \times 7$. We now clearly distinguish the localization transition which occurs for $k = k_c \approx 10$.

When $k < k_c$, the complex eigenstates of the random Hamiltonian have non-proportional real and imaginary parts. They appear as a random distribution of local currents with no defined orientation although they fulfill the conservation rule at each node of the 3D lattice. In fact the global current is not strictly zero for a finite size unit cell since the local current is random. Then, the global current is expected to decrease as a function of this size s proportionally to $L^3 p_{lc}/\sqrt{s^3}$ where L is the correlation length of this current glass. Assuming that this property is fulfilled for all the eigenstates at the Fermi energy, the Kubo—Greenwood formula then yields a finite conductivity proportional to the average of $L^3 p_{lc}$. We have not studied the correlation length L of this glass of current which corresponds to the mean free path of the quantum particle in this eigenstate. However, we can guess that the crossover in the variation of $|\mathbf{p}(k)|$ (see Figure 21a) indicates the value of k at which the size s of the unit cell becomes roughly proportional to $L(k)$. For $k > k_c$ we observe that the real and imaginary parts of the eigenstate which we

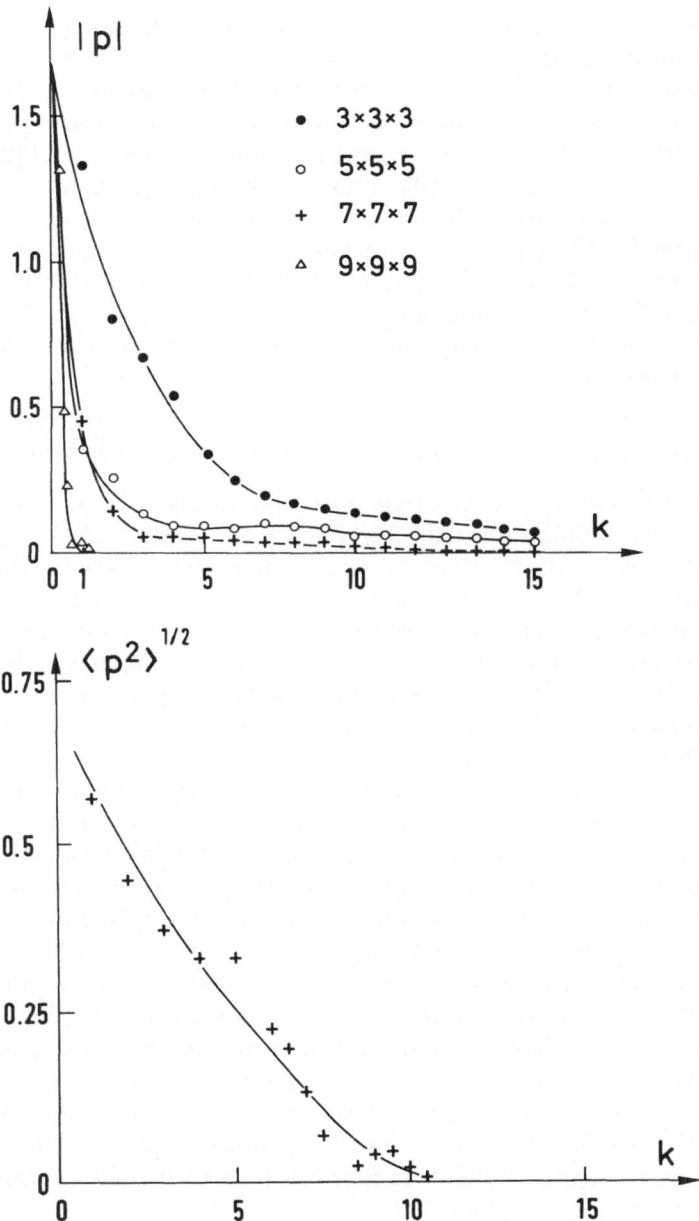

Fig. 21a. Modulus of the momentum $P(k)$ versus k of an eigenstate for a wave vector with components $q_1 = 0.37895 \times \pi\, q_2 = 0.56364 \times \pi$ and $q_3 = 0.55521 \times \pi$ in a periodic three-dimensional system with a cubic unit cell of size s with s^3 random on-site energies. The sizes s are $3 \times 3 \times 3, 5 \times 5 \times 5, 7 \times 7 \times 7, 9 \times 9 \times 9$. Note that for large size system, the global momentum goes to zero. *Figure 21b.* Variation of the square root of the quadratic mean value of the local momentum averaged over six random samples for a periodic system $7 \times 7 \times 7$.

continuously follow becomes proportional. Then no local current is carried and the eigenstate becomes localized.

In summary, the numerical test which have been performed show that *the localization transition in a quasiperiodic system and in a random potential are very different* despite both system can produce localization and mobility thresholds. The extended states in quasiperiodic systems are quasi-Bloch waves with a finite momentum. They undergo a TBA at a certain critical point which is qualitatively analogous to the breaking of a Kam torus and where their momentum vanishes. By contrast, the extended states in random 3D-systems have no well defined wave vector and a global zero momentum. However they exhibit finite local momenta with disordered orientations. The root mean square of these momenta vanishes at the localization transition.

4. The Transition by Breaking of Analyticity in One-Dimensional Peierls Chains

After having studied the effect of an incommensurate external potential on a periodic lattice (Section 2) and next the effect of an incommensurate lattice on the eigenstates of an electron (Section 3), we now study the self-consistent problem in one dimension, namely the ground state of Peierls chains. The Peierls instability originates essentially from the existence of an *electron—phonon coupling* in conducting systems where the propagation of the electrons in quasi-one-dimensional. (But note that similar lattice instabilities may exist for systems of electrons propagating in higher dimensions providing a strong enough electron—phonon coupling.)

We briefly explained the mechanism of this Peierls instability in the introduction to this paper. If we assume that the electronic density is frozen, the lattice can be considered as being submitted to a periodic potential proportional to this electronic density. Because this situation has strong analogies with the FK model where the role of the potential amplitude k is replaced by the electron—phonon coupling parameter, we conjectured that a transition by breaking of analyticity could exist in incommensurate Peierls chains at some value of the electron—phonon coupling. Otherwise, the quasiperiodic potential due to the lattice distortion can produce a localization transition of these electrons again beyond some value of the electron—phonon coupling, as we saw in the previous section. Then, considering the TBA, on the same basis as a usual symmetry breaking, we conjectured that the TBA could not occur separately in different parts of the incommensurate system and therefore that the TBA of the lattice should coincide with the TBA of the electrons [60].

Indeed, on one hand, once the analyticity of the hull function of the periodic lattice distortion is broken, the electrons become submitted to a quasiperiodic on-site potential which also has a *discontinuous hull function*. According to the previous section, one expects that all the quasi-Bloch eigenstates of the electrons break up. On the other hand, if the electronic waves at the Fermi energy are not quasi-Bloch waves, these eigenstates are not *continuous functions of the phase α* and then the hull function of the electron density can be expected to be a discontinuous function like for the self dual model in the localized regime (see Section 3.2). Therefore, the periodic potential which acts on the lattice being discontinuous, the analyticity of the hull function of the periodic lattice distortion must be broken. We concluded that the TBA is a *collective phenomenon* which concerns the properties of the whole system.

In the self dual model, the localization transition occurs when the largest gap in the electronic density becomes comparable to the band width of the free electron system. By contrast, when approaching the breaking of analyticity of the periodic lattice distortion, the electrons tends to localize, while the tendency to the localization of the electrons tends to break the analyticity of the PLD. Consequently, this cooperative effect should strongly reduce the critical value of the electron phonon coupling at which the TBA occurs. Indeed in the Holstein model which will be studied, the TBA is obtained when the electronic gap is at most 10% of the band width of the free electron system. As for the FK model, this critical value is expected to become smaller for incommensurability ratios close to simple rationals.

The effect of small Coulomb electron—electron interactions, not included in the models studied in the following, should be taken into account for the breaking of analyticity of the CDW. (Note that for strong enough electron—electron interaction, a spin density wave should appear instead of a CDW.)

In the nonanalytic incommensurate CDW, the localized electrons can be viewed as polarons because their charge density is necessarily associated to a lattice distortion due to the electron—phonon coupling. But if the Coulomb interaction between the electrons is not too strong, at 0 K this electronic state must be occupied by two electrons with opposite spins. The resulting object is called a *bipolaron*. Therefore, we suggested interpreting the nonanalytic charge density waves as an array of lattice pinned bipolarons (In [60], in the context of a spinless electron model, only the word polaron was used). The nonanalytic CDW then appears as a phase where the electrons pair up in the *real space*. It is interesting to note that, on the basis of a completely different approach, similar hypotheses were formulated by Alexandrov and Ranninger[61] and Alexandrov *et al.* [62] for systems with strong electron—phonon coupling. However, their physical assumptions and also their conclusions are not the same and will be discussed in the next (concluding) section.

This conjecture of breaking of analyticity in Peierls chains has been the motivation of the numerical study of two simple one-dimensional models [17, 18, 19] which are commonly used. These two models have been considered as being very different, in order to suggest that the existence of a TBA is not qualitatively sensitive to the details of the model (although it is quantitatively sensitive). The existence of a TBA with the same qualitative features as in the FK model, has been confirmed on these two models.

For sake of simplicity, these models assume that:

1 — The atoms of the lattice are *classical particles*. This assumption is physically valid as soon as the quantum fluctuations of the atomic positions become

negligible with respect to the effective distortion of the lattice due to the instability.

2 — The electrons are *non-interacting fermions* which essentially propagate in *one dimension*. (The fact that the Coulomb interaction between the electrons be neglected, prevents the possible formation of spin density waves.)

The effect of these assumptions on the TBA will be studied in further works.

The energy at 0 K of these two models decomposes into a classical elastic part and a quantum electronic part

$$\Phi(\{u_i\}) = \Phi_{\text{elast.}}(\{u_i\}) + \Phi_{\text{electr.}}(\{u_i\}) \tag{59}$$

$\{u_i\}$ are atomic displacements. $\Phi_{\text{elast.}}(\{u_i\})$ is the elastic energy cost for an atomic distortion determined by $\{u_i\}$. In both model the electronic energy $\Phi_{\text{electr.}}(\{u_i\})$ is defined by

$$\Phi_{\text{electr.}}(\{u_i\}) = 2 \sum_{\mu \text{ occ}} E_\mu(\{u_i\}) \tag{60}$$

where $E_\alpha(\{u_i\})$ are the eigenenergies of the electrons in the distorted lattice. The eigenstates μ occ in (60) have an eigenenergy smaller than the Fermi energy E_F and are occupied by a pair of electrons.

4.1. THE HOLSTEIN MODEL

In the Holstein model, the elastic energy corresponds to a dispersionless optical phonon branch:

$$\Phi_{\text{elast.}}(\{u_i\}) = \sum_i \frac{1}{2} m\omega_0^2 u_i^2 \tag{61a}$$

where ω_0 and m are the constant frequencies and masses of identical oscillators located at each site i. The electronic eigenenergies $E_\mu(\{u_i\})$ are given by the tight binding eigenequation

$$-T\Psi_{n+1}^\mu - T\Psi_{n-1}^\mu + \lambda u_n \Psi_n^\mu = E_\mu(\{u_n\})\Psi_n^\mu \tag{61b}$$

where T is the electronic exchange coupling between two neighbouring sites and λ is the electron—phonon coupling. By taking $2T$ as the energy unit, and redefining

$$u_n = \mathbf{u}_n \sqrt{\frac{2T}{m\omega_0^2}} \tag{62a}$$

$$k = \lambda \sqrt{\frac{2}{Tm\omega_0^2}} \tag{62b}$$

the *classical* Holstein model contains k only as a parameter. The total energy (59) becomes a function of the new variables $\{\mathbf{u}_i\}$

$$\Phi(\{\mathbf{u}_i\}) = 2TF(\{\mathbf{u}_i\}) \tag{63a}$$

$F(\{\mathbf{u}_i\})$ is a dimensionless quantity given by

$$F(\{\mathbf{u}_i\}) = \sum_{i=1}^{N} \frac{1}{2} \mathbf{u}_i^2 + \sum_{\mu=1}^{\zeta N} \mathbf{E}_\mu(\{\mathbf{u}_n\}) \tag{63b}$$

where $\mathbf{E}_\mu(\{\mathbf{u}_n\})$ are the eigenvalues in increasing order of the equation

$$\mathbf{H}|\Psi\mu\rangle_n = -\Psi_{n+1}^\mu - \Psi_{n-1}^\mu + k\mathbf{u}_n\Psi_n^\mu = \mathbf{E}_\mu(\{\mathbf{u}_n\})\Psi_n^\mu. \tag{63c}$$

The numerical problem then becomes to minimize the functional $F(\{\mathbf{u}_i\})$ which involves only two independent parameters k and ζ. In practice, we studied Peierls chains submitted to periodic boundary conditions $\Psi_N^\mu = \Psi_0^\mu$ with N atoms. Since the number $\zeta_r N$ of pairs of electrons on the chain (with $0 < \zeta < 1$) must be an integer, the rational number ζ_r will be chosen to be as close as possible to a given irrational number ζ. The well-known optimal choice is that ζ_r belongs the sequence of rational truncations of the continued fraction expansion of ζ (which are called 'best approximations' to ζ).

We mostly studied the best approximations of the irrational number $\zeta_{g'} = (3 - \sqrt{5})/2$ which have the form F_{n-2}/F_n where F_n is the Fibonacci sequence. Because of the electron—hole symmetry, this model can be mapped onto the same model with ζ_g' replaced by $\zeta_g = 1 - \zeta_g' = (\sqrt{5} - 1)/2$ which is the inverse golden mean. This number ζ_g' has been chosen because the sequence of denominators F_n of its best approximations has the smallest rate of divergency among all possible irrational numbers. Therefore, for a given order n in the approximation of ζ_g', the size F_n of the system which we have to study is the smallest possible. We mostly studied the commensurate systems where $\zeta_r = 21/55$, $34/89$ and $55/144$. The results for these three systems are almost the same and therefore satisfactorily approach the properties of the infinite incommensurate system.

Our method for minimizing (63b) is the same as the one which we used for finding the ground states and the metastable states of the FK model. From an initial configuration $\{\mathbf{u}_j(0)\}$, we performed the numerical integration of the equation

$$\frac{d\mathbf{u}_i(\tau)}{d\tau} = -\frac{\partial F(\{\mathbf{u}_j\})}{\partial \mathbf{u}_i} = -\left(\mathbf{u}_i + \sum_{\mu=1}^{\zeta N} \frac{\partial \mathbf{E}_\mu(\{\mathbf{u}_j\})}{\partial \mathbf{u}_i}\right) = -(\mathbf{u}_i + k\rho_i) \tag{64a}$$

where

$$\rho_i = \sum_{\mu=1}^{\zeta N} |\Psi_i^\mu(\{\mathbf{u}_j\})|^2 \tag{64b}$$

is the electron density at site i. The finite system is submitted to periodic boundary conditions and contains s atoms and r pairs of electrons. For large enough values of τ, $\{\mathbf{u}_j(\tau)\}$ converges either to a ground state or to a metastable configuration of the model depending on the value of k and on the initial configuration as we shall see in the following. The configurations which are obtained are good approximations of the incommensurate ground state with the electron pair concentration ζ'_g.

Before studying the many body problem, let us examine the single bipolaron. Within the continuous approximation, it can be easily calculated. We have $-\partial F(\{u_j\})/\partial u_i = 0$ which yields $u_i = -k|\Psi_i|^2$ for the unique occupied eigenstate. Then the eigenstate is given in the continuous limit by:

$$-\frac{\partial^2 \Psi(x)}{\partial x^2} - 2\Psi(x) - k^2\Psi(x)|\Psi(x)|^2 = \mathbf{E}\Psi(x) \tag{65a}$$

which is valid for small k (large bipolaron). The normalized electronic eigenstate of the bipolaron is given by

$$\Psi(x) = \frac{k}{2\sqrt{2}} \frac{1}{\cosh\left(\dfrac{k^2 x}{4}\right)}. \tag{65b}$$

For small k, the formation of a localized bipolaron with respect to the configuration with an extended eigenwave corresponds to an energy gain of

$$\Delta \approx T \frac{k^4}{24} \tag{65c}$$

(in the initial model (61)). The inverse of the characteristic extension of the bipolaronic electronic density is in spacing units $\gamma = 1/\xi = k^2/2$. When $k = \sqrt{2}$, the bipolaron halfwidth is equal to the lattice spacing. This value 1.414 also roughly corresponds to the critical value $k_c \cong 1.58$ which is observed for the TBA for a good irrational number ζ.

For larger k, the continuum approximation for calculating the single bipolaron is no longer valid. In the limit of large k, the bipolaron becomes localized on a single site. By perturbation around the large k limit, the energy gain for forming a bipolaron is comparable to having a pair of extended electrons with no lattice distortion cf. (86d)

$$\Delta \approx 2T\left(\frac{k^2}{2} - 2 + \frac{4}{k^2} + \ldots\right). \tag{65d}$$

4.2. THE FRÖHLICH–SSH MODEL

In the Fröhlich–SSH model, the elastic part of the Hamiltonian is determined by acoustic phonons

$$\Phi_{elast.}(\{u_i\}) = \sum_i \frac{1}{2} mc^2(u_{i+1} - u_i - b)^2 \tag{66a}$$

where c is the sound velocity in the free chain and m is the atomic mass. The electronic eigenenergies $E_\mu(\{u_i\})$ which determine the electronic energy is given by the tight binding eigenequation where now the intersite exchange coupling (instead of the on-site energy) is modulated by the lattice distortion

$$-[T - \lambda(u_{n+1} - u_n)]\Psi^\mu_{n+1} - [T - \lambda(u_n - u_{n-1})]\Psi^\mu_{n-1}$$

$$= E_\mu(\{u_n\})\Psi^\mu_n \tag{66b}$$

b in (66a) is a parameter which we choose for convenience such that the equilibrium position of the chain assumed to be undistorted (with $u_{i+1} - u_i =$ cste), is obtained for $u_{i+1} - u_i = 0$. This condition yields:

$$b = \frac{4\lambda \sin \pi\zeta}{\pi m\omega^2} . \tag{66c}$$

After defining

$$u_n = \mathbf{u}_n \sqrt{\frac{2T}{mc^2}} \tag{67a}$$

this classical model also depends of a unique parameter k defined as

$$k = \lambda \sqrt{\frac{2}{Tmc^2}} . \tag{67b}$$

We get

$$\Phi(\{\mathbf{u}_i\}) = 2TF(\{\mathbf{u}_i\}). \tag{68a}$$

The numerical problem is to minimize the function

$$F(\{\mathbf{u}_i\}) = \sum_{i=1}^{N} \frac{1}{2} (\delta_n - \mathbf{b})^2 + \sum_{\mu=1}^{\zeta N} \mathbf{E}_\mu(\{\delta_n\}) \tag{68b}$$

where $\mathbf{E}_\mu(\{\delta_n\})$ is given by the eigenequation

$$\mathbf{H}|\Psi^\mu\rangle_n = -[1 - k\delta_{n+1}]\Psi^\mu_{n+1} - [1 - k\delta_n]\Psi^\mu_{n-1} = \mathbf{E}_\mu(\{\delta_n\})\Psi^\mu_n \tag{68c}$$

δ_n is the relative atomic displacement

$$\delta_n = \mathbf{u}_n - \mathbf{u}_{n-1} \tag{68c}$$

and

$$\mathbf{b} = b \sqrt{\frac{mc^2}{2T}}. \tag{68d}$$

Our numerical study is limited to the region in k where the exchange coupling

$[1 - k\delta_n]$ keeps a positive sign. When the Peierls distortion becomes too large so that this term becomes negative for certain values of n, the incommensurate CDW becomes unstable because the exchange coupling almost vanish at certain bounds of the chain. The numerical approach is the same as for the Holstein model. We numerically solve the equation

$$\frac{d\delta_i(\tau)}{d\tau} = -\frac{\partial F(\{\delta_j\})}{\partial \delta_i} = -\delta_i + \mathbf{b} + \sum_{\mu=1}^{N\zeta} \frac{\partial \mathbf{E}_\mu(\{\delta_j\})}{\partial \delta_i}$$

$$= -\delta_i + \mathbf{b} - \sum_{\mu=1}^{N\zeta} k(\Psi_{n+1}^{\mu*}\Psi_n^\mu + \Psi_n^{\mu*}\Psi_{n-1}^\mu) \tag{69}$$

which for large τ yields a metastable configuration or a ground state depending on k and on the initial configuration. We studied the same rational concentration $\zeta = 34/89$ for the electronic pairs as for the Holstein model. The model has 89 bounds and 34 pairs of electrons and is submitted to periodic boundary conditions.

As for the Holstein model, the bipolaron in the continuous limit (valid for small k) can be calculated. The electronic eigenfunction has the same form as (65b). For small k the bipolaron energy gain compared to an electronic extended state is $\Delta \approx 2Tk^4/3$. The inverse size of the bipolaron is $\gamma \approx 2k^2$. When k exceeds $1/\sqrt{2}$, the bipolaron size becomes comparable to the lattice spacing. But unlike the Holstein model, for larger k the bipolaron tends to localize on two consecutive sites with equal electronic density. Thus the bipolaron in the Fröhlich—SSH model do not localize as well as the bipolaron in the Holstein model which probably explain why the TBA in this second model is observed for a critical value $k_c \cong 1.215$ much larger than the expected value $1/\sqrt{2} \cong 0.707$.

4.3. NUMERICAL OBSERVATION OF THE TRANSITION BY BREAKING OF ANALYTICITY IN PEIERLS CHAINS

For both models, the conjecture for the existence of a TBA [60] has been confirmed [17, 18, 19]. For $\zeta = \zeta_g'$, the critical value of the electron—phonon coupling has been found to be located at

$$k = k_c(\zeta_g') \cong 1.575 \pm 0.01 \qquad \text{for the Holstein model} \tag{70a}$$

and

$$k = k_c'(\zeta_g') \cong 1.21 \pm 0.01 \qquad \text{for the Fröhlich—SSH model.} \tag{70b}$$

In both models, we found that for $k < k_c$, the obtained configuration is unique apart an arbitrary phase shift, whatever is the initial configuration which is used in Equation 64 or 69. Therefore, this configuration must be the ground state of the model considered and is *undefectible*. (We are in the situation schematized by Figure 9a.)

But note that this result has been obtained because all the electronic states have a double occupancy. The ground state is undefectible by *spinless* defects. If we assume that certain electronic states have single occupancy, localized defects with a single spin can be found. Their creation energy is about half the energy of the Peierls gap (middle gap states [63, 64]).

According to the Peierls prediction, the ground state exhibits an incommensurate periodic lattice distortion associated with a charge density wave (PLD—CDW). For the Holstein model, the PLD which corresponds to an on-site modulation, can be described with a 2π-periodic hull function $g(x)$ defined as

$$\mathbf{u}_i = g(2k_F i + \alpha) \tag{71a}$$

where $k_F = \pi r/s$ (r is the number of electron pairs and s the number of sites) and α is an arbitrary phase. For the Fröhlich—SSH model, this hull function $g(x)$ now corresponds to a bound modulation and is defined as

$$\delta_i = \mathbf{u}_{i+1} - \mathbf{u}_i = g(2k_F i + \alpha). \tag{71b}$$

In the commensurate case $\zeta = r/s$, $g(x)$ is only defined for s values on the interval $[0, 2\pi]$. But when s is large enough and reaches values such as $s = 55, 89$ or 144, this function appears to be sufficiently well defined and independent of s so that a transition by breaking of analyticity (cf. Figures 22 and 23) is clearly observable. This transition looks similar to the TBA in the FK model (which has a rigorous existence proof) and which is exhibited by Figures 6a and 6b, which have been numerically calculated.

However, in a Peierls chain this transition appears to be much sharper because the amplitude of the lattice distortion is nonlinear in k but proportional to $\exp(-4\pi/k^2)$ (in the half-filled case studied in the introduction). For $k < k_c(\zeta)$, $g(x)$ appears to be a smooth function close to a sine function. For $k > k_c(\zeta)$, we found that Equation 64 or 69 converges to many possible metastable configurations which depend on the initial state and which generally are chaotic. Rigorous results are now available and prove the existence of these chaotic metastable states for k large enough [91]. In this situation, we followed by continuity, the 'analytic' ground state obtained for $k < k_c(\zeta)$ and we found that the metastable configuration which is obtained by this method, always has a lower energy than any other metastable configuration obtained for the same value of k. Then we concluded that for $k > k_c(\zeta)$, this metastable configuration which continuously follows the undefectible ground state obtained for $k < k_c(\zeta)$ is still the ground state but which is now *defectible*. (We are in the situation of Figure 9b.) Since we assume in our model that only a pair of electrons can occupy an eigenstate, the defects of these nonanalytic ground states are *spinless*.

The hull function of these ground states defined by (71a) or (71b) is shown Figures 22 and 23. It is clear that for $k > k_c(\zeta) g(x)$ has many discontinuities having a close similarity to the behavior of the nonanalytic hull functions $f(x)$ in the FK model. In that model, it was proved rigorously [16, 41] that the hull function of nonanalytic incommensurate structures defined in (8b) is not only a discontinuous function with an infinite number of discontinuities but is a *discrete*

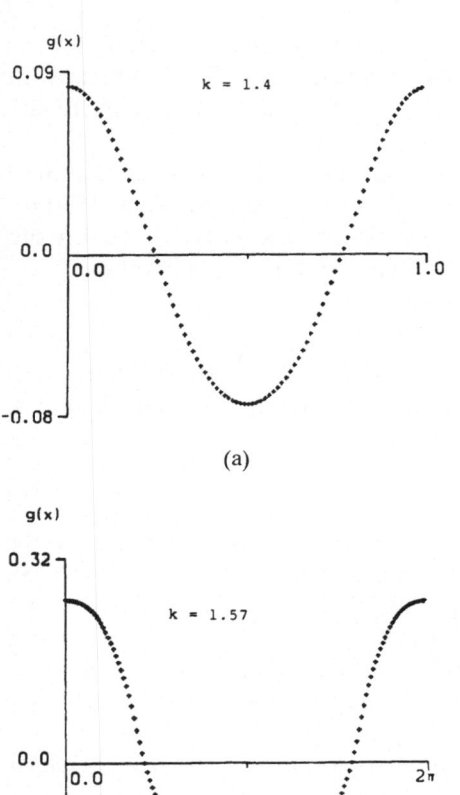

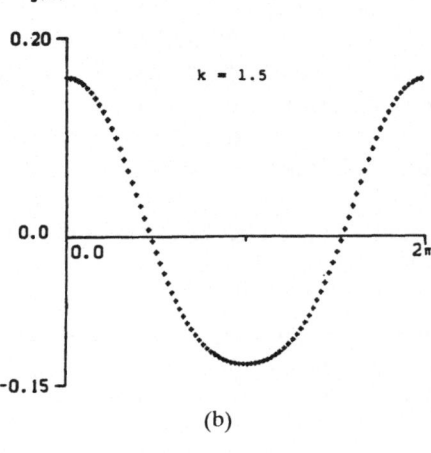

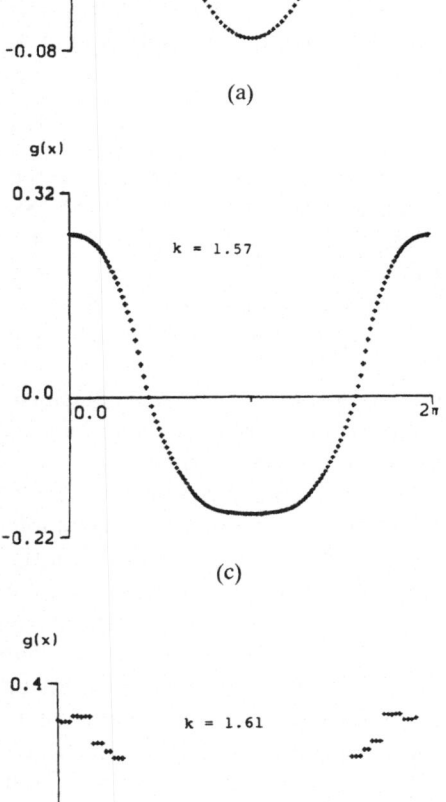

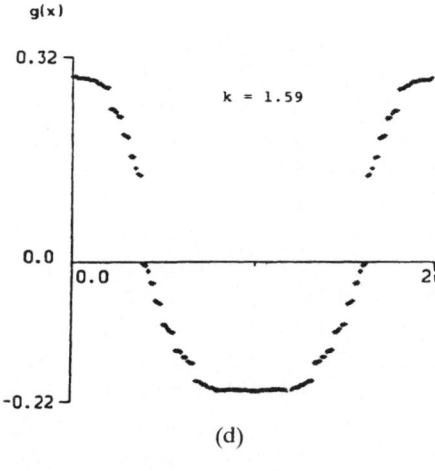

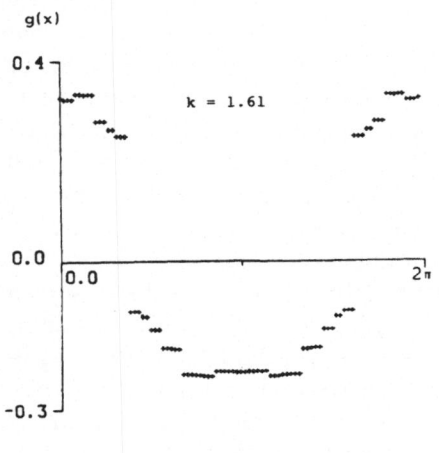

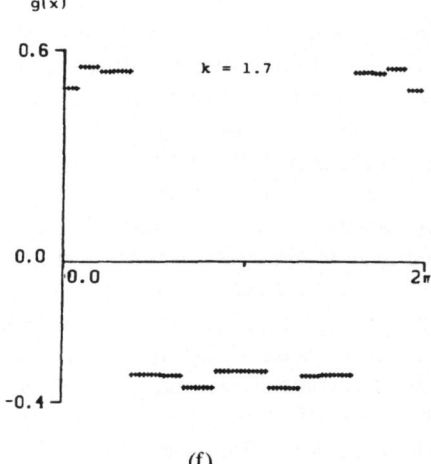

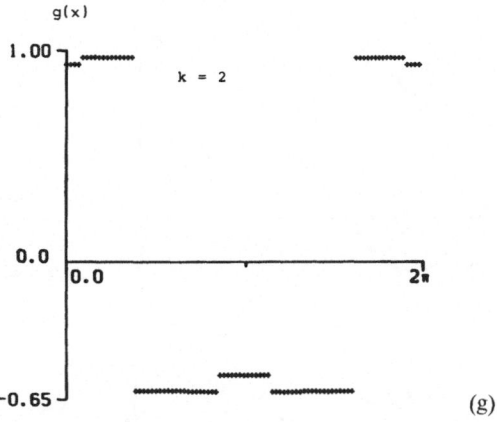

Fig. 22. Hull function $-g(x)$ of the periodic lattice distortion defined by (71a) for the Holstein model (61) for $k = 1.4$, 1.5 ($\zeta = 34/89$), $k = 1.57$, 1.59 ($\zeta = 55/144$) and k 1.61, 1.7 and 2.0 ($\zeta = 34/89$). Note that for $\zeta = 34/89$ and $\zeta = 55/144$, the limit of the infinite system $\zeta = (3 - \sqrt{5})/2$ concerning the aspect of hull function is practically reached. The TBA clearly occurs for $1.57 < k < 1.59$. Also note that the positive part of this curve corresponds to the sites occupied by bipolarons (for $k > k_c$).

function taking its values in a *zero measure* Cantor set. In other words, it can be written for $0 < x < 2\pi$ as an infinite series

$$f(x) = f_0 + \sum_n a_n Y(x - x_n) \tag{72a}$$

where $Y(x)$ is the step function which is zero for $x < 0$ and one for $x \geqslant 0$, $\{x_i\}$ is the set of discontinuity points of $f(x)$ on the interval $[0, 2\pi]$, a_n is the amplitude of this discontinuity of $f(x)$ at x_n. It is also proven that a_n is strictly positive for any n and f_0 is a constant. For the original FK model (and other modified FK models with simple periodic potentials having a single maximum per period), this set is generated from the knowledge of a unique discontinuity x_0 by the formula

$$x_n = x_0 - 2\pi n\zeta \text{ modulo } 2\pi. \tag{72b}$$

For $i \to \pm\infty$, a_n behaves as $\exp(-\gamma|n|)$ where γ is the inverse coherence length of the considered incommensurate ground state, the k-variation of which is shown for example in Figure 11.

Therefore, by analogy with the FK model, we expect that similar properties hold for the hull function $g(x)$ of the nonanalytic incommensurate Peierls chains which we study here. In the limit where ζ is a true irrational number, this number of discontinuities of $g(x)$ is expected to become infinite. The numerical observations suggest that the set of discontinuity points x_n of $g(x)$ is also generated by formula (72b) from the knowledge of a unique discontinuity point x_0. But, unlike $f(x)$, which is monotonous increasing, $g(x)$ is 2π-periodic. Because $\sum_n a_n = 0$,

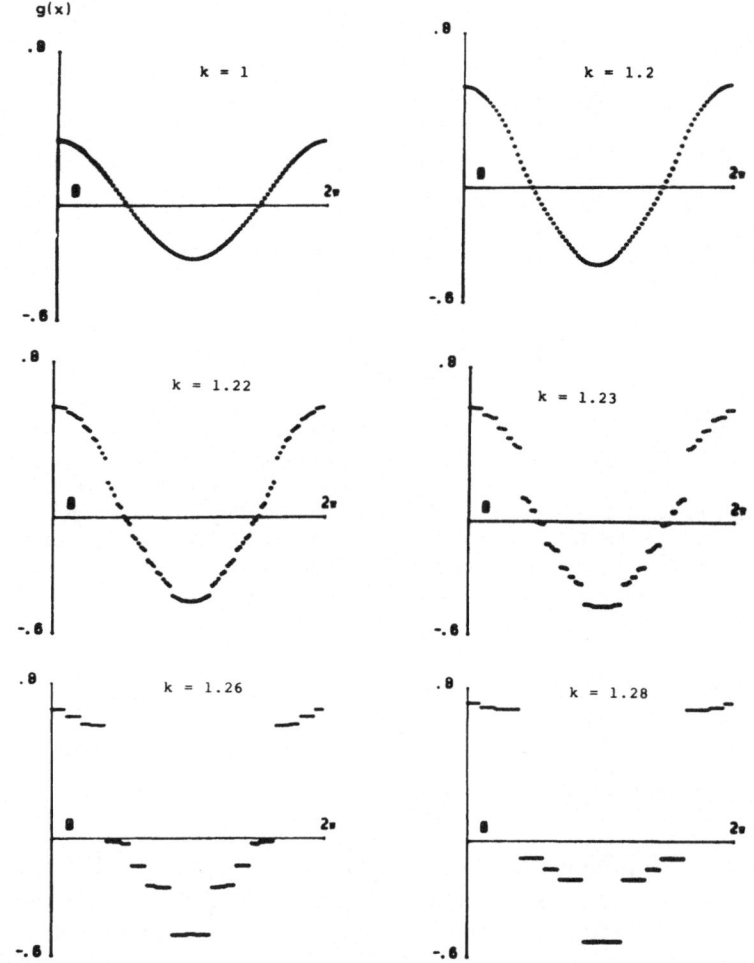

Fig. 23. Hull function $-g(x)$ of the periodic lattice distortion defined by (71b) for the Fröhlich–SSH mode (66) for $k = 1.0$, 1.20, 1.22, 1.23, 1.26 and 1.28 ($\zeta = 34/89$). The TBA clearly occurs for $1.20 < k < 1.22$.

$g(x)$ can be written as

$$g(x) = g_0 - \sum_n b_n \chi(x - x_n) \tag{73a}$$

where $\chi(x)$ is 2π-periodic and is defined on the interval $[0, 2\pi]$ by

$$\chi(x) = Y(x) - Y(x + 2\pi\zeta) \tag{73b}$$

$\chi(x)$ is equal to 1 on the interval $[0, 2\pi\zeta]$ and vanishes on the remaining part of the 2π-period. $a_n = b_{n-1} - b_n$ is the amplitude of the discontinuity of $g(x)$ at $x = x_n$ and g_0 is a constant which will be shown to be zero. In Figures (22) and (23), we have chosen the phase α in order that $g(x)$ appears to be symmetric ($g(x) = g(-x)$). This choice of the phase α corresponds to chosing $x_0 = \pi(1 - \zeta)$. b_0 is the coefficient with the largest modulus in the sequence $\{b_n\}$ and the symmetry of $g(x)$ implies that $b_n = b_{-n}$. Consequently, the incommensurate ground state of the Holstein model can be described with the form

$$\mathbf{u}_i = \sum_j b_j \sigma_{1+j} \tag{74a}$$

where

$$\sigma_j = \chi[(2j + 1)\pi\zeta - \pi] \tag{74b}$$

is a pseudospin which is zero or 1. (Note that this form (74) can also be obtained in the FK model but for the hull function $h(x)$ of the bound modulation $\delta_i = \mathbf{u}_{i+1} - \mathbf{u}_i = h(i\zeta + \alpha)$ where $h(x) = f(x + \zeta) - f(x)$. Then the pseudospins σ_i, which are defined, correspond to those defined in [38].

This form (74) suggests that $\mathbf{u}_i$ could depend linearly on an arbitrary sequence $\{\sigma_j\}$. Then, the pseudospins which have value 1, can be interpreted as determining the location of the *centre* of a localized bipolaron, the extension of which is described by the sequence $\{b_n\}$. In the simplest words, σ_i is 1 when there is a bipolaron at site i and 0 when there is no bipolaron. Indeed, we shall show in the following that for the Holstein model at large k coupling, an approximate Hamiltonian can be written for these bipolaron variable but unlike the FK model, we did not find a particular variation of this Holstein, where this Hamiltonian would be exact.

In both models, we note that for large value of k the hull function $g(x)$ tends to be a single square function. Then the sequence $\{b_n\}$ becomes zero for $n \neq 0$ which means that the bipolaron extension is reduced to a single site. By contrast, when k approaches $k_c(\zeta)$, the sequence $\{b_n\}$ goes slowly to zero and then the extension length of the bipolaron diverges.

The same decomposition (74) of the hull function $g(x)$ holds for the Fröhlich— SSH model for the bound modulation $\{\delta_i\}$ but then the pseudospin sequence determines the bounds i which are occupied ($\sigma_i = 1$) or unoccupied ($\sigma_i = 0$) by a bipolaron.

For the Holstein model, we first checked that the values x_i at which $g(x)$ has a visible discontinuity are indeed generated by (72b) from the knowledge of a unique discontinuity point x_0. The sequence b_i has been calculated from the discontinuity amplitudes a_i by $a_i = b_{i-1} - b_i$. It has been found that this sequence goes to zero for large positive and negative i and is positive for all nonnegligible b_i. When there is no bipolaron at all $u_i \equiv 0$ and then $g_0 = 0$ in (73a) and (74a). This observation suggests that this sequence is proportional to the probability density for the electron of a localized single bipolaron $\{\Psi_i\}$ at site 0 and $b_i = k|\Psi_i|^2$ and that the incommensurate structure just consists in a distribution of these bipolarons at the sites of a quasiperiodic sublattice superimposed on the initial lattice. A numerical calculation of a bipolaron is shown ref. 96.

In the Fröhlich–SSH the discontinuity points x_i are also generated by (72b) from the knowledge of a unique discontinuity point x_0. This sequence b_i is found to converge to zero for large $|i|$ but is not found to be positive all i, but note that in this model δ_i is not proportional to the electron density. (This property should have been checked for the hull function of the electron density, which we did not calculate.)

As for the FK model, by choosing the $\{\sigma_i\}$ with a different rule, such as (74b), one can expect to obtain metastable configurations of the Peierls chain but with higher energy than the ground state. This picture provides the physical interpretation of the chaotic metastable configuration which, indeed, we did find numerically. We chose in advance a chaotic configuration in the Holstein model, by starting the numerical integration of (64) with an initial state were the pairs of electrons are localized on a given distribution of single sites which can be random or not. (This is obtained by choosig $u_i(0) = -k^2\sigma_i$ where $\sigma_i = 1$ or 0 whether the site is occupied or not.) The concentration ζ of pairs of electrons is kept the same as for the ground state. For a value of k larger than some critical value k_0 ($\{\sigma_i\}$) > $k_c(\zeta)$ depending on the initial configuration $\{\sigma_i\}$, Equation 64 (or Equation 69) converges to a metastable configuration where the electronic density has kept its maxima at the same sites (or bound) as in the intial state.

Figures 24a and 24b shows that the space Fourier spectrum $S(q)$ defined as

$$S(q) = \left| \sum_n \exp(iqn)\mathbf{u}_n \right|^2 \tag{75}$$

may be very different for two chaotic metastable configurations of the Holstein model. The broadening of the incommensurate satellites essentially depends on the number of defects which are introduced in the ground state spin sequence $\{\sigma_i\}$. In the case of Figure 24a, the satellite broadening for the chaotic configuration is weak and its energy is slightly larger than those of the incommensurate ground state. In the case of Figure 24b, the disorder of the chaotic configuration is much more important with a broad Fourier spectrum which makes the satellites undistinguishable. As expected, the energy of this configuration is much larger than those of the incommensurate ground state.

Our numerical studies suggest that the energy density of states of these defects is gapless as for the nonanalytic incommensurate ground states of the FK model where this property is proved. These chaotic configurations can also be described by a random distribution of localized discommensurations. The description by random sequences of 'words' should be applicable for determining the thermodynamical properties at low temperature of these 1D Peierls chains. The reader is referred to [20, 28, 29] where this new method has been applied for making a hierarchical classification of the chaotic configurations of 1D nonanalytic incommensurate FK models, according to their energies and then determining the low temperature specific heat, the static structure factor, etc. of this model.

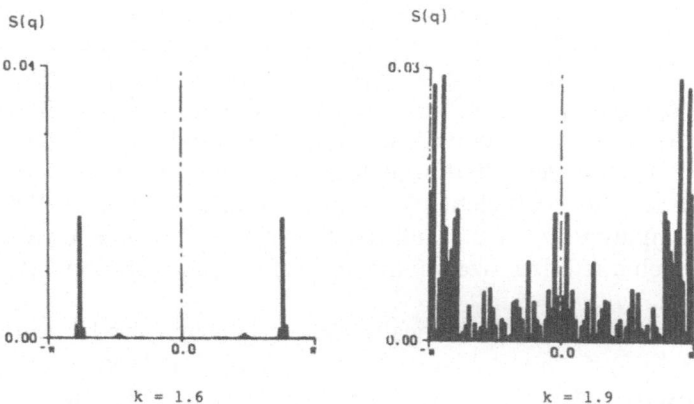

Fig. 24. Fourier spectrum $S(q)$ defined by (72) for two examples of chaotic configurations in the Holstein model for $k > k_c(\zeta)$ for $\zeta = 34/89$ and $k = 1.9$. In the first example, the Fourier spectrum is similar to those of the corresponding ground state but with slightly broaden peaks. In the second example, the incommensurate satellites are undistinguishable.

We also did few numerical investigations for the Holstein model with a different band filling ζ and again found the TBA with the same characteristic features at

$$k = k_c(\zeta_1) = 1.58 \pm 0.01 \quad \text{for} \quad \zeta_1 = (5 - \sqrt{5})/10 \cong 34/115 \qquad (76a)$$

and

$$k = k_c(\zeta_2) = 1.56 \pm 0.01 \quad \text{for} \quad \zeta_2 = (7 - \sqrt{5})/22 \cong 34/149 \qquad (76b)$$

This study confirms the existence of a TBA for other irrational concentrations of electron pairs and in addition shows a small sensitivity of critical value of $k_c(\zeta)$ to the variation of ζ. However, in principle we expect a variation of $k_c(\zeta)$ similar to that of the FK model which is schematized Figure 14. Studies on very large systems will be necessary to detail this variation but as we already said, for rational values of $\zeta = r/s$ with s large, the sharp crossover which replaces the TBA, can be interpreted physically as a true transition.

4.4. CRITICAL BEHAVIOR AT THE TBA OF PEIERLS CHAINS

Some of the physical quantities which are critical at the TBA in the FK model, can be calculated for these two models of Peierls chains. These quantities are also found to be critical with apparently the same critical exponent as in the FK model at the same incommensurability ratio ζ.

4.4.1. *Coherence Length*

The inverse coherence length has the same definition as for the FK model. For the Holstein model, we fix an atom (for example the atom 0 at $\mathbf{u}_0 + \varepsilon_0$) at a position slightly different from its position in a given ground state and then we allow the other atoms to relax from their initial position in this ground state by using Equation 64 or 69. In the limit configuration, the atom at site n is displaced at $\mathbf{u}_n + \varepsilon_n$ and we found that for $k > k_c(\zeta)$, ε_n behaves as $\exp(-\gamma|n|)$ for large n where $\gamma = 1/\xi$ is the inverse coherence length. For $k < k_c(\zeta)$, the motion of all the atoms of the chain are of comparable magnitude and the relaxed configuration is still a ground state described with the same hull function (71a) but with a different phase α. This coefficient $\gamma(k)$ is also the rate of decay of the sequence of discontinuity amplitude a_i of the hull function $g(x)$ (71a) or equivalently it is the inverse of the characterize size of the bipolaron configuration $\{b_i\}$ defined by (74a).

$\gamma(k)$ has been calculated numerically for the Holstein model and found to be critical at $k = k_c(\zeta)$ as shown Figure 25 for $\zeta = \zeta'_g$. For large k, $\gamma(k)$ behaves in the incommensurate case similarly to the coherence length of a single bipolaron but for smaller k, it vanishes as

$$\gamma(k) \propto (k - k_c(\zeta'_g))^\nu \tag{77a}$$

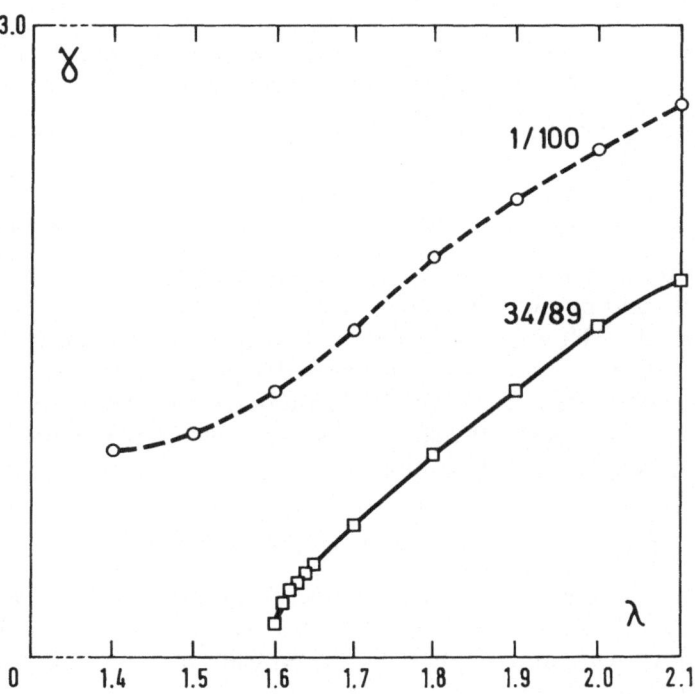

Fig. 25. Inverse coherence length $\gamma(k) = 1/\xi$ (in inverse lattice spacing units) versus k for the Holstein model (full line $\zeta = 34/89$, dashed line $\zeta = 1/100$). The second curve corresponds to a single bipolaron present in the system.

with an exponent

$$\nu \cong 0.95 \pm 0.05 \tag{77b}$$

which agrees reasonably with the similar exponent obtained at the TBA of the incommensurate ground state of the FK model with the equivalent incommensurability ratio $\zeta_g = 1 - \zeta'_g$.

Similar calculations can be done for the Fröhlich—SSH model. A perturbation of the bound length δ_0 by ε_0 produces a perturbation of the bound length δ_n by $\varepsilon_n \propto \exp(-\gamma|n|)\varepsilon_0$. This coefficient $\gamma(k)$ is found to be also critical at $k = k'_c(\zeta'_g)$ with practically the same critical exponent $\nu \cong 1.00 \pm 0.1$.

4.4.2. Peierls—Nabarro Energy Barrier

Again, the Peierls—Nabarro energy barrier is calculated in the same way as for the incommensurate ground state of the FK model. when $k > k_c(\zeta)$, when the phase α of the incommensurate ground state is equal to x_i, where x_i corresponds to a discontinuity of the hull function $g(x)$ (this sequence of discontinuities is generated by (72b)), the atomic configuration has two possible determinations. One is obtained from the other by the motion of a *single bipolaron*. The location of the center of this bipolaron is given by the atomic site i, the phase of which $i\zeta + \alpha$ corresponds to the largest discontinuity in the hull function $g(x)$. Then by fixing this atom i at positions $\mathbf{u}_i$ which are in between its two possible locations

$$\mathbf{u}_i^- = g^-(i\zeta + \alpha) < \mathbf{u}_i < \mathbf{u}_i^+ = g^+(i\zeta + \alpha) \tag{78}$$

and by letting the other atoms of the chain relax to their equilibrium position, one finds that the resulting configuration changes continuously between the two equivalent ground states. However, the energy of the chain does not remain a constant but has a maximum for $\mathbf{u}_i = \frac{1}{2}(\mathbf{u}_i^+ + \mathbf{u}_i^-)$ (see Figure 26). This energy barrier is well known for the motion of defects in discrete system and is called Peierls—Nabarro energy barrier. The relative height E_{PN} of this barrier (compared to the ground state energy) measures the strength of the pinning of the incommensurate CDW—PLD by the lattice.

The calculation of this physical quantity in the Holstein model as a function of k for $\zeta = \zeta'_g E_{PN}(k)$ is shown in Figure 27. Note that this PN energy barrier is very close to those of a single bipolaron. We again found a critical behavior

$$E_{PN}(k) \propto (k - k_c(\zeta'_g))\psi \tag{79a}$$

with an exponent

$$\psi \cong 2.90 \pm 0.10. \tag{79b}$$

The same calculation has been done for the Fröhlich—SSH model and yields the same critical behavior with the same critical exponent as (79) within the error bars.

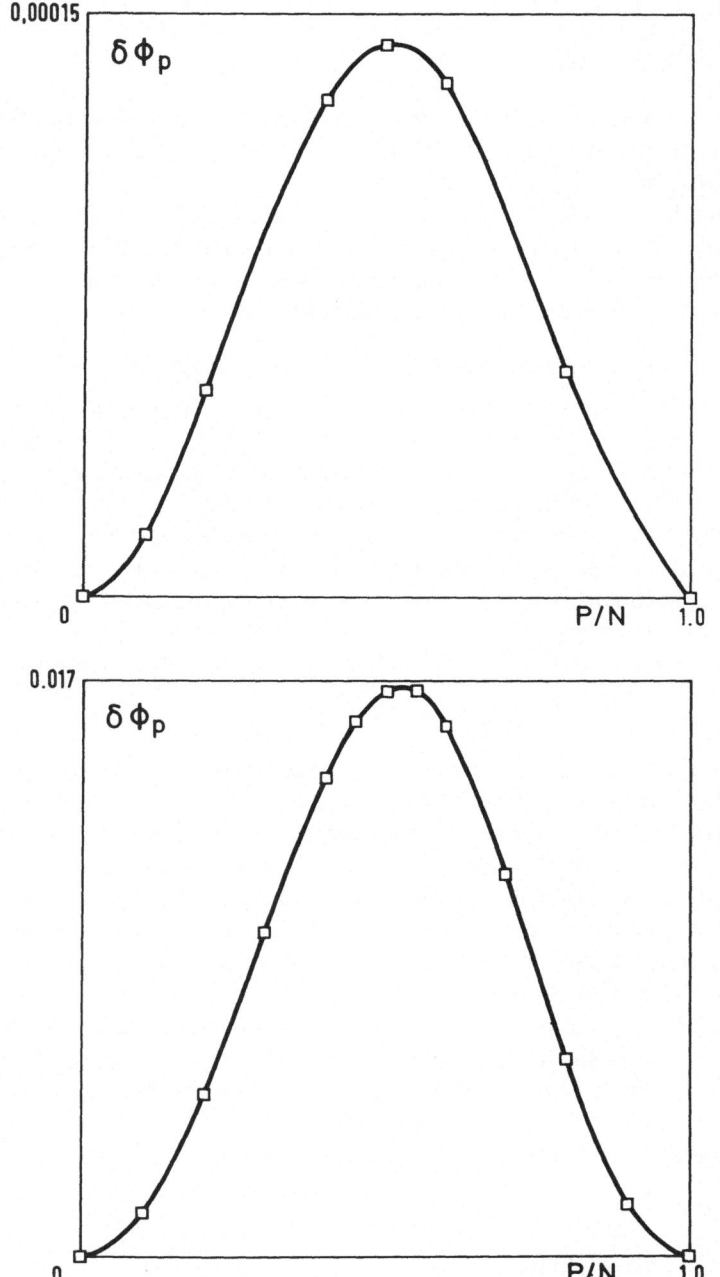

Fig. 26. Two examples of the energy variation of the Peierls chain for $k > k_c(\zeta)$ as a function of the atomic displacement through the largest gap of the hull function $k = 1.6$ (Figure 26a), $k = 1.7$ (Figure 26b). (Holstein model with $\zeta = 34/89$). The maximum of this function is the Peierls—Nabarro energy barrier E_{PN}.

4.4.3. *Phason Gap and Phonon Spectrum*

The classical phonon spectrum at absolute zero of the incommensurate ground states has been numerically calculated and an estimation for the intensity distribution of the dynamical response function $S(q, \omega)$ has been done. For the Holstein model, we perform a second-order expansion of the total energy $F(\{\mathbf{u}_i + \varepsilon_i\})$ (63a) around a ground state configuration $\{\mathbf{u}_i\}$:

$$F(\{\mathbf{u}_i + \varepsilon_i\}) - F(\{\mathbf{u}_i\}) \cong \frac{1}{2} \sum_{m.n} \frac{\partial^2 F(\{\mathbf{u}_i\})}{\partial \mathbf{u}_m \partial \mathbf{u}_n} \cdot \varepsilon_m \varepsilon_n$$

$$= \frac{1}{2} \sum_m \varepsilon_m^2 + \frac{1}{2} \sum_{\mu \, \text{occ}} \sum_{m,n} \partial^2 E^\mu(\{\mathbf{u}_i\})/\partial \mathbf{u}_m \partial \mathbf{u}_n \cdot \varepsilon_m \varepsilon_n. \quad (80a)$$

Using standard perturbation theory one readily obtains

$$\frac{\partial^2 E^\mu(\{\mathbf{u}_i\})}{\partial \mathbf{u}_m \partial \mathbf{u}_n} = -k \frac{\partial |\Psi_m^\mu(\{\mathbf{u}_i\})|^2}{\partial \mathbf{u}_n} \quad (80b)$$

and

$$\frac{\partial \Psi_n^\mu(\{\mathbf{u}_i\})}{\partial \mathbf{u}_m} = -k \sum_{\mu \neq \mu'} \frac{\Psi_m^\mu \Psi_m^{\mu'*}}{E_\mu - E_{\mu'}} \Psi_n^{\mu'} \quad (80c)$$

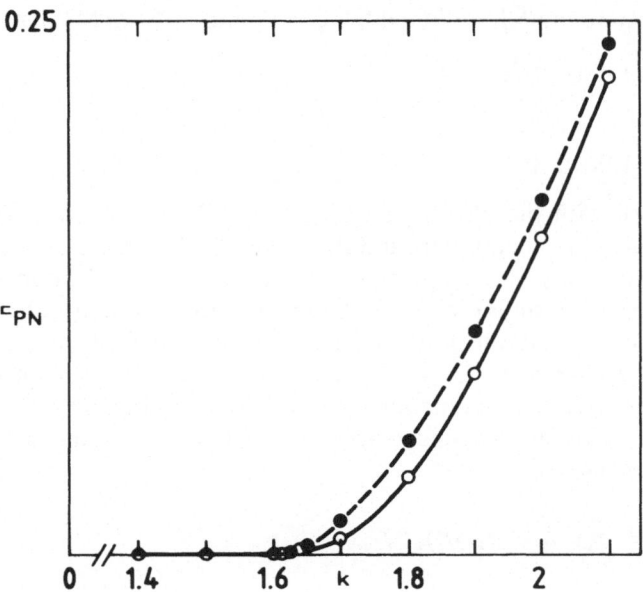

Fig. 27. Variation of the Peierls—Nabarro energy barrier E_{PN} versus k for the Holstein model, for $\zeta = 34/89$ (full line) and for $\zeta = 1/100$ (dotted line).

which yields

$$D_{m,n} = \frac{\partial^2 F(\{\mathbf{u}_i\})}{\partial \mathbf{u}_m \partial \mathbf{u}_n}$$

$$= \delta_{m,n} + k^2 \sum_{\mu \text{ occ}} \sum_{\mu' \text{ empt}} \left(\frac{\Psi_n^{\mu*} \Psi_m^{\mu} \Psi_m^{\mu'*} \Psi_n^{\mu'}}{E_\mu - E_{\mu'}} + \text{c.c} \right) \tag{80d}$$

where μ corresponds to the occupied states and μ' to the empty states, c.c is the complex conjugate of the first term in the parenthesis. Since the mass for the atoms in initial system (61) is m, the small motion equation for the variable ε_i is

$$\frac{d^2 \varepsilon_i}{dt^2} = \omega_0^2 \sum_j D_{i,j} \varepsilon_j. \tag{81}$$

The dynamical matrix $D_{i,j}$ has been numerically calculated from the known electronic eigenstates and they diagonalized. Then, we obtained the full set of eigenfrequencies ω_μ and their corresponding normalized eigenmodes $\{\varepsilon_n^\mu\} = |\mu\rangle$. The smallest eigenfrequency ω_G, if not zero, is the phason gap.

As expected, this phason gap is zero for $k < k_c(\zeta)$ and finite for $k > k_c(\zeta)$. When $k < k_c(\zeta)$ we checked that the eigenmode associated with the eigenzero frequency does correspond to the phase transition of the incommensurate PLD. For k going to $k_c(\zeta)$ by upper values, a critical behavior is observable (cf. Figure 28). For the Holstein model, we found for $\zeta = \zeta_g'$ and $k > k_c(\zeta_g')$ that

$$\omega_G(k) \propto (k - k_c(\zeta_g'))^\chi \tag{82a}$$

with an exponent

$$\chi \cong 1.00 \pm 0.10 \tag{82b}$$

which agrees well with the exponent 1.02 at the TBA of the incommensurate FK chain at the equivalent incommensurability ratio ζ_g. We have also plotted on the same curve (Figure 28), the frequencies of the two localized phonon modes of a single bipolaron as a function of k. Unexpectedly, there are two modes, the frequencies of which cross one another for a value of k not far from $k_c(\zeta_g')$.

The classical dynamical structure factor of this model can be calculated by assuming low temperature fluctuations in the harmonic regime of the eigenmodes of Equation 81. The modes observable by neutron scattering are the 'normal' modes $|q\rangle$ defined as

$$|q\rangle = \frac{1}{\sqrt{N}} \sum_n \exp(iqn)|n\rangle \tag{83a}$$

where $|n\rangle$ is the normalized mode localized at site n ($\varepsilon_n = 1$ and $\varepsilon_m = 0$ for $m \neq n$). These normal modes $|q\rangle$ can be written as a linear combination of the normalized eigenmodes ε^{μ} of the dynamical matrix $D_{i,j}$

$$|q\rangle = \sum_{\mu} U_{q,\mu} |\mu\rangle \tag{83b}$$

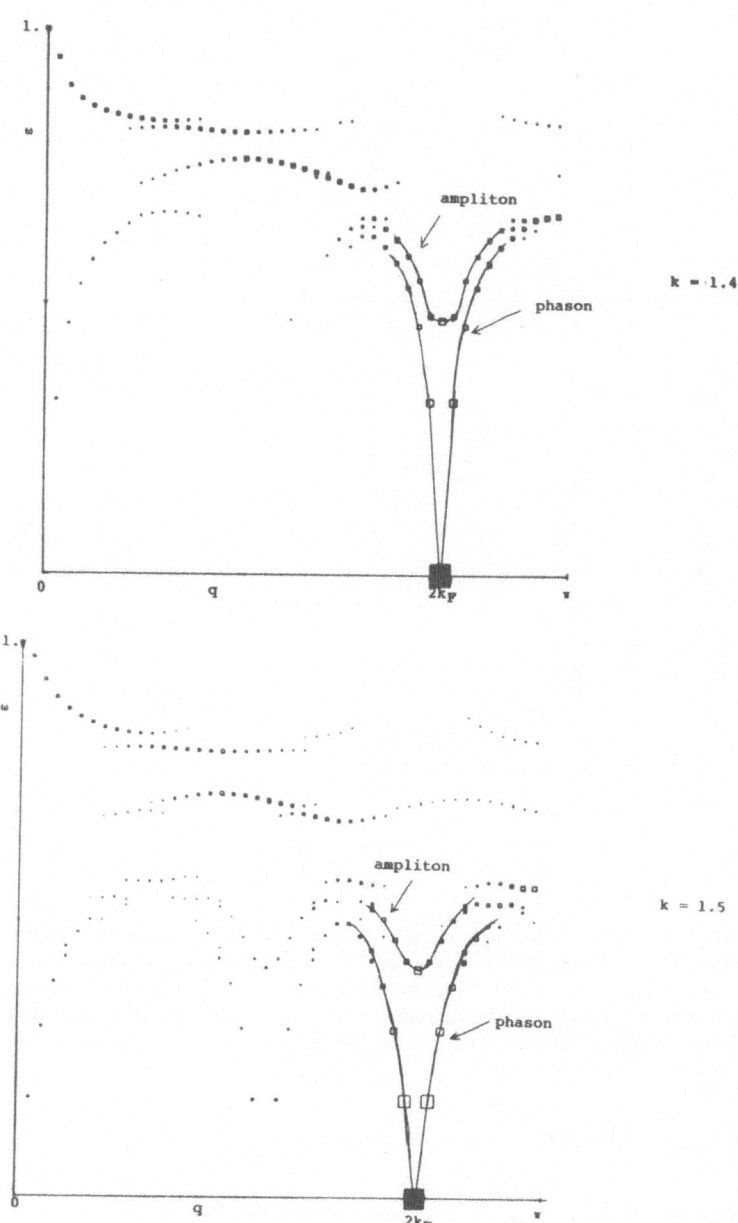

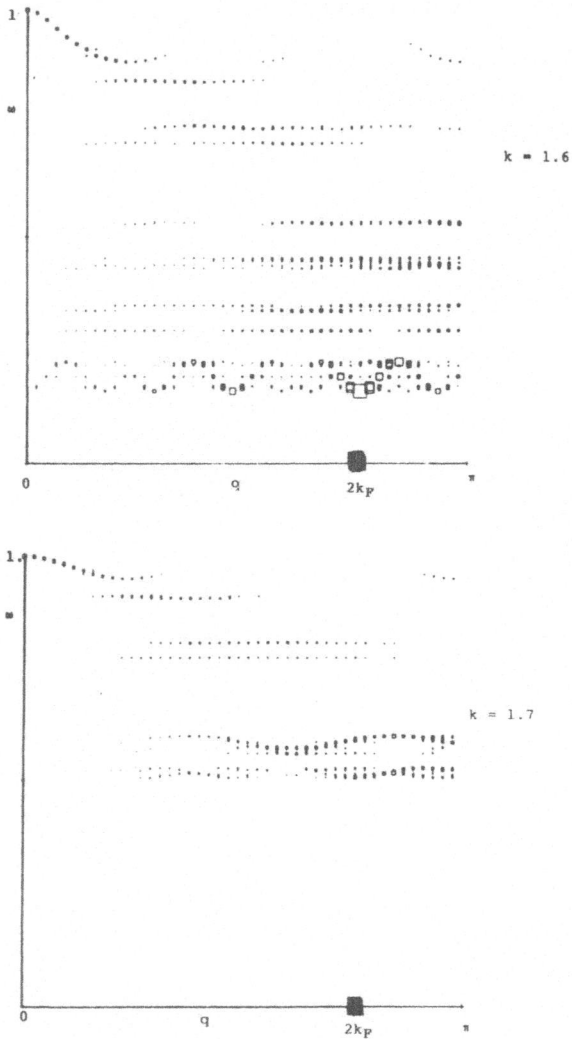

Fig. 28. Intensity distribution of the dynamical structure factor $S(q, \omega)$ in the plane (q, ω) for the Holstein model and $\zeta = 34/89$ at 0 K within an harmonic and classical approximation ($k = 1.4$ Figure 28a, $k = 1.5$ Figure 28b, $k = 1.6$ Figure 28c, $k = 1.7$ Figure 28d, $k = 1.8$ Figure 28e). Note that the zero gap phason branch and the ampliton branch appear to be rather well defined for $k < k_c(\zeta)$ (Figure 28a and 28b) while they sharply disappear for $k > k_c(\zeta)$ (Figure 28c—e).

where $U = \{U_{q, \mu}\}$ is a unitary matrix. The Fourier transform

$$S(q, \omega) = \int \langle \varepsilon_q(\tau)\varepsilon_q(\tau + t)\rangle_\tau \exp(i\omega t)\, d\tau \tag{83c}$$

of the time correlation function of the amplitude $\varepsilon_q(t)$ of this normal mode is the sum of the contributions coming from the time independent amplitudes ε_μ $\exp(i\omega t)$ of the orthogonal eigenmodes $|\mu\rangle$

$$S(q, \omega) = \sum_\mu |U_{q,\mu}|^2 \varepsilon_\mu^* \varepsilon_\mu \frac{1}{2} [\delta(\omega - \omega_\mu) + \delta(\omega - \omega_\mu)]. \tag{83c}$$

By choosing

$$\varepsilon_\mu^* \varepsilon_\mu = \frac{1}{\omega_\mu^2} \tag{83d}$$

one obtains a distribution of intensity proportional to the dynamical response factor $S(q, \omega)$ at finite temperature. Figure 29 represents this distribution of intensity in the plane q, ω. Since there are s normal modes $|q_p\rangle$ and s eigenfrequencies ω_μ for a finite system of s atoms, $S(q, \omega)$ is the sum of s^2 Dirac

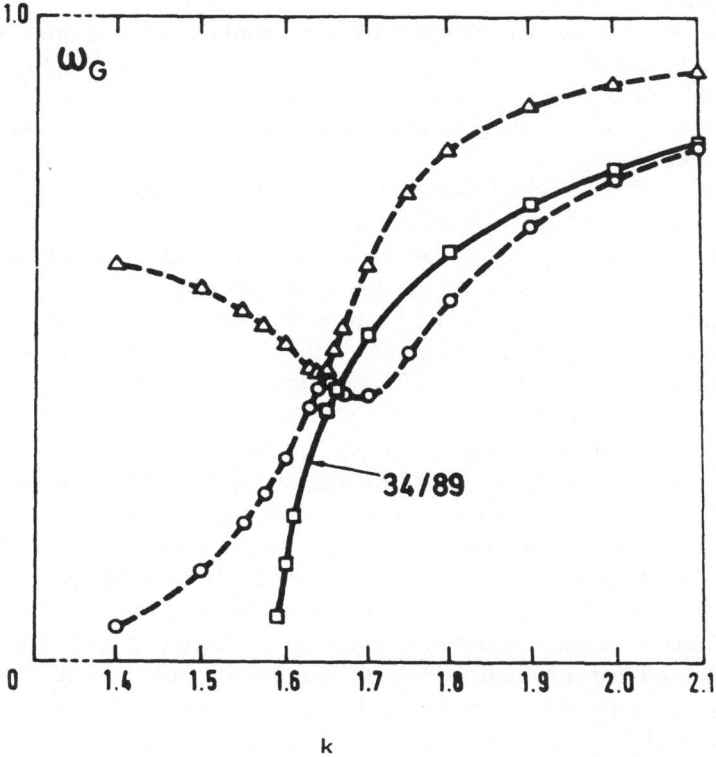

Fig. 29. Phonon gap ω_G versus k (with atomic unit mass) measured on Figure 28 (full line $\zeta = 34/89$, dashed line $\zeta = 1/100$). Note, incidently, that the single bipolaron in the Holstein model has two localized phonon modes. Their variation curves intersect for k close to $k_c(\zeta)$.

functions. We represented the Dirac function at (q_p, ω_μ) by a square centred at this point the area of which is proportional to the intensity of this Dirac function given by (83c) and (83d). When this intensity is negligible (a square smaller than the line thickness), nothing is plotted.

Then Figure 29 allows one to distinguish, for $k < k_c(\zeta)$, a well-defined phason branch (and the corresponding ampliton branch) which starts linearly at $\omega = 0$ from $q = 2\pi\zeta$. We have not been able to study the critical behavior of this phason velocity when k goes to $k_c(\zeta)$ via lower values but it is reasonable to believe that its behavior is the same as in the FK model with the same critical exponent. However, we clearly see by comparing the figures obtained for $k = 1.4$ and $k = 1.5$ that the phason velocity (which is the slope of these phason branch) softens significantly.

When k increases beyond $k_c(\zeta)$, the dynamical response function $S(q, \omega)$ suddenly changes and no well-defined phason branch appears. Moreover, for larger values of k, the spectrum only exhibits a weakly softened optical phonon branch which tends to be the flat optical phonon branch of the Holstein model without electrons.

The same study has been performed for the Fröhlich–SSH model. We do not reproduce the details of the analytical calculations which are similar to those of the Holstein model. First, we found the same critical behavior for ω_G and the same critical exponent χ as for the Holstein model for the FK model at equivalent incommensurability ratio.

The intensity distribution of the dynamical response function $S(q, \omega)$ (calculated for the initial variables u_i) exhibits, at all values of k, an acoustic branch (cf. Figure 30). This is not surprising, since the phonon mode which couples to the electrons has been chosen to be an acoustic phonon. For $k < k_c(\zeta)$ we note the existence of the phason branch which cannot be confused with harmonics of the acoustic phonon branch (also distinguishable on most figures) because it has a different and larger slope. The phason velocity also softens when approaching $k_c(\zeta)$ from below. The ampliton branch is also observable. As soon as the TBA is crossed, one sees as for the Holstein model a sudden change in $S(q, \omega)$ where the phason branch disappears. A rather well-defined and flat phonon branch occurs.

In summary, these numerical observations confirms that the physical picture for the TBA in Peierls chains, is closely similar to the physical picture of the TBA in the FK model. Particularly, these results support the notion that in Peierls chains for $k > k_c(\zeta)$, the CDW and the PLD are well described by an incommensurate array of bipolarons which is *strongly* pinned by the lattice. (Note that the values which we obtained for PN energy barrier are comparable to the characteristic energies of the model and therefore will play a role in the thermodynamic properties of the CDW systems at finite temperature as soon as k exceeds $k_c(\zeta)$.)

The picture of a single bipolaron is a lattice deformation and a pair of electrons localized in the potential well formed by this lattice deformation. As soon as there are many bipolarons, this picture apparently becomes valid no longer. We will see in the next section that the eigenfunctions of the electrons in the models which we have studied *are not localized close to a single bipolaron site* but have some

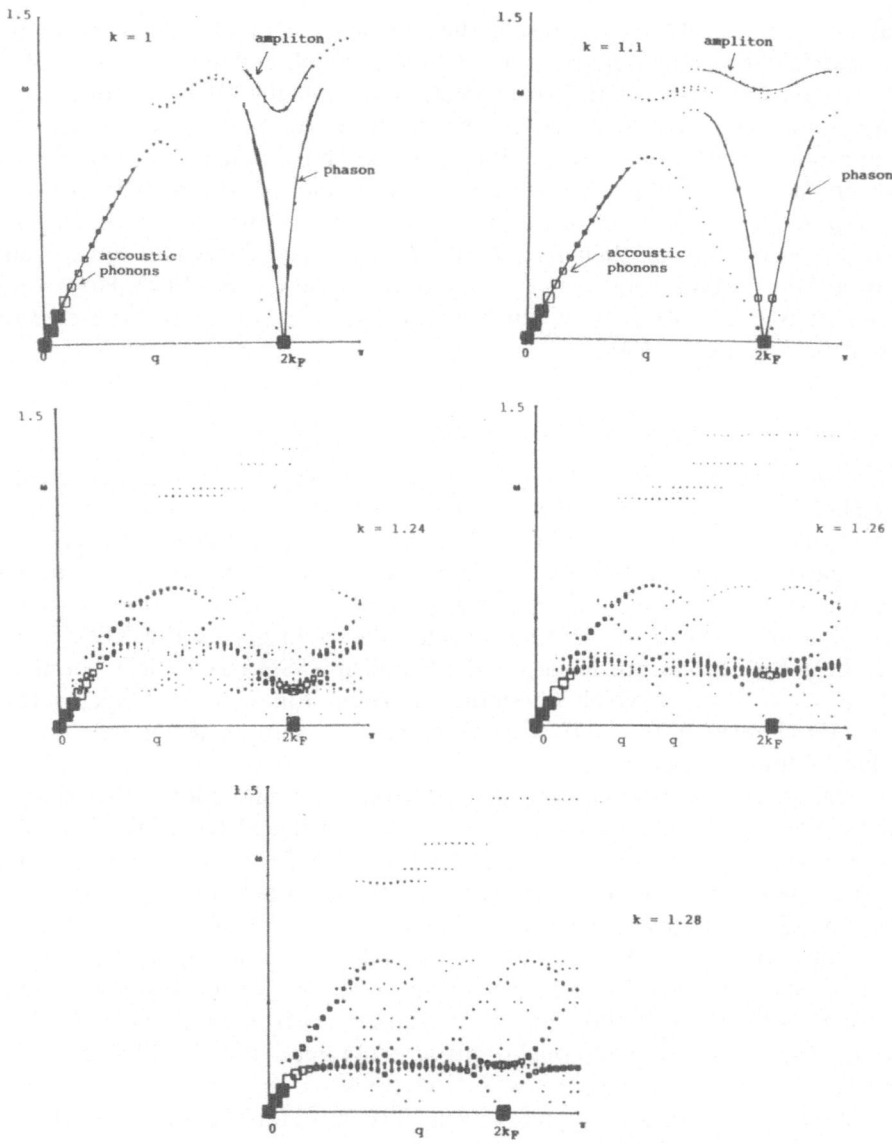

Fig. 30. Same as Figure 28 but for the SSH model ($k = 1.0$ Figure 30a, $k = 1.1$ Figure 30b, $k = 1.24$ Figure 30c, $k = 1.26$ Figure 30d and $k = 1.28$ Figure 30e). Note that there exists, for all values of k, a well defined acoustic branch. As on Figure 28, the phason and the ampliton branch are well defined for $k < k_c(\zeta) \cong 1.215$ and sharply disappear for $k > k_c(\zeta)$.

intermittent behavior which apparently knocks down the interpretation that was conjectured at the beginning of this section. In fact, this is an artefact of our initial numerical interpretation because, according to well-known principles in quantum mechanics, it is possible to represent the global state of the electrons in many ways, essentially by using unitary transformations which globally leave the space of occupied eigenstates invariant. For example, although the Wannier functions are not eigenstates, they can be used for an exact representation of the electronic states in a filled band, instead of the Bloch states. The simple model which we will develop in the section after this will confirm that, indeed, the picture of a lattice gas of *bipolarons with a localized pair of electrons* turns out to be physically correct. These bipolarons then behave as *classical particles* which form a lattice gas and in their ground states generate, the nonanalytic CDW—PLD. However, we think that it is useful to present our numerical results concerning the electronic eigenwaves electrons in CDW.

4.5. ELECTRONIC BEHAVIOR AT THE TBA

We first briefly discuss the numerical results concerning the energy density of states (EDS) of the electrons. Because of the incommensurate lattice modulation, we expect that the EDS exhibits infinitely many gaps. In fact below the TBA, only the main gap is clearly visible. At the TBA, there is *no sharp change* in the EDS, but we see that apart the main gap, many other gaps become more visible (cf. Figures 31 and 32). For the Holstein model, the main gap at the TBA is about 10% of the total band width while for the Fröhlich—SSH model, it is about 30%. Clearly, the system sizes which we studied were far too small (89 eigenstates) to detect subtle changes in the EDS corresponding, for example, to the occurrence of a singular continuous spectrum.

The change in the behavior of the electronic eigenfunctions also does not appear as sharp as the change of the hull function of the PLD—CDW at the TBA and does not allow one to detect the TBA. Figure 33 shows for the Holstein model the spatial density of some electronic wave functions (1) at the lowest eigenenergy, (2) at the lower edge of the Peierls gap. In the first case, the tendency to localization clearly appears just above the TBA while, in the second case, the effect of the TBA is very smooth. Again, the small size of the systems which we studied does not allow us to conclude if the electronic eigenwaves really localize or not at the transition. Similar conclusions are obtained for the SSH model (cf. Figure 34).

The lattice quasiperiodic potential being discontinuous beyond the TBA, the existence of quasi-Bloch waves is impossible because the perturbation theory of plane waves diverges at the lowest order (see Section 3). However, the assumption that the electronic spectrum is singular continuous for the nonanalytic incommensurate phases can be supported. In the limit of large electron phonon coupling k, the hull function of the PLD in the Holstein model becomes almost a square function:

$$\mathbf{u}_i \cong -k[\text{int}((i+1)\zeta + \alpha) - \text{int}(i\zeta + \alpha)]. \tag{84}$$

Equation (63b) with such a square quasiperiodic potential has been studied by

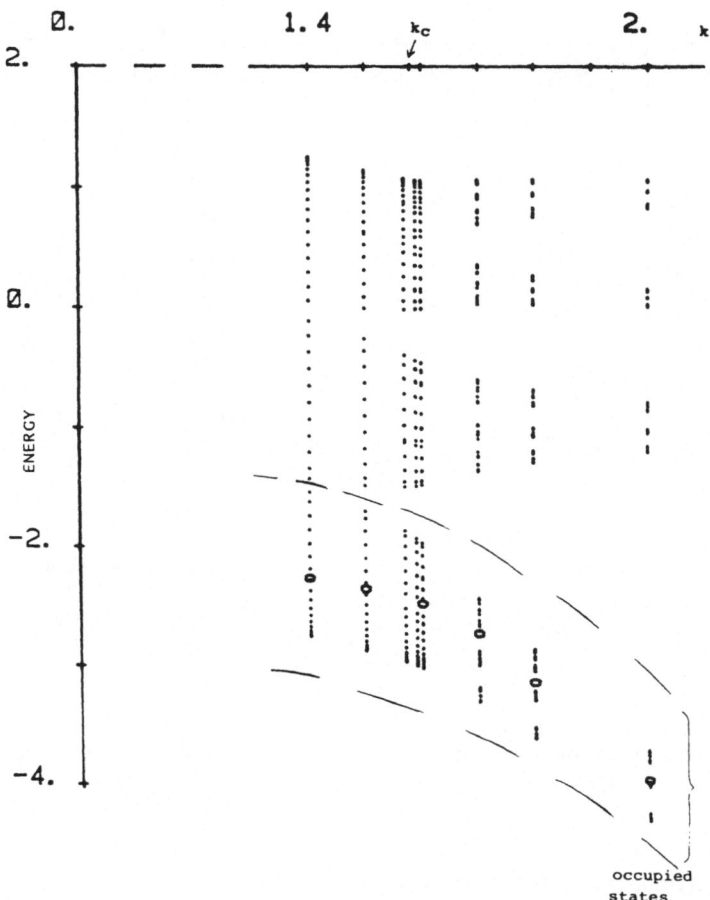

Fig. 31. Distribution of the electronic eigenenergies as a function of k for the Holstein model with 34 electron pairs and 89 atoms.

Kohmoto *et al.* (1983) for $\zeta = (\sqrt{5} - 1)/2$. It was shown that this model does not exhibit any quasi-Bloch waves but strong arguments were found for a singular continuous spectrum. Despite the fact that the eigenwaves are not truly localized, our study at the end of Section 3 suggests that the eigenwaves in such a potential do not carry any finite momentum, which is the physically essential property.

By contrast, for the chaotic configurations the lattice potential for the electrons, we did indeed find that the electronic states are well localized, in agreement with the theoretical predictions which assert that all electronic states are localized in a random one-dimensional potential.

Therefore, although our numerical calculations apparently cannot provide any definite conclusion, our belief is that in the nonanalytic incommensurate ground

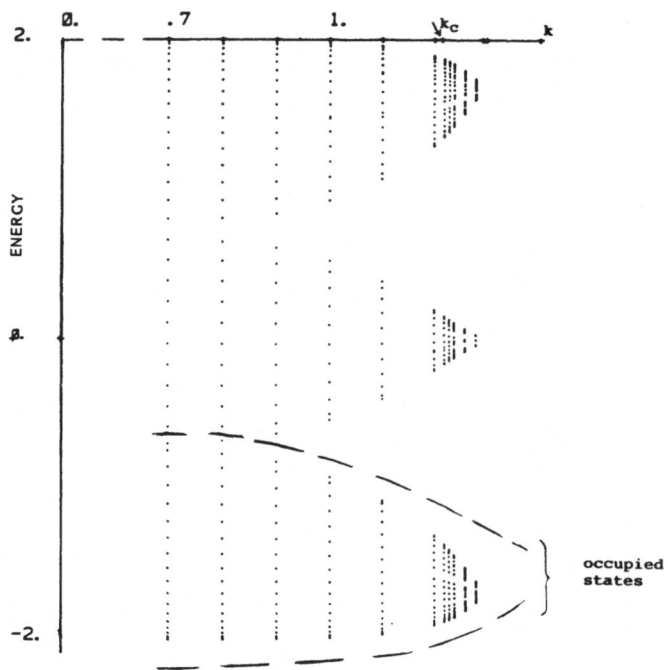

Fig. 32. Same as Figure 25 for the Fröhlich—SSH model for the same number of electrons and of atomic sites.

state, the electronic wave functions are neither exponentially localized for quasi-Bloch waves. They form a singular continuous spectrum for the ground states of nonanalytic CDW. In that case, the eigenstates also have the essential property of the localized eigenstates, namely they cannot carry any finite electric current. Further studies need to be done to confirm this conjecture.

4.6. A CLASSICAL LATTICE GAS MODEL FOR THE HOLSTEIN MODEL IN THE LARGE ELECTRON—PHONON COUPLING LIMIT

This model is constructed by analogy with the piecewise parabolic model (cf. Section 2.5) which approximates the sine FK model in the large coupling limit. This modified model does not exhibit any analytic incommensurate structures but yields a qualitatively good description of the *nonanalytic* incommensurate structures (with a lattice pinning of the incommensurate configurations, a complete devil's staircase and infinitely many metastable chaotic configurations.).

In the limit of large electron—phonon coupling, the energy of the metastable configurations of the Holstein model can be described as a function of the

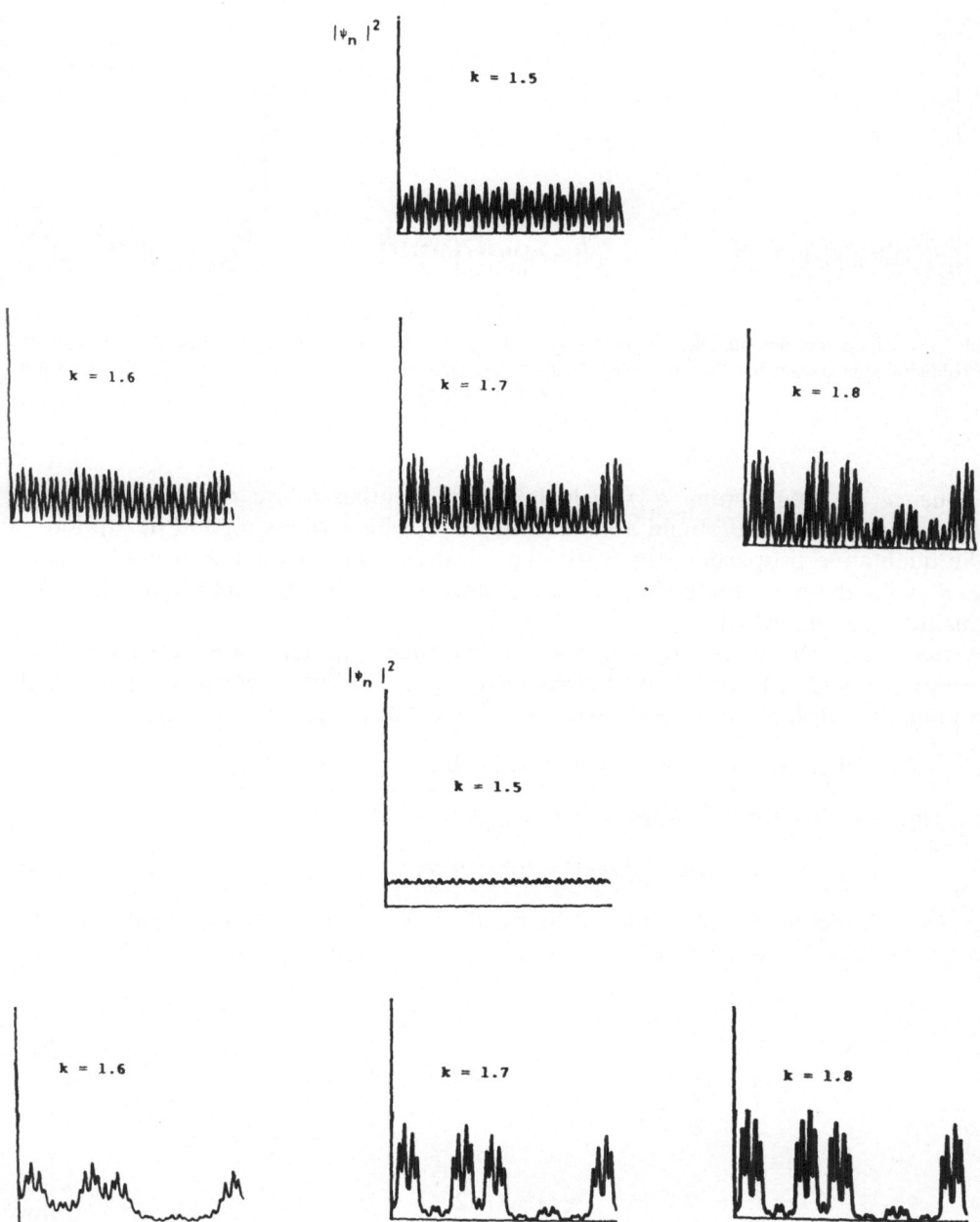

Fig. 33. Electronic density $|\Psi_n|^2$ versus n ($n = 1, 2, \ldots, 89$) of the electronic eigenfunction for the Holstein model ($\zeta = 34/89$) (a) density for the lowest energy state ($k = 1.5, 1.6, 1.7, 1.8$) (b) density for the eigenstate at the lower edge of the Peierls gap (same values of k).

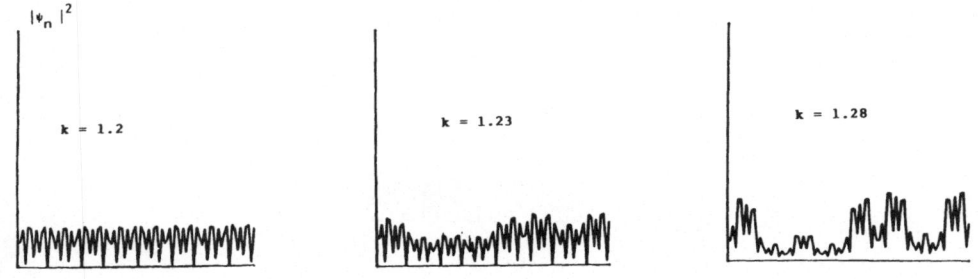

Fig. 34. Electronic density $|\Psi_n|^2$ versus n ($n = 1, 2, \ldots, 89$) of the electronic eigenfunction for the Fröhlich—SSH model for the eigenstate at the lower edge of the Peierls gap for $k = 1.20$, 1.23 and 1.28 ($\zeta = 34/89$).

sequence of pseudospins σ_i which determine whether a site i is occupied by a bipolaron or not. We obtain a lattice gas model which yields a good desription of the qualitative properties which we numerically observed for the ground-state as well as for the metastable chaotic states, although we have not yet obtained a good quantitative agreement.

Let us first study the Holstein model with a single bipolaron which is located for example at site p. In the limit of large k, the eigenstate of the electronic pair which is bound to this bipolaron tends to be localized at this unique site p

$$\Psi_p(p) = 1 \quad \text{and} \quad \Psi_n(p) = 0 \quad \text{for} \quad n \neq p \tag{85a}$$

which yields the atomic positions by using (64a) at 0th order

$$u_p = -k \quad \text{and} \quad u_n = 0 \quad \text{for} \quad n \neq p. \tag{85b}$$

Then Equation (63b) yields, to first order, in the perturbation expansion with respect to $1/k$, the normalized eigenstate

$$\Psi_n^{(p)} = \sqrt{\frac{1 - x^2}{1 + x^2}} \, x^{|n - p|} \tag{86a}$$

with

$$x = \frac{-k^2 + \sqrt{k^4 + 4}}{2} \cong \frac{1}{k^2} - \frac{2}{k^6}. \tag{86b}$$

The eigenenergy of the localized electron is

$$E_0 = -\sqrt{k^4 + 4} \cong -k^2 - \frac{2}{k^2} + \cdots \tag{86c}$$

and Equation 64a yields the first order correction to the atomic positions (85b)

$$\mathbf{u}_p \cong -k + \frac{2}{k^3} + \ldots \quad \text{and}$$

$$\mathbf{u}_n \cong -k . k^{-4|n-p|} + \ldots \quad \text{for} \quad n \neq p. \tag{86d}$$

For large k, the total energy of the system with a single bipolaron is

$$U_{\text{bip.}} \cong -\frac{k^2}{2} - \frac{4}{k^2} + \ldots . \tag{86e}$$

For large k, the Peierls—Nabarro energy barrier for a single bipolaron can be also calculated by assuming that when moving a bipolaron from site p to site $p + 1$, the energy of the system reaches its maximum when the electronic probability density is equally distributed on two sites

$$\Psi_p = \frac{1}{\sqrt{2}}, \quad \Psi_{p+1} = \frac{1}{\sqrt{2}} \quad \text{for} \quad n \neq p \quad \text{and}$$

$$\Psi_n = 0 \quad \text{for} \quad n \neq p \quad \text{and} \quad n \neq p+1. \tag{87a}$$

A similar calculation as above yields the energy $U_{\text{bip.unst.}} = -k^2/4 - 1 + \ldots$ of this unstable bipolaron and, consequently, the PN energy barrier $E^0_{\text{PN}} = U_{\text{bip.unst.}} - u_{\text{bip.}}$ of a single bipolaron is

$$E^0_{\text{PN}} \cong \frac{k^2}{4} - 1 + \ldots . \tag{87b}$$

According to our numerical observations, a many-bipolaron configuration can be described by a sequence of pseudospins $\sigma_i = 0$ or 1. For large k, the atomic positions are given at 0th order by

$$\mathbf{u}_i = -k\sigma_i. \tag{88a}$$

In order to give some idea of the pseudospin Hamiltonian which can be obtained, let us calculate the energy of a pair of bipolarons as a function of their distance. Within the assumption that the energy of a many-bipolaron system is only given by the sum of all bipolaron pair energies, one obtains an approximate pseudospin Hamiltonian $U(\{\sigma_i\})$ which yields correct qualitative properties for the Holstein model. But this form is not quantitatively correct because as for the FK model this pseudospin Hamiltonian should also involve multispin interactions at all orders. In fact, one expects that when k approaches $k_c(\zeta)$ from above, the role of these energy terms involving n-bipolaron interactions becomes essential for producing the TBA.

Let us consider of a pair of bipolarons located at $i = p$ and $i = q$ respectively, with $m = q - p > 0$. The atomic configuration at the lowest order is given by

$$\mathbf{u}_p = \mathbf{u}_q = -k \quad \text{and} \quad \mathbf{u}_i = 0 \quad \text{for} \quad i \neq p \quad \text{and} \quad q \tag{88b}$$

and the eigenwave functions of the two electron pairs for the atomic configuration
(88b) have the form

$$\Psi_n = A[x^{|n-p|} + \sigma x^{|n-q|}] \tag{89}$$

where $\sigma = +1$ and -1 for the symmetric eigenstate and the antisymmetric
eigenstate, respectively. A is a normalization factor. By checking that (89) is an
eigensolution of the electron eigenequations of the Holstein model, it comes out
that the electron eigenenergies E fulfill the equation

$$E_\sigma^2 = 4 + k^4 \left(1 + \sigma \cdot \left(-\frac{E_\sigma + \sqrt{E_\sigma^2 - 4}}{2} \right)^m \right)^2 \tag{90}$$

with the condition $E_\sigma < 0$. For $m = +\infty$, the two bipolarons are far apart, and
Equation (90c) yields the eigenenergy E_0 (86c) of the localized electron of a single
bipolaron. For finite m, it turns out that the eigenenergies of the symmetric and
antisymmetric states expand as

$$E_\sigma = E_0 \left(1 + \frac{\sigma}{k^{2m}} - \frac{m}{k^{4m}} + \ldots \right). \tag{91a}$$

Since the symmetric and antisymmetric eigenstates are both occupied, the part
of the electronic eigenenergy of the bipolaron pair which depends on their distance
m is

$$U_{\text{electr.}}(m) \cong \frac{2m}{k^{4m-2}} \tag{91b}$$

which shows that the electronic contribution to the interaction between two
bipolarons is *repulsive*.

The elastic contribution to the energy of the bipolaron pair which depends on
the distance can also be calculated at the lowest order by using the form (89)

$$U_{\text{elast.}}(m) \cong \frac{m+1}{k^{4m-2}} \tag{92}$$

which shows that the interaction between two bipolarons is also *repulsive*.
Neglecting the multispin interactions which involve more than two spins, the
simplest approximation for the energy of an arbitrary distribution of bipolaron $\{\sigma_i\}$
in the initial Holstein model is a lattice gas model with Hamiltonian:

$$\Phi(\{\sigma_j\}) = U_{\text{bip.}} \sum_i \sigma_i + \sum_{i,j} J_{i-j} \sigma_i \sigma_j \tag{93a}$$

where

$$J_m \cong T \frac{3m+1}{k^{4m-2}}. \tag{93b}$$

The calculations of Alexandrov, Ranninger and Robaskiewicz (1986) can be
applied specifically to this Holstein model without electron—electron interactions

in the adiabatic limit. The coupling constant for the nearest bipolaron interaction which they obtained is $v = 2T^2/\Delta$ (in their notations $S_i^z = \sigma_i - \frac{1}{2}$). Turning back to our notations, one finds $\Delta = \lambda^2/m\omega_0^2 = k^2 T/2$ and $v = 4T/k^2 = J_1$ which shows that our calculation and theirs agree concerning the nearest neighbor interaction. But although the calculation of these authors has the advantage of taking the quantum lattice effect into account, their calculation truncates the bipolaron interaction beyond the nearest neighbor ($J_m = 0$ for $m > 1$). In the absence of quantum terms, the drastic consequence of that approximation is that:

— *First* all possible incommensurate ground states (CDW) in their model are eliminated,
— *Second* for a non-half-filled band, the ground state of the bipolaron system is highly degenerate (with finite entropy). In that condition, the coexistence of 'bipolaronic superconductivity' and charge ordering, which they claim to exist in this kind of model, could be an artefact of their approximation. This problem will be discussed again in the next section.

For $k \gtrsim 1.15$ (and therefore in the nonanalytic regime) J_m is a convex function of m for $m > 0$. The ground state of this one-dimensional lattice gas model with the energy (93) [66, 67, 68] is an incommensurate structure described by the sequence of pseudospins

$$\sigma_i = \chi(2\pi i\zeta + \alpha) \tag{94}$$

ζ is the concentration per site of bipolarons and α is an arbitrary phase. $\chi(x) = \mathrm{int}(x + \zeta) - \mathrm{int}(x)$ is a 2π-periodic function which is unity on the interval $]0, 2\pi\zeta]$ and zero on the interval $]2\pi\zeta, 1]$. This result agrees with the numerical experiments and their interpretation which has been presented in the beginning of this section (see (74b)). Within this approximation, the sequence b_i defined in (74a) is given by

$$b_i \cong k|\Psi_i^{(0)}|^2 = kx^{2|i|} \cong k/k^{|4i|} \tag{95}$$

where $\{\Psi_i^{(0)}\}$ is the wave function of the electron localized on a single bipolaron at site 0. The inverse coherence length $\gamma = 1/\xi$ is the length that characterizes the extension of the probability density associated with this wave function. Note that the same lattice gas model holds for the incommensurate filling of the electron band as well as for the commensurate case. This method for finding this model for classical large electron—phonon coupling is not restricted to this Holstein model in one dimension; it could also be used for general models with any dimensions.

In summary, we have proved in this section the existence of a phase transition between two regimes for incommensurate Peierls chains with a classical lattice which occurs as a function of the electron—phonon coupling. These two regimes have qualitatively different properties and can be interpreted as follows:

(1) for $k < k_c(\zeta)$, (small electron—phonon coupling regime), the electronic states are essentially quasi-Bloch waves and the CDW—PLD have a hull function close to a pure sine function. The CDW—PLD has a zero gap sliding mode (phason) corresponding to its phase rotation which carry on electric current.

(Fröhlich conductivity) and does not accept any metastable (spinless) phase defects. This is the standard picture for CDW in a perfect system.

(2) for $k > k_c(\zeta)$ (large electron—phonon coupling regime), the PLD—CDW can be described as an array of bipolarons which are localized in the real space. In the ground state, these bipolarons form an *ordered* incommensurate structure, while the metastable chaotic configurations correspond to disordered distribution of these bipolarons. These bipolarons are pinned to the lattice and an energy barrier is required for their motions from one lattice site to another one. The CDW appears as a *lattice gas of bipolarons*.

Finally, let us suggest that the TBA is not a phenomenon which is purely restricted to 1D Peierls systems but should exist in models with a strong enough electron—phonon interaction with any number of dimensions and where the atomic lattice is taken as being *classical*, although the CDW structure becomes much more complex.

5. Future Prospects: Quantum Lattice Effects and Thermal Effects

All the studies presented at the beginning of this paper assumed that the atoms are classical particles and that the system is at absolute zero. In the first part of this section, we discuss the quantum lattice effects and present arguments that, in three dimensional models, the classical model remains qualitatively valid only for nonanalytic incommensurate CDW. We argue that the analytic incommensurate CDWs are unstable with respect to quantum lattice fluctuations and the system could becomes superconducting. Therefore, the real CDW systems should always have nonanalytic incommensurate ground states and be insulating at 0 K.

In the second part of this concluding section, we discuss the thermal fluctuations of the system and distinguish two limiting ideal behaviors: 1 — for large electron—phonon coupling k, the system is in an order-disorder regime. It has no underdamped phason but a diffusive dynamics. The CDW transition at the critical temperature T_c is described as the melting transition of an incommensurate lattice gas. 2 — For k close to and above $k_c(\zeta)$, the critical value at which super-conducting transition should occur, the physical behavior becomes displacive and approaches the standard behavior predicted in most of the literature on CDW. There is a phason mode with a small gap and, above the critical temperature, the system exhibits a soft mode which almost vanishes. Real systems range somewhere in the intermediate regime between these two limiting cases. Anomalous nonlinear conductivity, observed in many real CDW compounds, is explained on the basis of the assumption that real CDW are nonanalytic incommensurate structures.

The studies presented here are at a preliminary stage and are still in progress. We explain here essentially our physical ideas and suggest a new approach and new programs of work which, although not easy, seems to be very promising. More results concerning the thermal fluctuations are available in the Ph.D. dissertation of P. Quemerais [1].

5.1. QUANTUM LATTICE EFFECTS

The quantum fluctuations of the atomic positions in models with electron—phonon interactions have been taken into account in the calculations of Alexandrov *et al.* [62]. They have also shown that for large electron—phonon coupling, the electrons forms bipolarons and proposed to describe the system as a quantum lattice gas model with a spin $\frac{1}{2}$ Hamiltonian. The spin operator S_i^z at site i is $+\frac{1}{2}$ if this site is occupied by a bipolaron and $-\frac{1}{2}$ if it is empty. The bipolaron interacting energy only involves S_i^z operators while the quantum fluctuations of the lattice are described in this Hamiltonian by extra terms proportional to the operator $S_i^+ S_j^-$ which exchange a bipolaron from site i to site j.

Using their formalism, we found arguments for *3-dimensional* incommensurate uniaxial CDW (i.e. with a wave vector modulation in the direction of a crystal axis) suggesting that approaching the TBA from above the CDW becomes unstable with respect to quantum lattice fluctuations. The long range order of the CDW is expected to be destroyed at another transition *occuring at* $k_c^* > k_c$. Let us note that quantum fluctuations were already recognized to be important for sliding CDW ('analytic') by Bardeen [86]. However, this instability was not expected.

If this is true, *only nonanalytic CDWs could exist in nature*. When the long range order of the CDW is destroyed at 0 K by quantum effects, a new order in 'reciprocal space' is naturally expected in 3-dimensions to occur that is *superconductivity* ordering. We also found arguments for ruling out the mixed states between bipolaronic superconductivity and (nonanalytic) charge ordering predicted by Alexandrov *et al.* [62].

As a consequence, the physical and plausible scenario which is left (and takes the place of the classical TBA scenario) is that for simple electron—phonon coupled systems in the *absence of strong electron—electron interaction*, there are two kinds of possible phases:

For 3D systems with a small electron—phonon coupling $k \ll k_c^*$, the system starts to be a BCS superconducting phase which continuously turns into bipolaronic nature when approaching k_c^*, while the Cooper pairs tend to localize (but without any well-defined CDW). We conjecture that when the quantum fluctuations of the atomic lattice are taken into account at 0 K (and, physically, they must be taken into account), the TBA is replaced by a transition from a *Superconducting* state to a *Nonanalytic* CDW state.

For large electron—phonon coupling $k_c^*(\zeta) < k$ the system is in a *Nonanalytic* CDW—PLD ordering which corresponds to the localization in real space of the Cooper pairs which now become classical bipolarons. In the commensurate case $\zeta = \zeta_r = r/s$, $k_c(\zeta_r) = 0$ but $k_c^*(\zeta_r) > 0$ although for small commensurability order s, the critical electron—phonon coupling could be zero. A similar conclusion has also been recently obtained by Nasu [68] for a half-filled band 2D model (where $k_c(\frac{1}{2}) = 0 < k_c^*(\frac{1}{8})$) but for a different model (with only electron—electron interactions) and with a different approach. To fix the ideas, we focus here essentially on the quantum Holstein model (in 1, 2 and 3 dimensions) although further generalisation of our arguments to other models should be possible.

5.1.1. *A Quantum Lattice Gas Model for the Holstein Model in the Large Electron—Phonon Coupling Limit*

Using the formalism of fermion and boson operators, the Hamiltonian of the Holstein model where the kinetic energy term $p_i^2/2m$ is included, becomes $H + \Sigma_i \hbar\omega_0/2$ where

$$H = - \sum_{\langle i.j \rangle.\sigma} Tc_{i.\sigma}^+ c_{j.\sigma} + U \sum_i \left(\sum_\sigma c_{i.\sigma}^+ c_{i.\sigma} \right) (a_i^+ + a_i) +$$

$$+ \sum_i \hbar\omega_0 a_i^+ a_i \tag{96a}$$

$\langle i.j \rangle$ denotes nearest neighboring sites and

$$U = \lambda \left(\frac{\hbar}{2m\omega_0} \right)^{1/2} \tag{96b}$$

a_i^+ and a_i are the creation and annihilation phonon operators at site i, respectively defined as

$$u_i = \left(\frac{\hbar}{2m\omega_0} \right)^{1/2} (a_i^+ + a_i) \tag{97a}$$

and

$$p_i = i \left(\frac{\hbar m\omega_0}{2} \right)^{1/2} (a_i^+ - a_i). \tag{97b}$$

Then we obtain a form to which the theory of Alexandrov *et al.* [62] can be applied in the limit of large electron—phonon coupling. When U is large, the kinetic part of the electrons in the Hamiltonian can be considered as a perturbation. Using their theory, this Hamiltonian (72) is transformed by approximate unitary transformations into the Hamiltonian of a quantum lattice gas

$$H = \sum_i \mu S_i^z + \sum_{\langle i.j \rangle} J_{i-j} S_i^z S_j^z - \sum_{\langle i.j \rangle} t_{i-j} (S_i^x S_j^x + S_i^y S_j^y) \tag{98}$$

where (S_i^x, S_i^y, S_i^z) are spin $\frac{1}{2}$ operators. $\langle i.j \rangle$ denotes nearest neighboring sites.

$$S_i^+ = S_i^x + iS_i^y = c_{i\uparrow}^+ c_{i\downarrow}^+ \tag{99a}$$

corresponds to the creation of a bipolaron at site i and

$$S_i^- = S_i^x - iS_i^y = c_{i\uparrow} c_{i\downarrow} \tag{99b}$$

corresponds to its annihilation. Close to the adiabatic limit (classical lattice), they found (in our notation)

$$J_1 = 4T/k^2 \tag{100a}$$

and

$$t_1 = (4T/k^2)\exp(-4g^2) \tag{100b}$$

where g is a dimensionless constant:

$$g^2 = U^2/(\hbar\omega_0)^2 = Tk^2/4\hbar\omega_0. \tag{101}$$

Beyond the nearest neighbour their calculation, which was done at lowest order, yields zero coupling constants for J_n and t_n ($n > 1$). In the classical limit, which has been discussed in the previous section, $g = +\infty$ and $t_1 = 0$. Then, this model becomes identical to the classical lattice gas model described in Subsection 4.6 where $\sigma_i = S_i^z + \frac{1}{2}$.

Starting from the classical calculation of a bipolaron described in the previous section, one can reproduce the result of Alexandrov $et\ al.$ [62] close to the adiabatic limit ($\hbar\omega_0 \ll -U_{\text{bip.}} \cong Tk^2$). The quantum operator for creating a bipolaron at site n, S_n^+ is defined as the boson operator:

$$S_n^+ = \left(\sum_i \Psi_i^{(n)}c_{i,\uparrow}^+\right)\left(\sum_i \Psi_i^{(n)}c_{i,\downarrow}^+\right) \times$$

$$\times \exp\left(-k\left(\frac{T}{\hbar\omega_0}\right)^{1/2}\sum_i |\Psi_i^{(n)}|^2(a_i^+ - a_i)\right). \tag{102}$$

By definition, this operator just creates a pair of electrons with opposite spins in the state (86a) and the corresponding lattice distortion. For calculating the bipolaron bandwidth, the Hamiltonian (96a) is projected into the subspace generated by

$$|n\rangle = S_n^+|O\rangle \tag{103a}$$

where $|O\rangle$ is the empty state with no ferminons and no bosons. We obtain

$$\langle m\,\boldsymbol{H}|n\rangle = A_{m-n}\exp(-4G_{m-n}^2) \tag{103b}$$

and

$$\langle m\,|\,n\rangle = D_{m-n}^2\exp(-4G_{m-n}^2) \tag{103c}$$

where

$$D_{m-n} = \sum_i \Psi_i^{(m)*}\Psi_i^{(n)} \tag{104a}$$

is the overlap of the electronic states of two bipolarons at site m and n

respectively. A_{m-n} is equal to

$$A_{m-n} = -2D_{m-n}\left(T\sum_{\langle i\rangle} D_{m-n-i} - UD_{m-n}\right) +$$

$$+ D_{m-n}^2 Tk^2 \sum_i |\Psi_i^{(m)}|^2 \cdot |\Psi_i^{(n)}|^2 \qquad (104b)$$

$\langle i \rangle$ denotes the bonds such that $n + i$ be the nearest neighbor of n, and

$$\exp(-4G_{m-n}^2) = \exp\left(-\frac{Tk^2}{2\hbar\omega_0}\sum_i (|\Psi_i^{(m)}|^2 - |\Psi_i^{(n)}|^2)^2\right) \qquad (104c)$$

is the coefficient which characterizes the quantum lattice effects. This coefficient vanishes in the adiabatic limit. For large k, $D_m \cong (m + 1)/k^{2m}$ ($m > 0$). At the order of $1/k^2 \exp(-4G_1^2)$, $\langle m | n \rangle = 0$ for $m \neq n$ and $\langle n | n \rangle = 1$. Then, the modulus of the off-diagonal term yields the half bipolaron band width

$$\langle n + 1 | \boldsymbol{H} | n \rangle \cong -\frac{4T}{k^2} \cdot \exp\left(-\frac{Tk^2}{\hbar\omega_0}\right) \qquad (105)$$

which is exactly the result of Alexendrov *et al.* for this specific model in the adiabatic regime ($t_1 = \langle n + 1 | \boldsymbol{H} | n \rangle$). Then $g^2 = G_1^2 \gg 1$.

5.1.2. *Effects of the Quantum Lattice Fluctuations on an Incommensurate CDW*

For describing many bipolaron structures $\{\sigma_n\}$ (with $\sigma_n = 0$ or 1), we propose to study a variational generalization of this description suggested by our numerical observations. We take us an approximation of Hamiltonian (96a), its projection into the subspace E spanned by the states

$$S^+(\{\sigma_n\})|O\rangle = \prod_n [\sigma_n S_n^+ + (1 - \sigma_n)]|O\rangle \qquad (106)$$

where $\{\sigma_n\}$ are arbitrary bipolaronic configurations. Instead of using the eigenstate of a single bipolaron, the coefficients $\{\Psi_i^{(n)}\} = \{\Psi_{i-n}^{(0)}\}$ involved in (102) are determined variationally by minimizing the ground state energy. (Note that we will have to take into account that these states are not orthogonal.) We have not the explicit form of the projection $\boldsymbol{H}$ of Hamiltonian $\boldsymbol{H}$ in the subspace of states E. However, due to its definition, it can be formally written as a function of spin $\frac{1}{2}$ operators (with $S_i^z = \sigma_i - \frac{1}{2}$) with S_i^z, S_i^+ and S_i^- operators. In the large k limit (small bipolarons), the Hamiltonian (99) obtained by Alexandrov *et al.* is an approximation of this Hamiltonian.

In the classical limit (large k and large $T/\hbar\omega_0$), we expect to recover the numerical results presented in the last section for a one dimensional model. The localized variational electronic state $\{\Psi_i^{(n)}\}$ yields approximately the coefficients b_i defined in (73a) by Equation (96) and the TBA of this model should appear in the fact that by decreasing k down to some value $k_c^{var}(\zeta)$ approximately equal to $k_c(\zeta)$, the localization length of $\{\Psi_i^{(n)}\}$ (which is the CDW coherence length ζ defined in the previous section) diverges. Beyond this limit for $k < k_c^{var}(\zeta)$, the variational form (106) for the ground state is not appropriate. In 2 or 3D models, a similar TBA should occur at some value of k: $k_c(\zeta)$, where ζ is now the band filling, although the pseudospin configuration in the nonanalytic case is generally very difficult to find.

Now to estimate the quantum tunnelling of the bipolarons close to this adiabatic limit, we consider two bipolaron configurations $C_m = \{\sigma_i\}$ and $C_n = \{\sigma_i'\}$ which are identical apart from the location of only one bipolaron which is either at site m for C_m or at site n for C_n, the exchange coefficients $\langle C_m | H | C_n \rangle$ and $\langle C_m | C_n \rangle$ are obtained by a similar calculation as for the single bipolaron calculation described above. Then, if the localization length of $\{\Psi_i^{(n)}\}$ diverges, the quantum coefficient (104c) associated with the tunnelling of a bipolaron between two sites m and n, is no longer small and *goes to unity*. Coefficient $\langle m | n \rangle$ goes to 1, and $\langle m | H | n \rangle$ goes to $-2(zT - U)$ where z is the number of neighbouring sites to a given site i). As expected, this variational calculation should become more and more inaccurate while approaching the TBA.

Nevertheless, the important physical remark which emerges from this calculation is that when k approaches k_c from above, the spin $\frac{1}{2}$ operators in the Hamiltonian S_i^x and S_i^y due to *the quantum lattice fluctuations cannot be neglected* and becomes comparable to or larger than the S_i^z terms *even very close to the classical lattice limit*. These terms, which flip the classical S^z spins, tends to destroy the S^z component ordering and to replace it by an ordering of the S_i^x and S_i^y components. This assertion is physically explained by the fact that the Peierls—Nabarro barrier which classically pins the bipolarons is collapsing when approaching k_c from above.

If the terms in the Hamiltonian which are due to the quantum lattice fluctuations (which yield the quantum 'kinetic energy' of the CDW) becomes comparable at $k = k_c^*$ to the 'classical terms' (which yield the potential energy of the classical CDW configuration). When k becomes smaller than k_c^*, the amplitude of the CDW distortion decreases and then it is reasonable to expect that the relative contribution to the energy of the CDW of the quantum lattice terms becomes larger than that of the classical lattice distortion, determined by k smaller than k_c^* with k_c^* *larger than* k_c. When the energy terms in the Hamiltonian due to the quantum lattice fluctuations are predominant, the model is said to be in the 'antiadiabatic' regime.

The antiadiabatic regime obtained for large k while $Tk^2/\hbar\omega_0$ is small (very small electronic band width or large phonon frequency) has been studied by Alexandrov *et al.* [62]. They found a superconducting state which they claimed to be a new kind of superconductivity. Unlike the BCS superconductivity which only involve the electrons close to the Fermi surface, this superconducting state involves all the electrons of the considered band. In the spin $\frac{1}{2}$ representation, this superconductivity is associated with the ordering of the S^x and (or) S^y component of the spins. Unlike the BCS superconductivity, this superconductivity was found to be gapless and has been called 'bipolaronic'. We think that the gapless character of this superconductivity is an artefact due to the truncation of the bipolaron interactions beyond the nearest neighbours. We believe that a BCS superconducting state can evolve to a bipolaronic superconducting state without crossing any transition line and that there is no qualitative differences between these two states although there are very large quantitative differences.

In our situation (large $T/\hbar\omega_0$), it is well known that at least for small k, the BCS theory applies and that this quantum Holstein model should be a BCS superconductor (in 3D). First, let us rewrite the Hamiltonian (96a) in reciprocal space

$$H = \sum_{q.\sigma} \varepsilon(q)c^+_{q.\sigma}c_{q.\sigma} + \frac{U}{\sqrt{N}} \sum_{q.q'.\sigma} c^+_{q+q'.\sigma}c_{q.\sigma}(a^+_{-q'} + a_{q'}) \tag{107a}$$

where

$$\varepsilon(q) = -2T \sum_\alpha \cos q_\alpha \tag{107b}$$

q_α are the components of q, N is the number of sites. The electron creation operator in the reciprocal space with wave vector q is defined as

$$c^+_{q.\sigma} = \frac{1}{\sqrt{N}} \sum_n e^{iqn}c^+_{n.\sigma} \tag{107c}$$

$c_{q,\sigma}$ is defined as the conjugate operator of $c^+_{q,\sigma}$. The same definition holds for the Fourier transform of the phonon creation operator and its conjugate. With a standard unitary transformation $\exp(S)$, the new Hamiltonian $H' = \exp(S)H \exp(-S)$ involves an effective electron–electron interaction of order U^2 (see for example [69] for the detailed calculation of H'). This operator S calculated to first order in U is:

$$S = \frac{U}{\sqrt{N}} \sum_{q.q'.\sigma} \left(\frac{c^+_{q+q'.\sigma}c_{q.\sigma}a^+_{-q'}}{\varepsilon(q+q') - \varepsilon(q) + \hbar\omega_0} + \frac{c^+_{q+q'.\sigma}c_{q.\sigma}a_{q'}}{\varepsilon(q+q') - \varepsilon(q) - \hbar\omega_0} \right). \tag{108}$$

The variational standard BCS ground state of the new Hamiltonian H' has the form

$$|G\rangle = \prod_q (u_q + v_q c^+_{q.\uparrow}c^+_{-q.\downarrow})|0\rangle \tag{109a}$$

where the coefficients u_q and v_q are determined variationally. The pair creation operator $c^+_{q.\uparrow}c^+_{q.\downarrow}$ only creates a pair of electrons with opposite spins and wave vectors q and $-q$ in the limit of a small electron–phonon coupling U (or k). In fact, the new Fermion creation operator

$$c_{q, \sigma}^+ = \exp(S)c_{q, \sigma}^+ \exp(-S) \tag{109b}$$

also creates a 'cloud' of pairs of electrons with opposite spins and different wave vectors q' and q'' and with their associated phonons. It is interesting to note that the bipolaron creation operator S_n^+ at site n defined by (102), has a similar property in the real space. It creates a single site bipolaron at site n only for large U (or k) only in the limit of a large electron–phonon coupling U (or k). But for smaller U, it also creates many pairs of electrons with opposite spins at different sites, together with their associated phonons. Within this variational formulation, *the BCS pair creation operator*

$$S_q^+ = c_{q, \uparrow}^+ c_{q, \downarrow}^+ \tag{109c}$$

with wave vector q just plays in *the reciprocal space*, a role dual of the *bipolaron creation operator* S_n^+ at site n (defined by (102)) in *the real space*. In principle, a better result must be obtained for the superconducting ground state if the unitary operator S in (109b) instead of being calculated perturbatively for small U, is also determined variationally. The same form for S as (109), but where the coefficients of the operators are found by minimization of the ground state energy, seems to be appropriate.

Since the lattice quantum terms becomes very important for $k < k_c^*$, it is not unreasonable to believe that all the analytic incommensurate CDW expected in the classical limit for $k < k_c < k_c^*$ are unstable with respect to the quantum lattice fluctuations and do not exist. Close to this adiabatic limit, the ground state should be well approximated by a BCS ground state with the form (109).

In contradiction to Alexandrov *et al.*, let us now present an argument suggesting that a nonanalytic CDW cannot coexist with superconductivity. For that purpose, let us consider for example a Hamiltonian with the form (99) given by Alexandrov *et al.* but where the coefficients J_{i-j} are positive and long range in the x-direction and negative and only between nearest neighbours in the perpendicular direction (ferromagnetic coupling). For example, for a 2 dimensional model

$$J_{n, 0} = K\eta^{|n|} \tag{110a}$$

and

$$J_{0, 1} = -J. \tag{110b}$$

All the other coefficients J are zero. When the quantum terms are switched off, the ground state is a nonanalytic incommensurate structure.

The quantum coefficients t_{i-j} are supposed to be nearest neighbour and equal to a positive constant t. Without quantum terms ($t = 0$), the ground state of this model in one dimension, is exactly known and given by $S_n^z + \frac{1}{2} = \sigma_n = \chi(n\zeta + \alpha) = 0$ or 1 with the characteristic function (95). For several dimensional models, the spin configuration $\sigma_{n, i}$ only depends on the i index. The ground state of the whole system is made of parallel chains identical to the one-dimensional ground state with a constant phase α: $S_{n,i}^z + 1/2 = \sigma_{n,i} = \chi(n\zeta + \alpha)$. The ground state is a *uniaxial* incommensurate structure with a wave vector which always remains along the x-axis and can represent a simplified model for a quasi-one dimensional CDW. These incommensurate ground states are obviously degenerate with respect to an arbitrary phase shift of the phase α of the modulation.

(Note that, at the present stage, we ignore the long range interaction of the bipolaron Hamiltonian associated with the quantum Holstein model. We only know that it can be written as a function of spin 1/2 operators and its ground state in the adiabatic limit is also unknown. We have chosen such a form for the sake of simplicity because in that case the exact classical ground state is known and the effect of the lattice quantum fluctuations can be analysed. In spite of that, the ideas presented for this simple case are generalisable to other quantum lattice gas model).

In one dimension, an infinitesimal phase shift around a discontinuity of the hull function $\chi(x)$, exchanges only one bipolaron from a site to a neighbouring site (see Sections 2 and 4) in this nonanalytic incommensurate ground state. The consequence is that the phase degeneracy of one-dimensional non-analytic incommensurate structures is raised by any small quantum term proportional to $S_i^+ S_{i+1}^- + S_i^- S_{i+1}^+$ and no nonanalytic incommensurate CDW with long range order can exist in principle in 1D models with quantum lattice fluctuations. (In the case of a commensurate CDW of order s, one finds that it remains stable up to a certain critical value of the quantum perturbation t. This critical value of t goes to zero as s goes to infinity.)

For a three dimensional model, a mean field calculation should become qualitatively exact. (We do not discuss the two dimensional case for which a mean fied approximation may be wrong.) The standard mean field calculation of the ground state of an arbitrary quantum spin $\frac{1}{2}$ Hamiltonian consists in finding the coefficients s_n at all sites n, for a quantum state $|G\rangle$ with the form

$$|G\rangle = \prod_n [s_n S_n^+ + (1 - s_n)]|O\rangle \qquad (111a)$$

which minimize the variational energy $\langle G|\mathbf{H}|G\rangle$. The solution is considered as a good approximation of the real quantum ground state. Then $\langle \sigma_n \rangle = s_n$ is the average pseudospin $\sigma_n = S_n^z + \frac{1}{2}$ at site n. Note that this form for the ground state appears now completely analogous in the real space with the BCS form (109) in the reciprocal space where S_n^+ is replaced by S_q^+ given by (109c).

For $t_{i-j} = 0$, s_n is given by the nonanalytic hull function (95) with a wave vector modulation in the x direction (a square function). The ground-state of the mean-field Hamiltonian $\langle G|\mathbf{H}|G\rangle$ for a finite quantum term t is not exactly known. It is reasonable to assume at least for small t, that the sequence s_n which minimises $\langle G|\mathbf{H}|G\rangle$, is quasiperiodic and is described by a hull function $s(x)$ with a wave vector in the x direction with length Q and an arbitrary phase α

$$s_{n,i} = s(Qn + \alpha). \qquad (111b)$$

It is then reasonable to conjecture two effects

1 — Because of the quantum lattice fluctuations, the wave vector Q of the modulation should be shifted and not exactly equal to $2k_F$ (if it is incommensurate with the lattice wave vector. However, this effect should be negligible for k much larger than k_c^* but it could become non-negligible when k is closely above k_c^*. We do not consider this effect as essential and, at the present stage, we assume that Q is constant as a function of k.

2 — Arguing on the universality of the concept of breaking of analyticity, we claim that for small t, the hull function $s(x)$ is a *discrete function* without any continuous part which can be written as function $g(x)$ in (73a) where $\chi(x - x_n) = 0$ or 1

$$s(x) = \sum_n B_n \chi(x - x_n) \tag{112}$$

where B_n are some coefficients which go to zero as n goes to infinity.

Note that, for larger t, there could exist a transition by breaking of analyticity in the mean field model with respect to the quantum parameter t or another transition. If this TBA exists, it could be numerically studied by fixing ζ' to a given irrational number (by choosing the field $\mu(t)$). The hull function $s(x)$ then can be calculated with the same gradient method used in the other models studied in the previous sections of this paper. Since a TBA for the same hull function of the CDW cannot occur twice, it must be interpreted as corresponding to the TBA which we proved to exist in the adiabatic limit.

Nevertheless, for small enough t, in view of our previous studies of the transition by breaking of analyticity in other models, it is highly probable that $s(x)$ *remains a discrete function* as it is for $t = 0$. In that case, a new bipolaron operator $S_n'^z$ is definable so that the form (106) where S_n^z is replaced by $S_n'^z$, can be used again for describing this mean field ground state. The mean field ground state has the form (111a) where s_n is given by a discrete function in (111b). Form (107) can be written again

$$|G\rangle = N \exp\left[\sum_n \frac{s_n}{1 - s_n} S_n^+ \right] |0\rangle \tag{113a}$$

where N is a normalization factor. The coefficient $s_n/(1 - s_n)$ is also a discrete function of $Qn + \alpha$ and can thus be written as:

$$s_n/(1 - s_n) = \sum_i F_i \chi(Q(n - i) + \alpha) = \sum_i F_i \sigma_{n-i}' \tag{113b}$$

where F_i is some set of constants (playing the role of b_i in (79)) which goes exponentially to zero for i going to infinity

$$\sigma_n' = \chi(Qn + \alpha) = 0 \quad \text{or} \quad 1 \tag{113c}$$

is a new set of pseudospins determining the incommensurate configuration. Then, the nonanalytic incommensurate ground state (111), takes the form

$$|G\rangle = N \prod_n [\sigma'_n S'^+_n + (1 - \sigma'_n)]|O\rangle \tag{114a}$$

where

$$S'^+_n = \exp\left(\sum_i F_i S^+_{n+i}\right) \tag{114b}$$

can be interpreted as the creation operator of a *localized 'quantum' bipolaron*.

Also, in the case of a uniaxial incommensurate CDW, coefficients F_i are not uniquely determined but a choice can be made in order that all coefficients $F_{i,j}$ vanish for $j \neq 0$ where i is the index component in the wave vector direction, and j represent the other transverse components. The coefficients F_i decay exponentially at infinity with a characteristic length ξ'. A similar situation is encountered when describing a filled band by the creation operators of electrons in the Wannier states.

(Note that $(S'^+_n)^2$ is not zero, unlike the initial operator S^+_n. It is perhaps more convenient to change the variational form proposed in (111a) in order to define more conveniently a quantum bipolaron operator S'^+_n with the useful property $(S'^+_n)^2 = 0$. The problem is then to find variationally the best unitary transformation of S^+_n such that the ground state has the form (114a)).

Nevertheless, these arguments show that if the incommensurate CDW ground state remains nonanalytic, it can be constructed as the product of *localized* quantum bipolaronic operators at the sites of an incommensurate sublattice determined by σ'_n in (113c). *In spite of the existence of quantum terms in the Hamiltonian, the electrons keeps forming a classically ordered structure of localized bipolarons.* The effect of the quantum lattice terms is just to increase the extension of the classical bipolarons which nevertheless remains finite.

(But note that this result does not necessarily hold for the first excited states of this ground state. In the case of a uniaxial incommensurate structure in the x direction and because of the periodicity of the ground state in the x and y directions, the lowest energy excitations have a gap in energy and are extended in the directions y and z perpendicular to the x direction and localized in the x direction.)

An ideal variational calculation should be able to determine *self-consistently* both the coefficients Ψ_i in (102) and F_i in (114b) by minimizing the energy of the initial quantum Hamiltonian. The two characteristic lengths ξ (the 'classical' length) and ξ' (the 'quantum' length) are *a priori* different. Note for example that in the adiabatic limit (large $T/\hbar\omega_0$) that ξ' vanishes while ξ is finite. The effective extension of the 'quantum' bipolaron is the largest of these two lengths $\sup(\xi, \xi')$. But we expect that the self-consistent calculation should ensure that ξ is always larger than ξ' because the variational electronic wave function, which determines

ξ, has to be calculated in the effective lattice distortion which already has the extension $\sup(\xi, \xi')$.

By decreasing the parameters k, in 3 dimensional models one expects a phase transition. Let us assume that it is a second order transition. Then, these two characteristic length ξ and ξ' should diverge simultaneously. This result is clear if the characteristic length ξ' first diverges because $\xi > \xi'$ but suppose that the localization length ξ first diverge. Then we have already noted that the amplitude t of the quantum terms in the lattice gas Hamiltonian should be strongly enhanced. If this transition is second order, a TBA for the hull function $s(x)$ must be reached and then the localization length ξ' should also diverge.

In the opposite regime (small k), we noted that the BCS form for the super-conducting state can be considered in some respect 'dual' (in the reciprocal space) of the variational form (110). For sake of simplicity, we conjecture that in the 3D quantum Holstein model, a unique phase transition takes place of the transition by breaking of analyticity obtained in the classical lattice case. It is a transition from a *non-analytic incommensurate CDW to a superconducting state*.

Let us note that if, conversely to our expectation, the stability of an analytic CDW state would persist in some domain of the parameter space in spite of the important quantum lattice terms which we have shown to necessarily exist, the possibility of having quantum tunnelling of the CDW suggested by Bardeen (1978, 1980) would become essential for the CDW transport property [86].

Within the variational description which we proposed, some kind of duality between a 3D superconducting state (represented in the reciprocal space by S_q^+ pair creation operators (109a)) for small U and a 3D nonanlytic incommensurate CDW (represented in the real space by S_n^{-2} operators (110)) for large U appears. This situation is nicely reminiscent of the duality property of the Harper Equation (31) which has been described in Section 3. In that model, a TBA corresponding to a transition from a regime with extended quasi-Bloch eigenstates (localized in the reciprocal space) to a regime with exponentially localized eigenstates in the real space has been found. More detailed investigations, both analytical and numerical, of some specific models with electron—phonon interactions are neces-sary. Particularly, it should be possible to check numerically the existence of this transition by assuming that the variational ground state of model (96a) can be found within a reasonable approximation with either the BCS form (109) or the bipolaronic form (111).

As we have already stated in this paper, recall that a commensurate CDW is also a nonanalytic CDW and that a transition to a superconducting state is also possible. However, if the commensurability order is small, the critical value of k_c^* might vanish (see e.g. [70, 71] for models in 1D with a half-filled band). For large commensurability order, the behavior of a commensurate CDW is physically the same as for a true incommensurate CDW.

At the present stage, our arguments for claiming that the transition between a nonanalytic CDW to a superconductor is a TBA are empirical and do not require very precise details on the Hamiltonian. They make use of the assumption that the concept of transition by breaking of analyticity tested on other simpler models can also be extended to more complex models involving similar conceptual problems. As for a standard symmetry breaking, the breaking of analyticity of a system has a global effect: *it simultaneously affects all the physical properties of the system*. This assertion was indeed confirmed by numerical experiments but only on 'semi-quantum' models in Section 4.

Finally, let us recall that many other problems are left in the first rough approach suggested here. The CDW structures can be more complex than simple uniaxial incommensurate structures and these CDWs are not all stable against quantum lattice fluctuations. In one dimensional systems, both the superconducting states and the nonanalytic incommensurate CDWs are unstable and the resulting structure should be conducting at absolute zero. In 3 dimensional systems, nonanalytic incommensurate CDW may be also unstable with respect to quantum fluctuations, for example when the wave vector of their modulation has three incommensurate components. In those cases, the system does not necessarily become superconducting although it is conducting at 0 K. It should be interesting to know what all these states are and if some of them correspond to the picture of a metal, of mixed valence or heavy fermions compounds etc . . . ?. Of course, understanding the temperature stability of the superconducting state close to the CDW transition at k_c^* is interesting for potential application to high T_c super-conductors.

5.2. THERMAL EFFECTS ON A NONANALYTIC INCOMMENSURATE CDW GROUND STATE

All the arguments developed in the previous section have supported the idea that real incommensurate CDW must have a nonanalytic hull function and can be described as *classical* incommensurate arrays of localized bipolarons super-imposed on the lattice. Let us recall once again that we work with the assumption that the direct (Coulomb) electron—electron interaction is negligible at the microscopic scale. Despite the fact that the concept of breaking of analyticity could also be useful in these cases (but needs further investigation), the following description does not apply unchanged to CDW systems where these interactions are important, and particularly when there is spin ordering and magnetism, or only important spin fluctuations.

We describe here our ideas on what could be the physical effects of the thermal fluctuations on the CDW. We assume in this subsection that we are in the adiabatic limit, where the classical model is valid. In that situation, the most important low temperature excitations of the nonanalytic incommensurate CDW do not correspond to electrons jumping above the Peierls gap but are *configurational excitations*. These excitations corresponds to vacancies and interstitials in the bipolaron configurations and also to phase defects and vortices. In one-dimensional nonanalytic incommensurate CDW, all the configurational excitations can be described as phase defects (or discommensurations) corresponding to arbitrary phase differences or by 'word breaking' leading in both cases to the same hierarchical classification of their energies [28, 29].

For three dimensional models, the melting transition of the bipolaron sublattice corresponds to the critical temperature T_c at which the CDW disappears. However, it is essential to distinguish the two limit regimes of the model according to the fact that the characteristic temperature T_{PN} of the Peierls—Nabarro barrier defined by $k_B T_{PN} = E_{PN}$, is much larger than T_c, or much smaller.

5.2.1. Order—Disorder CDW

The first situation is obtained in the limit of a large electron—phonon coupling k. We have calculated, in Section 4, this PN energy in (87b) for the 1D Holstein model which *diverges* as k^2 when k diverges. The same behavior is easily proved by similar calculations for a classical Holstein model at any dimension. The energy for breaking a bipolaron into two free polarons (which both have a spin $\frac{1}{2}$) is half the energy of a bipolaron which is also approximately the PN energy barrier. Otherwise, the characteristic energy for the bipolaron interactions J given by (93b) goes to zero as $1/k^2$.

Therefore, in the limit of large k, the system can be described, for all temperatures smaller than T_{PN} by a classical lattice gas model of stable bipolarons well located at the lattice sites. The corresponding Ising model has a fixed magnetization corresponding to a fixed concentration of electrons. As for the stability with respect to quantum lattice fluctuations, we expect that the stability of the CDW with respect to thermal fluctuations, depends on the incommensurability ratio of the wave vector. In the case of a uniaxial incommensurate CDW (only one component of the wave vector of the CDW is incommensurate), a phase transition should exist in 3 dimensions. Because the PN barrier is very high and mostly locates the bipolarons at the lattice sites, this CDW transition is of the order—disorder type and occurs at a temperature $T_c \ll T_{PN}$ comparable to the nearest neighbour bipolaron interaction energies $k_B T_c \propto J_1$.

As shown by mean field calculations on simple lattice gas models [1], the wave vector of the CDW can exhibits a significant temperature dependence (as a devil's staircase) in the ordered phase for $T < T_c$. At low temperature, the variation curve becomes almost constant with an essential singularity at 0 K. At higher temperature, the melting of a nonanalytic incommensurate lattice gas appears to be rather complex and depends on the model. First order transitions are not excluded and depends on the details of the bipolaron interactions.

It is interesting to note that when k goes to infinity, the electronic gap also diverges as k^2 but that the CDW transition temperature goes to zero as $J(k) \propto 1/k^2$. Clearly the well-known standard mean field calculation for the critical temperature of a CDW (see formula (5)) breaks down in that limit. (However note that the Holstein model is peculiar because the phonon branch has no dispersion. In a more general case with some phonon dispersion, the bipolarons do not become localized on a single site for large k and then the interaction between bipolarons does not drop to zero for large k as well as the critical temperature T_c).

The lattice dynamics structure factor $S(q, \omega)$ of the system does not exhibit any underdamped phason mode but a diffusive central peak associated to the bipolaron diffusion. $S(q, \omega)$ should exhibit a critical narrowing at T_c corresponding to the appearance of the CDW lattice modulation peak.

5.2.2. Displacive CDW

This is the critical regime close to the TBA, the size of the bipolaron increases and the PN barrier tends to vanish. At the ideal limit only $k = k_c^*$ (when the amplitude of the CDW also tends to zero), one should reach the standard behavior of a sliding incommensurate CDW with Fröhlich conductivity.

At 0 K, the crossover from the order—disorder regime to the displacive regime occurs when the characteristic energy of the bipolaron interactions J becomes comparable to or larger than the Peierls—Nabarro barrier E_{PN}. Then the configurational defects of the CDW become much thicker than the unit cell and have a soliton-like behavior. The CDW cannot be described properly by classical lattice gas model because the PN barrier is not high enough for locating the bipolarons at the lattice sites. That situation is well-known for standard displacive structural phase transitions (see e.g. [72—74]) and has again been found in incommensurate structures [42].

A simplified model which could help for a qualitative understanding of some aspects of this regime is a classical gas model where the interacting particles can be at any continuous position and are submitted to a weak periodic potential representing the lattice pinning potential. The ground state of this model is supposed to be an incommensurate structure. In the limit of a large potential and at low enough temperature, a lattice gas model is recovered. Let us emphasize that such models have not the same properties as a model of coupled 1D FK chains in d dimensions because the connectivity of the bipolaron lattice is not given a priori. Vacancies and interstitials which are not permitted in the d-dimensional FK model, play a crucial role for the melting of the CDW structure. However, this model suffers the defect of not being self-consistent because the lattice pinning potential should depend on the distribution of the bipolarons when the temperature varies. It also neglects the phonon fluctuations of the underlying lattice which become very important when approaching the CDW transition.

By analogy with the displacive structural transitions, one expects that the displacive CDW transitions are driven by a soft mode. In the ideal displacive limit, above T_c, there should exist a soft mode the frequency of which goes to zero at T_c. The wave vector of this soft mode is the same as the wave vector of the CDW. Below T_c in the ordered CDW phase, this soft mode splits into the phason (at zero frequency) and the ampliton modes.

In fact in a real system which cannot be strictly at the displacive limit, this phonon frequency never strictly vanish at T_c (incomplete softening). Close to T_c and for q in the vicinity of the CDW wave vector, the structure of the dynamical factor $S(q, \omega)$ becomes overdamped with a growing central peak. Below T_c, this soft mode splits into the phason mode but this one has a small gap and the ampliton mode. The phason structure remains overdamped with a central peak that is broad in frequency (with a width comparable to the soft mode frequency) and narrow in q. As the temperature decreases, and if the CDW structure is in perfect thermal equilibrium (no quenching), the dynamical structure factor should evolve and become underdamped. The phason gap should increase and at 0 K reach, the small pinning frequency which the incommensurate CDW should have in its ground state. The central peak associated with the bipolaron diffusion should get narrower in q and ω, and disappear at 0 K.

In contrast with the order—disorder case, close above the critical temperature the diffusive central peak disappears. The fluctuations of the local lattice distortions which are essentially due to the phonons are fast and average to zero. The bipolaron structure is then swept out in the high temperature phase which returns to a standard metallic regime.

Instead of this behavior, one might believe that, within a mean field description, there could exist a TBA as a function of temperature at T_P in the regime where the CDW is ordered. For $T_P < T < T_c$ where T_P is some critical temperature, the CDW might recover an analytical hull function and thus a sliding mode with a strictly zero phason gap. Then the soft mode should strictly reach the zero frequency at T_c. We do not believe that this situation occurs except at the perfect displacive limit. However, the answer to this question is an open problem.

Let us also mention that the wave vector of the CDW should also vary in the ordered phase as a function of the temperature but much less than in the order—disorder case. At the ideal displacive limit only, it becomes independent of the temperature.

In the intermediate regime between the order—disorder and the displacive limit, the phonon softening at T_c is weaker and the phason gap larger than in the displacive case. The central peak exists in a larger domain in temperature above T_c. While approaching the order—disorder limit, the soft phonons becomes less and less dependent on temperature and the central peak becomes well distinguished from the soft phonon peaks.

5.2.3. Ohmic Conductivity of a Nonanalytic CDW

At 0 K, a nonanalytic CDW ground state is always pinned by the lattice. Except very close to the displacive limit, this pinning energy E_{PN} which is a microscopic energy comparable to a thermal energy (typically the CDW transition temperature ~ 100 K) remains *by far larger by several order of magnitude*, than the largest microscopic energies which could be involved in real experiments with a macroscopic electric field. *Consequently, at 0 K, all CDW compounds should be insulating.* Indeed, most real materials with a CDW are insulators at 0 K.

One of the very few exceptions to this rule is given by $NbSe_3$ which remains conducting at 0 K (see Monceau [75]). But this system has peculiarities because its structure is formed by three kinds of parallele and weakly coupled conducting chains. A CDW structure appears only on two of these three kinds of chains. The chains of the third kind, do not exhibit any CDW structure but it is also interesting to note that at low temperature (and under pressure), the system becomes super-conducting.

At finite temperature, the CDW motion under an electric field is essentially due to the thermal motion of the bipolarons above the PN barrier. However, there is an important qualitative difference according to the fact that the CDW is of the order—disorder type or of the displacive type.

In this order—disorder regime, the motion of the bipolarons corresponds to a single particle diffusion in a quasiperiodic potential which is *thermally activated* with a large characteristic energy roughly equal to the PN energy. For a large range of temperature from zero to T_{PN}, the ohmic conductivity of the system increases with the temperature. It has roughly the behavior of the conductivity of an insulator apart an anomaly at T_c where the bipolaron mobility should sharply increases due to the melting of the bipolaron superstructure. The order—disorder CDWs remain bad conductors at all temperatures.

By contrast, close to the displacive limit, the PN energy barrier is small and the bipolaron are strongly coupled one with each other. The motion of the bipolarons becomes more collective. The conductivity is also thermally activated but the eleectric transport is due to the diffusion of the configurational defects of the CDW (bipolaron vacancies and interstitial) which extends over several unit cell of the crystal and thus are very mobile. The thermal activation energy involves the energy for creating these configurational defects which is comparable to the thermal energy $k_B T_c$ associated with the CDW transition temperature (and thus much smaller than in the order disorder case).

Above the transition temperature, since the CDW modulation disappears, the bipolarons are destroyed by the thermal fluctuations and the system returns to a metallic regime. New results concerning quantum fluctuations of CDW are available in ref. 96 and in preparation.

5.2.4. *Nonlinear Electric Conductivity of a Nonanalytic CDW*

At 0 K, the phase of the CDW is pinned to the lattice. By increasing the temperature from zero, the effective pinning should decrease because of the thermal diffusion of the bipolaron and remains finite up to the temperature T_c (or T_P) at which it should vanish. However, in practice this pinning always remains very strong and prevents a *global uniform motion* of the CDW under the effect of the macroscopic electric field applied in real experiments.

This lattice pinning naturally allows the existence of phase domains in the CDW structure which are *metastable* and do not require any impurities or defects for pinning the CDW phase. In each domain, the CDW has a constant phase pinned by the lattice, which is different of those of the neighbouring domains.

It is well known that these domain walls have an electric charge if they are not parallel to the phase gradient. This situation is analogous to that of *the domain walls in ferroelectrics*. Because the Coulomb energy cost involved in the formation of a charged domain wall with area S diverges as $S^{3/2}$ and thus becomes very great for macroscopic surfaces, the domains are always parallel to the polarisation axis in order to be electrically neutral. For the same reason, in CDW systems the phase domain wall are electrically neutral on the macroscopic scale. (There are obviously thermal microscopic wall fluctuations which involve some electric charges.) The consequence for quasi-one dimensional conductors is that the domain walls must be *parallel to the conducting chains.* (Note that although the Coulomb interactions between the electrons could sometimes be neglected on the microscopic scale compared to the other electronic forces, they necessarily play a role in the domain formation.)

Up to the end of this paper we focus on the quasi-one dimensional conductors such as the molybdenum bronze compounds to which most of this book is devoted. (But nonlinear conductivity could also exist in two dimensional CDWs or three dimensional CDWs, if any.) Since the CDW is incommensurate, the phase shift between two phase domains can be very small and thus the energy of such a phase wall per unit surface can be also very small. Consequently, in a real system, the phase domain wall creation is much easier than in a ferroelectric and many phase domain walls should exist in nonanalytic CDW. Since they are mostly parallel to the chain axis, this effects should produce a systematic transverse broadening of the satellites peaks observable by X-ray or neutron scattering. It is also reasonable to expect that the width of these peaks decreases at lower temperature while the wall kinetics are still sufficiently fast because the wall free energy increases. At very low temperature, the wall structure should become quenched because the wall motion is due to the thermal diffusion of the bipolarons, which then becomes very slow.

It is well known that the application of an electric field to a ferroelectrics cannot globally reverse the polarisation because of the huge energy required for causing the polarization to vanish. However, in practice, the polarization of a ferroelectric is readily reversed by the creation and the propagation of walls. The situation is the same for a pinned CDW. The electric field cannot globally depin the CDW but it can move the walls in order to extend the domains where the phase is the largest at the expense of the domains where the phase is the smallest.

These domain walls of a nonanalytic CDW can easily be moved under the effect of an electric field, as for a ferroelectric. The motion of these neutral phase domain walls produces an extra electric depolarization current which adds to the Ohmic electric current. In a ferroelectric, this depolarization current is transitory and ends when the sample becomes a ferroelectric monodomain. The situation is very different in a nonanalytic CDW because the phase has no upper bound, unlike the polarization of a ferroelectric which only has two possible values. A nonanalytic CDW in an electric field can generate indefinitely new phase domain walls, producing a depolarization current in a stationary regime.

However, the new phase domain walls which are permanently produced must also be eliminated. This could be done at the edge of the sample. But one now has to take into account the fact that in practice *the current density is non-constant in any sample.* It is also well known that the big defects, such as dislocations etc., can pin or slow the wall motion down. To fix the ideas, let us consider the border between a region with a constant electric current and a region with no electric current. The phase domain walls must slow down and stop because of the opposite electric field-induced electric charges which are generated by the excess of the depolarisation current at this border. If no phase domain walls disappear, the nucleation of new phase domain walls cannot continue because of the energy cost of the wall interactions (which are repulsive) or equivalently because of the phase distortion. The sample is then highly polarized but the extra current due to the phase wall motion stops.

However, beyond a certain depinning field E_T, (the observed electric threshold in experiments), the destruction of phase domain wall becomes possible and starts around few nucleations centers such as big defects (dislocations, grain boundaries). The phase domain walls must annihilated by a packet with a total phase shift which is necessarily 2π. When this annihilation occurs, vortex line defects are produced and their motion is parallel to the conducting chains. In order to have narrow band noise, the number of these nucleation centers has to be of some units only. Of course, this number sharply depends on the sample. (On this point, there is a similarity between our interpretation and earlier theories developed by Gor'kov [76], Batistic et al. [77] and also by Ong, Verma and Maki [78].) There is a sudden relaxation of the CDW stress which produces a voltage drop and the nucleation of new phase domain walls can start again, as well as the extra current. After a short time, the same process starts again around the same nucleation centers. The whole system can reach a stationary regime in time, which is periodic, quasi-periodic, or chaotic.

If the number of these nucleation centers in the sample is a few units, a narrow band voltage noise occurs and also regular patterns in the phase wall motion could form in the bulk of the sample. If the number of nucleation centers is large ('dirty sample'), and/or if the electric current is very nonuniform the voltage noise becomes broad band. The geometry of the sample as well as its quality (only a few nucleation centers) plays an important role for the occurrence of narrow-band noise. The change in the density and the number of walls after various conductivity experiments should induce memory effects. The effective lattice pinning increases at low temperature which allows quenching of metastable configurations. More details about this theory are given in Quemerais [1].

In real systems, the wall mobility should be much higher in the displacive CDW than in the order–disorder CDW (because of a smaller PN energy barrier). Since nonlinear conductivity can be measured only when E_T is not too large, we expect that the real systems where it is observed corresponds to nonanalytic CDW close enough to the diplacive regime.

The phenomenological model used for impurity pinning [79, 80, 81] give many physical features similar to those due to a lattice pinning. Therefore, it appears to be difficult to discriminate between these two interpretations on the basis of the only conductivity experiments. The discrimination could arise because the two theories differ concerning the temperature effects. In our interpretation, and unlike the theory based on impurities, or defects, the *temperature of the system plays an intrinsically important role* because the wall motions, nucleations and annihilations are produced at the microscopic scale by the hopping of bipolarons, which is a thermally-activated process. The global mobility of the CDW has to be strongly temperature dependent and this fact is confirmed by the experiments. Other experiments, like dielectric response experiments ranging from the infrared to radio frequency, could also be better interpreted in terms of bipolaron pinning frequency and phase domain wall motions.

The ideas presented or suggested in this paper concerning various problems concerning the CDW suggest an approach drastically different than the other earlier approaches. There is a tremendous number of new problems and quantitative calculations which need to be examined in order to confirm that this new approach based on the concept of breaking of analyticity is useful and perhaps necessary for a deeper understanding of some aspects of the real systems.

Acknowledgements

We thank D. Grempel, P. Nozieres and S. Robaszkiewicz for useful discussions while visiting the theoretical group of Institut Laue—Langevin in 1986 where the writing of this paper was started. We also thank the physicists of the experimental groups working on CDW compounds at the CNRS laboratories and at the University of Grenoble, especially P. Monceau, C. Schlenker, J. Dumas, C. Berthier and J. Lajzerovicz. We obtained a great benefit from the knowledge of many of their results prior to publication and/or of many useful discussions. One of us (S.A) also thanks A. Bishop and D. Campbell for their hospitality at the Los Alamos National Laboratory and P. Vogel for useful discussions. R. Schrieffer and W. Kohn are also acknowledged for their hospitality at the ITP where the writing of this paper was completed.

This work was supported in part by the National Science Foundation under Grant PHY82-17853 and by funds from NASA.

References

1. P. Quemerais, 'Une nouvelle approche pour l'étude des composés à ondes de densité de charge: Conséquences de la brisure d'analyticité'. Thèse de doctorat, Université de Nantes, 1987.
2. F. Nabarro, *The Theory of Crystal Dislocations*, Oxford University Press, 1967.
3. R. E. Peierls, *Quantum Theory of Solids*, Oxford University Press, 1955, p. 108.
4. H. Fröhlich, *Proc. Roy. Soc.* **A223**, 296 (1954).
5. I. S. Gradshteyn and I. M. Ryshik, *Table of Integrals, Series and Products*, Academic press, New York, 1965.
6. D. Feinberg and J. Ranninger, *J. Phys.* **C16**, 1875—1885 (1983).
7. S. A. Brazovskii, X. X. Dzyaloshinskii, and I. M. Krichever, *Sov. J. Phys. JETP* (Engl. trans.), **56**, 212 (1982).
8. S. Aubry, 'Structures et instabilités', Lecture Notes, BEG—ROHU Summer School (C. Godreche, Ed.), Editions de Physique, 1986, pp. 73—193, and S. Aubry in *Intrinsic Stochasticity in Plasmas* (G. Laval and D. Gresillon, Eds.), Editions de Physique, 1979, pp. 63—83.
9. J. Bardeen, L. N. Cooper, and J. R. Schrieffer, *Phys. Rev.* **108**, 1175 (1957).
10. P. Monceau (Ed.) *Electronic Properties of Quasi-One-Dimensional Materials*, Vol. II, Reidel, 1985, pp. 139—268.
11. C. M. Varma and A. L. Simons, *Phys. Rev. Lett.* **51**, 138 (1983).
12. P. A. Lee, T. M. Rice, and P. W. Anderson, *Phys. Rev. Lett.* **31**, 462 (1973).
13. S. Aubry in *Symmetries and Broken Symmetries* (N. Boccara, Ed.), IDSET, Paris, 1981, pp. 313—322.
14. S. Aubry and L. de Sèze, *Festkörperprobleme*, Vol. XXV, 1985, pp. 59—69.

15. S. Aubry and P. Y. le Daeron, *Physica* **8D**, 381—422 (1983).
16. S. Aubry, *Physica* **8D**, 250—258 (1983).
17. P. Y. le Daeron, Ph.D. Diss., Univ. Paris Sud (Orsay), 1983.
18. P. Y. le Daeron and S. Aubry, *J. Phys.* **C16**, 4827—4838 (1983).
19. P. Y. le Daeron and S. Aubry, *J. Physique* **44**, C3, 1573—1577 (1983).
20. F. Vallet, Ph.D. Diss., Univ. Pierre et Marie Curie (Paris), 1986.
21. F. Vallet, S. Aubry, and L. de Sezel, unpublished, 1985.
22. M. Peyrard and S. Aubry, *J. Phys.* **C16**, 1593—1608 (1983).
23. S. Coppersmith and D. Fisher, *Phys. Rev.* **B28**, 2567 (1983).
24. L. de Sèze and S. Aubry, *J. Phys.* **C17**, 389—403 (1984).
25. L. Kadanoff, *Phys. Rev. Lett.* **47**, 1641 (1981).
26. S. Schenker and L. Kadanoff, *J. Stat. Phys.* **27**, 631 (1982).
27. J. Greene, *J. Math. Phys.* **20**, 1183 (1979).
28. F. Vallet, R. Schilling, and S. Aubry, *Europhys. Lett.* **2**, 815—822 (1986).
29. F. Vallet, R. Schilling, and S. Aubry, *J. Phys.* **C21**, 67—105 (1988) and R. Schilling and S. Aubry, *J. Phys.* **C20**, 4881—4889 (1987).
30. D. Escande, *Phys. Rep.* **121**, 165—261 (1985).
31. D. Escande and F. Doveil, *Bull. Am. Phys. Soc.* **25**, 987 (1980).
32. S. Aubry, *Solid State Sciences* **47**, 126—143 (1983).
33. V. I. Pokrovsky, *Solid State Commun.* **26**, 77 (1978).
34. S. R. Sharma, B. Bergsen, and B. Joos, *Phys. Rev.* **B29**, 6335—6340 (1984).
35. S. Aubry, *Ferroelectrics* **24**, 53 (1980).
36. S. Aubry, *Seminar on the Rieman Problem, Spectral Theory and Complete Integrability, 1978—1979* (D. and G. Chudnovsky, Eds.), Lecture Notes in Mathematics, Vol. 925, 221 (1982).
37. V. I. Pokrovsky, *J. Physique* **42**, 761 (1981).
38. S. Aubry, *J. Phys.* **C16**, 2497 (1983).
39. S. Aubry, *J. Physique Lett.* **44**, L247 (1983).
40. S. Aubry and G. Andre, *Ann. Israel Phys. Soc.* **3**, 133 (1980), and *The Physics of Quasicrystal* (P. J. Steinhardt and S. Ostlund, Eds.), World Scientific, Singapore, 1987, pp. 554—595.
41. S. Aubry, P. Y. Le Daeron, and G. Andre, unpublished preprint (1982).
42. K. Parlinski, *Phys. Rev.* **B35**, 8680—8695 (1987).
43. J. B. Sokoloff, *Phys. Reports* **126**, 189—244 (1985).
44. B. Simon, *Adv. Appl. Math.* **3**, 463 (1982).
45. E. J. Dinaburg and Ya. G. Sinaï, *Func. Anal. Appl.* **9**, 279 (1976).
46. D. R. Hofstadter, *Phys. Rev.* **B14**, 2239 (1976).
47. D. J. Thouless, *J. Phys.* **C5**, 77—81 (1972).
48. D. J. Thouless and Q. Niu, *J. Phys.* **A16**, 1911—1919 (1983).
49. M. Wilkinson, *Proc. Roy. Soc., London* **A391**, 305—350 (1985).
50. R. E. Prange, D. R. Grempel, and S. Fishman, *Phys. Rev. Lett.* **53**, 1592 (1984).
51. F. Delyon and D. Petritis, *Commun. Math. Phys.* **103**, 441 (1986).
52. M. Kohmoto, L. P. Kadanov, and C. Tang, *Phys. Rev. Lett.* **50**, 1870 (1983).
53. S. Ostlund, R. Pandit, H. J. Schnellhuber, and E. Siggia, *Phys. Rev. Lett.* **50**, 1873 (1983).
54. N. F. Mott and E. A. Davis, *Electronic Processes in Non-Crystalline Materials* (2nd Edn.), Clarendon Press (1979).
55. F. Delyon, Y. E. Levy, and B. Souliiard, *Phys. Rev. Lett.* **55**, 618—621 (1985).
56. G. Andre and S. Aubry, unpublished results presented at a poster session of the Rome Conference on Disordered Systems and Localization, May, 1981.
57. S. Aubry, C. Godreche, and J. M. Luck, *J. Stat. Phys.* **51**, 1033—1075 (1988).
58. S. Aubry, C. Godreche, and J. M. Luck, *Europhys. Lett.* **4**, 639—643 (1987).
59. P. W. Anderson, *Phys. Rev.* **109**, 1492 (1958).
60. S. Aubry in *Bifurcation Phenomena in Mathematical Physics and Related Topics* (C. Bardos and D. Bessis, Eds.), Reidel, 1980, pp. 163—184.
61. A. S. Alexandrov and J. Ranninger, *Phys. Rev.* **B24**, 1164 (1981).
62. A. S. Alexandrov, J. Ranninger, and S. Robaszkiewicz, *Phys. Rev.* **B33**, 4526—4542 (1986).

63. W. P. Su, J. R. Schrieffer, and A. J. Heeger, *Phys. Rev. Lett.* **25**, 1968 (1979) and *Phys. Rev.* **B22**, 2099 (1980).
64. B. K. Chakraverty, *Phys. Rev.* **B34**, 5287 (1986).
65. S. Aubry, *Solid St. Sci.* **8**, 254—277 (1978) (Eds. A. Bishop and T. Schneider).
66. J. Hubbard, *Phys. Rev.* **B17**, 494 (1978).
67. V. I. Pokrovsky and G. L. Uimin, *J. Phys.* **C11**, 3535 (1978).
68. K. Nasu, *Phys. Rev.* **B35**, 1748—1763 (1987).
69. C. Kittel, *Quantum Theroy of Solids*, John Wiley & Sons, 1976, p. 150.
70. J. E. Hirsch and E. Fradkin, *Phys. Rev.* **B27**, 4302—4316 (1983).
71. E. Fradkin and J. E. Hirsch, *Phys. Rev.* **B27**, 1680—1697 (1983).
72. S. Aubry, Ph.D. Diss., Univ. Paris VI (1975).
73. S. Aubry, *J. Chem. Phys.* **62**, 3217—3229 (1975).
74. S. Aubry, *J. Chem. Phys.* **64**, 3392—3402 (1976).
75. P. Monceau (Ed.), *Electronic Properties of Inorganic Quasi-One-Dimensional Materials*, Vol. II, Reidel, 1985, pp. 139—268.
76. L. P. Gor'kov, *Sov. Phys. JETP* **59**, 1057 (1984).
77. I. Batistic, A. Bjelis, and L. P. Gor'kov, *J. Physique* **45**, 1049—1059 (1984).
78. N. P. Ong, G. Verma, and K. Maki, *Phys. Rev. Lett.* **52**, 663—666 (1984).
79. H. Fukuyama and P. A. Lee, *Phys. Rev.* **B17**, 535 (1978).
80. P. A. Lee and T. M. Rice, *Phys. Rev.* **B19**, 3870 (1980).
81. P. B. Littlewood, *Phys. Rev.* **B33**, 6694—6708 (1986).
82. D. Feinberg and J. Dumas, *Europhys. Lett.* (1986), and Yamada Conference on *Physics and Chemistry of Quasi-One-Dimensional Conductors* (Lake Kawagushi, Japan), to appear in *Physica* **BC**.
83. W. I. MacMillan, *Phys. Rev.* **B14**, 1496 (1976).
84. K. A. Chao, R. Riklund, and G. Wahlström, *J. Phys.* **A18**, L403—L404 (1985).
85. P. Bak and V. L. Pokrovsky, *Phys. Rev. Lett.* **47**, 958 (1982).
86. J. Bardeen, *Phys. Rev. Lett.* **42**, 1498 (1979), and *Phys. Rev. Lett.* **45**, 1978 (1980).
87. G. Grüner and A. Zettl, *Phys. Rep.* **119**, 119 (1985).
88. D. Jerome and H. J. Schultz, *Adv. Phys.* **31**, 799 (1982).
89. C. Schlenker in *Physics in Disordered Materials* (D. Adler, Ed.), Plenum, 1985.
90. C. Schlenker and J. Dumas in *Crystal Chemistry and Properties of Materials with Quasi-One-Dimensional Structures* (R. Monceau, Ed.), Reidel, 1986, pp. 135—177.
91. S. Aubry in collaboration with G. Abramovici, J. P. Gosso, J. L. Raimbault and P. Quemerais, several papers will describe new exact results concerning the FK model and the Holstein model.
92. J. Fröhlich, T. Spencer, and P. Witten, *Localization for a Class of One-Dimensional Quasi-Periodic Operators*, preprint (1988).
93. V. Chulaevsky and F. Delyon, *Purely Absolutely Continuous Spectrum for Almost Mathieu Equation*, preprint (1989).
94. Ya. Sinaï, *J. Stat. Phys.* **46**, 861 (1987).
95. C. Oguey and S. Aubry, unpublished (1988), pp. 342—363.
96. S. Aubry and P. Quemerais in *Singular Behavior and Non Linear Dynamics*, Eds. St. Pnevmatikos, T. Bountis and Sp. Pnevmatikos World Scientific (1989).

IMPERFECTIONS OF CHARGE DENSITY WAVES IN BLUE BRONZES

DENIS FEINBERG

Centre National de la Recherche Scientifique,
Laboratoire d'Etudes des Propriétés Electroniques des Solides,
associated with Université Joseph Fourier,
B.P. 166, 38042 Grenoble Cedex, France.

JACQUES FRIEDEL

Laboratoire de Physique des Solides, Université Paris-Sud, Centre d'Orsay, Bât. 510,
91405 Orsay Cedex, France.

1. Introduction

Lattice modulations are usually coupled with charge density waves (CDW) in crystals which are conductors in the higher temperature range where the lattice modulations are unstable [1, 2, 3] (Figure 1).

The local phase of these modulations can be shifted under the influence of a strong enough force. This can be due to a change of equilibrium wavelength λ of the modulation, due to a change of temperature or of pressure. In the usual case where a CDW is present, this force can also be due to an externally applied electric field E: the motion of the wave at an average speed v then produces a Fröhlich current [4] of average intensity

$$j = n_s \, e \, v \tag{1}$$

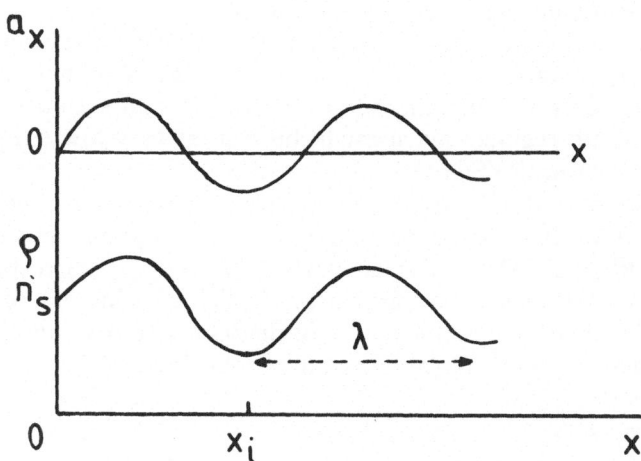

Fig. 1. Atomic displacement a_x and charge density ρ of a simple sinusoidal lattice modulation and its coupled charge density wave.

C. Schlenker (Ed.), Low-Dimensional Electronic Properties of Molybdenum Bronzes and Oxides,
407—448.
© 1989 *Kluwer Academic Publishers.*

if n_s is the average condensed charge per unit volume carried by the CDW, such that

$$n_s = \frac{1}{\lambda} \int_{x_i}^{x_i + \lambda} \rho \, \mathrm{d}x \qquad (2)$$

where ρ is the local conduction electron density of the CDW (Figure 1).

Equating the work of the configurational force F exerted per unit volume on the modulation to the electric work due to E gives, parallel to the chain axis x,

$$F_x = E \, n_s \, e \qquad (3)$$

In the limit of slow motion where the viscous friction due to electronic or phononic excitations can be neglected, the frictional force that can oppose such motion arises from the periodicity of the lattice or from couplings with lattice defects [5]. The lattice modulations become distorted, owing to such couplings, especially if they move; and some of these distortions can help the modulations to move further.

The discussion of these questions is expected to be rather involved, as in the simpler but similar question of dislocation motion in crystals [6]. The case of blue bronzes [7] offers a particularly attractive field to test possible models:

(i)   One has indeed a CDW coupled with a low-temperature lattice modulation, leading to Fröhlich conductivity.

(ii) The lattice modulation develops on linear chains which seem to be only weakly coupled [8]; and it only deviates weakly from a pure sinusoidal variation. Its 'phason' mode is then sharply distinct from its 'amplitudon' mode at low speed, where the CDW can be safely assumed to keep a rigid amplitude, with only local changes of its phase.

(iii) The lattice modulation has a wavelength λ which varies widely and continuously with temperature T over a large range of temperatures where it behaves obviously as incommensurate [8]. At low temperatures, it approaches a commensurate regime; and it might be that, at very low temperatures, it is exactly commensurate (Figure 2) [8, 9].

(iv) Because coherent atomic displacements are involved over large distances, the slow motion of the lattice modulation can be treated classically, and the coupling with the CDW assumed perfect. Indeed the corresponding conditions are well fullfilled, showing that the typical coupling energy Δ is respectively smaller than the kinetic energy of the bare electrons and larger than the one of condensed electrons (involving the Fröhlich mass):

$$\frac{\hbar^2}{M \lambda^2} < \Delta < E_F$$

where, in the blue bronzes, the Fermi energy is $E_F \cong 1$ eV, the single-particle gap of the CDW is $\Delta \cong 0.075$ eV and the Fröhlich mass is $M \cong 10^2$ to $10^3 \, m$ (m = electron mass).

(v)   Much is already known of the Fröhlich current, of its pinning by imperfec-

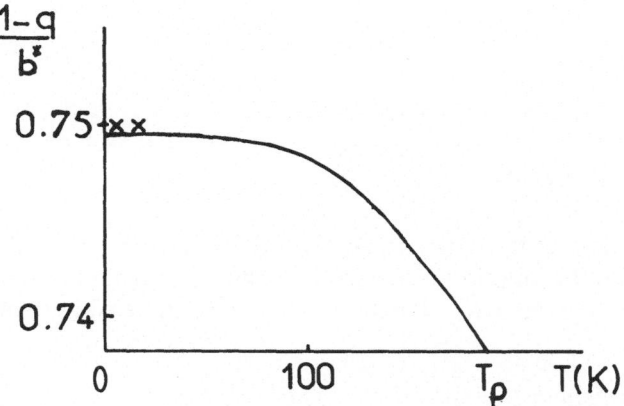

Fig. 2. Variation with temperature T of the wave vector c of the lattice modulation in blue bronzes [8], in units of the reciprocal lattice vector b^* along the chains. The crosses depict the commensurate phase which seem to be observed sometimes at low temperatures.

tions or possibly by the lattice, of its non-uniformity across large single crystals [10—11] and the corresponding non-uniformity of the lattice modulations. [13, 14, 51, 53]

Our purpose is perforce modest. Restricting ourselves to the case considered of *nearly sinusoidal charge density wave* on *weakly-coupled parallel chains*, we will try and analyse the nature and properties of static imperfections of the CDW and their possible couplings with lattice defects. With this background knowledge, we will comment on known 'elastic' and 'plastic' responses to electric excitations. In the limit of slow motion considered here, the CDW can be considered as classical waves, with a fixed amplitude but variable phase. We shall furthermore restrict ourselves to temperatures small enough with respect to the Peierls transition temperature T_p for critical fluctuations to be neglected.

2. Properties of Static Imperfections of CDWs

Three possible cases apply to blue bronzes and must be discussed in turn. They are of increasing complexity (Figure 2).

2.1. INCOMMENSURATE CDW

In the high temperature regime of blue bronzes, the equilibrium period λ of the CDW is obviously far from any submultiple mb/n of the lattice period b of the chains (m, n integers) where the CDW could be appreciably blocked [8]. The lattice modulation and ensuing CDW can then be expected to be *near* a simple sinusoidal wave moving in a continuum. In this approximation, the electronic density of the wave is (Figure 1):

$$r \cong n_s [1 + A \cos(qx + \phi)] \tag{4}$$

where

$$q = 2\pi/\lambda. \tag{5}$$

and

$$A \cong \frac{\Delta}{E_F} \ln \frac{E_F}{\Delta}. \tag{6}$$

As explained above, the low-energy distortions considered in this paper affect the phase ϕ but not the amplitude A, which is assumed constant in space and time.

Imperfections of decreasing dimensionalities and increasing complexity can be considered [15].

2.1.1. *Long Wavelength Distortions; the 3D Elastic Limit*

If ϕ varies slowly in space in a perfect crystal, we can treat the crystal as a continuum and the energy stored by this distortion reduces, to lowest order in grad ϕ, to an elastic energy [16, 17]

$$U_{\text{ela}} = \iiint \frac{1}{2} \left[K_x \left(\frac{\partial \phi}{\partial x} \right)^2 + K_y \left(\frac{\partial \phi}{\partial y} \right)^2 + K_z \left(\frac{\partial \phi}{\partial z} \right)^2 \right] d\tau. \tag{7}$$

We can introduce a local *displacement* of the CDW

$$u_x = \phi/q \tag{8}$$

parallel to the chains and equivalent to the phase shift ϕ. The corresponding *strain* tensor [e] of the CDW has the following nonzero components [18]:

$$e_{xx} = \frac{1}{q} \frac{\partial \phi}{\partial x}$$

$$e_{xy} = e_{yx} = \frac{1}{2q} \frac{\partial \phi}{\partial y} \tag{9}$$

$$e_{xz} = e_{zx} = \frac{1}{2q} \frac{\partial \phi}{\partial z}.$$

If we introduce a *stress* tensor [σ] of the CDW, related to [e] by *elastic constants* K_x, K_y, K_z such that

$$\sigma_{xx} = q^2 K_x e_{xx}$$

$$\sigma_{xy} = \sigma_{yx} = q^2 K_y e_{xy} \tag{10}$$

$$\sigma_{xz} = \sigma_{zx} = q^2 K_z e_{xz}$$

the elastic energy density (7) is indeed $\frac{1}{2}$ [σ] [e].

More precisely, $e_{xx} = \delta\lambda/\lambda$ is a compression or a dilatation along the chains (Figure 3a), while e_{xy}, e_{xz} which are related to the inclination θ (Figure 3) measure the shear of the CDW (Figure 3b). The continuous lines pictured in Figure 3

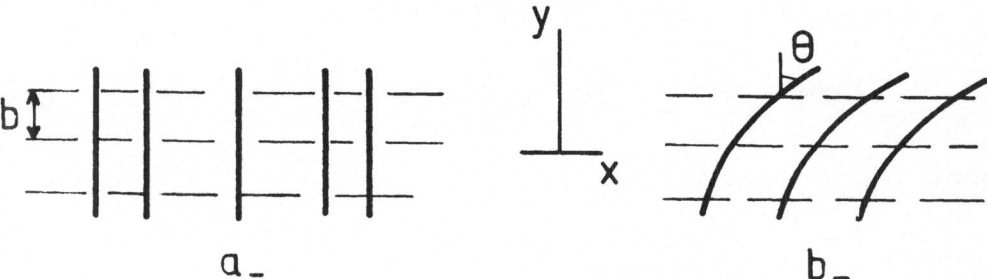

Fig. 3. Variable strains of a CDW: a — compression, dilatation; b — shear. Continuous lines: surfaces of equal argument; dotted lines: chains.

represent surfaces of equal argument

$$\Phi = \phi + qx = 2n\pi \tag{11}$$

(n integer).

Dilatation e_{xx} produces an electric charge density with an average (measured over distances large compared with λ)

$$\delta\rho = (n_s e)\, e_{xx} \tag{12}$$

while shear components e_{xy}, e_{xz} are neutral. $K_x = \hbar v_F / 4\pi s$, where s is the chain section and v_F the Fermi velocity. It is related to the strong intrachain interactions, while K_y and K_z are related to interchain interactions, known to be much weaker in blue bronzes. A very specific property of compounds such as blue bronzes is therefore that shear of the CDW is much easier than its compression or dilatation. As an order of magnitude, we can probably take

$$10^{-2} K_x \cong K_y \cong K_z \cong K_\perp. \tag{13}$$

The elastic distortions described here can be maintained statically in a perfect crystal by the application of an external force, such as the one due to an electric field.

Let a local force F per unit volume be applied to the CDW. It is a configurational force, coupled to the local phase ϕ or equivalent displacement u_x of the CDW parallel to the chains. Thus only F_x produces work; and, because $u_y = u_z = 0$, the only condition of equilibrium is

$$\sum_i \frac{\partial \sigma_{ix}}{\partial i} + F_x = 0 \qquad i = x, y, z \tag{14}$$

or

$$K_x \frac{\partial^2 \phi}{\partial x^2} + K_y \frac{\partial^2 \phi}{\partial y^2} + K_z \frac{\partial^2 \phi}{\partial z^2} + \frac{F_x}{q} = 0. \tag{15}$$

If the force is due to an electric field E applied parallel to the chains, Equation (3) applies. It is equivalent to say that the electrical potential due to **E** is

$$U_{\text{ele}} = -\iiint \frac{n_s e}{q} \, \phi E \, d\tau = -\iiint \mathbf{P} \, \mathbf{E} \, d\tau \tag{16}$$

where

$$P = \frac{n_s e}{q} \, \phi \tag{17}$$

is the local *polarisation*, parallel to x.

Two simple examples will be used below in the context of surface pinning. In the first one, the CDW is longitudinally compressed along the chain direction (Figure 4a). In the second case, surfaces of constant argument ϕ bend in the x direction, between two planar surfaces parallel to $x\,z$, where they are pinned (Figure 4b).

Equations (14) and (15) then lead to

$$\phi(x) = \frac{n_s e}{q} \frac{E}{2K_x} \, x(L - x) \quad \text{for case a} \tag{18}$$

$$\phi(y) = \frac{n_s e}{q} \frac{E}{2K_y} \, y(L' - y) \quad \text{for case b.} \tag{19}$$

These distortions produce an electric susceptibility

$$\chi = \frac{dP}{dE} \tag{20}$$

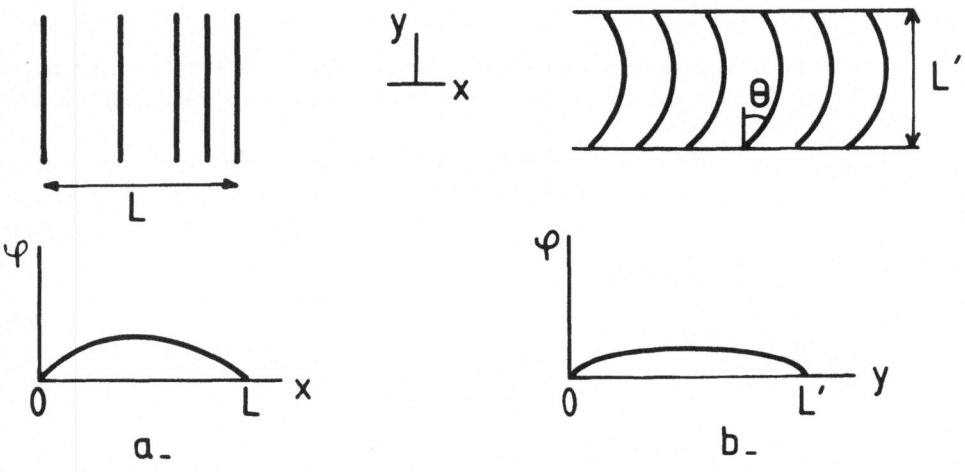

Fig. 4. Simple elastic distortions under an electric field. a — Thin sample normal to the chains; b — thin slice parallel to the chains. Surface pinning in both cases. Continuous lines: surfaces of constant argument $\Phi = 2n\pi(n = \text{integer})$.

which is

$$\chi = \left(\frac{n_s e}{q}\right)^2 \frac{L^2}{K_x} \quad \text{in case a} \tag{21}$$

$$\chi = \left(\frac{n_s e}{q}\right)^2 \frac{L'^2}{K_y} \quad \text{in case b.} \tag{22}$$

As $K_x \gg K_y$, it is seen that, for $L' \cong L$, the susceptibility of case b is much larger than in case a: surface pinning is more effective for surfaces normal to the chains than for surfaces parallel to them. This can also be expressed in terms of a *coherence length* ξ. In the present case, this is the minimum crystal size L or L' over which ϕ changes by, say, π under an applied force F_x. (18) and (19) give

$$\xi_i \cong 4\pi \sqrt{\frac{K_i}{F_x \lambda}} . \tag{23}$$

2.1.2. *Short Wavelength Distortions. 2D Ridges and Walls [15]*

If ϕ is allowed to vary more rapidly in space, the crystal structure can play some role. This can occur under large stresses. Because we consider here incommensurate CDWs, we can to a good approximation neglect the atomic structure of the chains; but, as these chains are only weakly interacting, specific effects due to the lattice periodicities normal to the chains can be expected.

Consider for instance a shear configuration such as in Figure 4b. One can expect, on an atomic scale, the lateral variations of ϕ, normal to the chains, to concentrate somewhat between the chains, more than on the chains. The surfaces of constant argument ϕ, Equation (11), should then exhibit some deviations from the smooth variations pictured so far, with a tendancy to present cylindrical *ridges* parallel to the chains joined by steps of atomic height across the chains (Figure 5a). Such ridges can extend into cylindrical *walls* parallel to the chains (Figure 5b). Ridges and walls are obviously neutral, while steps carry the Fröhlich charge density ρ (Equation 4).

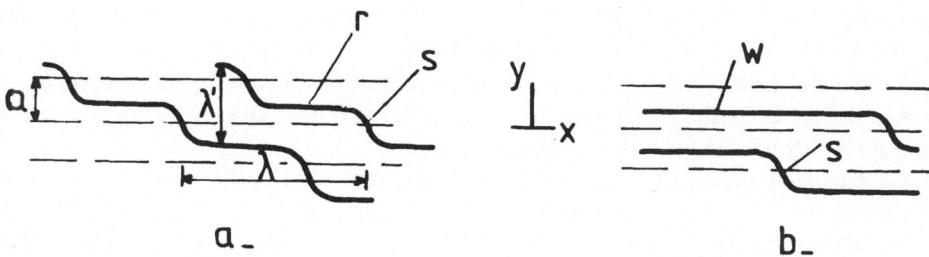

Fig. 5. Effect of lattice periodicity across the chains: a — ridges or steps s; b — walls w. Continuous lines: surfaces of constant argument $\Phi = 2n\pi$ (n = integer); broken lines: chains.

Being incommensurate, the steps move without friction along the chains. This is not true of the motion of the ridges or walls, normal to the chains, if their period λ' normal to the chains is commensurate with the interchain distance a: a sizeable lattice friction should then oppose such motion.

More quantitatively, the effect discussed here is related to the fact that the lateral constants K_y, K_z vary somewhat with y and z. To the lowest order, we can take

$$K_y \cong K_y^0 \left(1 + \beta \cos \frac{2\pi}{a} y \right) \tag{24}$$

where $|\beta| \ll 1$, and a similar expression for K_z. Taking, for a pure shear,

$$\phi \cong \frac{2\pi y}{\lambda'} - \phi_1 \sin ky \tag{25}$$

with $|\phi| \ll 1$, one finds, according to (7), a density of elastic energy

$$\frac{1}{2} K_y^0 \left[\frac{4\pi^2}{\lambda'} + \frac{1}{2} k^2 \phi_1^2 - \frac{2\pi}{\lambda} \beta k \phi_1 \overline{\cos \frac{2\pi y}{a} \cos ky} \right].$$

This is minimised if

$$k = 2\pi/a \tag{26}$$

and

$$\phi_1 = \frac{b}{2\lambda'} \beta. \tag{27}$$

The undulations have the wavelength expected across the chains. Ridges and walls will have an appreciable lattice friction if λ' is simply commensurate with a.

2.1.3. *1D Perfect Dislocations of the CDW*

This is another type of imperfections to be considered.

The only quantised translation symmetry of a CDW is the period λ parallel to the chains. This is necessarily the elemental Burgers vector of such perfect dislocations [17, 18]. Dislocations with multiple Burgers vectors $n \lambda$ (n integer $\neq \pm 1$) are necessarily unstable, as the long-range elastic energy is proportional to $n^2 \lambda^2$, which is larger than $|n| \lambda^2$, the corresponding energy term when the multiple dislocation splits into $|n|$ elementary lines.

Simple cases of screw dislocations (line L parallel to λ) and edge dislocations (line L normal to λ) can be considered [17] (Figures 6, 7, 8). It is obvious that screw dislocations only involve shears, and thus have smaller energies than edge ones, which involve also compressions and dilatations. More general orientations of straight dislocation lines are expected to give energies between these two extremes.

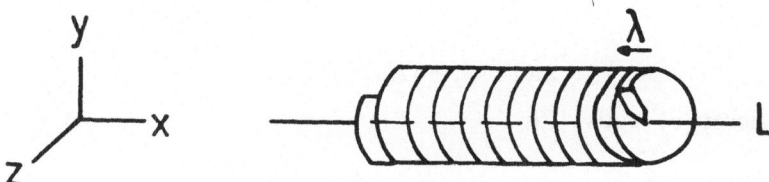

Fig. 6. Screw dislocation line L of a CDW.

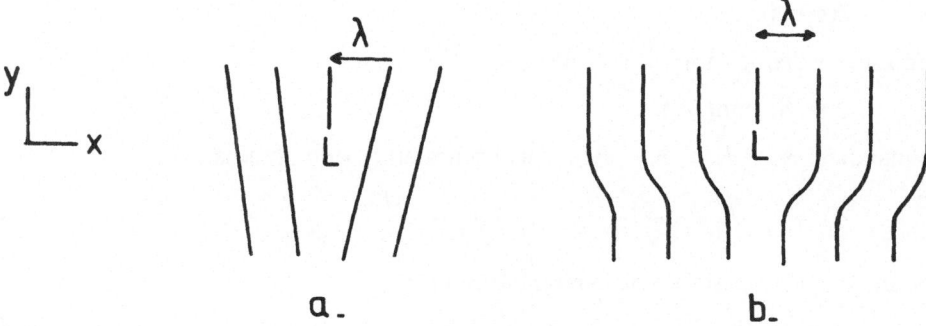

Fig. 7. Edge dislocation line L of a CDW. $a - \xi_\perp \gg a$; $b - \xi_\perp \cong a$.

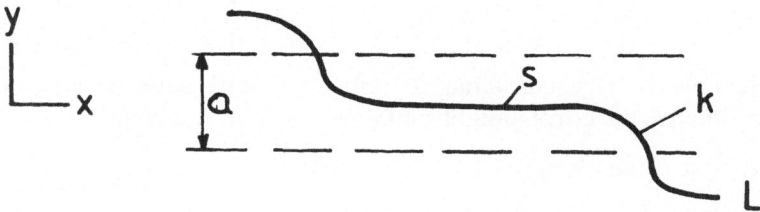

Fig. 8. Kinked dislocation line L on a glide plane parallel to the chains (S screw part, k kink).

These perfect dislocations can also be considered as line singularities of the phase ϕ. The Burgers vector $\mathbf{B}$ is defined by integration around a Burgers circuit C enclosing the line:

$$\mathbf{B} = \oint_c \operatorname{grad} \mathbf{u} \, dr = \frac{\mathbf{x}}{q} \oint_c \nabla\phi \, dr \tag{28}$$

where $\mathbf{x}$ is the unit vector along the chain axis. As the phase slip has to be a multiple of 2π, the Burgers vector is a multiple of $\lambda \, \mathbf{x} = \boldsymbol{\lambda}$.

If we scale the coordinates according to

$$\tilde{x} = x, \quad \tilde{y} = \alpha_y \, y, \quad \tilde{z} = \alpha_z \, z \tag{29}$$

with

$$\alpha_y = (K_x/K_y)^{1/2} \quad \text{and} \quad \alpha_z = (K_x/K_z)^{1/2} \tag{30}$$

the elastic energy (7) reads [17]

$$\int d\tilde{\tau} \sqrt{\frac{K_y K_z}{2}} (\text{grãd } \phi)^2 = \int d\tau \frac{K_x}{2} (\text{grãd } \phi)^2 \tag{31}$$

and the equilibrium condition (14) gives, in the absence of applied force,

$$\tilde{\Delta}\phi = 0. \tag{32}$$

For a screw dislocation, the solution is

$$\phi = \text{Arctan}(\tilde{z}/\tilde{y}). \tag{33}$$

If, for instance, $K_y = K_z = K_\perp$ (cylindrical symmetry), one obtains

$$\sigma_{xy} = -\frac{q K_\perp}{2} \frac{z}{y^2 + z^2}, \qquad \sigma_{xz} = \frac{q K_\perp}{2} \frac{y}{y^2 + z^2}. \tag{34}$$

For an edge dislocation, one can take similarly

$$\phi = \text{Arctan}(\tilde{y}/\tilde{x}). \tag{35}$$

This is due to the fact that the elastic equilibrium involving deformations only along $\hat{x}$, edge dislocations are topologically identical to screw ones. Hence

$$\sigma_{xx} = -q \, \alpha_\perp^3 K_\perp \frac{y}{x^2 + \alpha_\perp^2 y^2}, \qquad \sigma_{xy} = \frac{q \, \alpha_\perp \, K_\perp}{2} \frac{x}{x^2 + \alpha_\perp^2 y^2}. \tag{36}$$

In cylinders with their axis along L and inner and outer radii $\tilde{\xi}$ and $\tilde{R}$ in reduced coordinates, the corresponding energies are, per unit length of L,

$$E_s = \pi E_\perp \ln (\tilde{R}/\tilde{\xi}) \tag{37}$$

$$E_e = \pi \sqrt{K_x K_\perp} \ln (\tilde{R}/\tilde{\xi}). \tag{38}$$

For a straight dislocation where the line L makes with λ, and thus the axis x of the chains, an angle ψ, one finds in a similar way the more general formula

$$E(\psi) = \pi K_\perp \left[\frac{K_\perp}{K_x} \sin^2 \psi + \cos^2 \psi \right]^{-1/2} \ln (\tilde{R}/\tilde{\xi}). \tag{39}$$

As for lattice dislocations [6] $\tilde{R}$ can be taken as a representative size of the perfect CDW around the dislocation considered, i.e. the distance to a surface or to another dislocation of opposite sign, appropriately scaled according to (29). $\tilde{\xi}$ is a core radius which measures the minimum size below which the present analysis fails. By analogy with superfluidity, it is possible that ξ_x, measured along the x axis, is a coherence length of the CDW amplitude, below which it tends to vanish. If we treat the energy change related to this variation of amplitude of the CDW as due to an outside force, the reasoning leading to equation (23) gives for the coherence length $\xi_\perp$ measured normal to the chains

$$a_\perp \, \xi_\perp \; \cong \; \xi_x$$

thus an isotropic $\tilde{\xi}$. Such very rough estimates suggest that, for blue bronzes,

$$0.1 \, \xi_x \; \cong \; \xi_\perp \; \cong \; a. \tag{40}$$

The fact that $\xi_\perp \cong a$ and $\xi_x \gg b$, the unit length along a chain, follow, whatever the details, from the weak interchain and strong intrachain couplings. But better founded estimates than (40) would be appropriate.

We conclude that screw dislocations should have small isotropic cores, of size of the order of the interchain distance a. Edge dislocations should have very anisotropic cores, rather widely split along the glide plane containing L and λ (Figure 7b, not 7a).

As in lattice dislocations, this analysis neglects small corrective terms due to relaxation of surface stresses. The energies (37), (38), (39) also neglect the core energy. These two corrections reduce $\tilde{\xi}$ a little, without changing these estimates significantly, as long as $\tilde{R} \gg \tilde{\xi}$. As in superfluids, the gap should go to zero in the middle of the core, which should be more nearly metallic.

The very large difference in core behavior between screws and edges should have some consequences on the stability and mobility of these dislocations [6].

Screw dislocations, with small cores, should have a very different core energy, depending whether the line sits on a chain or between chains. They should therefore have a definite tendency to sit along the one of those two configurations which has the lowest core energy. And a nearly screw dislocation should be transformed spontaneously into a zigzag form, with screw parts related by *kinks* of nearly edge character and length of order a, so that each screw part can sit along a minimum of its Peierls potential (Figure 8).

Definite predictions can also be made about the mobility of these dislocations, due to the differences in Peierls friction associated with differences in splitting.

Thus edge dislocations should *glide* more easily than screw dislocations. The same would apply to the special case of kinks, Figure 8, which should easily glide along the screws and leave them straight under small stresses. Screws could only move under rather high stresses, or at high enough temperatures for double kinks to be produced thermally in the thermal vibrations of screws (Figure 9).

Edge dislocations should also *glide* parallel to λ more easily than *climb* normal to λ, a motion which most probably requires a variation of the splitting of the core, depending whether the line L is on a chain or between chains.

A further phenomenon must be taken into account in such a climb: *free charges emission or absorption* [15, 19]. This is the equivalent, for dislocations of CDW, of the point defects emission or absorption for the climb of lattice dislocations [6]. It is clear that the 'up climb' of the edge dislocation, Figure 7, will remove part of one period of the CDW, carrying a charge density $n_s \lambda e$ per unit area. When these excess emitted charges have been dispersed, the charge neutrality will have been

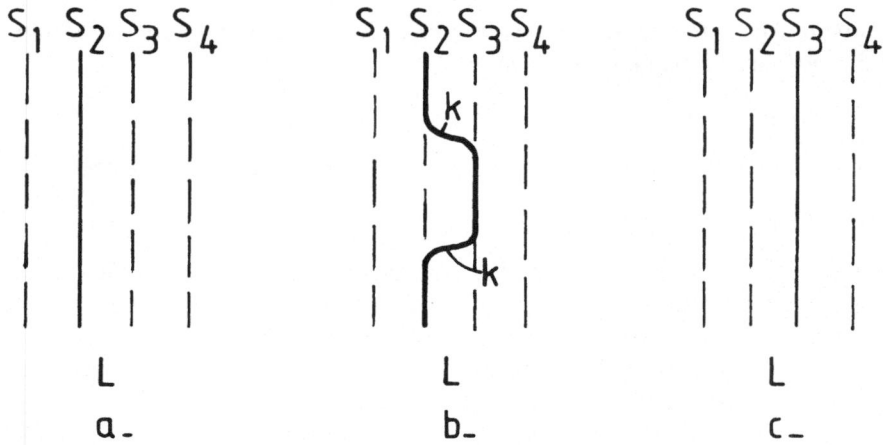

Fig. 9. Glide of a screw L by glide of thermally activated double kinks k. S_i: positions of stable Peierls potential.

preserved by the concomittant glide of the remaining periods of the CDW, to fill the empty space created by the up climb. In such a motion the absorption of already present positive holes is possible, in parallel with the emission of free electrons, just as for lattice dislocations, the same climb can occur by emission of interstitials or absorption of vacancies. The 'down climb' of the same dislocation, Figure 7, will, in a similar way, create free holes or absorb free electrons. At low temperatures where no carriers exist initially, the climb is submitted to a supplementary solid friction due to the creation of free carriers by climb. The energy δ required to create one free carrier is related to the gap energy Δ of the CDW, and the corresponding frictional force is

$$F_\delta = n_s \lambda \delta \qquad (41)$$

per unit length of dislocation.

It is worth pointing out that the force acting on these dislocations and inducing their motion is of purely elastic origin and due to the local distortions of the CDW along the dislocation lines. It is the Peach and Koehler force per unit length

$$\mathbf{F}_{PK} = \lambda [s] \times \mathbf{L} \qquad (42)$$

where $[\sigma]$ is the local stress tensor of the CDW. $\mathbf{L}$ being the unit vector along the line L, $\mathbf{F}_{PK}$ is as expected normal to L. There is *no* direct coupling of an external electric field with the dislocations, because they are neutral. However, as explained above, the electric field produces a volume force on the CDWs which deforms them; and these strains in turn will produce a force on dislocations. Examples will be given below. Also the dislocation line carries an electric dipole which can couple with the gradient of an electric field (Figure 21).

2.1.4. *Disclinations and Point Singularities of CDWs*

To be complete, one must consider, as in smectics, other low dimensions singularities of the order parameter, with rotational symmetry: disclinations and point singularities.

In the absence of an applied volume force, such singularities, solutions of Equation (32), have amplitudes which can decrease continuously to zero: they are not quantised, and therefore not stable.

The problem is however different *under an applied volume force*, such as due to an electric field. These singularities are very sensitive to such a force and can indeed be stabilised by it under suitable boundary conditions.

Equation (32) is then replaced according to (15) and, in reduced coordinates, by

$$\tilde{\Delta}\phi + f = 0 \tag{43}$$

where

$$f = \frac{F_x}{qK_x} = \frac{n_s e E}{q K_x}. \tag{44}$$

Solutions of lowest energy where ϕ depends on a radial (reduced) distance $\tilde{r}$, have a long-range behavior given by

$$\phi = -\frac{f\tilde{r}^2}{2\gamma} \tag{45}$$

where $\gamma = 2$ for a disclination and $\gamma = 3$ for a singular point. The corresponding surfaces of constant argument $\Phi = \phi + qx$ are respectively elliptical cylinders and ellipsoids centred on the x axis (at $x = \gamma q/f$) and having radii in reduced coordinates, of

$$\tilde{R} = \frac{2\gamma q}{f} \left(1 - \frac{\Phi \phi}{2\gamma q^2} \right)^{1/2} \tag{46}$$

as pictured schematically Figure 10.

The corresponding energies are, from (7) and (16),

$$U_{ela} + U_{ele} = \left(\frac{1}{2} - 1 \right) \pi f^2 \sqrt{K_x K_y} \frac{\tilde{R}^4}{8}$$

$$= -\frac{\pi f^2}{16} \sqrt{K_x K_y} \tilde{R}^4 \tag{47}$$

per unit line for the disclination, and

$$U_{ela} + U_{ele} = \left(\frac{1}{3} - 1 \right) 2\pi f^2 \sqrt{K_y K_z} \frac{\tilde{R}^5}{15}$$

$$= -\frac{4\pi}{45} f^2 \sqrt{K_y K_z} \tilde{R}^5 \tag{48}$$

for the total energy of the singular point.

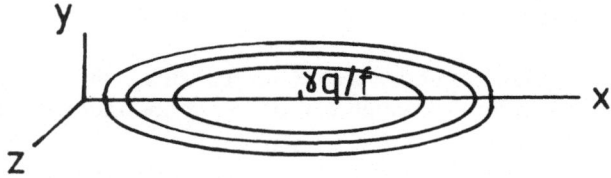

Fig. 10. Line or point singularity with rotational symmetry. Continuous lines: surfaces of constant argument $\Phi = 2n\pi(n = \text{integer})$.

These energies are negative: the electric field stabilises them. If boundary conditions allow, the disclination is, as expected, more stable than the singular point for the same volume. However both configurations are less stable, for suitable boundary conditions, than those of Figure 4a and b, which have, respectively, energies per unit area (parallel to yz and xz, respectively)

$$U_{\text{ela}} + U_{\text{ele}} = -\frac{f^2}{24} K_x L^3 \tag{49}$$

and

$$U_{\text{ela}} + E_{\text{ele}} = -\frac{f^2}{24} \frac{K_x^2}{K_y} L'^3. \tag{50}$$

It is therefore clear that disclinations or singular points could only by produced over small volumes, where the surface term due to boundary conditions overcomes the volume terms discussed here.

2.2. NEARLY COMMENSURATE CDW [15]

This case occurs when the equilibrium wavelength λ of the lattice modulations is very near to the wavelength λ_c of a simple commensurate modulation, with an expected sizeable lattice friction. This case is most probably encountered in blue bronzes at rather low temperatures (Figure 2).

It is then useful to consider the *lattice of discommensurations* [20, 21] which describe the deviations of the incommensurate and distorted CDW from the perfect commensurate modulations. This is equivalent to take for the basic crystal structure not the atomic period a along the chains but the period λ_c of the commensurate structure. The equilibrium period λ_d of the lattice of discommensurations is given by

$$\frac{1}{\lambda_d} = \left| \frac{1}{\lambda} - \frac{1}{\lambda_c} \right|. \tag{51}$$

Everything which was stated above for the lattice of modulations still holds qualitatively for the lattice of discommensurations. The only quantitative differences come from the stronger deviations expected from a pure sinusoidal spatial variation of the CDW in its equilibrium configuration, due to the large values of λ_d and stronger effects of the underlying lattice along the chains. Thus Equation (7) should be replaced by a corrected form which, in its simplest form, reads

$$U'_{ela} = U_{ela} - V_p \cos p(qx + \phi) \tag{52}$$

where

$$p = \lambda_d / \lambda_c \tag{53}$$

and the discommensurations have, at equilibrium, a classical soliton form. Their interactions along the chain axis are then smaller than in the previous case: K_x is smaller, but still large enough to preserve the incommensurability.

At very low temperatures, where the blue bronzes becomes insulators, the equilibrium distance λ_d between discommensurations seems large. The problem of their electrostatic stability arises: the nearly commensurate phase might well then be replaced by a commensurate one along the chains. There is indeed some experimental evidence, reported in Figure 2, that the CDW become commensurate below 20 K or so [8] in some samples.

2.3. COMMENSURATE CDW

There is no experimental evidence at very low temperatures of strong deviations from the basic sinusoidal modulations observed at higher temperatures. A commensurate phase, if it exists in blue bronzes, is then expected to be only weakly coupled with the lattice.

In such 'weakly commensurate phase', one expects isolated *discommensurations* to be produced and moved under not very high electric fields. This is what remains of the elastic distortions of the CDW under electric field of the incommensurate phases, when the Peierls friction on the motion of the (isolated) discommensurations is large enough to make the phase commensurate.

These discommensurations have a Burgers vector equal to $\pm$ **b**, the elementary period of the chains. They can be analysed as in Figure 5 in *charged steps* and *neutral ridges* or *walls*, the charge of each step being $\pm n_s eb$ per chain cut by the step. These discommensurations can appear as surfaces of finite area bordered by loops of *partial dislocations* of the CDW, with Burgers vectors $\pm$ **b** (Figure 11a). Also the perfect dislocations of the commensurate CDW will have a tendency to split into a collection of partial dislocations bordered by discommensurations (Figure 11b).

These details have been analysed elsewhere [15]. They should be taken into account in a detailed analysis of the very low 'plasticity' of CDW, if it is true that it becomes commensurate in blue bronzes.

Two observations made in this temperature range fit with this hypothesis:

— a higher critical electric field [22] is expected in this regime, because the 'elastic' distortions of CDW under small electric fields are blocked by the friction of commensuration.

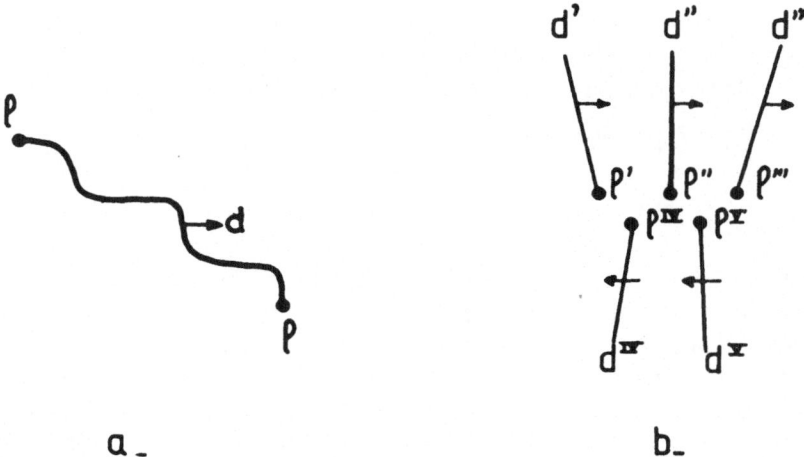

Fig. 11. Discommensurations *d* and partial dislocations *l*: a — dislocation loop *l*; b — perfect dislocation of Burgers vector 5**b**.

— lateral walls of discommensurations are subjected to no force due to the electric field. They could then subsist easily in the low-temperature commensurate phase, the break the CDW into domains of small lateral sizes, normal to the chains. This seems the case, especially after the passage of a Fröhlich current [13, 14].

In the absence of firm evidence, we shall however *not* consider the commensurate case any further.

3. Interaction of CDW with Lattice Defects

By breaking the periodicity of the underlying lattice, lattice defects tend to couple locally with specific values of the argument $\phi = qx + \phi$ of the CDW. This coupling can have two opposite effects:

— it usually tends to pin down the CDWs and their defects such as dislocations or discommensurations; this introduces a characteristic friction which prevents the apparition of a Fröhlich current below a critical electric field [5, 17, 23].
— it might in some cases help to create defects under an applied field, and thus lower the critical electric field.

A similar ambivalence exists for the plastic properties due to the motion of lattice dislocations [6].

We shall review briefly three types of lattice defects, which often should add their frictions: surfaces, dislocations and point defects.

3.1. SURFACES AND INTERFACES

The local change in atomic structure along a free surface or an interface is expected to couple fairly effectively with the argument of the CDW. In chain compounds such as blue bronzes, a surface cutting the chains is expected to be more effective than a surface parallel to the chains.

An especially simple case would then be provided by a cylindrical crystal, cut with atomically smooth surfaces parallel and perpendicular to the chains (Figure 12a). The discussion of Figure 4 a then applies.

In general however, surfaces and interfaces are rough on a scale comparable with the coherence lengths (Figure 12b). Two complications have to be taken into account, which work in opposite directions: the roughness of the lateral surfaces blocks the argument of the CDW, which then bends in a way similar to figure 4b, if the crystal is a long enough cylinder; this raises the surface friction. However the surface roughness, both on frontal and lateral surfaces, produces local irregularities and stress-concentrations of the CDW under an applied field. This lowers the surface pinning.

In a quantitative discussion, one must distinguish 'brute force' processes from those due to the multiplication of dislocation loops.

3.1.1. *Brute Force Processes* [24, 25]

In the *simple* case of Figure 12a, the CDW will start moving when the local stresses induced by the electric field near the two end surfaces, at $x = 0$ and L, become larger than the local pinning stress e_c caused on the CDW by these surfaces. From (15) and (18), and in terms of a critical strain e_c such that $\sigma_c = q^2 K_x e_c$,

$$\frac{F_x L}{2q^2 K_x} = \frac{n_s e E_c L}{2q^2 K_x} \geq e_c. \tag{54}$$

The critical field E_c is inversely proportional to the length L of the crystal; it depends on e_c, i.e. the nature of the surface or interface involved, and can only be estimated by a microscopic computation.

In the more general case of Figure 12b, the lateral pinning induces a frictional

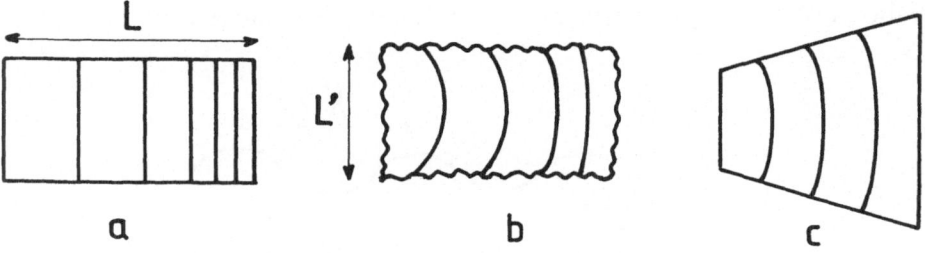

Fig. 12. Blocking of CDW by surfaces and interfaces: a — simple case (smooth surfaces); b, c — general cases (rough surfaces and lateral surfaces not parallel to the chains).

resistance to brute force processes until the CDW reach a critical bending angle θ_c at the surfaces such that $e_c = \frac{1}{2} \tan \theta_c$ (Figure 12b, c). In the case of free surfaces for instance, the motion of the CDW involves the creation of *surface charges* wherever the lateral surfaces are not parallel to the chains. Thus if these surfaces make an *angle* with the chains (Figure 12c), uniform charges are created which soon correspond to large electrostatic energies and totally block the motion of the CDW near the surfaces except if these are made conductive by an appropriate surface treatment. Charges of opposite signs are created by the motion of the CDW on *rough* lateral surfaces with a distance of order λ', the wavelength of the roughness, if this is smaller than the wavelength λ of the CDW (*fine scale roughness*, Figure 13). In both cases, pinning is expected to be very effective, leading to large values of θ_c.

There is actually a limitation to the strength of lateral surface pinning: for large angles θ, the continuum approximation used here fails, and an analysis of the CDW in steps and ridges must be made (Figure 5). Above a certain critical angle θ_c, the CDW turns into a wall w parallel to the pinning surface or interface, of increasing mismatch (Figure 14). Within approximation (24), θ_c is given by

$$\cos \theta_c = 1 - |\beta|. \tag{55}$$

One can say that a cylinder of CDW is 'punched' into a static medium. Equations (15) and (19) then give

$$\frac{F_x L'}{q^2 K_x} \geqslant \frac{1}{2} \tan \theta_c = e_c \tag{56}$$

which again fixes a critical field inversely proportional to a characteristic length L'. Similar conditions would obtain for interfaces with fine scale roughness such as smooth grain boundaries.

In the opposite case of *large scale roughness* ($\lambda' \gg \lambda$), local stress concentrations are produced which favour the multiplication of dislocations, as presently described.

3.1.2. *Dislocation Multiplication*

In parallel with the brute force processes, one should also consider processes where the CDW glides in a progressive way, by dislocation multiplication.

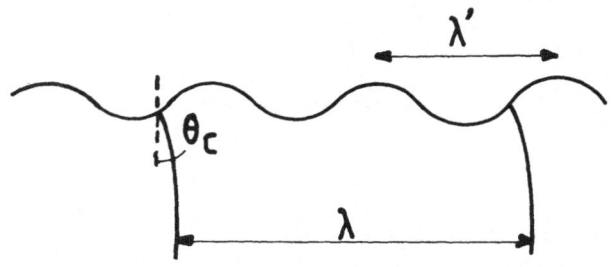

Fig. 13.   Critical angle θ_c for fine scale (free) surface blocking of the CDW.

Fig. 14. Large angle lateral pinning of the CDW.

Let us first point out that the usual multiplication mechanism met for lattice dislocations, i.e. *the activation of Frank—Read sources* [26], *is not very likely in blue bronzes*: as there is only one type of possible Burgers vectors ($\pm\lambda$), the dislocations lines cannot build a stable three-dimensional Frank network (Figure 15a); and the other possibility, of successive loops strongly pinned by impurities or precipitates in different glide planes would require a somewhat unlikely geometry (Figure 15b).

The only processes to be considered are therefore by *nucleation and growth of dislocation loops* [19].

Let us first consider the simple case of Figure 12a. Under the stress concentrations at either end of the crystal, edge dislocation loops can nucleate, thus introducing or removing small disk-like portions of a period λ of the CDW (Figure 16a). Equating the work $\sigma_{xx} 2\pi R\lambda\, dR$ of the Peach and Koehler force due to the external stresses to the change in elastic energy $d(2\pi R\, E_e)$ when the radius R of the disk varies by dR, we obtain, using (9), (10), (15) and (38),

$$\pm q\, K_x\, \lambda \left(\frac{\partial \phi}{\partial x} \right)_{0,\,L} 2\pi R\, dR = d\, [2\pi^2\, R\, \sqrt{K_\perp K_x}\, \ln\,(\tilde{R}/\tilde{\xi})].$$

With ϕ given by (18), the condition of equilibrium is

$$F_x = \frac{2\pi\, \sqrt{K_\perp K_x}}{R\, L\, \lambda}\, \ln\, \frac{e\,\tilde{R}}{\tilde{\xi}}\,, \qquad (\ln e = 1). \tag{57}$$

The critical field corresponds to the smallest nucleus where this 'elastic' analysis

Fig. 15. Frank—Read sources: a — loop of a Frank network (impossible); b — successive loops pinned by impurities or precipitates in different glide planes (unlikely).

Fig. 16. Nucleation and growth of dislocation loop under an applied field: a — dislocation climb; b
— dislocation glide.

holds, or $\tilde{R} \cong \tilde{\xi}$ and

$$F_x \cong \frac{2\pi \sqrt{K_x K_\perp}}{\xi_\perp L \lambda} \cong \frac{2\pi K_x}{\xi_x L \lambda}.$$ (58a)

A similar argument has been given recently by Maki [27]. As explained above, the disk of edge dislocation will actually *climb*, and the corresponding period of the CDW disappear or develop, only if the Peach and Koehler force due to the electric field compensates not only the increase in elastic energy but also *two frictional forces*: the rather large one due to the Peierls friction against climb, and the force (41) due to the emission of free carriers.

In the more general case of Figure 12b, the pinning by the lateral surfaces can be similarly relieved by the nucleation and *glide* of dislocations loops of mixed edge and screw character (Figure 16b). In the case of fine scale roughness depicted in Figure 13, an analysis similar to the preceding one shows that the critical nuclei are loops elongated along the x axis in the ratio $\sqrt{K_x/K_\perp}$. The critical nucleation field is given by a formula very similar to (58a), where L is replaced by L'. The extension of such loops by glide parallel to the x axis involves no friction due to free carrier emission; and the Peierls friction is also reduced, especially for the edge part. Large scale roughness helps the nucleation of dislocation loops on both types of surfaces: roughly speaking, a roughness of wavelength $\lambda' \gg \lambda$, and hence $\lambda' > \xi$, will induce the nucleation under low fields of loops of size λ'; their growth will only require the force given by (57) with ξ replaced by λ', and a similar expression for loop on lateral surface.

In conclusion, brute force or dislocation nucleation and growth will prevail in the simple case of Figure 12a, depending on the strength of the surface or interface pinning. But, as for lattice dislocation nucleation in plasticity, the two criteria are *not* expected to be very different. Thus (54) and (58a) give, for the ratio of critical fields for dislocation multiplication and brute force,

$$\frac{E^{dm}}{E^{bf}} \cong \frac{1}{4\pi e_c} \frac{\lambda}{\xi_x}$$ (58b)

and a similar relation holds for lateral surfaces. In usual cases where surface pinning is strong enough for e_c to be sizeable (e.g. typically 10^{-1}), the dislocation multiplication process will prevail as $\xi_x > \lambda$ in the incommensurate phase, except perhaps in the low temperature nearly commensurate case. The critical voltage for dislocation multiplication

$$e\,E^{dm}\,L \;\cong\; \frac{2\pi\,K_x}{n_s\,\xi_x\,\lambda} \tag{59}$$

is then expected, for blue bronzes with $n_s = 3 \times 10^{21}$ cm^{-3}, $K_x \cong 10^5$ eV cm^{-1}, $\lambda \cong 24$ Å and $\xi_x \cong 100$ Å, to be of order 1 meV. This estimate is reasonable and leads, for critical fields $E_c \cong 100$ mV cm^{-1} to a typical length $L \cong 100$ μm. The above discussion takes account only of elastic deformations of the CDW and neglects the role of amplitude variations which enter solely the calculation of the microscopic critical strain e_c. This simplified analysis nevertheless provides an essential feature of boundary effects in pinning. While, for large sample sizes L, the critical field, due to bulk pinning, is independent of L, for small sizes it should vary as L^{-1} (Figure 17a). This is true for both contact and lateral surface pinning and whether depinning is dominated by brute force effects or dislocation nucleation. This result is consistent with a number of previous works. First, in the phase-slip model of Gor'kov, and Batistic et al. [24, 25], corresponding here to the so-called brute force situation, where the phase slips as a whole, the relevant quantity triggering the slipping event is the gradient $\partial\phi/\partial x$, which fixes a critical voltage (rather than a critical field). The fact that $E_T \sim L^{-1.23}$, rather than L^{-1} in [25], actually comes from the details of boundary effects and amplitude variations. Secondly, the relevance of a critical voltage V_T at contacts was shown by Saint-Lager et al. [28] in NbSe$_3$ short samples (Figure 17b) and by Mihaly et al. in o-TaS$_3$ [52]. The above estimated value $V_T \sim 1$ meV is consistent with the experimental ones, which should be strongly dependent on the contact geometry. Finally it was recently shown by Gill [29] that in narrow samples of NbSe$_3$ the critical field is proportional to the ratio C/A of the perimeter to area of the sample cross section, thus to L'^{-1} if L' is a typical transverse length.

If surfaces are rough on a fine scale, the increase in friction due to lateral pinning is expected to be more easily relieved than the frontal pinning, either by the punching process described Figure 15 or by multiplication of gliding dislocation loops of mixed character, more easily formed, as explained, than the climbing edge loops on the end surfaces. *The critical lateral distances L' can therefore be definitely smaller than L:* cylindrical parts of the CDW of sizes somewhat larger than these values of L, L', can move independently from each other, creating at most either lateral walls of mismatch or pile up of glide dislocations along such lateral walls.

Finally large-scale surface roughness helps the nucleation of dislocations and lowers the critical fields.

In this discussion, we have only considered the most simple surface and interface geometries and athermal processes. It should be pointed out that more complex boundary conditions might apply than those of Figure 12. One might wonder for instance whether, under large fields, the configuration reached could ressemble locally part of the singular configurations pictured Figure 10. A possible example is given Figure 18. However this is very *unlikely*: as soon as the CDW

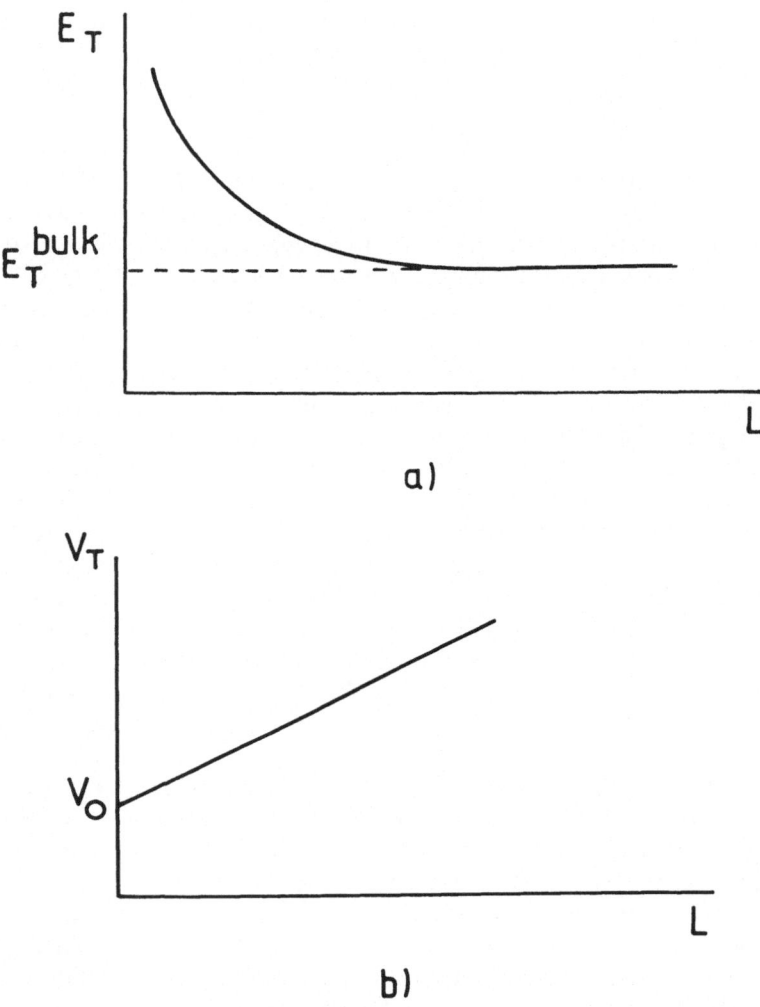

Fig. 17. a — Variation of the critical field E_T with a typical sample size L for both frontal or lateral pinning. Bulk pinning dominates at large sizes while $E_T \sim L^{-1}$ for small L. b — variation of the critical voltage V_T for a sample of length L. A critical voltage V_0 due to contact depinning is required for small L.

makes with the x axis a sizeable angle θ, the corresponding stresses are expected to be relieved either by the punching process of Figure 15 or by the multiplication of glide dislocations of Figure 16b.

Indeed *the CDW cannot deviate too much from its equilibrium state*, if only because of the possibility of stress relief by dislocation multiplication. Thus one deduces easily from (58) that the maximum local change of CDW wavelength at

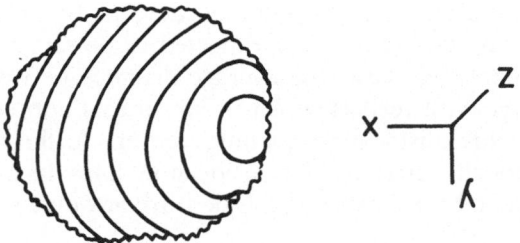

Fig. 18. Unlikely surface pinning under large fields.

the end surfaces of Figure 12a is given by

$$\frac{\delta\lambda}{\lambda} \cong \frac{\lambda}{4\pi\,\xi_x} \tag{60}$$

while the maximum shear on the lateral surfaces of Figure 12b is given by

$$e_c = \tan\theta_c \cong \sqrt{\frac{K_x}{K_\perp}}\,\frac{\lambda}{4\pi\,\tilde{\xi}} \tag{61}$$

where $\tilde{\xi}$ is an average of ξ_x and $\xi_\perp$. As $\xi_x > \lambda$, these deviations should remain modest, at least at temperatures where thermal fluctuations are negligible.

Thermal activation can also help nucleating dislocation loops. This is most easily seen from a computation of the energy of the loops of minimum size. For a pure edge loop, of diameter $\xi_\perp$, the critical energy is (from (37, 38))

$$U_e = 2\pi\,\xi_\perp\,E_e \simeq 2\pi^2\sqrt{K_x K}\,\xi_\perp.$$

For a loop of mixed edge and screw character,

$$U_s \cong 2E_s\,\xi_{||} + 2E_c\,\xi_\perp = 2\pi(K_\perp\,\xi_{||} + \sqrt{K_x K_\perp}\,\xi_\perp).$$

As $(\xi_\perp/\xi_{||})^2 \cong K_\perp/K_x$, $U_e \cong U_s$. With the estimates above, these critical energies are of the order of 10^{-2} to 10^{-1} eV. Thermal creation under fields much less than (59) are then possible at temperatures less than T_p.

3.2. LATTICE DISLOCATIONS

Dislocation networks which build low angle boundaries are expected to play much the same role as the interfaces just discussed. Indeed it is expected that, in large single crystals, this is the main reason why, near the critical field, the Fröhlich current only develops in parts of the sample. The geometry of these current-carrying parts, expected to be long cylinders parallel to the chains, and its relation to the mosaic structure of the crystal should be more systematically and quantitatively studied.

More random dislocations, as produced in work-hardening, certainly produce some friction on the motion of CDWs and their defects. By analogy with the case of magnetic walls, this friction is however expected to be rather modest: except in the very nearly commensurate case, the average distance between lattice dislocations remains very large with respect to λ (or λ_d), so that appreciable piling up of CDW occur against any dislocation pinning center; furthermore, pinning by dislocations will be mostly effective for dislocation lines locally parallel to the CDW, a rather rare event. More studies in this field are however required.

3.3. POINT DEFECTS

In blue bronzes, impurities such as Fe, V or the substitution alkalis (Rb) have little influence on the critical fiels [30]. W seems to have a stronger effect [31]. Nonstoichiometry usually produces vacant alkaline ions sites, which can play a fundamental role in the nature of the CDW: these vacancies create a charged defect which can either shift the Fermi level in the metallic phase from its position for the stoichiometric formula $A_{0.3}MoO_3$, where $2\,k_M = 0.75\,b^*$ leads to the commensurate phase; it can also become localised as positively charged electronic defects. Finally electronic defects such as Mo^{5+} do exist; these defects introduce levels in the Peierls gap and contribute to conductivity; they are paramagnetic and observed by EPR [32]. Other strong defects such as O or Mo vacancies can be created by irradiation [33].

The possible couplings of such point defects with CDW and their dislocations have given rise to many studies. Only some general points will be stressed, in the absence of systematic studies with samples doped in a controlled way.

3.3.1. *Interactions with CDW*

The exact nature of the coupling of a given point defect with the modulations of a CDW or with the discommensurations of nearly commensurate CDW must be studied case by case.

However, following Mott and Nabarro's classical study on (lattice) dislocations interacting with point defects [34], one can distinguish cases of strong and weak pinning [17].

In strong pinning, each point defect locally blocks the argument $\Phi = qx + \phi$ of the wave, and usually also alters locally its amplitude. For more or less random defects, these local pinnings are not coherent; the corresponding frustration is relieved by elastic distortions of the waves (Figure 19a).

In weak pinning, each point defect is individually unable to pin the argument of the wave effectively. But cooperative effects can take place, where the wave is locally blocked by fluctuations of large enough concentrations of point defects (Figure 19b).

It is then tempting to state that Fe, V and substitution alkalis are weak pinning defects.

However the analogous case of (lattice) dislocations clearly shows that a full study of the effects of concentration, temperature and heat treatment must be made before a sharp distinction can be made.

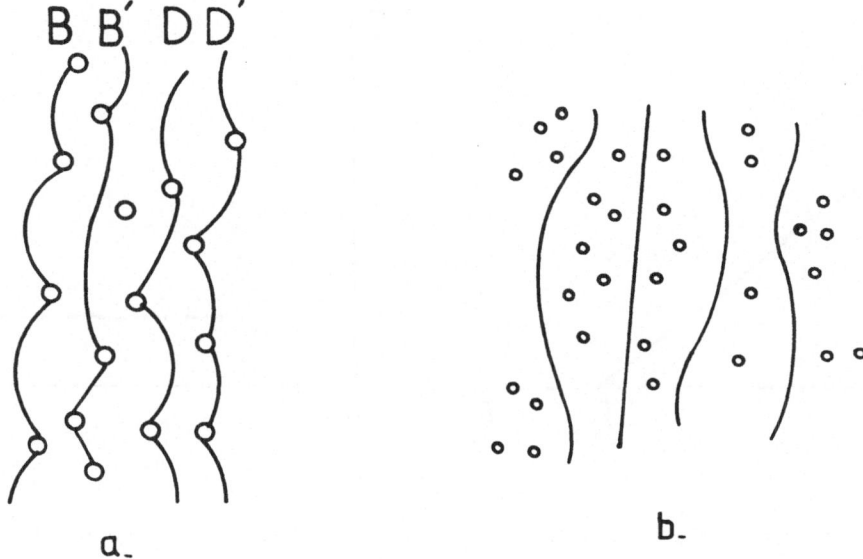

Fig. 19. Pinning of CDW by point defects: a — strong; b — weak (continuous lines surfaces of constant argument Φ).

— It is clear that the same point defects which are moderately strong pinning points at small concentrations can become weak pinning points at concentrations large enough for their average distance to be less than the coherence length over which elastic distortions of the waves are effective.

— Strong pinning is, in essence, strongly localised on an atomic volume. Thermal vibrations of the waves can then be very effective in freeing the waves from pinning points under moderate applied fields: a thermally activated creep of the CDW should develop, where the CDW make successive (incoherent) jumps from one pinning point to the next, as sketched in Figure 19a (B to B', D to D'). The electric field necessary to obtain a given average speed of the CDW should then strongly decrease with increasing temperature; and above a critical temperature Tc, the blocking should disappear (Figure 20a), at least if T_c is much below the critical temperature T_P for the Peierls distortion. A different regime, not discussed here, should obtain in the opposite case ($T_c \gg T_P$).

— Weak pinning is much smaller at low temperatues, but because the thermally activated motion of the waves involves jump over large regions with many impurities, it is much less effective and the field necessary to move the CDW is much less temperature sensitive (Figure 20b). As a result, it can dominate in the higher temperature range.

— In both cases, heat treatments can alter the pinning properties. Indeed one expects a maximum pinning effect in the range of temperatures where point defects can diffuse in the crystal fast enough to optimise their coupling with the

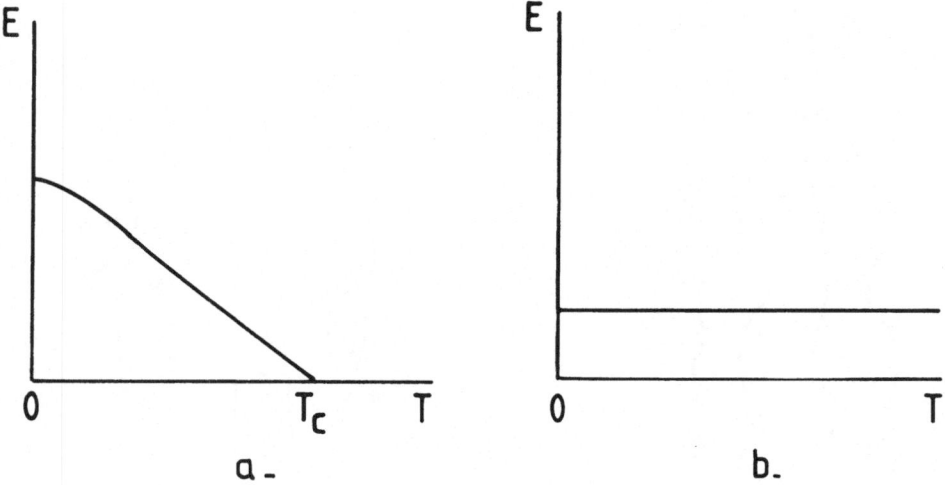

Fig. 20. Expected thermal variation of the field E for a given speed of creep of the CDW: a —
strong pinning; b — weak pinning.

CDW when it is undistorted and at rest, but not fast enough to follow them
when they move in a Fröhlich motion.

3.3.2. *Interaction with Dislocations of the CDW*

We can distinguish the interaction due to the short-range coupling of a point
defect with the local argument of the CDW, and an eventual long-range coupling
mostly due to electrostatic effects.

The short-range interaction potential is written

$$W = -v \cos (qx + \phi). \tag{62}$$

For a screw dislocation along the x axis,

$$\phi = \text{Arctan } \frac{z}{y} = \theta.$$

The interaction force has only a tangential component (Figure 21).

$$F_y = -\frac{v}{r} \sin(qx + \theta). \tag{63}$$

For an edge dislocation, $\phi = \text{Arctan } (\tilde{y}/x)$, thus

$$F_x = -v\left(Q - \frac{\alpha_\perp y}{x^2 + \alpha_\perp^2 y^2} \right) \sin(qx + \phi)$$

$$F_y = -v \frac{x}{x^2 + \alpha_\perp^2 y^2} \sin(qx + \phi) \tag{64}$$

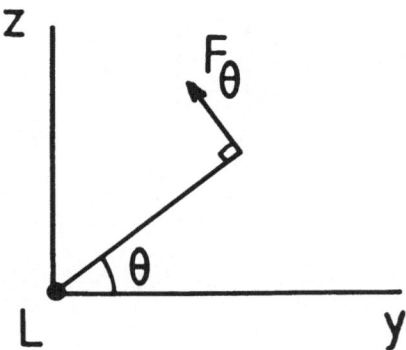

Fig. 21. Short range coupling with a screw dislocation L.

F_θ and F_x oppose the glide of the dislocations, F_y its climb. Owing to their non-oscillatory nature when the distance of the point defect, moving along $\hat{y}$, varies with respect to the dislocation line, F_θ and F_y will attract point defects towards at least some parts of the dislocation lines, thus building Cottrell clouds if the point defects are mobile enough.

The long-range Coulomb interactions of point defects with dislocations can only arise with edge dislocations, and in the low-temperature ranges where blue bronzes are insulators.

The upper and lower parts of an edge dislocation have indeed opposite charges which can attract the local charge due to a point defect. According to (12) and (35), the local electrical density is

$$\delta\rho = -\frac{n_s e}{q} \frac{\alpha_\perp y}{x^2 + \alpha^2 y^2}. \tag{65}$$

Because of its dipolar nature, it can only attract impurities at fairly small ranges (Figure 22).

Conversely, one can ask what is the equilibrium solution of an initially perfect CDW when a charged point defect is introduced. A possible solution is to build an edge dislocation loop centred on the point (Figure 23). It is, however clear without recourse to a detailed computation, that the dislocation loop is only stable if the charge of the point defect is large; and even then, the dipolar nature of the electric charge carried by the dislocation prevents its radius R to be much larger than atomic distances. This is therefore *not* an effective nucleation process for edge dislocations in perfect CDWs.

A similar analysis would hold for a loop of partial dislocations, limiting a disk of discommensuration, in the case of commensurate phases. The energies involved being rather small, the dislocation loops can then be more easily stable and of somewhat larger sizes. This would be the natural extension of the amplitude solitons studied on independent chains [35, 36, 37].

Finally it has been recently suggested by Zawadowski [38] that a disclination

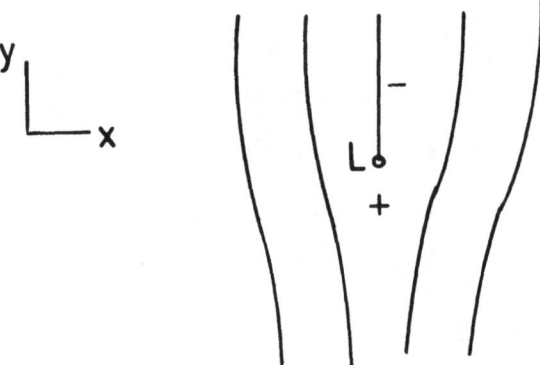

Fig. 22.   Electrical dipole of an edge CDW dislocation.

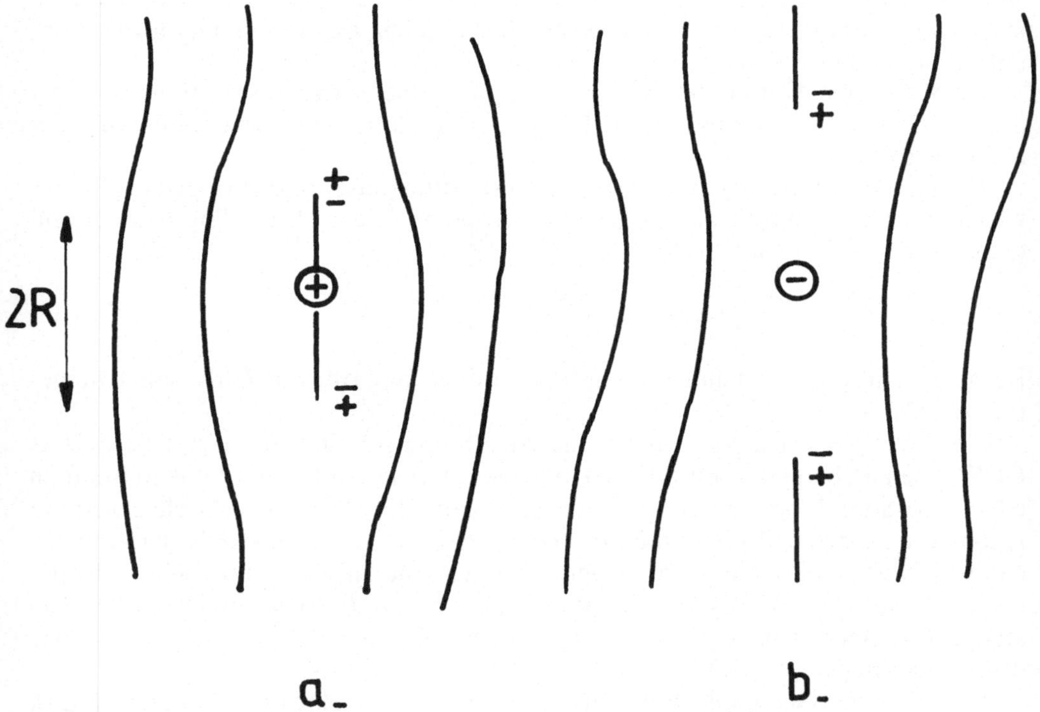

2R

a.   b.

Fig. 23. Dislocation loop associated with a charged point defect: a — positive defect; d — negative
defect.

such as pictured Figure 10 could be stabilised by the presence of a strong pinning impurity at its center.

There are clearly more studies needed on these various possibilities.

4. Elastic and Anelastic Responses of CDW

After having studied the nature of CDW imperfections, their coupling to lattice imperfections and their reactions to applied electric fields, we are now able to discuss the global properties of the blue bronzes to an applied field. We shall distinguish the 'elastic' responses below the critical field for Fröhlich current and the 'plastic' ones above such critical field. In the elastic regime local motion of the CDW under alternating fields gives rise to anelastic internal friction processes.

4.1. ELASTIC EQUILIBRIUM OF A CDW UNDER AN ELECTRIC FIELD

Starting from the free energy for the phase field, which, in reduced coordinates, can be written

$$F = \iiint \left[\frac{1}{2} K_x (\nabla \phi)^2 - \frac{n_s e}{q} E\phi + V(\mathbf{r}, \phi) \right] d\tau, \tag{66}$$

one obtains the equilibrium equation

$$K_x \Delta\phi + \frac{n_s e}{q} E - \frac{\partial}{\partial \phi} V(\mathbf{r}, \phi) = 0. \tag{67}$$

Here $V(\mathbf{r}, \phi)$ represents all the interactions which tend to pin the CDW, i.e. the surfaces and interfaces, the dislocations, the point defects and possibly the discrete lattice.

The incoherence of these various pinnings leads to an equilibrium configuration which, in zero field, is distorted and not unique. However, if the sample has been cooled down in the absence of an applied electric field, it is reasonable to expect the equilibrium phase $\phi_0(\mathbf{r})$ reached to be near to the configuration of minimum energy, and free from polarisation. Thus

$$K_x \Delta\phi_0 = \frac{\partial V(\mathbf{r}, \phi)}{\partial \phi_0} \tag{68}$$

with

$$\int \phi_0(\mathbf{r}) \, d\tau = 0. \tag{69}$$

If now we apply a small electric field E, the phase becomes

$$\phi = \phi_0 + \phi' \tag{70}$$

with

$$K_x \, \Delta\phi' = M(\mathbf{r}) \, \phi' - \frac{n_s e}{q} \, E \tag{71}$$

where

$$M = \frac{\partial^2 \, V(\mathbf{r}, \phi)}{\partial\phi^2} \tag{72}$$

is an inhomogeneous spring constant for phase distortion. Apart from the simple commensurate CDW where $V = -V_p \cos p\phi$ and thus M is a constant, the solution of (71) is complex. It can be written

$$\phi' = - \frac{n_s e}{q} \, E \, g(\mathbf{r}) \tag{73}$$

where

$$(K_x \, \Delta - M) \, g(\mathbf{r}) = 1. \tag{74}$$

The strain tensor $[e] = q^{-1} \, \nabla\phi$ is then inhomogeneous, as well as the polarisation density (parallel to the chain axis)

$$p = \frac{n_s e}{q} \, \phi \tag{75}$$

(75) leads to the electric susceptibility

$$\chi_e = P_e/E = \left(\frac{n_s e}{q} \right)^2 \frac{1}{v} \int g(\mathbf{r}) \, \mathrm{d}\tau \tag{76}$$

where v is the volume of the sample.

The solutions $g(\mathbf{r})$ of the nonlinear Equation (74) are not unique. Indeed, they can contain a number of singularities: walls, dislocations, disclinations, singular points. However, as was explained above, such singularities should be rather rare in a sample cooled down in the absence of an electric field. They are furthermore at least as strongly pinned down by defects as the underlying CDW (except in the strongly commensurate case). The contribution of their motion under field to the electrical susceptibility at low fields can therefore be safely neglected.

Equations (21) and (22) are then particular cases of expression (76), for surface pinning of regular CDWs, without such singularities. Similar expressions for strong and weak pinning by random impurities should be worked out. An order of magnitude reasoning can however be given.

If we Fourier analyse M and g,

$$M = \sum_{k \neq 0} M_{\mathbf{k}} \cos(\mathbf{k}\mathbf{r} + \eta_{\mathbf{k}}) \tag{77}$$

$$g = \sum_{\mathbf{k}} g_{\mathbf{k}} \cos(\mathbf{k}\mathbf{r} + \phi_{\mathbf{k}}) \tag{78}$$

Equation (74) leads to

$$\sum_k M_k g_k \cos(\eta_k - \phi_k) = -2 \tag{79}$$

and

$$K_x k^2 g_k \cos(\mathbf{kr} + \phi_k) + g_0 M_k \cos(\mathbf{kr} + \eta_k) +$$
$$\sum_{k'} \frac{M_{k'} g_{k-k'}}{2} \cos(\mathbf{kr} + \eta_k + \phi_{k'-k}) +$$
$$+ \sum_{k'} \frac{M_{k'} g_{k'-k}}{2} \cos(\mathbf{kr} + \eta_{k'} - \phi_{k-k'}) = 0. \tag{80}$$

With $\phi_k = \eta_{k'}$, this leads to an approximate expansion

$$g \cong g_0 + \sum_{k \neq 0} g_k \cos(\mathbf{kr} + \eta_k)$$

with

$$g_k \cong -g_0 \frac{M_k}{k^2 K_x}$$

and

$$\sum_k M_k g_k = -2.$$

Hence

$$g_0 \cong 2K_x \bigg/ \sum_k k^{-2} M_k^2. \tag{81}$$

If $k_L = 2\pi/L$ is the predominating wavelength of pinning,

$$g_0 \cong \frac{8\pi^2}{L^2} \frac{K_x}{M_{k_L}^2}$$

and

$$\chi_e \cong \left(\frac{n_s e}{q}\right)^2 \frac{8\pi^2}{L^2} \frac{K_x}{M_{k_L}^2}. \tag{82}$$

As expected, the strong pinning case gives a susceptibility much smaller than mere surface pinning, because the coupling strength M_k will be of the order of K_x/L^2 leading to an expression analogous to (21) to (22), but the characteristic distance L will be much smaller than the size of the crystal. Weak impurity pinning, if alone, will on the other hand allow much larger susceptibilities as, for the same concentrations, L is much larger and M_k smaller.

4.2. ANELASTIC RESPONSE

The elastic motion of a CDW induced by a small applied oscillating field will induce some *viscous* internal friction, due to energy losses by collisions of the moving CDW with phonons and with free electronic carriers.

In the absence of friction, the usual Lagrangian of the incommensurate case is, in reduced coordinates,

$$\mathcal{L} = \int \frac{1}{2} K_x \left[\frac{1}{c^2} \left(\frac{\partial \phi}{\partial t} \right)^2 - (\tilde{\nabla} \phi)^2 \right] d\tau \tag{83}$$

where

$$c = v_F (m/M)^{1/2} \tag{84}$$

is the longitudinal phonon velocity, v_F the Fermi velocity and

$$M = m[1 + 4\Delta^2/\lambda_{ep} \omega_q^2 \hbar^2] \tag{85}$$

the Fröhlich mass (λ_{ep} is the electron—phonon coupling constant and ω_q a phonon frequency). One gets from (83) the equation of motion, the propagation equation of elastic waves in the CDW medium:

$$\Delta \phi - \frac{1}{c^2} \frac{\partial^2 \phi}{\partial t^2} = 0. \tag{86}$$

In original coordinates, one must separate such a 'phason' into a longitudinal wave with velocity c_x and transverse waves with velocity $c_\perp$, with

$$c_x = c$$
$$c_\perp = c(K_\perp / K_x)^{1/2}. \tag{87}$$

In the presence of an alternating applied field of low pulsation ω and of viscous friction, Equations (71) and (86) are replaced by

$$K_x \left(\Delta \phi' - \frac{1}{c^2} \frac{\partial^2 \phi'}{\partial t^2} \right) = M(\mathbf{r}) \phi' - \frac{1}{q} \left[n_s e E_0 \cos \omega t - \frac{B \partial \phi'}{q \partial t} \right]. \tag{88}$$

In this equation, the inertial term in $\partial^2 \phi'/\partial t^2$ is negligible at low pulsations ω.

First, in the absence of applied field ($E_0 = 0$), ϕ' has an exponential decrease with time

$$\phi' = \phi'_0 e^{-t/\tau}$$

with

$$K_x \Delta \phi' - M(\mathbf{r}') \phi' + \frac{B}{q^2 \tau} \phi' = 0. \tag{89}$$

In the weak perturbation limit, the Fourier component ϕ'_{k_L} of ϕ' related to the average pinning length L has a relaxation time τ related to B by

$$B \cong K_x q^2 K_L^2 \tau. \tag{90}$$

Correlatively, under an alternating field ($E_0 \neq 0$), equation (88) leads for the same Fourier components to an internal friction

$$\tan \delta \propto \frac{\omega \tau}{1 + \omega^2 \tau^2} \tag{91}$$

and an electrical susceptibility due to the distortion of the CDW

$$\chi(\omega) = \frac{\chi_e(0)}{1 + i\omega\tau} \tag{92}$$

where the static value $\chi_e(0)$ is given by (82).

Some features of dielectric relaxation of CDW in blue bronzes and some other compounds can be qualitatively described by the above analogy with internal friction in solids. While, at high frequencies, the elastic response dominates, the anelastic effects are observed at low frequencies. The peak in the imaginary of $\chi(\omega)$ is similar to the expected peak in internal friction. Moreover the decrease in $\chi(\omega)$ as the impurity concentration n_i increases, evidenced in bronzes [39] can be understood by observing that $L \propto n_i^{-1/3}$ for strong pinning (resp., n_i^{-1} for weak pinning), thus χ_e should decrease roughly as $n_i^{-2/3}$ (resp., n_i^{-2}). The relaxation time should also decrease with increasing impurity concentration, as $n_i^{-2/3}$ (resp., n_i^{-2}) according to (90).

More systematic studies of the effect of strong impurities, for instance by irradiating the sample, would be useful to check the above picture. It should however be stressed that the temperature dependence of the relaxation time τ seems to show an activation energy [40]. This has been attributed to screening by normal carriers. It could also be possible due to the activation necessary for walls to overcome the lattice friction in the nearly commensurate (low temperature) range. It might also be due to a pinning effect by some impurities, a situation often met in internal friction of lattice dislocations. In such a case, the susceptibility should be amplitude dependent, and the corresponding anelasticity should properly belong to the next chapter on plasticity.

5. Plastic Properties of CDW

By this term, we cover the establishment of a Fröhlich current and the large hysteresis in electrical properties associated with it.

As in the plastic properties of crystals under mechanical stresses, the application of a strong electric field on a CDW can strongly distort it and make it explore configurations which are rather far, in a configurational space, from the initial configuration obtained by cooling down under zero field: e.g. strong Fröhlich currents, large remanent polarisations, amplitude-dependent internal friction. All these properties, which are also met not only in usual mechanical plasticity but in magnetic hysteresis [41] or in the hysteresis of type II superconductors [42] are clear signs of a strong *solid* friction.

5.1. AMPLITUDE-DEPENDENT INTERNAL FRICTION. APPROACH TO CRITICAL CURRENT

For increasing amplitudes E_0 of an alternating electric field, one expects the CDW to become locally and temporarily unpinned from some of its internal pinning points — mostly impurities but possibly some interfaces such as polygonised walls. The successive pinning (1) then unpinned (2, 3) configurations pictured schematically Figure 24 a produce a hysteresis loop (Figure 24b) and a characteristic supplementary internal friction which appears above a critical field E_{0c} and then decreases with increasing amplitude E_0 (Figure 24c).

As E_0 increases, an increasing number of such unpinning events will occur, each producing an infinitesimal contribution such as in Figure 14c. The sum result will be an excess internal friction increasing with the amplitude E_0, with a form which will depend on the strength of pinning, the spatial distribution of the pinning objects and — for strong pinning points — temperature. Correlatively there should be a frequency-dependent electrical susceptibility which should increase with amplitude [6, 43].

For dc fields E well above those just referred to, CDWs are progressively freed from all impurities.

This *progressive freeing* from strong impurity pinning centers is qualitatively pictured Figure 25. Due to the necessary distribution of impurity distances and ensuing impurity pinning strengths, some yielding will not be homogeneous but lead under increasing electric field to an increasing average pinning length $\hat{L}(E)$. A *catastrophic depinning* occurs at a certain field, leading to $\hat{L}$ of order of the size of the sample or of its constituent grains.

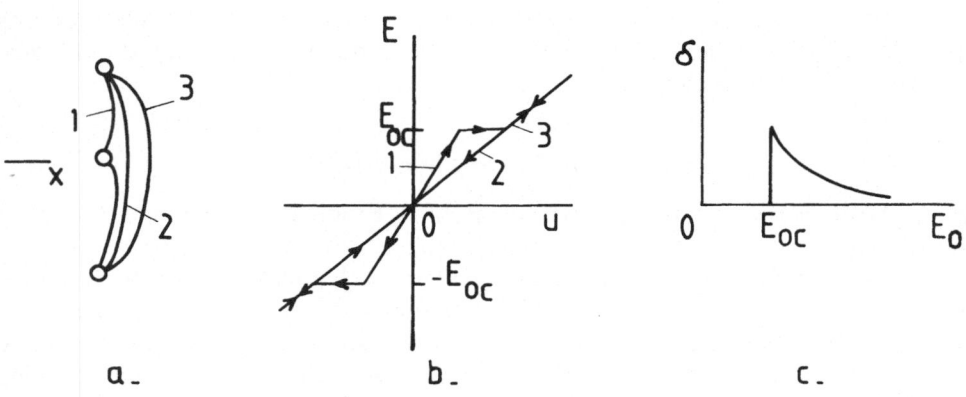

Fig. 24. Hysteresis loop (b) and internal friction δ(c) due to an unpinning process of a CDW from a local pinning point.

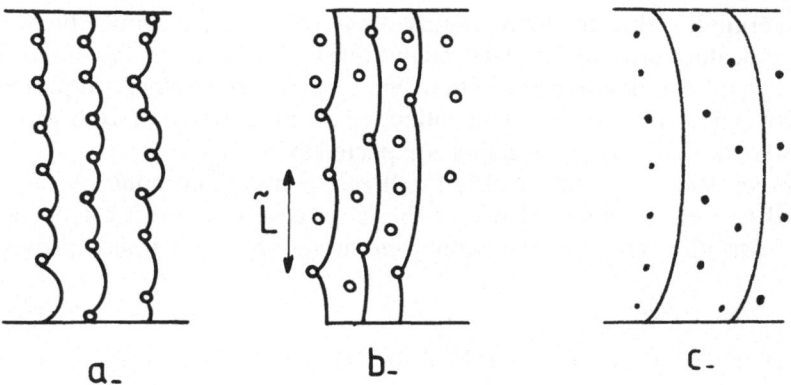

Fig. 25. Various configurations obtained in increasing field. The pinning length $\check{L}(E)$ increases up to the size of the crystal.

One can give a *very* schematic description of such a phenomenon in the case of strong pinning, assuming for simplicity a one dimensional configuration.

Let us assume a distribution $p(e_c)$ of pinning strengths. At a given field, all centres stronger than strength $e_c(E)$ hold, while the others have yielded. The average distance between the former is $\check{L}(E) = 1/n(e_c)$ where

$$n(e_c) = n_i \int_{e_{c(L)}}^{\infty} p(e_c)\, \mathrm{d}e_c \tag{93}$$

$e_c(E)$ is determined by the following equality

$$\frac{n_s e\, E\, \check{L}(E)}{q^2\, K_x} = e_c(E). \tag{94}$$

As E increases, $\check{L}(E)$ tends to diverge and bulk depinning obtains when $e_c(E)$ equals some maximum strength e_c^* or $\check{L}(E)$ becomes of the order of the sample size. For example, with a square distribution $p(e_c) = 1/e_c^M (e_c < e_c^M)$, one gets

$$e_c\left(1 - \frac{e_c}{e^M}\right) = \frac{n_s e\, E}{q^2\, K_x\, n_i}. \tag{95}$$

The critical field E_T is reached when the first member is maximum, e.d. $e_c^* = \frac{1}{2}e_c^M$, thus

$$E_T^{\text{bulk}} = \frac{q^2\, K_x}{n_s e}\, \frac{e_c^M}{4}\, n_i \qquad \check{L}(E_T) = 2/n_i. \tag{96}$$

The susceptibility, given by an expression similar to (18) in this simplified picture, increases as $(\check{L}(E))^2$, as well as the relaxation time (Equation 90). Such an increase and softening of the low-frequency responses has actually been observed as precursor effects to depinning [44]. One never observes a true divergence of the static susceptibility due to depinning as obtained above. It would be worthwhile verifying if scaling laws are obeyed below threshold, between the static dielectric constant ε_0 and the average relaxation time. In the above very simplified model, one would expect ε_0/τ to be independent of E. In a more realistic three dimensional model, the same kind of scaling is expected to hold.

The above discussion only works for strong pinning and could be qualitatively transposed to weak pinning. However the latter case is more involved due to the inherent frustration and the outcoming hierarchy of length scales present in the problem.

5.2. CRITICAL FIELD FOR FRÖHLICH CURRENT

When the applied field is large enough to practically free the CDW from impurity pinning (Figure 25), a Fröhlich current can establish itself if, over such a macroscopic region, the CDW is not blocked by surface or interface pinning.

Two rather different regimes can therefore occur:

1 — Surface pinning weaker than impurity pinning. The critical field should be that for impurity depinning, as given very approximately by (96). The corresponding variations of $\tilde{L}(E)$, ε_0 and τ near to $E_c = E_T^{bulk}$ are given in Figure 26.

2 — Surface pinning stronger than impurity pinning. Above the impurity depinning field E_1, one should observe a temporary current producing a stronger polarisation, but a Fröhlich current only above the larger surface depinning

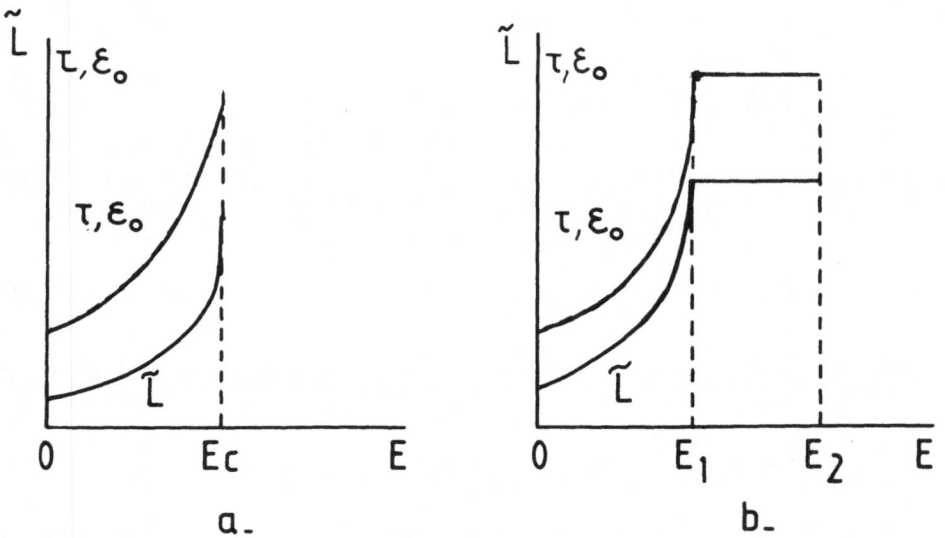

Fig. 26. Schematic variation with E of the pinning length $\tilde{L}$, static dielectric constant ε_0 and average relaxation time τ.

field E_2. There is some evidence that this second regime can be observed [29]. The mechanisms then involved (by brute force or more likely, in general, by dislocation multiplication) have been discussed above; they can be due to frontal or to lateral pinning. The variations of $\tilde{L}$, ε_0 and τ with E are given in Figure 25b.

In all cases, there are strong indications that the Fröhlich current is not homogeneous, but concentrated in some parts of the crystal while others stay static. This is surely due to an inhomogeneous repartition of the pinning defects.

These *inhomogeneities* can be on the scale of the inhomogeneities of the Fröhlich current. One expects, for instance, that electrodes placed laterally on the side of a sample will distort the CDW and concentrate the current near the surface. One also observes cases where the current is concentrated on cylinders parallel to the chains but much smaller in diameter than the sample itself, suggesting that for instance internal polygonised boundaries prevent the CDW from moving in other parts. Figures 27 and 28 depict some cases of this sort.

But a more subtle and short-range inhomogeneity of point defects, on the scale λ of the CDW itself, seems also to be possible, when a heat treatment allows the point defects to take their most stable position with respect to an undistorted CDW [45]. Such a local rearrangement of impurities not only stabilises the CDW by increasing its critical temperature T_c. But it should impede the Fröhlich current, if the impurities are not able to diffuse very fast. As a result, the critical field for Fröhlich current, when it is due to impurity pinning, should present a peak at an intermediate temperature where the point defects are mobile enough to adapt to the presence of a CDW but not mobile enough to follow its motion. Such a peak seems indeed to be observed around 100 K in blue bronzes [7]. Because, in a region where the CDW has become unpinned from its impurity modulation, this tends to disperse, it is easier for the region which has slipped to go on slipping than for slip to propagate elsewhere, one expects such pinning to produce a yield point, i.e. the critical field necessary to continue the current should be less than the initial, critical field. It is finally expected that such local Fröhlich current should produce some 'hardening': the inhomogeneous current should produce defects

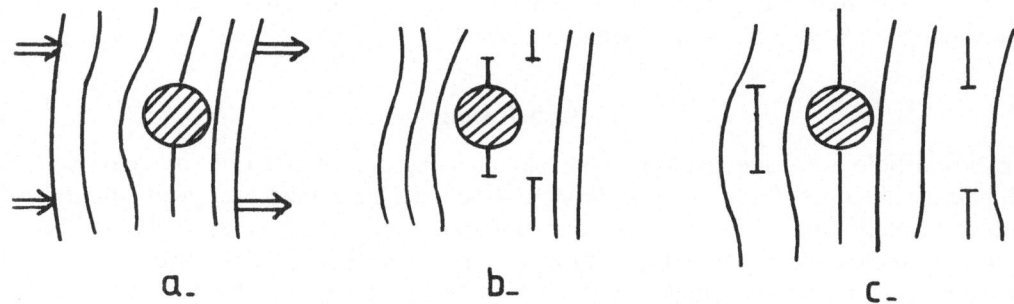

Fig. 27. Example of bypassing of a strong pinning center by a shear plastic process.

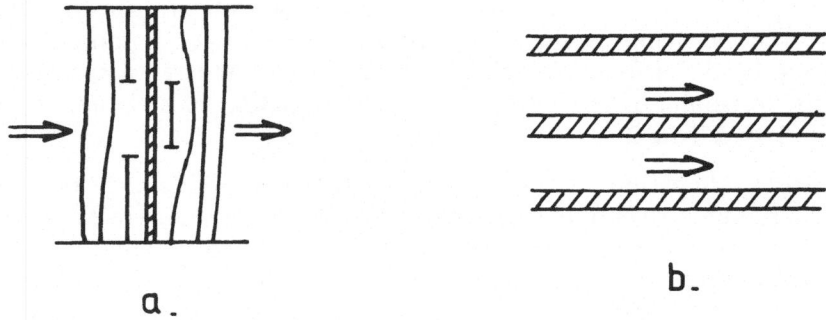

Fig. 28. Phase-slip center (a) and pinned longitudinal domains.

(CDW dislocations or walls) which, by multiplying, should make a further current more difficult. One should therefore expect successive nucleations and propagations of localised regions of Fröhlich current, each accompanied by a characteristic yield point. These characteristic behaviours are observed in blue bronzes [7]. They remind one of very similar behaviours in the plastic properties of solid solutions [6] (blue brittleness of steel, Portevin—Le Châtelier effect of substitutional alloys).

It is quite clear that more microscopic observations are required in this field.

5.3. REMANENT POLARISATION

When, after a Fröhlich current has been produced the electric field is switched off, a remanent polarisation usually remains in the sample, which can be suppressed only by heat treatment.

Such a polarisation clearly measures the deviation of the CDW from the initial locally distorted but unpolarised state. It can have two different origins:

— a mere distortion of the CDW, as in Figure 4, which is kept in place by impurity pinning.
— remanent singularities (walls, dislocations, etc.), again pinned by impurities.

It is clear that the second type of process is more effective than the first in producing the large polarisations often observed.

6. Conclusions

In this article we have exploited the classical picture of a CDW in quasi-one-dimensional compounds. This approach which neglects macroscopical quantum effects, widely studied by Bardeen and coworkers [46], is expected to be valid at low time scales and at the temperatures of interest here, i.e. ~10—100 K in blue bronzes. Furthermore, at higher temperature, fluctuations should alter the present picture of a rigid '*electronic solid*'. An essential feature of CDWs is indeed their internal rigidity due to ordering in real space, as stressed by Anderson [47]. Neglecting long-range Coulomb interactions present in the low-temperature insulating state of blue bronzes and some other compounds [3], the deformations of pinned CDWs under an electric field can be usefully compared to mechanical deformations of crystals under external stresses.

Starting from the phase description of elastic deformations of CDWs in the continuum model, we have sketched the basis for a theory of elasticity and plasticity of CDWs. Amplitude variations necessarily occur at strong pinning points, at topological defects or in phase-slip events. For sake of simplicity, they are not considered here. The distorted CDW is defined by strain and stresss fields. Due to the uniaxial nature of CDWs deformations along the chain axis, internal strains split into longitudinal ones related to an amplitude modulation of the wave vector, and transversal ones, producing a rotation of the wave vector. These deformations occur respectively as compressions (dilatations) or shear of the CDW 'crystal'. They originate from pinning forces providing solid friction phenomena encountered when an electric field is applied to a pinned CDW. The description has been specialized to systems with weakly coupled chains inducing a strong anisotropy of the stiffness constants of the CDW. This seems to be the case for blue bronzes and some other compounds. The limitations of the continuum model come naturally in the extreme limit $\xi_\perp \sim a_\perp$, where some particular lattice effects are expected (Section 2.1.2).

A description of topological defects of CDWs has been given, including walls, dislocations and disclinations. The anisotropy of the CDW superlattice has important consequences as far as the core structure, the energy and the mobility of such defects are concerned. In particular they are pinned at low fields by lattice friction and impurities and should be as metastable as dislocations in ordinary crystals (with obviously different temperature scales, due to the much smaller energies involved). The most likely mechanism for their nucleation is by strain concentration at strong pinning centers under applied field. Interfaces and surface pinning are in this respect very efficient as sources of CDW dislocations, especially in the shear configuration. They release locally strong pinning by processes more economical than brute force mechanisms involving phase slip on the whole pinning area. The argument is quite similar to that involved in the comparison of plastic glide of crystal by pure shear or by dislocation glide [6].

We have described the response of a CDW to an applied electric field of increasing intensity. Electric fields act directly only on longitudinal strains and, as in solids, the forces that work in the elastic and plastic deformations are internal elastic forces reflecting the system's rigidity. For low fields, the phase configuration remains close to the equilibrium one, and the response is inelastic, e.g. linear and reversible on time scales which depend on the pinning length (internal friction). For larger fields, the phase is still strongly pinned by surfaces, interfaces or grain boundaries; but weak pinning allows it to explore configurations far from equilibrium. As a consequence low and thermally activated time scales occur in the dielectric response, and a metastable polarisation is built, directly related to the net deformation of the CDW. Such a non-linear and highly metastable response can be considered as plastic. Nucleation of defects could also occur in this intermediate field regime, which notably increases the remanence.

Depinning mechanisms are complex and may involve catastrophic processes leading to a strong increase of both low-frequency response and relaxation times. In thin samples, pinning by surfaces is expected to be dominant and to determine the threshold for Fröhlich conductivity. In that case precursor effects linked to bulk impurity depinning should be observed and also especially high and very metastable polarisations.

Depinning from surfaces, interfaces and more generally any kind of strong pinning centre can be achieved without reaching the true critical strength which can be very high, provided that dislocation loops can nucleate and allow to bypass the obstacle. More generally dislocation and defect motion allow one part of the CDW to glide past another. They provide the clue for both inhomogeneities of CDW velocities and the fact that for moderate fields above threshold only part of the CDW is depinned. They could also be a cause of broad-band noise emission.

The relationship with other domains of condensed matter physics is worth to explore apart from the link with crystalline elasticity, developed in this article. The topological properties of CDW defects are very close to those of liquid crystals, especially smectics [48]. Interesting information could also be gained from the comparison with vortices in type II superconductors where solid friction is a crucial phenomenon [42]. Finally the core structure of CDW dislocations presents analogies with that of dislocations of vortex lattice in flowing nematic liquid crystals [49].

All the arguments developed here rest on a few microscopic parameters, among which are the constants K_x, K_y, K_z or the related coherence lengths. A better estimation of these parameters and of the ensuing core structure of CDW dislocations is required in the low temperature range, both experimentally and theoretically, especially by taking proper account of the interchain coupling. Moreover the role of long range Coulomb interactions in compounds like blue bronzes or TaS_3 is also of importance and can stiffen notably the CDW lattice at very low temperatures [50]. Finally a better understanding of the microscopical mechanism for surface and interface pinning is also highly required. Direct surface probes may be extremely useful to this respect. More generally local probes measuring locally the degree of deformation of CDWs are necessary to improve the general picture given above, which is actually closer to reality than the old but nevertheless useful single particle model for CDWs dynamics.

Acknowledgements

The authors have benefitted from useful discussions with S. Aubry, S. Barisic, A. Bjelis, B. K. Chakraverty, J. Dumas, L. Kubin, J. Lajzerowicz, P. Monceau, J. P. Pouget, M. Renard and C. Schlenker.

References

1. R. E. Peierls, *Quantum Theory of Solids* (Oxford University Press 1955) p. 108.
2. J. Solyom, *Adv. in Physics* **28**, 201 (1979).
3. For reviews see *Electronic Properties of Inorganic Quasi-one-Dimensional Compounds*, P. Monceau (Ed.) (Reidel, Dordrecht 1985); G. Grüner, A. Zettl, *Phys. Rept.* **119**, 117 (1985); J. C. Gill, *Contemp. Phys.* **27**, 37 (1986).

4. H. Fröhlich, *Proc. Roy. Soc.* **A223**, 296 (1954).
5. P. Lee, T. M. Rice, and P. W. Anderson, *Solid State Comm.* **14**, 703 (1974).
6. J. Friedel, *Dislocations* (Pergamon Press, Oxford, 1964).
7. For a review see C. Schlenker, J. Dumas in *Crystal Chemistry and Properties of Materials with Quasi-One-Dimensional Structures*, J. Rouxel (ed.) (D. Reidel, Dordrecht, 1986) p. 135, and chapters in the present volume.
8. J. P. Pouget, C. Noguera, A. H. Moudden, and R. Moret, *J. Physique* **A6**, 1731 (1985).
9. R. M. Fleming, L. F. Schneemeyer, and D. E. Moncton, *Phys. Rev.* **B31**, 899 (1985).
10. P. Segransan, A. Janossy, C. Berthier, J. Marcus, and P. Butaud, *Phys. Rev. Lett.* **56**, 1854 (1986).
11. C. Berthier and P. Segransan, NATO ASI on *Low Dimensional Conductors and Superconductors*, L. Caron, D. Jérome (Eds.), Magog (Canada), August 1986, to be published (Plenum).
12. J. H. Ross, Z. Wang and C. P. Slichter, *Phys. Rev. Lett.* **56**, 663 (1986).
13. R. M. Fleming, R. G. Dunn, and L. F. Schneemeyer, *Phys. Rev.* **B31**, 4099 (1985).
14. T. Tamegai, K. Tsutsumi, S. Kagoshima, Y. Kanai, H. Tomozawa, M. Tani, Y. Nogani, and M. Sato, *Solid State Comm.* **56**, 13 (1985).
15. J. Friedel, Ref. 11.
16. W. L. MacMillan, *Phys. Rev.* **B12**, 2042 (1975).
17. P. A. Lee and T. M. Rice, *Phys. Rev.* **B19**, 3970 (1979).
18. J. Dumas and D. Feinberg, *Europhys. Lett.* **2**, 555 (1986) and *Physica* **143B**, 111 (1986).
19. N. P. Ong, G. Verma, and K. Maki, *Phys. Rev. Lett.* **52**, 663 (1984); N. P. Ong and K. Maki, *Phys. Rev.* **B32**, 6582 (1985).
20. W. L. MacMillan, *Phys. Rev.* **B14**, 1496 (1976).
21. P. Bak, *Rep. Progr. Phys.* **45**, 587 (1982).
22. L. Mihaly and G. X. Tessema, *Phys. Rev.* **B37**, 5858 (1986).
23. H. Fukuyama and P. A. Lee, *Phys. Rev.* **B17**, 535 (1978).
24. L. P. Gor'kov, *Zh, Eksp. Teor. Fiz.* **86**, 1818 (1984) transl. *Sov. Phys. JETP* **59**, 1057 (1984).
25. I. Batistic, A. Bjelis, and L. P. Gor'kov, *J. Physique* **45**, 1049 (1984).
26. F. C. Frank, *Adv. Physics* **1**, 51 (1952).
27. K. Maki, *Physica* **143B**, 59 (1986).
28. M. C. Saint-Lager, Thesis, Université Scientifique, Technologique et Médicale de Grenoble; P. Monceau, M. Renard, J. Richard and M. C. Saint-Lager, *Physica* **143B**, 64 (1986).
29. J. C. Gill, to be published.
30. L. F. Schneemeyer, F. J. Di Salvo, S. E. Spengler and J. V. Waszczak, *Phys. Rev.* **B30**, 4297 (1984) and *Mol. Cryst. Liq. Cryst.* **125**, 41 (1985).
31. J. Dumas, R. Buder, J. Marcus, C. Schlenker, and A. Janossy, *Physica* **143B**, 183 (1986).
32. J. Y. Veuillen, R. Chevalier, J. Marcus, and C. Schlenker, *Physica* **143B**, 186 (1986).
33. H. Mutka and S. Bouffard, *J. Physique Lett.* **45**, L729 (1984).
34. N. F. Mott and F. R. N. Nabarro, *Proc. Phys. Soc.* **52**, 86 (1940).
35. S. A. Brazowski, *Sov. Phys. JETP* **51**, 342 (1980).
36. K. Sasaki and K. Maki, *Phys. Rev.* **B33**, 2665 (1986).
37. B. K. Chakraverty, *Phys. Rev.* **B34**, 5287 (1986).
38. A. Zawadowski, personal communication; I Tüttö and A. Zawadowski, *Phys. Rev.* **B32**, 2449 (1985).
39. R. J. Cava, L. F. Schneemeyer, R. M. Fleming, P. B. Littlewood, and E. A. Rietman, *Phys. Rev.* **B32**, 4088 (1985).
40. R. J. Cava, R. M. Fleming, P. Littlewood, E. A. Rietman, L. F. Schneemeyer, and R. G. Dunn, *Phys. Rev.* **B32**, 3228 (1984).
41. M. Kléman in *Magnetism of Metals and Alloys* (Les Houches, Winter School 1980) edited by M. Cyrot (North Holland, 1982) p. 535.
42. D. Saint-James, G. Sarma, and E. J. Thomas, *Type II Superconductivity*, (Pergamon Press, 1969).
43. A. Granato and K. Lücke, *J. Appl. Phys.* **27**, 583 (1956).
44. R. J. Cava, R. M. Fleming, R. G. Dunn, E. A. Rietman and L. F. Schneemeyer, *Phys. Rev.* **B30**, 7290 (1984).
45. P. Lederer, J. P. Jamet, and G. Montambaux, *Mol. Cryst. Liq. Cryst.* **121**, 99 (1985); *Ferroelectrics* **66**, 25 (1986).

46. J. Bardeen and J. R. Tucker in *Charge Density Waves in Solids* G. Hutiray and J. Solyom (Eds.), *Lectures Notes in Physics*, vol. 217 (Springer, Berlin, 1985), p. 155; J. Bardeen, *Physica* **143B**, 14 (1987) and refs. therein.
47. P. W. Anderson, *Basic Notions of Condensed Matter Physics*, (Benjamin Cummings, 1984) p. 159, and Ref. 5.
48. M. Kléman, *Dislocations* 1984, Colloque International du C.N.R.S. Aussois (Ed. du C.N.R.S., 1984) p. 1.
49. R. Ribotta and A. Joëts, *J. Physique* **47**, 739 (1986).
50. S. Barisic, Ref. 11 and *Mol. Cryst. Liq. Cryst.* **119**, 413 (1985).
51. S. E. Brown, L. Mihaly, and G. Grüner, *Solid State Comm.* **58**, 231 (1986).
52. G. Mihaly, Gy. Hutiray, and L. Mihaly, *Solid State Comm.* **48**, 203 (1983).
53. J. C. Gill, Ref. 46, p. 377.

INDEX

4d electron distribution 53

amplitudon 98, 109, 133, 164

band structure 2, 120, 140, 144, 179, 241
bipolaron 33, 37, 295, 363
blue-black bronzes 55, 59
breaking of analyticity 295
Burgers vector 414

chaotic metastable configurations 304
charge density wave (CDW) 88, 159, 160
　coupling between 94, 104, 138, 417
　effective mass 286
　instabilities 16
　response function 90, 117, 125
　transport 165, 180
coherence length 413
Cole–Cole plot 266
compression 410
configurational force 411
correlation length 92, 117, 125, 135, 149
critical strain 427
critical voltage 427
critical wave vector 119, 139
crystal structure 5, 12, 51
crystallographic shear 77

damping 286
dielectric function 261, 264, 270
dielectric relaxation 260, 263, 266, 277, 439
disclinations 418
discommensuration 168, 304, 420
　advanced 304
　delayed 304
displacement 410
distortions 410
dual eigenequation 331

elastic anomalies 151
elastic constants 410
elastic properties 118, 177
electric field induced disorder 129
electric susceptibility 412, 436, 439
electron–hole polarizability 88
electron paramagnetic resonance 207
electron–phonon coupling 90, 100, 128, 136, 148
exponentially localized eigenwave 331

far infrared 289
Fermi surface 88, 106, 144, 160
fluctuations 101, 262

Frank–Read sources 425
Frenkel–Kontorova model 301
friction 408, 421, 422, 430
　internal 438, 439
　solid 440
　viscous 438
Fröhlich 165
Fröhlich current 165, 407, 442
Fukuyama, Lee and Rice model 167, 262

ground state
　defectible 359
　undefectible 358

Hall effect 228
hardening 443
hull function 303
hydrogen molybdenum bronzes 62
hysteresis 181, 197, 440

impurities 11, 166, 178, 184, 273
incommensurate 409
incommensurate systems 295
ion channeling 204

kinks 417
Kohn anomaly 95, 117, 131, 135

$La_2Mo_2O_7$ 30
lattice dislocations 429
lattice distortion 94, 146, 163, 296
lattice friction 414
lattice gas bipolarons 384
lattice gas model 378

magnetic susceptibility 11, 18, 176, 231
magnetoresistance 224
memory effects 181, 201
metastability 197, 263, 279
Mössbauer effect 204

nesting wave vector 89, 124, 145
noise voltage 181, 191
nonlinear conductivity 165, 400
nuclear magnetic resonance 209

optical reflectivity 242
order parameter 95, 126, 149

pair creation operator 391
Peach and Koehler force 418, 425
Peierls chains 365

449

Peierls friction 426
Peierls transition 6, 114, 151, 160, 173
Peierls–Nabarro barrier 367
Peierls–Nabarro potential 295
phase diagram 142
phase dislocations 168, 414, 421
    Burgers vector 417
    climbing 417, 432
    edge 414
    gliding 417, 432
    loops 425
    nucleation 425
    screw 414
phase-slip 427
phason 98, 109, 132, 164, 259, 300, 369
photoemission spectroscopy 179
pinning 166, 303
    random 262
    strong 430
    surface 442
    uniform 260
    weak 430
plasticity 426, 439
point singularities 418
polarisation 412, 436
    remanent 444
polarization 202, 260, 282
Portevin–Le Châtelier effect 444
preparation 4, 50, 73
pressure effects 25
pseudospin 363

quantum transport 244
quasi-Bloch wave functions 328

red bronze 26, 54, 67
relaxation time 439, 441
ridges 413

screening 260, 286, 289
shear 410
single-particle model 261
sodium hydrated molybdenum bronzes 64
specific heat 23, 24, 177, 237
strain tensor 410, 436
stress tensor 410
structural modulation 127
substitutional disorder 129
superconductivity 247
superstructure 142, 146
surface friction 423

thermopower 22, 221
Thouless formula 331
threshold field 165, 182, 259, 424
    critical 424, 425, 441
titanium bronze 139
transport properties 173, 212
tungsten bronzes 38

vanadium bronzes 32

walls 413
washboard frequency 166, 279

yield point 443
yielding 440

Zachariasen formula 53, 59

Physics and Chemistry of Materials
with Low-Dimensional Structures

Previously published under the Series-title:
PHYSICS AND CHEMISTRY OF MATERIALS WITH LAYERED STRUCTURES

1. R.M.A. Lieth (ed.): *Preparation and Crystal Growth of Materials with Layered Structures.* 1977   ISBN 90-277-0638-7
2. F. Lévy (ed.): *Crystallography and Crystal Chemistry of Materials with Layered Structures.* 1976   ISBN 90-277-0586-0
3. T. J. Wieting and M. Schlüter (eds.): *Electrons and Phonons in Layered Crystal Structures.* 1979   ISBN 90-277-0897-5
4. P.A. Lee (ed.): *Optical and Electrical Properties.* 1976   ISBN 90-277-0676-X
5. F. Hulliger: *Structural Chemistry of Layer-Type Phases.* Ed. by F. Lévy. 1976   ISBN 90-277-0714-6
6. F. Lévy (ed.): *Intercalated Layered Materials.* 1979   ISBN 90-277-0967-X

Published under:
PHYSICS AND CHEMISTRY OF MATERIALS WITH LOW-DIMENSIONAL STRUCTURES
SERIES A: LAYERED STRUCTURES

7. V. Grasso (ed.): *Electronic Structure and Electronic Transitions in Layered Materials.* 1986   ISBN 90-277-2102-5
8. K. Motizuki (ed.): *Structural Phase Transitions in Layered Transition Metal Compounds.* 1986   ISBN 90-277-2171-8

PHYSICS AND CHEMISTRY OF MATERIALS WITH LOW-DIMENSIONAL STRUCTURES
SERIES B: QUASI-ONE-DIMENSIONAL STRUCTURES

B1. P. Monceau (ed.): *Electronic Properties of Inorganic Quasi-One-Dimensional Compounds.* Part I: Theoretical. 1985   ISBN 90-277-1789-3
B2. P. Monceau (ed.): *Electronic Properties of Inorganic Quasi-One-Dimensional Compounds.* Part II: Experimental. 1985   ISBN 90-277-1800-8
B3. H. Kamimura (ed.): *Theoretical Aspects of Band Structures and Electronic Properties of Pseudo-One-Dimensional Solids.* 1985   ISBN 90-277-1927-6
B4. J. Rouxel (ed.): *Crystal Chemistry and Properties of Materials with Quasi-One-Dimensional Structures.* A Chemical and Physical Synthetic Approach. 1986   ISBN 90-277-2057-6

Discontinued.

PHYSICS AND CHEMISTRY OF MATERIALS WITH LOW-DIMENSIONAL STRUCTURES
SERIES C: MOLECULAR STRUCTURES

C1. I. Zschokke (ed.): *Optical Spectroscopy of Glasses.* 1986   ISBN 90-277-2231-5
C2. J. Fünfschilling (ed.): *Relaxation Processes in Molecular Excited States.* 1989   ISBN 07923-0001-7

Discontinued.

Physics and Chemistry of Materials
with Low-Dimensional Structures

9. L.J. de Jongh (ed.): *Magnetic Properties of Layered Transition Metal Compounds.* 1990
ISBN 0-7923-0238-9
10. E. Doni, R. Girlanda, G. Pastori Parravicini and A. Quattropani (eds.): *Progress in Electron Properties of Solids.* Festschrift in Honour of Franco Bassini. 1989
ISBN 0-7923-0337-7
11. C. Schlenker (ed.): *Low-Dimensional Electronic Properties of Molybdenum Bronzes and Oxides.* 1989
ISBN 0-7923-0085-8
12. R. H. Friend (ed.): *Conducting Polymers.* 1990 (forthcoming)

KLUWER ACADEMIC PUBLISHERS – DORDRECHT / BOSTON / LONDON